Handbook of Optical Design

OPTICAL ENGINEERING

Founding Editor

Brian J. Thompson

University of Rochester
Rochester, New York

1. Electron and Ion Microscopy and Microanalysis: Principles and Applications, *Lawrence E. Murr*
2. Acousto-Optic Signal Processing: Theory and Implementation, *edited by Norman J. Berg and John N. Lee*
3. Electro-Optic and Acousto-Optic Scanning and Deflection, *Milton Gottlieb, Clive L. M. Ireland, and John Martin Ley*
4. Single-Mode Fiber Optics: Principles and Applications, *Luc B. Jeunhomme*
5. Pulse Code Formats for Fiber Optical Data Communication: Basic Principles and Applications, *David J. Morris*
6. Optical Materials: An Introduction to Selection and Application, *Solomon Musikant*
7. Infrared Methods for Gaseous Measurements: Theory and Practice, *edited by Joda Wormhoudt*
8. Laser Beam Scanning: Opto-Mechanical Devices, Systems, and Data Storage Optics, *edited by Gerald F. Marshall*
9. Opto-Mechanical Systems Design, *Paul R. Yoder, Jr.*
10. Optical Fiber Splices and Connectors: Theory and Methods, *Calvin M. Miller with Stephen C. Mettler and Ian A. White*
11. Laser Spectroscopy and Its Applications, *edited by Leon J. Radziemski, Richard W. Solarz, and Jeffrey A. Paisner*
12. Infrared Optoelectronics: Devices and Applications, *William Nunley and J. Scott Bechtel*
13. Integrated Optical Circuits and Components: Design and Applications, *edited by Lynn D. Hutcheson*
14. Handbook of Molecular Lasers, *edited by Peter K. Cheo*
15. Handbook of Optical Fibers and Cables, *Hiroshi Murata*
16. Acousto-Optics, *Adrian Korpel*
17. Procedures in Applied Optics, *John Strong*
18. Handbook of Solid-State Lasers, *edited by Peter K. Cheo*
19. Optical Computing: Digital and Symbolic, *edited by Raymond Arrathoon*
20. Laser Applications in Physical Chemistry, *edited by D. K. Evans*
21. Laser-Induced Plasmas and Applications, *edited by Leon J. Radziemski and David A. Cremers*

22. Infrared Technology Fundamentals, *Irving J. Spiro and Monroe Schlessinger*
23. Single-Mode Fiber Optics: Principles and Applications, Second Edition, Revised and Expanded, *Luc B. Jeunhomme*
24. Image Analysis Applications, *edited by Rangachar Kasturi and Mohan M. Trivedi*
25. Photoconductivity: Art, Science, and Technology, *N. V. Joshi*
26. Principles of Optical Circuit Engineering, *Mark A. Mentzer*
27. Lens Design, *Milton Laikin*
28. Optical Components, Systems, and Measurement Techniques, *Rajpal S. Sirohi and M. P. Kothiyal*
29. Electron and Ion Microscopy and Microanalysis: Principles and Applications, Second Edition, Revised and Expanded, *Lawrence E. Murr*
30. Handbook of Infrared Optical Materials, *edited by Paul Klocek*
31. Optical Scanning, *edited by Gerald F. Marshall*
32. Polymers for Lightwave and Integrated Optics: Technology and Applications, *edited by Lawrence A. Hornak*
33. Electro-Optical Displays, *edited by Mohammad A. Karim*
34. Mathematical Morphology in Image Processing, *edited by Edward R. Dougherty*
35. Opto-Mechanical Systems Design: Second Edition, Revised and Expanded, *Paul R. Yoder, Jr.*
36. Polarized Light: Fundamentals and Applications, *Edward Collett*
37. Rare Earth Doped Fiber Lasers and Amplifiers, *edited by Michel J. F. Digonnet*
38. Speckle Metrology, *edited by Rajpal S. Sirohi*
39. Organic Photoreceptors for Imaging Systems, *Paul M. Borsenberger and David S. Weiss*
40. Photonic Switching and Interconnects, *edited by Abdellatif Marrakchi*
41. Design and Fabrication of Acousto-Optic Devices, *edited by Akis P. Goutzoulis and Dennis R. Pape*
42. Digital Image Processing Methods, *edited by Edward R. Dougherty*
43. Visual Science and Engineering: Models and Applications, *edited by D. H. Kelly*
44. Handbook of Lens Design, *Daniel Malacara and Zacarias Malacara*
45. Photonic Devices and Systems, *edited by Robert G. Hunsperger*
46. Infrared Technology Fundamentals: Second Edition, Revised and Expanded, *edited by Monroe Schlessinger*
47. Spatial Light Modulator Technology: Materials, Devices, and Applications, *edited by Uzi Efron*
48. Lens Design: Second Edition, Revised and Expanded, *Milton Laikin*
49. Thin Films for Optical Systems, *edited by François R. Flory*
50. Tunable Laser Applications, *edited by F. J. Duarte*
51. Acousto-Optic Signal Processing: Theory and Implementation, Second Edition, *edited by Norman J. Berg and John M. Pellegrino*
52. Handbook of Nonlinear Optics, *Richard L. Sutherland*
53. Handbook of Optical Fibers and Cables: Second Edition, *Hiroshi Murata*

54. Optical Storage and Retrieval: Memory, Neural Networks, and Fractals, *edited by Francis T. S. Yu and Suganda Jutamulia*
55. Devices for Optoelectronics, *Wallace B. Leigh*
56. Practical Design and Production of Optical Thin Films, *Ronald R. Willey*
57. Acousto-Optics: Second Edition, *Adrian Korpel*
58. Diffraction Gratings and Applications, *Erwin G. Loewen and Evgeny Popov*
59. Organic Photoreceptors for Xerography, *Paul M. Borsenberger and David S. Weiss*
60. Characterization Techniques and Tabulations for Organic Nonlinear Optical Materials, *edited by Mark Kuzyk and Carl Dirk*
61. Interferogram Analysis for Optical Testing, *Daniel Malacara, Manuel Servín, and Zacarias Malacara*
62. Computational Modeling of Vision: The Role of Combination, *William R. Uttal, Ramakrishna Kakarala, Sriram Dayanand, Thomas Shepherd, Jagadeesh Kalki, Charles F. Lunskis, Jr., and Ning Liu*
63. Microoptics Technology: Fabrication and Applications of Lens Arrays and Devices, *Nicholas F. Borrelli*
64. Visual Information Representation, Communication, and Image Processing, *Chang Wen Chen and Ya-Qin Zhang*
65. Optical Methods of Measurement: Wholefield Techniques, *Rajpal S. Sirohi and Fook Siong Chau*
66. Integrated Optical Circuits and Components: Design and Applications, *edited by Edmond J. Murphy*
67. Adaptive Optics Engineering Handbook, *edited by Robert K. Tyson*
68. Entropy and Information Optics, *Francis T. S. Yu*
69. Computational Methods for Electromagnetic and Optical Systems, *John M. Jarem and Partha P. Banerjee*
70. Laser Beam Shaping: Theory and Techniques, *edited by Fred M. Dickey and Scott C. Holswade*
71. Rare-Earth-Doped Fiber Lasers and Amplifiers: Second Edition, Revised and Expanded, *edited by Michel J. F. Digonnet*
72. Lens Design: Third Edition, Revised and Expanded, *Milton Laikin*
73. Handbook of Optical Engineering, *edited by Daniel Malacara and Brian J. Thompson*
74. Handbook of Imaging Materials, *edited by Arthur S. Diamond and David S. Weiss*
75. Handbook of Image Quality: Characterization and Prediction, *Brian W. Keelan*
76. Fiber Optic Sensors, *edited by Francis T. S. Yu and Shizhuo Yin*
77. Optical Switching/Networking and Computing for Multimedia Systems, *edited by Mohsen Guizani and Abdella Battou*
78. Image Recognition and Classification: Algorithms, Systems, and Applications, *edited by Bahram Javidi*
79. Practical Design and Production of Optical Thin Films: Second Edition, Revised and Expanded, *Ronald R. Willey*

80. Ultrafast Lasers: Technology and Applications, *edited by Martin E. Fermann, Almantas Galvanauskas, and Gregg Sucha*
81. Light Propagation in Periodic Media: Differential Theory and Design, *Michel Nevière and Evgeny Popov*
82. Handbook of Nonlinear Optics: Second Edition, Revised and Expanded, *Richard L. Sutherland*
83. Polarized Light: Second Edition, Revised and Expanded, *Dennis Goldstein*
84. Optical Remote Sensing: Science and Technology, *Walter G. Egan*
85. Handbook of Optical Design: Second Edition, *Daniel Malacara and Zacarias Malacara*

Additional Volumes in Preparation

Nonlinear Optics: Theory, Numerical Modeling, and Applications, *Partha P. Banerjee*

Handbook of Optical Design

Second Edition

Daniel Malacara
Zacarias Malacara
Centro de Investigaciones en Optica, A.C.
León, Mexico

MARCEL DEKKER, INC. NEW YORK • BASEL

Library of Congress Cataloging-in-Publication Data
A catalog record for this book is available from the Library of Congress.

The first edition of this book was published as *Handbook of Lens Design* (Marcel Dekker, Inc., 1994).

ISBN: 0-8247-4613-9

This book is printed on acid-free paper.

Headquarters
Marcel Dekker, Inc.
270 Madison Avenue, New York, NY 10016
tel: 212-696-9000; fax: 212-685-4540

Eastern Hemisphere Distribution
Marcel Dekker AG
Hutgasse 4, Postfach 812, CH-4001 Basel, Switzerland
tel: 41-61-260-6300; fax: 41-61-260-6333

World Wide Web
http://www.dekker.com

The publisher offers discounts on this book when ordered in bulk quantities. For more information, write to Special Sales/Professional Marketing at the headquarters address above.

Copyright © 2004 by Marcel Dekker, Inc. All Rights Reserved.

Neither this book nor any part may be reproduced or transmitted in any form or by any means, electronic or mechanical, including photocopying, microfilming, and recording, or by any information storage and retrieval system, without permission in writing from the publisher.

Current printing (last digit):
10 9 8 7 6 5 4 3 2 1

PRINTED IN THE UNITED STATES OF AMERICA

Preface to the Second Edition

The first edition of this book was used by our students in a lens design course for several years. Taking advantage of this experience, this second edition has been greatly improved in several aspects.

Most of the material in the original second chapter was considered quite important and useful as a reference. However, to make an introductory course on lens design more fluid and simple, most of the material was transferred to the end of the book as an Appendix. In several other sections the book was also restructured with the same objective in mind.

Some of the modifications introduced include the clarification and a more complete explanation of some concepts, as suggested by some readers. Additional material was written, including additional new references to make the book more complete and up to date. We will mention only a few examples. Some gradient index systems are now described with greater detail. The new wavefront representation by means of arrays of gaussians is included. The Delano diagram section was enlarged. More details on astigmatic surfaces with two different curvatures in orthogonal diameters are given.

We would like to thank our friends and students who used the previous edition of this book. They provided us with many suggestions and pointed out a few typographical errors to improve the book.

Daniel Malacara
Zacarias Malacara

Preface to the First Edition

This is a book on the general subject of optical design, aimed at students in the field of geometrical optics and engineers involved in optical instrumentation. Of course, this is not the first book in this field. Some classic, well-known books are out of print however, and lack any modern topics. On the other hand, most modern books are generally very restricted in scope and do not cover important classic or even modern details.

Without pretending to be encyclopedic, this book tries to cover most of the classical aspects of lens design and at the same time describes some of the modern methods, tools, and instruments, such as contemporary astronomical telescopes, gaussian beams, and computer lens design.

Chapter 1 introduces the reader to the fundamentals of geometrical optics. In Chapter 2 spherical and aspherical optical surfaces and exact skew ray tracing are considered. Chapters 3 and 4 define the basic concepts for the first- and third-order theory of lenses while the theory of the primary aberrations of centered optical systems is developed in Chapters 5 to 7. The diffraction effects in optical systems and the main wave and ray methods for lens design evaluation are described in Chapters 8, 9, and 10. Chapters 11 to 17 describe some of the main classical optical instruments and their optical design techniques. Finally, Chapter 18 studies the computer methods for optical lens design and evaluation.

In conclusion, not only is the basic theory treated in this book, but many practical details for the design of some optical systems are given. We hope that this book will be useful as a textbook for optics students, as well as a reference book for optical scientists and engineers.

We greatly acknowledge the careful reading of the manuscript and suggestions to improve the book by many friends and colleagues. Among these many friends we would like to mention Prof. Raúl Casas, Manuel Servín, Ricardo Flores, and several of our students. A generous number of members of the research staff from Optical Research Associates provided a wonderful help with many constructive criticisms and suggestions. Their number is large and we do not want to be unfair by just mentioning a few names. We also acknowledge the financial support and enthusiasm of the Centro de Investigaciones en Optica and its General Director Arquímedes Morales. Last, but not least, the authors greatly acknowledge the encouragement and understanding of our families. One of the authors (D.M.) especially thanks his sons Juan Manuel and Miguel Angel for their help with many of the drawings and the word processing of some parts.

Daniel Malacara
Zacarias Malacara

Contents

Preface to the Second Edition   *iii*
Preface to the First Edition   *v*

1. Geometrical Optics Principles   **1**
    1.1 Wave Nature of Light and Fermat's Principle   1
    1.2 Reflection and Refraction Laws   9
    1.3 Basic Meridional Ray Tracing Equations   12
    1.4 Gaussian or First-Order Optics   18
    1.5 Image Formation   22
    1.6 Stop, Pupils, and Principal Ray   25
    1.7 Optical Sine Theorem   29
    1.8 Herschel Invariant and Image Magnifications   33
    1.9 Ray Aberrations and Wave Aberrations   34
    References   36

2. Thin Lenses and Spherical Mirrors   **39**
    2.1 Thin Lenses   39
    2.2 Formulas for Image Formation with Thin Lenses   42
    2.3 Nodal Points of a Thin Lens   44
    2.4 Image Formation with Convergent Lenses   45
    2.5 Image Formation with Divergent Lenses   47
    References   47

3. Systems of Several Lenses and Thick Lenses   **49**
    3.1 Focal Length and Power of a Lens System   49
    3.2 Image Formation with Thick Lenses or Systems of Lenses   51
    3.3 Cardinal Points   53
    3.4 Image Formation with a Tilted or Curved Object   55
    3.5 Thick Lenses   57
    3.6 Systems of Thin Lenses   60

	3.7	The Lagrange Invariant in a System of Thin Lenses	63
	3.8	Effect of Object or Stop Shifting	64
	3.9	The Delano y–$\bar{y}$ Diagram	67
		References	72
4.	**Spherical Aberration**	**73**
	4.1	Spherical Aberration Calculation	73
	4.2	Primary Spherical Aberration	77
	4.3	Aspherical Surfaces	86
	4.4	Spherical Aberration of Aspherical Surfaces	86
	4.5	Surfaces without Spherical Aberration	87
	4.6	Aberration Polynomial for Spherical Aberration	91
	4.7	High-Order Spherical Aberration	98
	4.8	Spherical Aberration Correction with Gradient Index	99
		References	101
5.	**Monochromatic Off-Axis Aberrations**	**103**
	5.1	Oblique Rays	103
	5.2	Petzval Curvature	109
	5.3	Coma	112
	5.4	Astigmatism	116
	5.5	Distortion	129
	5.6	Off-Axis Aberrations in Aspherical Surfaces	132
	5.7	Aberrations and Wavefront Deformations	135
	5.8	Symmetrical Principle	138
	5.9	Stop Shift Equations	139
		References	142
6.	**Chromatic Aberrations**	**145**
	6.1	Introduction	145
	6.2	Axial Chromatic Aberration	146
	6.3	Secondary Color Aberration	158
	6.4	Magnification Chromatic Aberration	160
		References	168
7.	**The Aberration Polynomial**	**171**
	7.1	Wave Aberration Polynomial	171
	7.2	Zernike Polynomials	175
	7.3	Wavefront Representation by an Array of Gaussians	180
	7.4	Transverse Aberration Polynomials	183
		References	189

Contents ix

8. Diffraction in Optical Systems — 191
- 8.1 Huygens–Fresnel Theory — 191
- 8.2 Fresnel Diffraction — 192
- 8.3 Fraunhofer Diffraction — 195
- 8.4 Diffraction Images with Aberrations — 200
- 8.5 Strehl Ratio — 202
- 8.6 Optical Transfer Function — 204
- 8.7 Resolution Criteria — 209
- 8.8 Gaussian Beams — 211
- References — 214

9. Computer Evaluation of Optical Systems — 217
- 9.1 Meridional Ray Tracing and Stop Position Analysis — 217
- 9.2 Spot Diagram — 219
- 9.3 Wavefront Deformation — 224
- 9.4 Point and Line Spread Functions — 230
- 9.5 Optical Transfer Function — 232
- 9.6 Tolerance to Aberrations — 234
- References — 236

10. Prisms — 239
- 10.1 Tunnel Diagram — 239
- 10.2 Deflecting a Light Beam — 239
- 10.3 Transforming an Image — 242
- 10.4 Deflecting and Transforming Prisms — 244
- 10.5 Nondeflecting Transforming Prisms — 248
- 10.6 Beam-Splitting Prisms — 251
- 10.7 Chromatic Dispersing Prisms — 253
- References — 257

11. Simple Optical Systems and Photographic Lenses — 259
- 11.1 Optical Systems Diversity — 259
- 11.2 Single Lens — 260
- 11.3 Spherical and Paraboloidal Mirrors — 270
- 11.4 Periscopic Lens — 281
- 11.5 Achromatic Landscape Lenses — 283
- 11.6 Achromatic Double Lens — 284
- 11.7 Some Catoptric and Catadioptric Systems — 285
- 11.8 Fresnel Lenses and Gabor Plates — 288
- References — 290

12. Complex Photographic Lenses — 291
12.1 Introduction — 291
12.2 Asymmetrical Systems — 292
12.3 Symmetrical Anastigmat Systems — 301
12.4 Varifocal and Zoom Lenses — 306
References — 312

13. The Human Eye and Ophthalmic Lenses — 315
13.1 The Human Eye — 315
13.2 Ophthalmic Lenses — 318
13.3 Ophthalmic Lens Design — 322
13.4 Prismatic Lenses — 328
13.5 Spherocylindrical Lenses — 329
References — 331

14. Astronomical Telescopes — 333
14.1 Resolution and Light Gathering Power — 333
14.2 Catadioptric Cameras — 337
14.3 Newton Telescope — 341
14.4 Reflecting Two-Mirror Telescopes — 342
14.5 Field Correctors — 357
14.6 Catadioptric Telescopes — 362
14.7 Multiple Mirror Telescopes — 365
14.8 Active and Adaptive Optics — 367
References — 369

15. Visual Systems, Visual Telescopes, and Afocal Systems — 373
15.1 Visual Optical Systems — 373
15.2 Basic Telescopic System — 376
15.3 Afocal Systems — 379
15.4 Refracting Objectives — 382
15.5 Visual and Terrestrial Telescopes — 393
15.6 Telescope Eyepieces — 398
15.7 Relays and Periscopes — 407
References — 412

16. Microscopes — 415
16.1 Compound Microscope — 415
16.2 Microscope Objectives — 421
16.3 Microscope Eyepieces — 430
16.4 Microscope Illuminators — 434
References — 437

17.	**Projection Systems**	**439**
	17.1 Slide and Movie Projectors	439
	17.2 Coherence Effects in Projectors	440
	17.3 Main Projector Components	441
	17.4 Anamorphic Projection	445
	17.5 Overhead Projectors	446
	17.6 Profile Projectors	447
	17.7 Television Projectors	448
	References	449
18.	**Lens Design Optimization**	**451**
	18.1 Basic Principles	451
	18.2 Optimization Methods	452
	18.3 Glatzel Adaptive Method	453
	18.4 Constrained Damped Least Squares Optimization Method	455
	18.5 Merit Function and Boundary Conditions	463
	18.6 Modern Trends in Optical Design	469
	18.7 Flow Chart for a Lens Optimization Program	470
	18.8 Lens Design and Evaluation Programs	470
	18.9 Some Commercial Lens Design Programs	472
	References	474
Appendix 1.	**Notation and Primary Aberration Coefficients Summary**	**477**
	A1.1 Notation	477
	A1.2 Summary of Primary Aberration Coefficients	480
Appendix 2.	**Mathematical Representation of Optical Surfaces**	**485**
	A2.1 Spherical and Aspherical Surfaces	485
	References	494
Appendix 3.	**Optical Materials**	**497**
	A3.1 Optical Glasses	497
	A3.2 Optical Plastics	500
	A3.3 Infrared and Ultraviolet Materials	500
	Bibliography	501
Appendix 4.	**Exact Ray Tracing of Skew Rays**	**503**
	A4.1 Exact Ray Tracing	503
	A4.2 Summary of Ray Tracing Results	513

A4.3	Tracing Through Tilted or Decentered Optical Surfaces	515
References		518

Appendix 5. General Bibliography on Lens Design — **521**

Index — *523*

1
Geometrical Optics Principles

1.1 WAVE NATURE OF LIGHT AND FERMAT'S PRINCIPLE

The nature of light is one of the most difficult concepts in modern physics. Due to its quantum nature, light has to be considered in some experiments as an electromagnetic wave, and in some others it has to be considered as a particle. However, in ordinary optical instruments we may just think of the light as an electromagnetic wave with an electric field and a magnetic field, mutually perpendicular, and both perpendicular to the path of propagation. If the light beam is plane (linearly) polarized, the electric and the magnetic fields have a constant fixed orientation, changing only in magnitude and sign as the wave propagates. The electric and magnetic fields are in phase with each other, as shown in Fig. 1.1. This is the simplest type of wave, but we may find more complicated light beams, where the electric and magnetic fields do not oscillate in a fixed plane. The different manners in which the fields change direction along the light trajectory are called polarization states. It is shown in any physical optics textbook that any polarization state may be considered as a superposition of two mutually perpendicular plane-polarized light beams. The type of polarization depends on the phase difference between the two components and on their relative amplitudes as explained in any physical optics textbook. The frequency v and the wavelength λ of this wave are related by the speed of propagation v as follows

$$\lambda v = v \qquad (1.1)$$

Light waves with different frequencies have different colors, corresponding to certain wavelengths in the vacuum. In lens design the frequencies (or corresponding wavelengths in the vacuum) for the solar Fraunhofer lines are used to define the color of the light. These lines are shown in the Table 1.1.

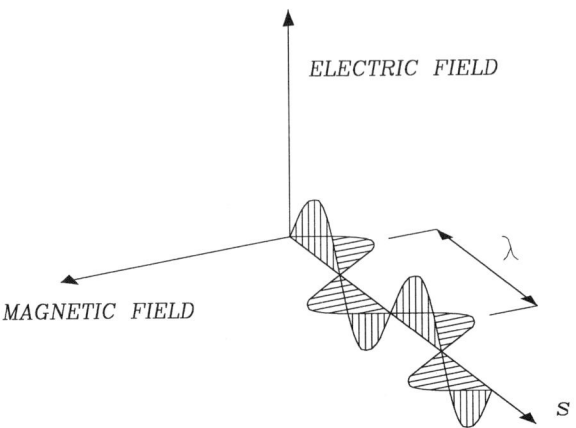

Figure 1.1 Electric and magnetic fields in an electromagnetic wave.

Table 1.1 Fraunhofer Lines and Their Corresponding Wavelengths

Line	Wavelength in nm.	Element	Color
i	365.01	Hg	UV
h	404.66	Hg	Violet
g	435.84	Hg	Blue
F'	479.99	Cd	Blue
F	486.13	H	Blue
e	546.07	Hg	Green
d	587.56	He	Yellow
D	589.29	Na	Yellow
C'	643.85	Cd	Red
C	656.27	H	Red
r	706.52	He	Red

Along the path of propagation of a light beam, the magnitude E of the electric field may be written as

$$E = A \exp i(ks - \omega t) = A \exp i(\phi - \omega t) \tag{1.2}$$

where A is the amplitude of the wave, k is the wavenumber, defined by $k = 2\pi/\lambda$, and ω is the angular frequency, defined by $\omega = 2\pi\nu$.

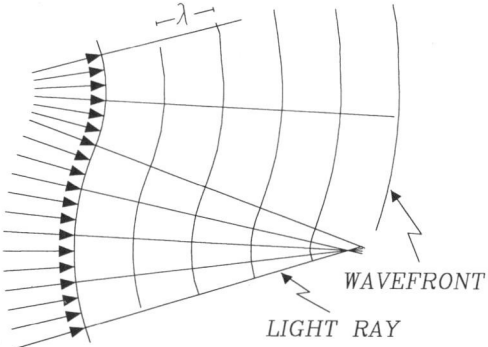

Figure 1.2 Light rays and wavefronts in an isotropic medium.

The wavelength is represented by λ and the frequency by ν. In this expression, s is the distance traveled along the light path, ϕ is the phase difference between the point being considered and the origin, and $\phi - \omega t$ is the instantaneous phase, assuming that it is zero at the origin for $t=0$. A wavefront in a light beam is a surface in space for which all points have the same instantaneous phase ϕ. Another equivalent definition given by Kidger (2001) is that a wavefront is a surface of constant optical path length, along the light path from a luminous point in the object. So, we may imagine on a light wave a family of surfaces in which the disturbance becomes a maximum at a certain time; i.e., the crests for the light waves. These surfaces are wavefronts and the distance between two consecutive wavefronts is the wavelength as illustrated in Fig. 1.2.

The speed of light in a vacuum is about 300,000 km/sec and it is represented by c. In any other transparent medium, the speed v is less than c (except in extremely rare conditions known as anomalous dispersion) and its value depends on the medium to be considered. The refractive index n for a material is defined as

$$n = \frac{c}{v} \tag{1.3}$$

For a given material, the refractive index n is a function of the light color (wavelength in the vacuum). As a general rule, this index decreases with increasing wavelength, as shown in Fig. 1.3 for two typical glasses. The index of refraction increases with the wavelength only in certain small spectral regions outside of the visible spectrum.

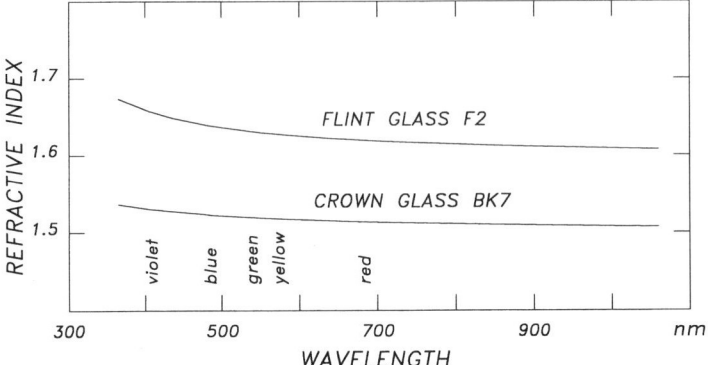

Figure 1.3 Refractive indices of a crown and a flint glass as a function of the wavelength.

Table 1.2 Refractive Indices for Some Optical Materials

Material	Refractive index
Vacuum	1.0000
Air	1.0003
Water	1.33
Fused silica	1.46
Plexiglass	1.49
Borosilicate crown	1.51
Ordinary crown	1.52
Canada Balsam	1.53
Light flint	1.57
Extra dense barium crown	1.62
Extra dense flint	1.72

The maximum sensitivity of the eye is for yellow light, with a peak at 555 nm. This is why the very common sodium doublet line D (589.3 nm) was originally chosen as the reference to measure refractive indices for visible light. Accuracy using this doublet, however, was found to be inconsistent. The more stable helium line d at 587.56 nm was adopted. Table 1.2 shows the refractive index at this wavelength for several transparent materials.

Although the formal definition of the refractive index is with respect to the vacuum, in practice it is measured and specified with respect to the air.

Using the definition for refractive index, the time t for light to go from a point $\mathbf{P}_1$ to another point $\mathbf{P}_2$ in an isotropic, homogeneous, or inhomogeneous medium is given by

$$t = \frac{1}{c} \int_{P_1}^{P_2} n \, ds \qquad (1.4)$$

where $ds^2 = dx^2 + dy^2 + dz^2$. It is convenient to define the optical path OP as

$$OP = \int_{P_1}^{P_2} n \, ds \qquad (1.5)$$

The direction in which a light beam propagates has been defined as a light ray. As we will see in Chap. 8, a ray cannot be isolated due to the phenomenon of diffraction. By using a diaphragm, we can try to isolate a single light ray, as shown in Fig. 1.4. This turns out to be impossible since when the aperture approaches the light wavelength, the light beam divergence increases. This effect is larger for smaller apertures. When an aperture or lens rim is large compared with the wavelength, the diffraction effects become less important and then we can approach the light ray concept with fair precision. The optics branch based on the concept of the light ray is known as *geometrical optics*.

An optically transparent medium is said to be homogeneous and isotropic if the light travels at the same speed in every direction inside the

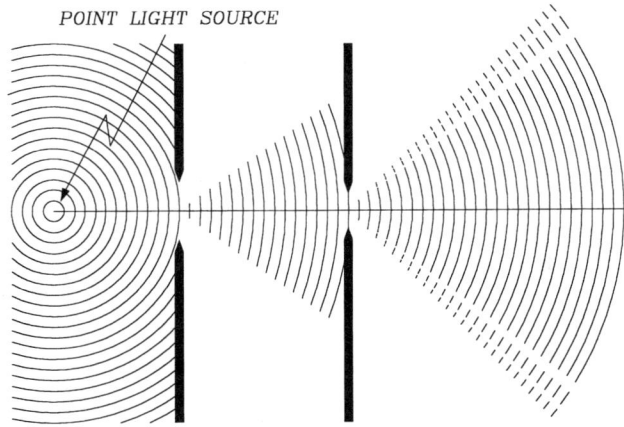

Figure 1.4 Unsuccessful attempt to isolate a single ray of light.

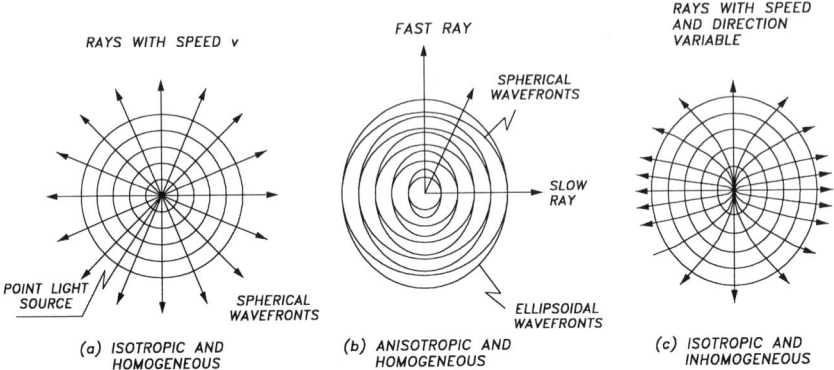

Figure 1.5 Wavefronts in different types of media: (a) isotropic and homogeneous; (b) anisotropic and homogeneous; (c) isotropic and inhomogeneous.

medium, independently of the orientation of the electric field (polarization), as shown in Fig. 1.5(a). A transparent medium is anisotropic (like in crystals) if the traveling velocity of the light is different for different orientations of the electric field (polarization state) of the wave, even if the traveling direction is the same, Fig. 1.5(b). Many crystals, like quartz or calcite, are anisotropic. In these materials, even if they are homogeneous (same refractive index for all points in the medium), depending on the polarization orientation, either a spherical or ellipsoidal wavefront is produced with a point light source. In this book we will consider only isotropic media. The medium is isotropic and inhomogeneous (like in gradient index glass to be described later or variable-density fluids) if the light speed depends on the direction of propagation, but not on the orientation of the electric field, Fig. 1.5(c).

Malus law—Equation (1.5) for the optical path may also be written in differential form as

$$\frac{dOP}{ds} = n \qquad (1.6)$$

where the OP is measured along any geometrical path ds. We define the *eikonal* φ as the optical path along trajectories always perpendicular to the wavefronts, related to the phase ϕ by $\varphi = \phi/k$. The *Malus law*, as illustrated in Fig. 1.6, states that in an isotropic medium, light rays are always perpendicular to the wavefront. We may mathematically state this law by means of the *eikonal equation*, which may be written as

$$|\nabla \varphi| = n \qquad (1.7)$$

Geometrical Optics Principles

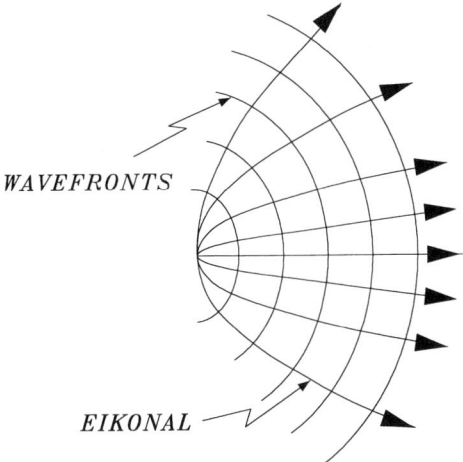

Figure 1.6 Propagation of wavefronts and light rays (eikonal).

As stated before, the Malus law is valid in homogeneous and inhomogeneous media but not in anisotropic media, like some crystals.

Fermat's Principle—This principle, which from the Malus law becomes a natural consequence, is the basis of all geometrical optics. It can be stated as follows:

"The path traveled by a light wave from a point to another is stationary with respect to variations of this path."

This is equivalent to saying that the time for the light to travel must be either the longest or smallest time or be stationary with respect to other trajectories. Figure 1.7 shows some examples for two cases, in which the light must go from point P_1 to P_2 after being reflected in a mirror. In inhomogeneous [Fig. 1.8(a)] or discontinuous [Fig. 1.8(b)] media there may also be several physically possible trajectories for the light rays. In this case the point P_1 is the object and the point P_2 is its image. The optical path along all of these trajectories from the object to the image is the same. This constant optical path is called, in Hamilton's theory of geometrical optics, the *point characteristic* of the system, because it depends only on the location of the initial and end points, not on the particular path.

1.1.1 Gradient Index of Refraction

The refractive index of glass can be made inhomogeneous on purpose by means of several experimental procedures. Then, we speak of a gradient index (GRIN) of refraction (Marchand, 1978; Moore, 1995). The most

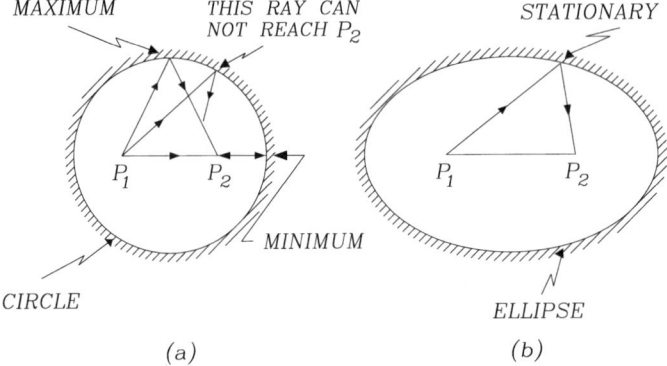

Figure 1.7 Illustration of Fermat's principle in a hollow sphere and a hollow ellipsoid.

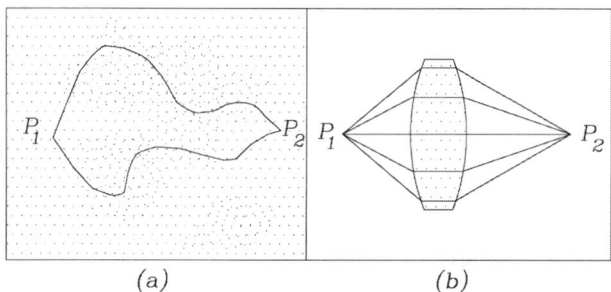

Figure 1.8 Optical path lengths from point P_1 to point P_2 are the same for any possible path.

common procedure for fabricating gradient index optical elements is by an ion-exchange process. They can be made out of glass, polymers, zinc selenide/zinc sulfide, and germanium. Gradient index optical fibers have also been made by a chemical vapor-deposition process. In nature, a gradient index frequently appears in the hot air above a road, creating a mirage.

The variation in the index of refraction or gradient index in a lens can be in the direction of the optical axis. This is called an *axial gradient index*. It can be in the perpendicular direction to the optical axis. This is a *radial gradient index*. It can also be symmetric about a point, which is the *spherical gradient index*. The spherical gradient is rarely used in optical components mainly because they are difficult to fabricate.

Geometrical Optics Principles

Gradient index lenses are very useful in optical instruments to correct many aberrations. Two popular examples of the use of gradient index optical elements are single lenses corrected for spherical aberration and imaging elements in endoscopes, as will be shown in later chapters.

1.2 REFLECTION AND REFRACTION LAWS

Reflection and refraction laws can be derived in a simple way using Fermat's principle, as follows.

1.2.1 Reflection Laws

The first reflection law states that the incident ray, the reflected ray, and the normal to the reflecting surface lay on a common plane. This law can be explained as a consequence from Fermat's principle.

The second law states that the magnitude of the reflected angle is equal to the magnitude of the incident angle. Consider Fig. 1.9, where a light ray leaves from point P_1 $(0, y_1)$ and reaches the point P_2 (x_2, y_2) after a reflection on a plane mirror at the point $P(x, 0)$. If the refractive index is n, the optical path from P_1 to P_2 is

$$OP = n\sqrt{x^2 + y_1^2} + n\sqrt{(x_2 - x)^2 + y_2^2} \tag{1.8}$$

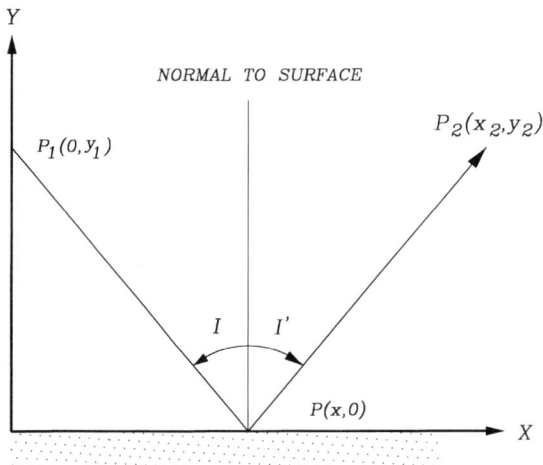

Figure 1.9 Derivation of the law of reflection.

Since this optical path must be an extremum, we set the condition:

$$\frac{dOP}{dx} = \frac{nx}{\sqrt{x^2 + y_1^2}} - \frac{n(x_2 - x)}{\sqrt{(x_2 - x)^2 + y_2^2}} = 0 \tag{1.9}$$

and from this last equation, we can easily see that

$$\sin I = -\sin I' \tag{1.10}$$

where the minus sign has been placed to introduce the convention that the angles I and I' have opposite signs because they are on opposite sides of the normal to the surface after reflection. Hence, we conclude that $I = -I'$, which is the second reflection law.

1.2.2 Refraction Laws

The first refraction law states that the incident ray, the refracted ray, and the refracting surface normal lie in a common plane. This law is also an immediate consequence from Fermat's principle.

The second refraction law, called also *Snell's Law*, can be derived from Fig. 1.10, where we can easily note that the optical path is given by

$$OP = n\sqrt{x^2 + y_1^2} + n'\sqrt{(x_2 - x)^2 + y_2^2} \tag{1.11}$$

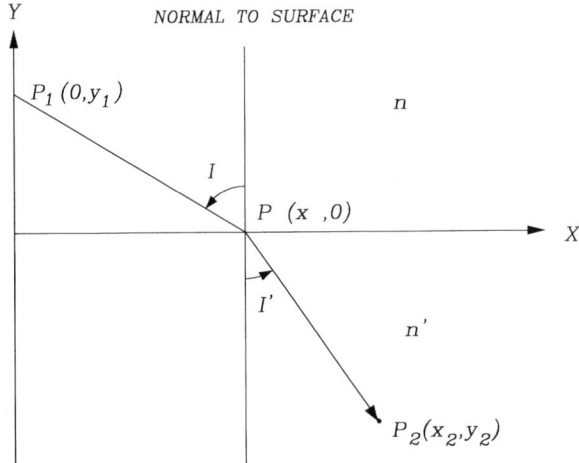

Figure 1.10 Derivation of the law of refraction.

Geometrical Optics Principles

By applying Fermat's principle, we impose the condition:

$$\frac{dOP}{dx} = \frac{nx}{\sqrt{(x^2+y_1)^2}} - \frac{n'(x_2-x)}{\sqrt{(x_2-x)^2+y_2^2}} = 0 \qquad (1.12)$$

where we can see that

$$n \sin I = n' \sin I' \qquad (1.13)$$

which is Snell's law. This relation becomes identical to the reflection law when the indices of refraction n and n' have the same magnitude but opposite sign. This fact is used to trace rays through optical systems with mirrors.

1.2.3 Vectorial Form of Refraction Laws

Frequently, it is not simple to apply Snell's law in three-dimensional space, especially after many reflections and refractions, when the light rays are not contained in a common plane. Then, it is a great advantage to use a vectorial form of the refraction law. This form may be easily derived with the help of Fig. 1.11. Let us define a vector $\mathbf{S}_1$ along the incident ray, with magnitude n, and a vector $\mathbf{S}_2$ along the refracted ray, with magnitude n'. Then, Snell's law may be written as

$$|\mathbf{S}_1| \sin I = |\mathbf{S}_2| \sin I' \qquad (1.14)$$

The refracted vector $\mathbf{S}_2$ is related to the incident vector $\mathbf{S}_1$ by

$$\mathbf{S}_2 = \mathbf{S}_1 - \mathbf{a} \qquad (1.15)$$

On the other hand, it is easy to see that the vector $\mathbf{a}$ is parallel to the vector $\mathbf{p}$ normal to the refracting surface, and that its magnitude is given by

$$\begin{aligned} |\mathbf{a}| &= \Gamma \\ &= (|\mathbf{S}_2| \cos I' - |\mathbf{S}_1| \cos I) \\ &= n' \cos I' - n \cos I \end{aligned} \qquad (1.16)$$

Thus, the final vectorial law of refraction is given by

$$\mathbf{S}_2 = \mathbf{S}_1 - \Gamma \mathbf{p} \qquad (1.17)$$

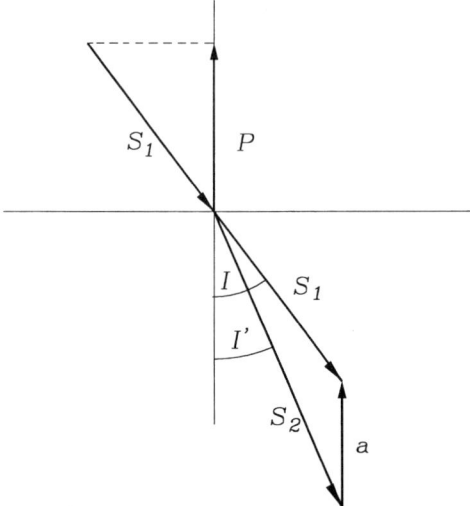

Figure 1.11 Derivation of the vectorial law of refraction.

where Γ is given by

$$\Gamma = n'\cos I' - n\cos I$$
$$= n'\left[\left(\frac{n}{n'}\cos I\right)^2 - \left(\frac{n}{n'}\right)^2 + 1\right]^{1/2} - n\cos I \qquad (1.18)$$

1.3 BASIC MERIDIONAL RAY TRACING EQUATIONS

A spherical refracting surface is the most common surface in optics. A plane surface may be considered a special case of a spherical surface, with an infinite radius of curvature. In a spherical refracting surface like the one shown in Fig. 1.12, we define the following parameters:

    1. *Center of curvature:* The center of an imaginary sphere that contains the refracting surface.

    2. *Radius of curvature:* The distance from the refracting surface to the center of curvature.

    3. *Vertex:* A point on the refracting surface, at the center of its free aperture. This definition assumes that the aperture is circular and centered and that the surface is spherical. More generally, if the surface is not spherical but has rotationally symmetry, the vertex is the point where the

Geometrical Optics Principles

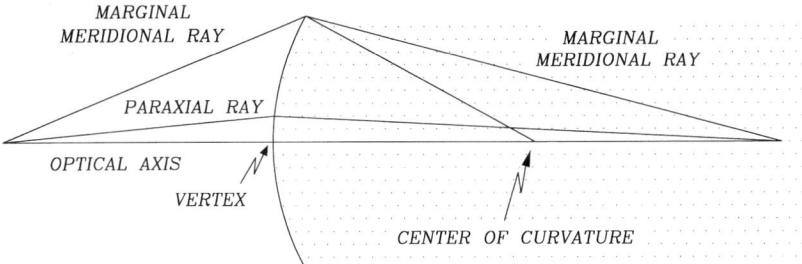

Figure 1.12 Some definitions in a refractive spherical surface.

axis of symmetry intersects the optical surface. Even more generally, we may say that the vertex is the local origin of coordinates to which the surface function is referred.

4. *Optical axis:* An imaginary straight line passing through the center of curvature and the vertex. For the case of nonspherical surfaces with rotational symmetry, the optical axis is the axis of symmetry.

According to their direction, rays incident on a refractive spherical surface are classified as follows:

1. *Meridional ray:* Any ray in a common plane with the optical axis, called the meridional plane. In this case, the surface normal and the refracted ray are also contained on the meridional plane.

2. *Oblique or skew ray:* Any nonmeridional ray. In this case, the ray is not in a common plane with the optical axis.

3. *Paraxial ray:* A meridional or skew ray that has a small angle with respect to the optical axis is a paraxial ray. However, in a more general way we can say that a paraxial ray is an approximation to a real ray, obtained by assuming valid small angle approximations.

In aberration theory, *axial, tangential,* and *sagittal rays* are also defined. Axial rays are meridional rays originating in an object point on the optical axis. Tangential rays are meridional rays originating in an off-axis object point, The meridional plane that contains the object point is called the *tangential plane.* On the other hand, the sagittal rays are skew rays contained in a single plane, called the *sagittal plane,* perpendicular to the tangential plane and containing the center of the entrance pupil (to be defined later in this chapter) of the optical system. These concepts will become more clear later when the tangential and sagittal planes are defined. To clarify these concepts, the reader is advised to see Fig. 7.6 in Chap. 7.

Meridional rays are used to trace rays through a spherical refracting surface. The behavior of meridional rays permits us to obtain many

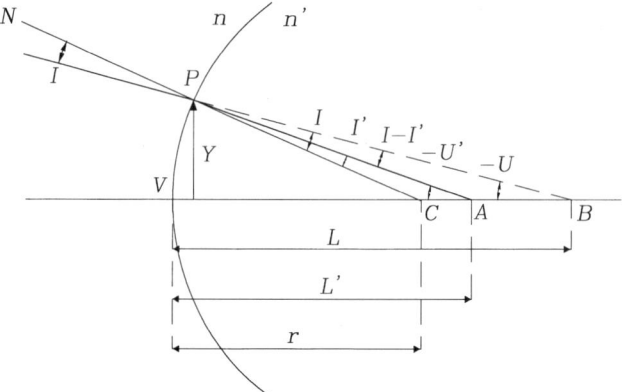

Figure 1.13 Meridional ray refracted at a spherical surface.

interesting properties of optical systems. Skew rays are mathematically more complex than meridional rays and their study is described in Appendix 4.

Figure 1.13 shows a spherical refracting surface and a meridional ray intersecting the surface at the point **P**. The surface normal at **P** is **N** and the curvature center is **C**.

A convention sign must be defined for all the parameters in Fig. 1.13. Such convention has to be consistent with most textbooks and commercial optical design programs. Unfortunately, there are many notations in books and the most widely used departs from the old definition by Conrady (1957). The sign convention used in this book, assuming that the light travels from left to right, is as follows, where primed quantities are used after refraction on the surface:

1. *Radius of curvature r:* Positive if the center of curvature is to the right of the vertex and negative otherwise. The curvature c is the inverse of the radius of curvature ($c = 1/r$).

2. *Angles U and U':* In agreement with analytic geometry, they are positive if the slope of the meridional ray is positive and negative otherwise. [Conrady (1957) and Kingslake (1965) use the opposite convention.]

3. *Angles I and I':* The angle of incidence I is positive if the ray arrives at the surface from left to right, below the normal, or from right to left above the normal. This angle is negative otherwise. The angle of refraction I' is positive if the ray leaves from the surface from left to right, above the normal, or from right to left below the normal. This angle is negative otherwise. This sign convention is illustrated in Fig. 1.14.

4. *Distances L and L':* L is the distance from the vertex of the surface to the intersection of the meridional ray before refraction (object) with the

Geometrical Optics Principles

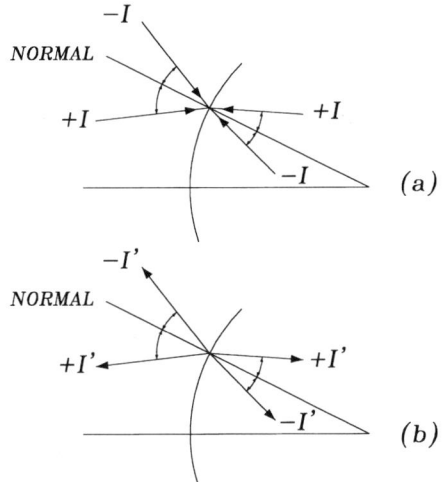

Figure 1.14 Sign convention for the angles of incidence and refraction.

optical axis. It is positive if this object is to the right of the vertex, and negative if it is to the left. L' is the distance from the vertex of the surface to the intersection of the meridional ray after refraction (image). It is positive if this image is to the right of the vertex, and negative if it is to the left. This rule is valid for the light traveling from left to right, as well as for light traveling from right to left.

5. *Thickness t:* Positive when the next surface in the optical system lies to the right of the optical surface being considered and negative if it lies to the left of it.

6. *Ray height Y:* It is positive if the ray crosses the optical surface above the optical axis and negative otherwise.

7. *Refractive index n:* It is positive if the light travels in this medium from left to right and negative if it travels in the opposite sense. The index of refraction changes its sign at any reflective surface, in order to be able to use the law of refraction on any reflection.

It is interesting to see that according to this convention, for any particular ray, not all three parameters L, Y, and U can be positive at the same time. Observing again Fig. 1.13, with a meridional ray and where negative parameters are indicated with a minus sign, we can apply sine law to the triangle **PCB**:

$$\frac{\sin I}{L-r} = \frac{-\sin U}{r} \tag{1.19}$$

and by using the same law to the triangle **PCA**:

$$\frac{\sin I'}{L' - r} = \frac{-\sin U'}{r} \tag{1.20}$$

Since triangles **PCB** and **PCA** both share a common angle, and since the sum for the internal angles in both triangles adds up to 180° it must be true that

$$I - U = I' - U' \tag{1.21}$$

and finally we write Snell's law:

$$n \sin I = n' \sin I' \tag{1.22}$$

From these relations, parameters r, n, and n' are fixed and known, while L, L', I, I', U, and U' are variables. Since we have four equations, all remaining variables can be calculated if any two of the three parameters L, I, U for the incident ray are specified.

An optical system is generally formed by many optical surfaces, one after the other. We have a *centered optical system* when the centers of curvature of all the surfaces lie on a common line called the *optical axis*. In these systems formed by several surfaces, all parameters relating to the next surface are represented by the subscript +1. Then, the transfer equations are

$$U_{+1} = U' \tag{1.23}$$
$$n_{+1} = n' \tag{1.24}$$

and

$$L_{+1} = L' - t \tag{1.25}$$

where t is the distance from the vertex of the surface under consideration to the vertex of the next surface.

1.3.1 Meridional Ray Tracing by the *L–U* Method

The equations in the preceding section have been described by Conrady (1957) and may be used to trace rays. This is the so-called L–U method, because the incident as well as the refracted rays are defined by the distances L and L' and the angles U and U'. Although these equations are exact, they

Geometrical Optics Principles

are never used in present practice to trace rays because they break down for plane and low curvature surfaces, and L and L' become infinite for rays parallel to the optical axis.

1.3.2 Meridional Ray Tracing by the Q–U Method

An alternative ray tracing method defines the meridional ray by the angle U and the perpendicular segment Q from the vertex of the surface to the meridional ray, as shown in Fig. 1.15. A line from **C**, perpendicular to the line Q, divides this segment into two parts. Thus, from this figure we may see that

$$\sin I = Qc + \sin U \tag{1.26}$$

where the curvature $c = 1/r$ has been used instead of the radius of curvature r. Then, from the refraction law in Eq. (1.22), we have

$$\sin I' = \frac{n}{n'} \sin I \tag{1.27}$$

and from Eq. (1.21):

$$U' = U - I + I' \tag{1.28}$$

From Eq. (1.26) we may obtain an expression for Q and, placing primes on this result, the value of Q' is obtained as

$$Q' = \frac{\sin I' - \sin U'}{c} \tag{1.29}$$

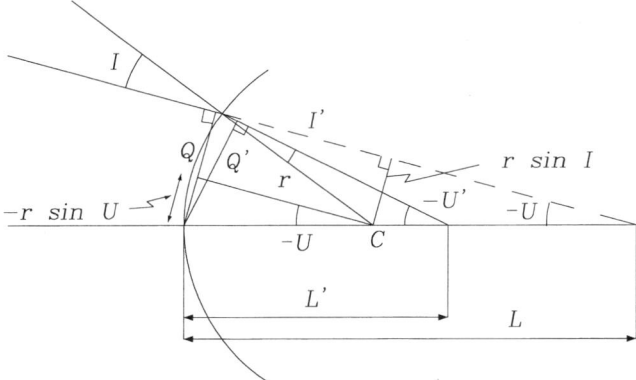

Figure 1.15 Meridional ray tracing by the Q–U method.

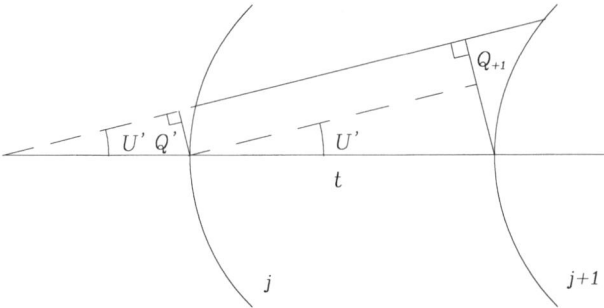

Figure 1.16 Derivation of a transfer relation for ray tracing of meridional rays.

In order to obtain the equivalent expression when the surface is flat ($c = 0$) we may see from Fig. 1.15 that $Q = -L \sin U$ and for flat surfaces $\tan U = Y/L$; thus, it is possible to show that an alternative expression for flat surfaces is

$$Q' = Q \frac{\cos U'}{\cos U} \tag{1.30}$$

The transfer equation is, as clearly illustrated in Fig. 1.16, as

$$Q_{+1} = Q' + t \sin U' \tag{1.31}$$

1.4 GAUSSIAN OR FIRST-ORDER OPTICS

First-order optics is formed by all ray tracing relations using a paraxial ray approximation. The location of the image for a given object position in an optical system using first-order optics may be found by means of the so-called Gauss formula, to be derived next. For this reason, gaussian optics is considered to be synonymous with first-order optics.

The Gauss formula is one of the main achievements of first-order optics and it can be derived from Eqs. (1.19) to (1.22). Before doing any paraxial approximations, we will work with the exact equations. From Eq. (1.19) we can obtain:

$$\frac{L}{r} = 1 - \frac{\sin I}{\sin U} = \frac{\sin U - \sin I}{\sin U} \tag{1.32}$$

from this we see that

$$\frac{r}{L} = 1 - \frac{\sin I}{\sin I - \sin U} \qquad (1.33)$$

and then, by multiplying both sides of this equation by n/r:

$$\frac{n}{L} = \frac{n}{r} - \frac{n}{r}\frac{\sin I}{\sin I - \sin U} \qquad (1.34)$$

and similarly, using Eq. (1.20):

$$\frac{n'}{L'} = \frac{n'}{r} - \frac{n'}{r}\frac{\sin I'}{\sin I' - \sin U'} \qquad (1.35)$$

we now subtract Eq. (1.35) from Eq. (1.34) and use Snell's law [Eq. (1.22)] to obtain:

$$\frac{n'}{L'} - \frac{n}{L} = \frac{n'-n}{r} + \frac{n \sin I}{r}\left[\frac{1}{\sin I - \sin U} - \frac{1}{\sin I' - \sin U'}\right] \qquad (1.36)$$

This relation is exact, but now we will develop the paraxial approximations. A paraxial ray approximation is taken by substituting the trigonometric functions $\sin U$ and $\sin I$ by the angles I and U in radians. These approximations are valid without a significant loss in precision if the angles I and U are very small. As explained before, first-order or gaussian optics is the branch of geometrical optics that uses only paraxial rays. The equations for first-order optics are obtained by replacing in the exact equations the following:

$$\sin I \Rightarrow i$$
$$\sin I' \Rightarrow i'$$
$$\sin U \Rightarrow u$$
$$\sin U' \Rightarrow u' \qquad (1.37)$$
$$L \Rightarrow l$$
$$L' \Rightarrow l'$$

obtaining from Eqs. (1.19) to (1.22):

$$\frac{i}{l-r} = \frac{-u}{r} \qquad (1.38)$$

$$\frac{i'}{l'-r} = \frac{-u'}{r} \qquad (1.39)$$

$$-u + i = -u' + i' \qquad (1.40)$$

$$ni = n'i' \qquad (1.41)$$

and the transfer equations (1.23) and (1.25) now are

$$u_{+1} = u' \tag{1.42}$$

and

$$l_{+1} = l' - t \tag{1.43}$$

Variables L and L' have been substituted by l and l' in order to distinguish exact values from paraxial approximations. Most of the lens and optical systems properties can be obtained with fair precision using first-order optics, except for monochromatic aberrations.

By approximating Eq. (1.36) for paraxial rays (first order) and using Eq. (1.40), we obtain finally the so-called Gauss formula:

$$\frac{n' - n}{r} = \frac{n'}{l'} - \frac{n}{l} \tag{1.44}$$

With this equation, we can obtain the distance l' from the refracting surface to the image, for a given l from the surface to the object. This distance l', so obtained, is independent of the incidence angle. From this we can conclude that, within first-order optics limits, a point object produces a point image.

The Gauss equation is so important that it has been obtained using many different approaches using ray as well as wave optics. A comparison of all these methods has been given by Greco et al. (1992).

Frequently, lens designers prefer to use a Gauss equation in terms of the angles u and u', instead of the distances l and l'. Then, this relation becomes, by using the curvature c instead of the radius r:

$$(n' - n)cy = -n'u' + nu \tag{1.45}$$

where the ray height y is related to the distances l and l' and the angles u and u' are

$$u = -\frac{y}{l} \tag{1.46}$$

and

$$u' = -\frac{y}{l'} \tag{1.47}$$

in accordance with our sign convention. Then, the transfer equation (1.43) is substituted by

$$y_{+1} = y + tu' \tag{1.48}$$

1.4.1 Paraxial Ray Tracing by y–nu Method

Meridional paraxial rays may be traced through an optical system by means of the following set of equations. They assume that the surface data (r, t, and n) are known, as well as the initial data for the light ray. These initial data in the y–nu method are the ray height y and the product nu of the refractive index n by the angle u.

The ray may be traced by the following relation derived from the Gauss equation (1.45):

$$[n'u'] = [nu] - (n' - n)yc \tag{1.49}$$

with the transfer equation (1.48) written as

$$y_{+1} = y + \frac{t[n'u']}{n'} \tag{1.50}$$

If the value of the angle of incidence is wanted, it may be computed with the formula [obtained from Eqs. (1.38) and (1.46)]:

$$i = yc + u \tag{1.51}$$

1.4.2 Delano's Relation

An interesting relation that relates the refraction of a paraxial ray with that of a marginal meridional ray has been found by Delano (1952). Let us consider a paraxial and a marginal ray as in Fig. 1.17. These rays do not necessarily originate at the same object point.

The perpendicular distances from the vertex of the optical surface to the incident and the refracted marginal rays are Q and Q', respectively. The perpendicular distances from the crossings of the incident and the refracted paraxial rays with the optical axis to the incident and refracted marginal rays are s and s', respectively. Thus, the marginal ray is defined by Q and U and the paraxial ray is defined by y and u. It is easy to see that

$$s = Q + l \sin U \tag{1.52}$$

Therefore, multiplying both sides of this expression by u and using Eq. (1.46):

$$su = Qu - y \sin U \tag{1.53}$$

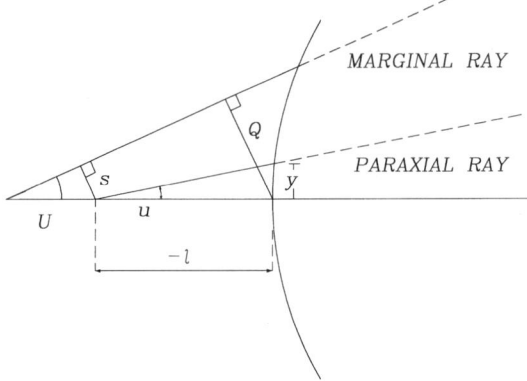

Figure 1.17 Derivation of Delano's relation.

Using Eq. (1.26), and Eq. (1.51), we obtain

$$snu = -yn \sin I + Qni \tag{1.54}$$

and in a similar manner, we may obtain for the refracted ray:

$$s'n'u' = -yn' \sin I' + Q'n'i' \tag{1.55}$$

Thus, subtracting Eq. (1.55) from Eq. (1.54) and using Snell's law [Eqs. (1.22) and (1.41)] we finally find that

$$s'n'u' = snu + (Q' - Q)ni \tag{1.56}$$

This is Delano's relation, which will be useful when studying the spherical aberration, as will be shown later.

1.5 IMAGE FORMATION

A refracting surface, a lens or a lens system establishes a one-to-one correspondence between a point in the object plane to a point in the image plane, when an image is formed. An image-forming system function is to refract (or reflect) light coming from a point in the object and send it to a single point in the image, as shown in Fig. 1.18.

Geometrical Optics Principles

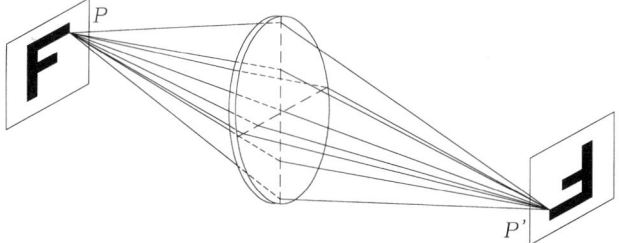

Figure 1.18 Image formation by an optical system.

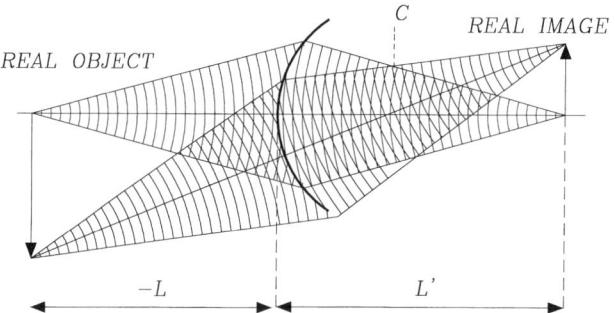

Figure 1.19 Formation of a real image with a real object.

An object, depending on its position with respect to the image-formation system, can be of two types:

1. *Real object:* An object is real when its distance L from the optical surface is negative; in other words, when the object is located to the left of the optical surface, as shown in Figs. 1.19 and 1.20. Conversely, when the light travels from the right to the left, the object is real when L is positive. A real object may be present when a real physical object or the image formed by another optical system is used.

2. *Virtual object:* An object is virtual when the distance L from the optical surface is positive; in other words, when the object is located to the right of the optical surface, as shown in Figs. 1.21 and 1.22. When the light travels from the right to the left, the object is virtual when L is negative. Let us consider another optical system located between the optical system and its image. This new optical system will change the image position, size, and perhaps its orientation. The image from the first optical system is the virtual object for the second system.

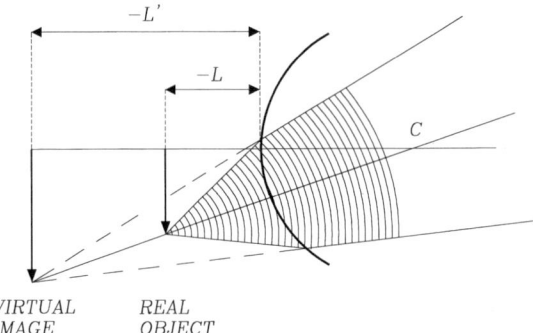

Figure 1.20 Formation of a virtual image with a real object.

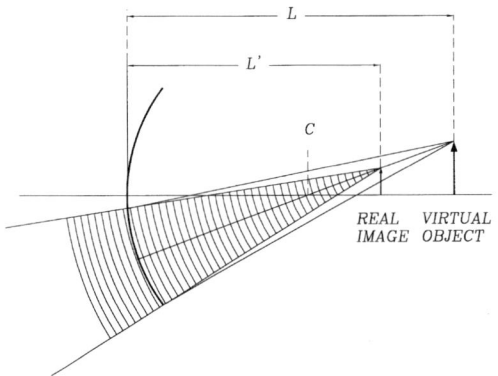

Figure 1.21 Formation of a real image with a virtual object.

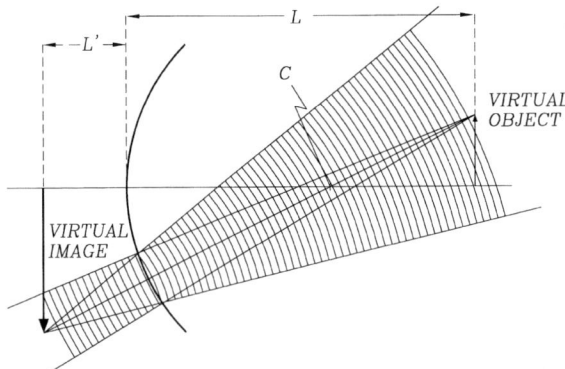

Figure 1.22 Formation of a virtual image with a virtual object.

As the object, the image can be both real and virtual as follows:

1. *Real image:* An image is real if the distance L' for the optical surface is positive; in other words, when the image is located to the right of the optical surface, as shown in Figs. 1.19 and 1.21. When the light travels from the right to the left, the image is real when L' is negative. A real image can be observed in either of two ways: by placing a screen where the image is formed, or by looking directly with the eye from a large distance from the place where the image is formed.

2. *Virtual image:* An image is virtual if the distance L' for the optical surface is negative; in other words, when the image is located to the left of the optical surface, as shown in Figs. 1.20 and 1.22. When the light travels from the right to the left, the image is virtual when L' is positive. When the rays emitted from a single point in the object are not convergent but divergent after passing through the optical surface or system, the image is virtual. The light beam will have an apparent diverging point, where the virtual image is formed. These images may be observed directly with the eye, but they may not be formed on a screen.

1.6 STOP, PUPILS, AND PRINCIPAL RAY

The refracting or reflecting surfaces in an optical system are not infinite in size, but limited, generally to a round shape. This finite transverse extension limits the beam of light passing through them. Let us consider a centered optical system. If the light beam entering this system comes from a point object on the optical axis, very likely only one of the surfaces will limit the transverse extension of the beam, as shown in Fig. 1.23. This limiting surface is called the *stop* of the system. If the stop is a diaphragm, we may think of it as a dummy refracting surface whose refractive indices are the same before and after the surface (diaphragm). The system stop may be at any surface. It need not be in the middle or at one end of the system; some optical surfaces are located before the stop and some others after it. If the stop is observed from the entrance of the system, it will be observed through the surfaces that precede it, changing its apparent size and position. This observed image of the stop is called the *entrance pupil*. If the stop is observed from the back of the system, it will be observed through the surfaces that are after it, changing again its apparent size and position. This observed image of the stop is called the *exit pupil*.

As shown in Fig. 1.23, of all meridional rays going from a point off-axis on the object plane, to the point on the image plane, only one passes through the center of the stop. This ray is the *principal ray,* defined as the

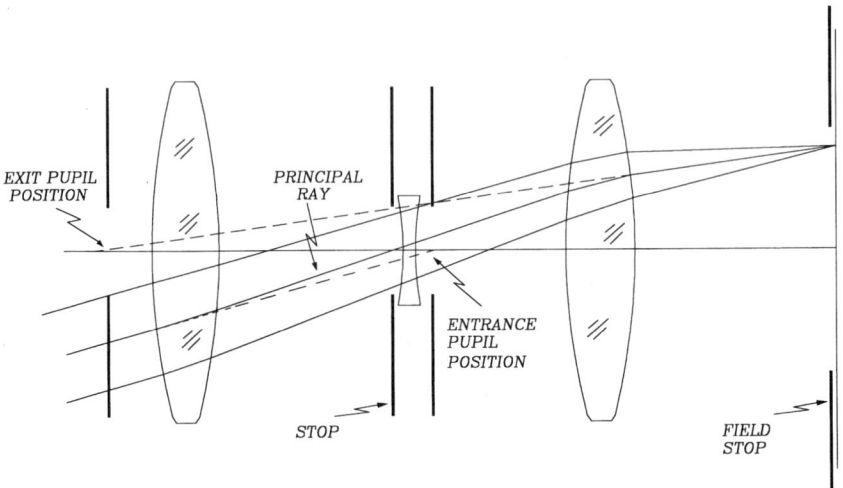

Figure 1.23 Definitions of principal ray, entrance pupil, and exit pupil.

ray that passes through the off-axis point object and the center of the stop. The intersection of the extension of the segment of the principal ray in the object space with the optical axis is the center of the entrance pupil. Similarly, the intersection of the extension of the segment of the principal ray in the image space with the optical axis is the center of the exit pupil.

An image of the stop can also be observed from any medium in the optical system, not only from the object or image media. As shown in Fig. 1.24 the real (or virtual) image of the stop is located at the point where the principal ray (or its extension) crosses the optical axis. This image of the stop is the *pupil* of that surface or medium.

All quantities referring to the principal ray are represented with a bar on top of the symbol; for e.g., $\bar{y}$ is the paraxial height of the principal ray and $\bar{u}$ is its paraxial angle with respect to the optical axis. By definition, the value of $\bar{y}$ is equal to zero at the stop. All quantities referring to the axial rays (meridional rays from a point object on the axis) are written without the bar.

The meridional ray heights at the pupil for the medium j are represented by Y_{pj} for the marginal rays or y_{pj} for the paraxial rays. The meridional ray heights at the entrance and exit pupils are represented by Y_{entr} and Y_{exit} for the marginal rays and y_{entr} and y_{exit} for the paraxial rays.

Summarizing, the stop is the aperture that limits the amount of light entering the optical system and its images are the pupils. The field stop, on the other hand, is located on the image plane and limits the image lateral extension, as shown in Fig. 1.23.

Geometrical Optics Principles

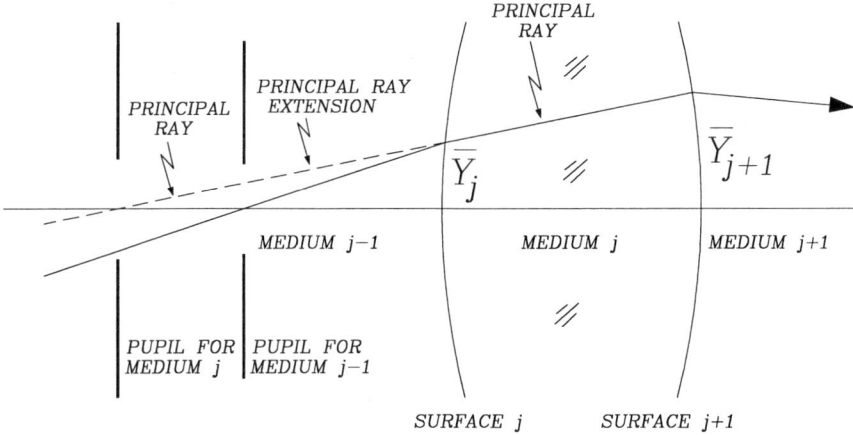

Figure 1.24 Location of the pupil of a surface in an optical system.

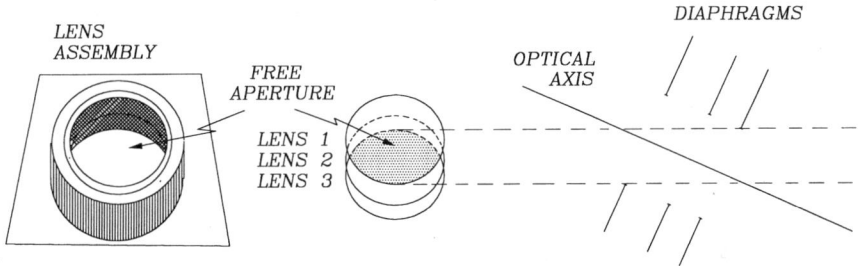

Figure 1.25 Vignetting in a lens.

If the light beam entering the system comes from an off-axis object point, as shown in Fig. 1.25, several surfaces may limit the transverse extension of the beam, producing an apparent aperture with a nearly elliptical shape. Then, the system is said to have *vignetting*. The vignetting effect appears only when the angle of incidence of the beam exceeds a certain limit. It is frequently desirable to avoid vignetting in a centered optical system, as shown in Fig. 1.26, to avoid excessive decreasing of the illuminance of the image at the edge of the field and to have a better control of the image analysis during the design stage. Some times, however, vignetting is introduced on purpose, to eliminate some aberrations difficult to correct.

The tangential and sagittal planes, defined previously in Section 1.3, may now be more formally defined. The tangential plane is a meridional

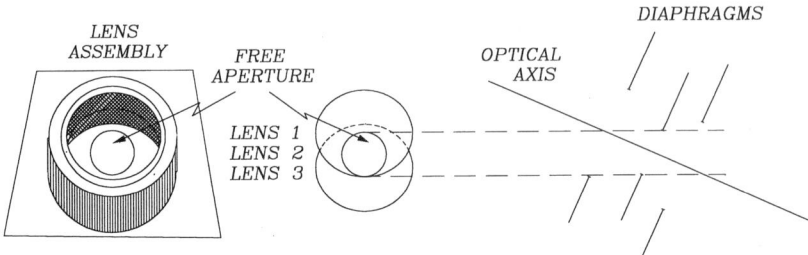

Figure 1.26 Stop size to avoid vignetting in a lens for a given off-axis angle.

plane that contains the principal ray (also off-axis point object). The sagittal plane is a plane perpendicular to the tangential plane, which contains the principal ray. As we may notice, there is a single common tangential plane for all media between two consecutive optical surfaces in a centered optical system. However, there is a sagittal plane for each medium, because the principal ray is refracted at each surface.

In order to trace the principal ray through an optical system we must know its direction in the object medium. This direction must be such that the principal ray passes through the center of the stop.

1.6.1 Telecentric Systems

A *frontal telecentric* system is one that has its entrance pupil placed at infinity. Since the stop (diaphragm) is at the back focal plane, the object must be at a finite distance to avoid forming the image on the focal plane. Let us consider the optical system in Fig. 1.27(a) where the principal ray is parallel to the optical axis, since the entrance pupil is at infinity. A small defocusing by a small change in the distance from the object to the system does not introduce any change in the magnification of the image. This property makes these systems useful for measuring systems where small defocusings do not introduce any errors.

A *rear telecentric* system has its exit pupil at infinity as in Fig. 1.27(b). The stop is at the front focal plane. The object may be at any distance from the system. In these systems a small defocusing by changing the distance from the optical system to the observing screen does not change the image size.

An optical system may be *simultaneously frontal and rear telecentric*, with both the object and the image at finite distances from the system. In this case the stop is in the middle of the system, at the back focal plane of the part of the system preceding the stop and at the front focal plane of the part of the system after the stop, as in Fig. 1.27(c).

Geometrical Optics Principles 29

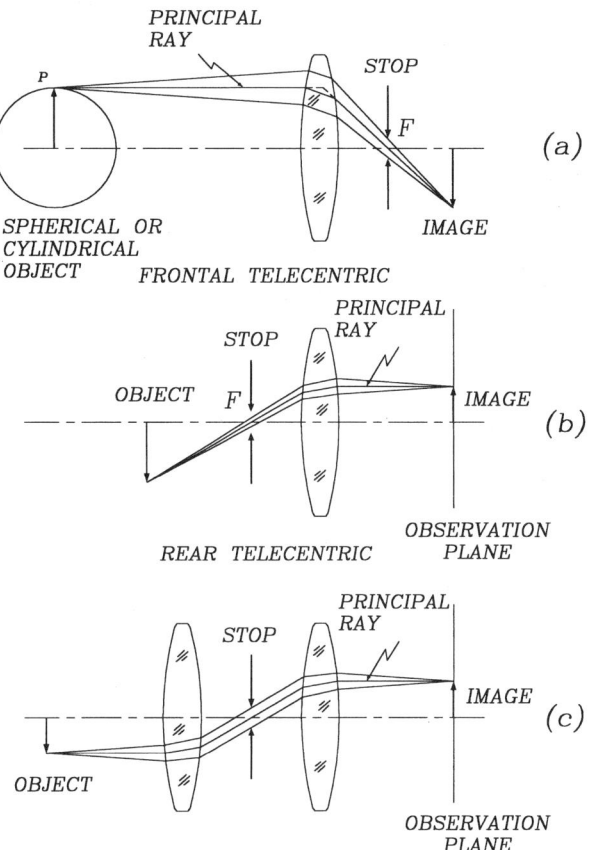

Figure 1.27 Telecentric lenses.

1.7 OPTICAL SINE THEOREM

This theorem was discovered almost simultaneously by Abbe and Helmholtz. Before studying it let us define the *auxiliary axis* of an optical surface as an imaginary straight line that passes through an off-axis point object (for that surface) and its center of curvature. Thus, every surface in a centered system has a different auxiliary axis. The sine theorem defines the ratio between the image size and the degree of convergence or divergence for the rays in the image plane. This theorem is derived with the help of Fig. 1.28. Let us consider an object point **O** with height H and the auxiliary axis. Also, let us assume that the image **O**′ is on the auxiliary optical axis

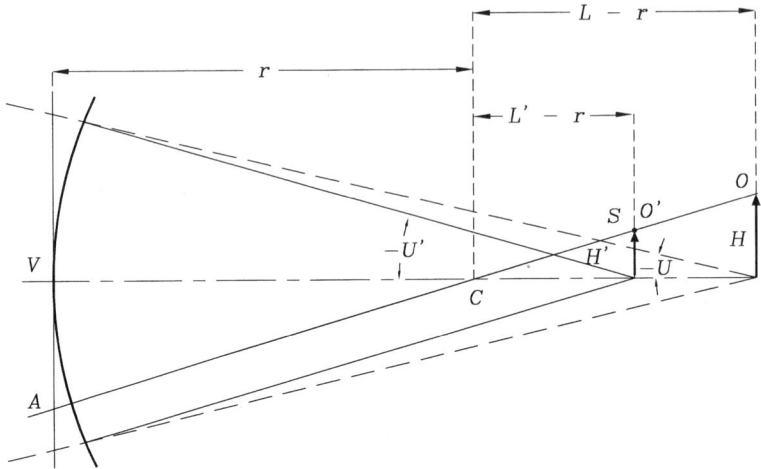

Figure 1.28 Optical sine theorem.

(this is true only for sagittal rays, as will be shown), with height H'. Then, we can see that

$$\frac{H'}{H} = \frac{L' - r}{L - r} \qquad (1.57)$$

By using now Eqs. (1.19) and (1.20) we can see that

$$nH \sin U = n'H' \sin U' \qquad (1.58)$$

Now, let us prove that the sagittal image **S** is on the auxiliary axis, by means of Fig. 1.29. The rays $\mathbf{T}_1$ and $\mathbf{T}_2$ are two tangential rays, passing through the upper edge and the lower edge of the entrance pupil, respectively. These two rays converge at a point **T** called the tangential focus, not necessarily at the auxiliary optical axis. Two sagittal rays $\mathbf{S}_1$ and $\mathbf{S}_2$ are symmetrically placed with respect to the auxiliary axis and converge at the point **S** called the sagittal focus. Due to the symmetry about the auxiliary axis, the point **S** is on this axis.

If the field is small (small H' compared with the radius of curvature) the sagittal focus **S** approaches the point **O′**. Thus, we may say this relation, known as the optical sine theorem, is strictly valid only for sagittal rays and, for relatively small off-axis displacements of the image, so that the distance $O'S$ may be neglected.

Geometrical Optics Principles

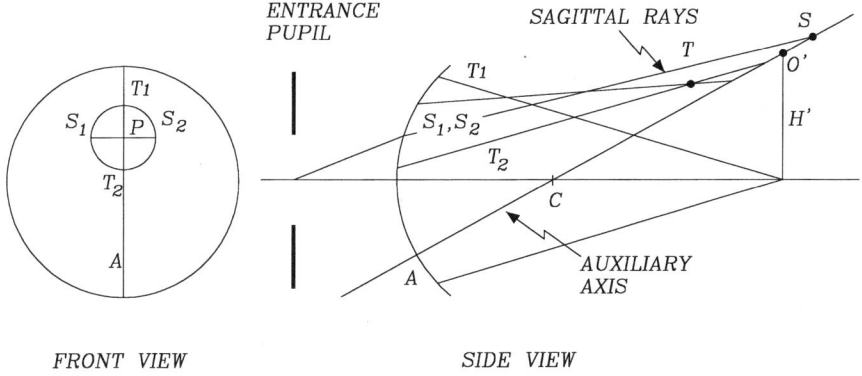

Figure 1.29 Proof that the sagittal image is located on the auxiliary axis.

It is important to notice that a small field is assumed, but not a paraxial approximation for the meridional marginal ray. This means that the angles U and U' may be large and the optical sine theorem is still valid.

The triple product $n H \sin U$ is said to be an optical invariant because any optical system formed with centered refracting and/or reflecting surfaces maintains its magnitude throughout all surfaces in the optical system.

Since the exit pupil is an image of the entrance pupil we can apply this theorem to these pupils and the principal ray by writing

$$n_1 Y_{\text{entr}} \sin \overline{U}_1 = n'_k Y'_{\text{exit}} \sin \overline{U}'_k \qquad (1.59)$$

where Y_{entr} and Y'_{exit} are the heights of the meridional ray at the entrance end exit pupils, respectively.

1.7.1 Lagrange Invariant

The paraxial approximation of the optical sine theorem is known as the *Lagrange theorem* and it is written as

$$\Lambda = hnu = h'n'u' \qquad (1.60)$$

where Λ is called the Lagrange invariant, since it has a constant value for all optical surfaces in the optical system. The sagittal image position calculated with the Lagrange theorem also falls on the auxiliary axis, but longitudinally displaced (if there is spherical aberration) to the paraxial focus plane. In the

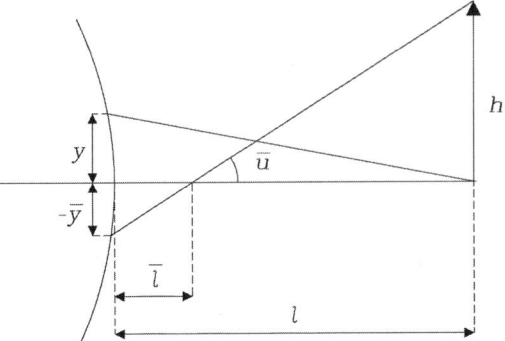

Figure 1.30 Derivation of an alternative form of the Lagrange invariant.

paraxial approximation there is no astigmatism, so the tangential image must be coincident with the sagittal image at the auxiliary axis. This image position is also known as the gaussian image. The Lagrange theorem may be physically interpreted as an energy conservation theorem, since the amount of light collected from the object by the optical system is directly proportional to the square of the angle u. Thus, the image illuminance of an extended object is independent of the distance from the lens to the object and depends only on the diameter of this lens and the distance to the image.

There is an alternative form of the Lagrange invariant, useful in aberration theory. To derive this form let us consider an optical system with a meridional ray and the principal ray as shown in Fig. 1.30. The object height h may be written as

$$h = (l - \bar{l})\bar{u} \tag{1.61}$$

thus, the Lagrange invariant may be written, by using Eq. (1.46) and its equivalent for the principal ray, as

$$\begin{aligned}\Lambda &= nu\bar{u}(l - \bar{l}) \\ &= n(\bar{y}u - y\bar{u}) = n'(\bar{y}u' - y\bar{u}')\end{aligned} \tag{1.62}$$

This form of the invariant may also be readily obtained from the Gauss equation (1.45) by writing it for the meridional ray as well as for the principal ray and then taking the ratio of the two equations.

Another invariant for the magnification of the pupils can be obtained from the paraxial approximation of Eq. (1.59) as follows:

$$n_1 Y_{\text{entr}} \bar{u}'_1 = n'_k \, y_{\text{exit}} \bar{u}'_k \tag{1.63}$$

1.8 HERSCHEL INVARIANT AND IMAGE MAGNIFICATIONS

The *lateral magnification* of an optical system is defined as

$$m = \frac{H'}{H} \tag{1.64}$$

where H is the object height and H' is the image height. Using the optical sine theorem this magnification may be written:

$$m = \frac{n \sin U}{n' \sin U'} \tag{1.65}$$

We see that the lateral magnification of an optical system depends on the magnitude of convergence U' of the axial rays from an object on the optical axis and also on the ratio n/n' of the indices of refraction. The paraxial approximation of this magnification is

$$m = \frac{h'}{h} = \frac{nu}{n'u'} \tag{1.66}$$

The *longitudinal magnification* $\overline{m}$ is defined as the ratio of a small longitudinal displacement $\Delta l'$ of the image and the corresponding displacement Δl of the object. By differentiating the Gauss equation (1.44) we obtain

$$\overline{m} = \frac{\Delta l'}{\Delta l} = \frac{n\, l'^2}{n'\, l^2} \tag{1.67}$$

but using relations in Eqs. (1.46) and (1.47) we find that

$$\overline{m} = \frac{\Delta l'}{\Delta l} = \frac{n\, u^2}{n'\, u'^2} \tag{1.68}$$

and then, rewriting this expression:

$$\Delta l\, n u^2 = \Delta l'\, n' u'^2 \tag{1.69}$$

This quantity $\Delta l\, n u^2$ has a constant value for all surfaces of the system, before and after refraction and is called the *Herschel invariant*. From relations (1.66) and (1.68) we may find that the two types of magnification are related by

$$\overline{m} = \frac{n'}{n} m^2 \tag{1.70}$$

Thus, we may see that the lateral magnification is equal to the longitudinal magnification only if the ratio of the indices of refraction n/n' is equal to the lateral magnification m.

Using the invariant in Eq. (1.63), the *paraxial angular magnification M* of an optical system, defined as the ratio of the slopes of the principal ray after and before being refracted by the optical system, can be written as

$$M = \frac{\overline{u}'_k}{\overline{u}_1} = \frac{n_1 y_{\text{entr}}}{n'_k y_{\text{exit}}} \tag{1.71}$$

The ratio $y_{\text{exit}}/y_{\text{entr}}$ is called the *pupil magnification*. If the object and image medium is air the angular magnification is equal to the ratio of the diameters of the entrance pupil to the exit pupil, i.e., equal to the inverse of the pupil magnification.

The optical invariants described in this chapter are not the only ones. There are some others, like the skew invariant described by Welford (1968).

To conclude this chapter let us now study an interesting relation, similar to an invariant, but not from surface to surface as the ones just described. Instead, this is a quantity whose value remains constant for any incident orientation and path of the refracted ray. This relation is easily found from Eq. (1.40) and Snell's law, given by Eq. (1.41), as follows:

$$\frac{u - u'}{i} = \frac{\overline{u} - \overline{u}'}{\overline{i}} = 1 - \frac{n}{n'} \tag{1.72}$$

Since this expression is a constant for any ray, it has the same value for the meridional and for the principal ray. Another interpretation of this relation is that the change in direction of the ray on refraction at the spherical surface $(u - u')$ is directly proportional to the angle of incidence i.

1.9 RAY ABERRATIONS AND WAVE ABERRATIONS

We have seen at the beginning of this chapter that in an isotropic medium the light rays are defined by the normals to the wavefront. Let us assume

Geometrical Optics Principles

that an almost spherical wavefront converges to a point in the image. If the wavefront is not exactly spherical, we say that the wavefront is aberrated. Let us now consider an aberrated wavefront with deformations $W(x, y)$ with respect to the reference sphere, which are related to the transverse aberrations $TA_x(x, y)$ and $TA_y(x, y)$ by

$$\frac{\partial W(x,y)}{\partial x} = -\frac{TA_x(x,y)}{r_W - W(x,y)} \qquad (1.73)$$

and

$$\frac{\partial W(x,y)}{\partial y} = -\frac{TA_y(x,y)}{r_W - W(x,y)} \qquad (1.74)$$

where r_W is the radius of curvature of the reference sphere. These exact expressions were derived by Rayces (1964). The plane where the transverse aberrations are measured contains the center of curvature of the reference sphere, as shown in Fig. 1.31. In general, the radius of curvature of the reference sphere r_W is much larger than the wave aberration $W(x,y)$. Then, with a great accuracy, enough for most practical purposes, we may approximate this expression by

$$\frac{\partial W(x,y)}{\partial x} = -\frac{TA_x(x,y)}{r_W} \qquad (1.75)$$

and

$$\frac{\partial W(x,y)}{\partial y} = -\frac{TA_y(x,y)}{r_W} \qquad (1.76)$$

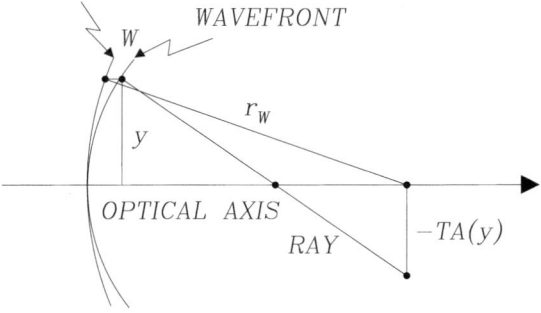

Figure 1.31 Relation between ray and wave aberrations.

If the transverse aberrations are known, the wavefront deformations may be calculated by integration of these aberrations, as follows:

$$W(x,y) = -\frac{1}{r_W} \int_0^x TA_x(x,y)\,dx \tag{1.77}$$

and

$$W(x,y) = -\frac{1}{r_W} \int_0^x TA_y(x,y)\,dy \tag{1.78}$$

More details on the fundamentals of geometrical optics may be found in the book by Herzberger (1963) and in the chapters by Hopkins and Hanau (1962) and Hopkins and Malacara (1988).

REFERENCES

Conrady, A. E., *Applied Optics and Optical Design*, Dover Publications, New York, 1957.

Delano, E., "A General Contribution Formula for Tangential Rays," *J. Opt. Soc. Am.*, **42**, 631–633 (1952).

Greco, V., Molesini, G., and Quercioli, F., "Remarks on the Gaussian Formula for the Refraction at a Single Spherical Interface," *Am. J. Phys.*, **60**, 131–135 (1992).

Herzberger, M., "Some Recent Ideas in the Field of Geometrical Optics," *J. Opt. Soc. Am.*, **53**, 661–671 (1963).

Hopkins, R. E. and Hanau, R., "Fundamentals of Geometrical Optics," in *Military Standardization Handbook: Optical Design, MIL-HDBK 141*, U.S. Defense Supply Agency, Washington, DC, 1962.

Hopkins, R. E. and Hanau, R., "First Order Optics," in *Military Standardization Handbook: Optical Design, MIL-HDBK 141*, U.S. Defense Supply Agency, Washington, DC, 1962.

Hopkins, R. E. and Malacara, D., "Applied Optics and Optical Methods," in *Methods of Experimental Physics, Geometrical and Instrumental Optics*, Vol. 25, D. Malacara, ed., Academic Press, San Diego, CA, 1988.

Kidger, M. J., *Fundamentals of Optical Design*, SPIE. The International Society for Optical Engineering, Bellingham, WA, 2001.

Kingslake, R., "Basic Geometrical Optics," in *Applied Optics and Optical Engineering*, Vol. I, R. Kingslake, ed., Academic Press, New York, 1965.

Marchand, E. W., *Gradient Index Optics*, Academic Press, New York, 1978.

Moore, D. T., "Gradient-Index Materials," in *CRC Handbook of Laser Science and Technology, Supplement 1: Lasers,* M. J. Weber, ed., 499–505, CRC Press, New York, 1995.

Rayces, J. L., "Exact Relation Between Wave Aberration and Ray Aberration," *Opt. Acta,* **11**, 85–88 (1964).

Welford, W. T., "A Note on the Skew Invariant of Optical Systems," *Opt. Acta,* **15**, 621–623 (1968).

2
Thin Lenses and Spherical Mirrors

2.1 THIN LENSES

A lens is a glass plate whose faces are spherical, concave, or convex and almost parallel at the center. Let us consider a beam of parallel light rays (collimated) arriving at the first lens face. If these rays converge to a point (focus) after being refracted by the lens, the lens is *convergent or positive*. If the rays diverge the lens is *divergent or negative*. (See Fig. 2.1.)

In the most common case, when the medium surrounding the lens is less dense (smaller index of refraction) than the lens material, a lens thicker at the center than at the edge is convergent, and a lens thinner at the center is negative. A thin lens may have different shapes, as shown in Fig. 2.2.

Since the lens has two spherical surfaces (a plane surface is a spherical one with an infinite radius of curvature), we define the *optical axis* as the line that passes through the two centers of curvature. If the lens has a plane surface, the optical axis is the line that passes through the center of curvature of the spherical surface and is perpendicular to the plane surface. Obviously, this axis passes through the thickest or thinnest part of the lens.

The *focus* of the lens is the point where the collimated beam of light converges to, or diverges from, after being refracted by the lens. The *focal length* is the distance from the thin lens to the point of convergence or divergence, being positive for convergent lenses and negative for divergent lenses. The *power P* of the lens is defined as the inverse of the focal length, as follows:

$$P = \frac{1}{f} \qquad (2.1)$$

A common unit for the power is the *diopter*, when the focal length is expressed in meters.

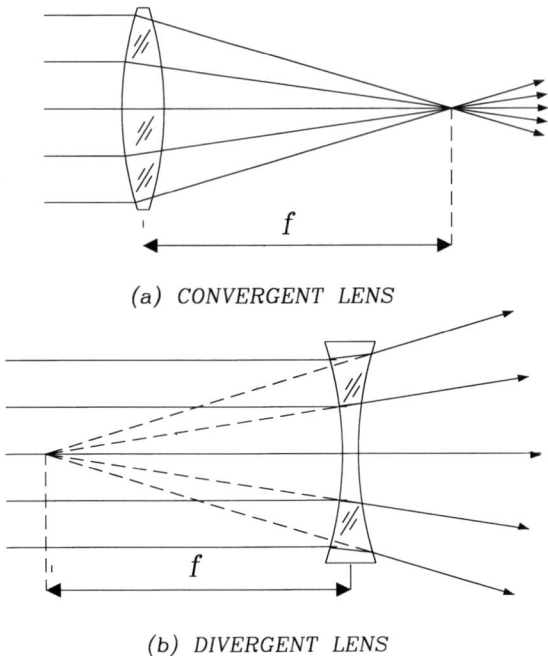

(a) CONVERGENT LENS

(b) DIVERGENT LENS

Figure 2.1 A convergent and a divergent lens: (a) convergent lens; (b) divergent lens.

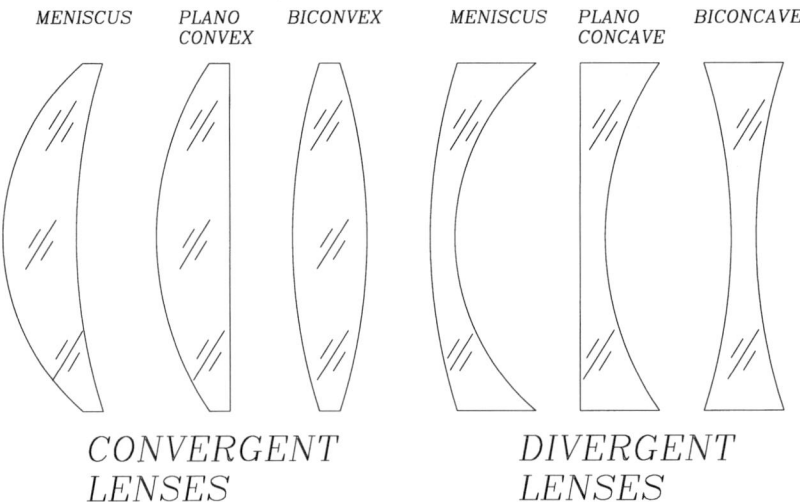

Figure 2.2 Possible shapes for thin lenses.

Thin Lenses and Spherical Mirrors

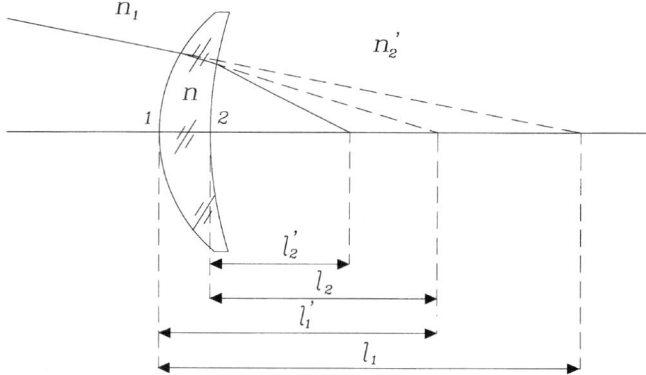

Figure 2.3 Ray refraction in a thin lens.

To understand how a lens refracts a beam of light let us apply the *Gauss law* (*Eq. 1.44*) to both surfaces of the lens, as in Fig. 2.3. For the first surface we may write

$$\frac{n'_1}{l'_1} - \frac{n_1}{l_1} = \frac{n'_1 - n_1}{r_1} \tag{2.2}$$

and for the second surface:

$$\frac{n'_2}{l'_2} - \frac{n_2}{l_2} = \frac{n'_2 - n_2}{r_2} \tag{2.3}$$

Using now transfer equations *(1.42)* and *(1.43)* with the thickness t equal to zero ($l_2 = l'_1$), and denoting the refractive index $n'_1 = n_2$ of the lens by n we find, after adding both equations, that

$$\frac{n - n_1}{r_1} + \frac{n'_2 - n}{r_2} = \frac{n'_2}{l'_2} - \frac{n_1}{l_1} \tag{2.4}$$

where l_1 is the distance from the object to the lens and l'_2 is the distance from the lens to the image. The distance l_1 is positive if the object is virtual and negative if it is real. The distance l'_2 is positive if the image is real and negative if it is virtual.

Given a thin lens, the left-hand side of Eq. (2.4) is a constant, so the right-hand side must also be a constant for all positions of the object and the image. A particular case of interest is when the object is at infinity

($1/l_1 = 0$) and l'_2 is equal to the focal length f', for the collimated light beam arriving at the lens from the left side. Then, from Eq. (2.4) we obtain

$$\frac{n'_2}{f'} = \frac{n - n_1}{r_1} + \frac{n'_2 - n}{r_2} = \frac{n'_2}{l'_2} \tag{2.5}$$

If the collimated beam of light enters the thin lens traveling from right to left, the convergence point is at the focus at the left of the lens, at a distance f, given by

$$\frac{n_1}{f} = \frac{n - n_1}{r_1} + \frac{n'_2 - n}{r_2} = -\frac{n_1}{l_1} \tag{2.6}$$

These two focal lengths for a lens with different object and image media, which are different from the focal length when the lens is surrounded by air, are related by

$$\frac{n'_2}{f'} = \frac{n_1}{f} \tag{2.7}$$

In the particular case in which the media before the lens and after the lens is air ($n_1 = n'_2 = 1$), the focal lengths are identical ($f = f' = f$) and have the value:

$$\begin{aligned}\frac{1}{f} &= (n-1)\left[\frac{1}{r_1} - \frac{1}{r_2}\right] \\ &= (n-1)\kappa = (n-1)(c_1 - c_2) \\ &= P = P_1 + P_2\end{aligned} \tag{2.8}$$

where κ is called the total lens curvature. This is the so-called *lens maker's formula*.

2.2 FORMULAS FOR IMAGE FORMATION WITH THIN LENSES

Some ray paths in the formation of images with a thin lens are shown in Fig. 2.4. Notice that the ray through the center of the lens is straight only when the media before and after the lens are the same. From Eqs. (2.5) and (2.6) we may find a relation for the positions of the object and the image as

Thin Lenses and Spherical Mirrors

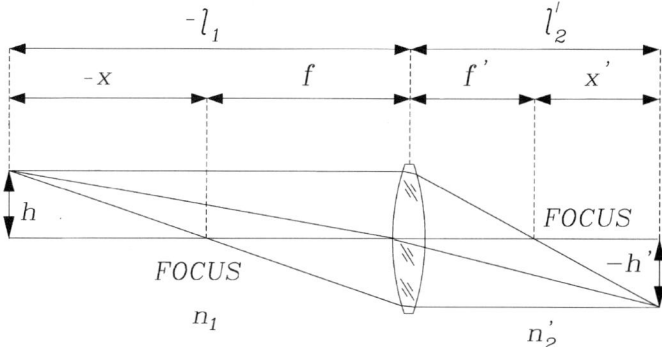

Figure 2.4 Image formation with a thin lens.

$$1 = \frac{f'}{l'_2} - \frac{f}{l_1} \tag{2.9}$$

Equivalently, if we define x as the distance from the focus f to the object, being positive if the object is to the right of this focus, as

$$x = l_1 + f \tag{2.10}$$

and x' as the distance from the focus f' to the image, being positive if the image is to the right of this focus, as shown in Fig. 2.4,

$$x' = l'_2 - f' \tag{2.11}$$

we may find

$$xx' = -ff' \tag{2.12}$$

which is known as *Newton's formula*.

If the lens is surrounded by air, from Eq. (2.9) we can write

$$\frac{1}{f} = \frac{1}{l'_2} - \frac{1}{l_2}$$

$$= \frac{1}{y}(u'_2 - u_1) \tag{2.13}$$

or

$$xx' = -f^2 \qquad (2.14)$$

The positions for the object and the image, defined by these relations, are said to be *conjugate* of each other.

The paraxial lateral magnification m may be found by any of the following relations, derived using Fig. 2.4 and Eqs. (2.7), (2.10), and (2.11):

$$\begin{aligned} m &= \frac{h'}{h} = \frac{n_1 l'_2}{n'_2 l_1} \\ &= \frac{f}{x} = -\frac{x'}{f'} \\ &= 1 - \frac{l'_2}{f'} = \left[1 + \frac{l_1}{f}\right]^{-1} \end{aligned} \qquad (2.15)$$

2.3 NODAL POINTS OF A THIN LENS

We have mentioned before that a ray passing through the center of the thin lens changes its direction after passing through the lens, unless the refractive indices of the media before and after the lens are the same. It is easy to see that the deviation of the central ray is given by

$$n_1 \sin \theta_1 = n'_2 \sin \theta_2 \qquad (2.16)$$

The nodal point **N** of a thin lens is defined as a point on the optical axis such that any ray entering the lens and pointing towards the nodal point, exits the lens without changing its direction. This nodal point position may be found from Fig. 2.3, with the condition $l'_2 = -l_1$. As shown in Fig. 2.5, using Eqs. (2.7) and (2.9), the distance $A = l'_2$ from the thin lens to the nodal point is given by

$$A = \left[1 - \frac{n_1}{n'_2}\right] f' \qquad (2.17)$$

As we may expect, if the thin lens is in air the nodal point is at the center of the lens.

Thin Lenses and Spherical Mirrors

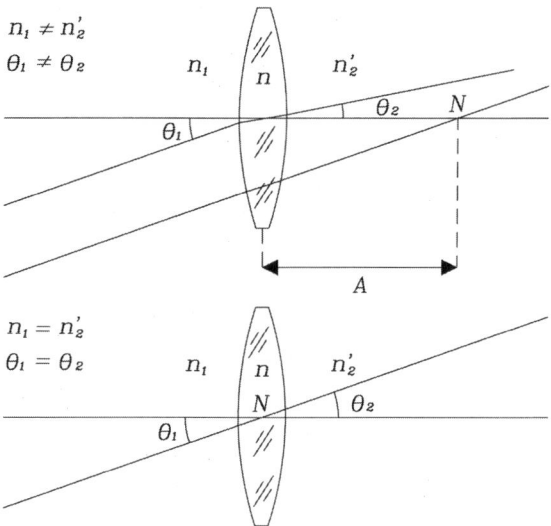

Figure 2.5 Location of the nodal point of a thin lens.

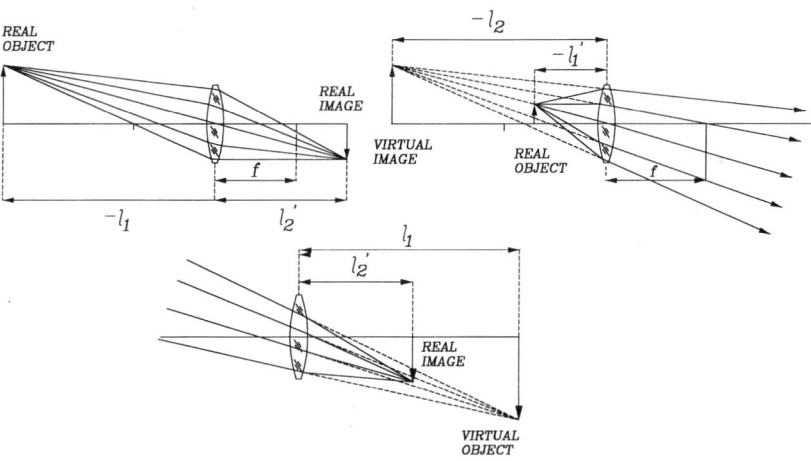

Figure 2.6 Image formation with a thin convergent lens.

2.4 IMAGE FORMATION WITH CONVERGENT LENSES

Figure 2.6 shows the three types of images that may be formed with convergent lenses. The image formation with convergent lenses may be studied by plotting in a diagram the values given by Eq. (2.13), as in Fig. 2.7.

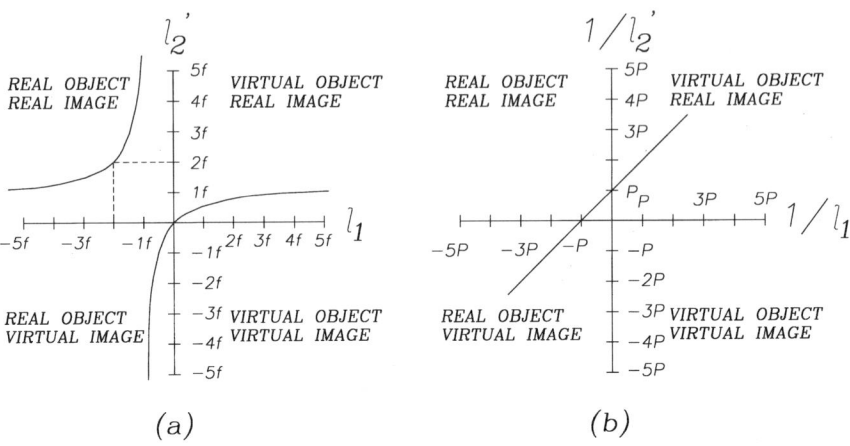

Figure 2.7 Diagram for image formation with convergent lenses.

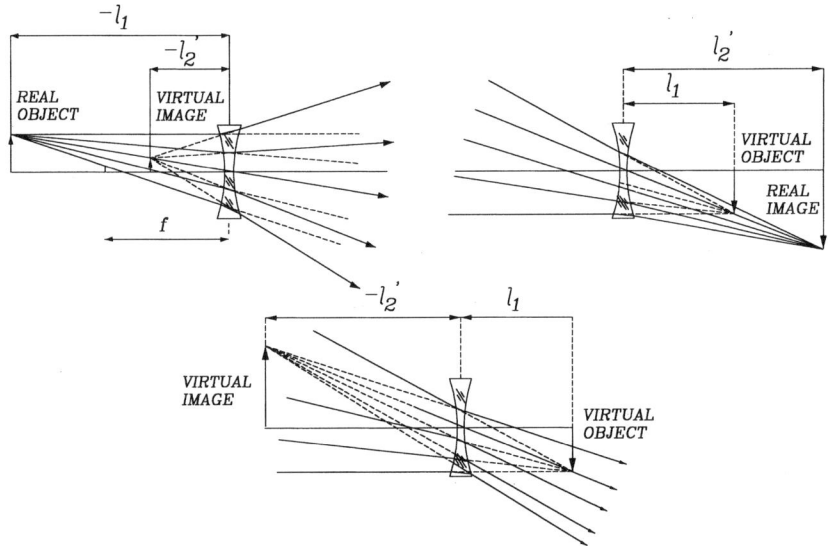

Figure 2.8 Image formation with thin divergent lenses.

As we see, it is not possible with a convergent lens to form virtual images with virtual objects. We may notice that when moving the object, near the focus f_1, there is a singularity. On one side of the focus the image is real and on the other side it is virtual.

Thin Lenses and Spherical Mirrors

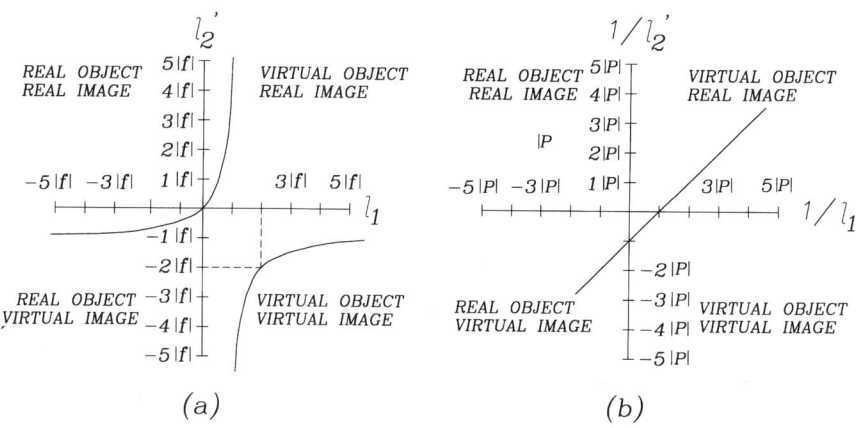

Figure 2.9 Diagram for image formation with diverging lenses.

2.5 IMAGE FORMATION WITH DIVERGENT LENSES

Figure 2.8 shows the three types of images that may be formed with divergent lenses. The image formation with divergent lenses may be studied in the same manner, with the diagram in Fig. 2.9. As we see, it is not possible with a divergent lens to form virtual images with real objects. Here, when moving a virtual object, near the focus f_2, there is a singularity. On one side of the focus the image is real and on the other side virtual. Additional details on the subject of first-order paraxial optics may be found in Hopkins and Hanau (1962a,b).

REFERENCES

Hopkins, R. E. and Hanau, R., "Fundamentals of Geometrical Optics," in *Military Standardization Handbook: Optical Design, MIL-HDBK 141*, U.S. Defense Supply Agency, Washington, DC, 1962a.

Hopkins, R. E. and Hanau, R., "First Order Optics," in *Military Standardization Handbook: Optical Design, MIL-HDBK 141*, U.S. Defense Supply Agency, Washington, DC, 1962b.

3
Systems of Several Lenses and Thick Lenses

3.1 FOCAL LENGTH AND POWER OF A LENS SYSTEM

The lateral magnification of a thick centered optical system in air, using the Lagrange theorem in Eq. (1.60) and the definition of lateral magnification for a distant object, is given by

$$m = \frac{u_1}{u'_k} = \frac{y_1}{l_1 u'_k} \tag{3.1}$$

assuming that the object and image media are the same (typically air).

The effective focal length of a thick lens or system of lenses is defined by

$$F' = -\frac{y_1}{u'_k} \tag{3.2}$$

hence, the lateral magnification with a distant object depends only on the effective focal length F, independently of the particular lens configuration. With this definition we may see from Fig. 3.1 that the effective focal length is the distance from the focal plane to an imaginary plane called the principal plane.

In general, there are two principal planes in any centered optical system, one for each orientation of the system. In Fig. 3.1 we have graphically defined the principal planes P_1 and P_2 and the effective focal lengths F and F'. It is interesting to notice that a system may be convergent and have a negative effective focal length, or divergent and have a positive effective focal length. This happens when the incident paraxial ray crosses the optical axis an odd number of times before reaching the focus.

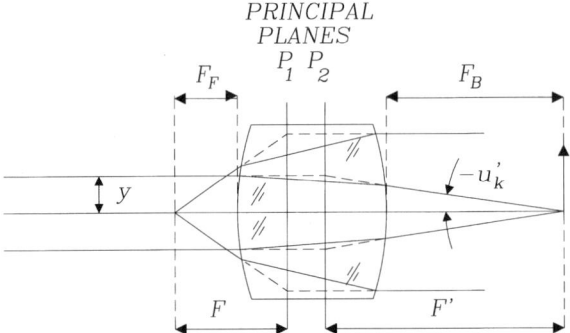

Figure 3.1 Diagram to illustrate the principal planes and the effective focal lengths.

Given a centered optical system, the effective focal length may be computed with the help of the Gauss equation (1.45). Summing this expression for a system of k surfaces we find that

$$-n'_k u'_k + n_1 u_1 = \sum_{i=1}^{k} y \left[\frac{n' - n}{r} \right] \tag{3.3}$$

then, using the definition of effective focal length, and making $u_1 = 0$, we find that

$$P = \frac{1}{F'} = \sum_{i=1}^{k} \frac{y}{y_1} \left[\frac{n' - n}{n'_k r} \right] \tag{3.4}$$

For the particular case of a system of thin lenses this expression becomes

$$P = \frac{1}{F} = \sum_{i=1}^{k} \frac{y}{y_1} \left[\frac{1}{f} \right] = \sum_{i=1}^{k} \frac{y_i}{y_1} P_i \tag{3.5}$$

where P_i is the power of the lens i. We see that the contribution to the total power of a surface or a thin lens in a system is directly proportional to the height y of the marginal ray on that surface or lens.

From Eq. (2.13) we can write the power of each individual lens as

$$P_i = \frac{1}{f_i} = \frac{1}{y}(u'_i - u_i) = \frac{1}{\bar{y}}(\bar{u}'_i - \bar{u}_i) \tag{3.6}$$

Systems of Several Lenses and Thick Lenses

The back focal length F_B and the front focal length F_F are also defined in Fig. 3.1.

3.2 IMAGE FORMATION WITH THICK LENSES OR SYSTEMS OF LENSES

Several important relations in thick optical systems may be found with only the definitions of effective focal length and Lagrange's theorem. To do this let us consider Fig. 3.2. In a first approximation, for paraxial rays we may write

$$\frac{u}{u'} = \frac{L'}{L} \tag{3.7}$$

and with the help of Lagrange's theorem we obtain the lateral magnification as

$$m = \frac{H'}{H} = \frac{nL'}{n'L} \tag{3.8}$$

This expression is analogous to Eq. (2.15) for thin lenses. From Fig. 3.2 we may obtain

$$\frac{-H'}{H} = \frac{L' - F'}{F'} \tag{3.9}$$

and

$$\frac{-H'}{H} = \frac{F}{-L - F} \tag{3.10}$$

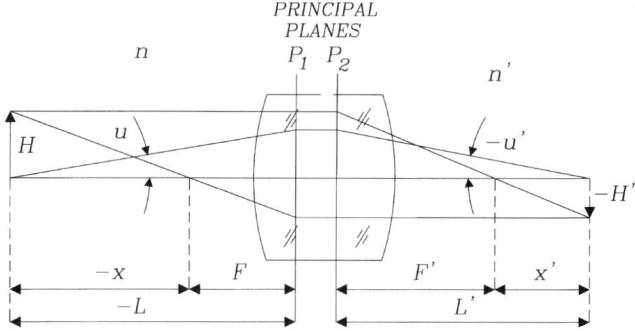

Figure 3.2 Image formation with a thick optical system.

We now define the distances $X = L + F$ and $X' = L' - F'$, where X' is positive if the image is to the right of the focus F' and X is positive if the object is to the right of the focus F. Then, we find again Newton's equation, as for thin lenses in Eq. (2.12),

$$XX' = -FF' \tag{3.11}$$

but we also may find, by equating Eqs. (3.9) and (3.10), that

$$1 = \frac{F'}{L'} - \frac{F}{L} \tag{3.12}$$

and from this expression:

$$L' - F' = -\frac{FL'}{L} \tag{3.13}$$

which, substituted into Eqs. (3.7) and (3.8), gives

$$\frac{n'}{F'} = \frac{n}{F} \tag{3.14}$$

The equivalent thin lens expression is Eq. (2.7). For the most common case when the object and the image refracting media are the same, the two focal lengths have the same value. Three possible exceptions are: (1) an underwater camera, with the object in water and the image in air, (2) an immersion microscope, where the object is in oil and the image in air, and (3) the human eye, where the object is in air and the image is in the eye's liquid.

Using now Eqs. (3.11) and (3.13) we find that

$$\frac{1}{F'} = \frac{1}{L'} - \frac{n}{n'L} \quad ; \quad \frac{1}{F} = \frac{n'}{nL'} - \frac{1}{L} \tag{3.15}$$

For the most common case of object and image media being the same we obtain

$$\frac{1}{F} = \frac{1}{L'} - \frac{1}{L} \tag{3.16}$$

whose thin lens is analogous to Eq. (3.14).

Systems of Several Lenses and Thick Lenses 53

3.3 CARDINAL POINTS

The nodal points of an optical system are two points on the optical axis, with the property that an incident ray pointing to nodal point N_1, after refraction comes out from the optical system pointing back to the nodal point N_2, and parallel to the incident ray. Let us consider Fig. 3.3 with a point light source S in the focal plane containing F. If rays R_1 and R_2 are emitted by S, after refraction they will come out from the optical system as rays R_3 and R_4, parallel to each other. The ray R_2 is selected so that it points to the nodal point N_1 so, by definition of nodal points, the ray R_4 will point back to the nodal point N_2, parallel to R_2. Since rays R_3 and R_4 are parallel to each other, the ray R_3 will also be parallel to R_2. The triangles SAN_1 and F_2P_2B are identical, hence the distances F_1N_1 and F_2P_2 are equal. Thus, we may write

$$F_1N_1 = F' \tag{3.17}$$

Then, the distance from the nodal point N_1 to the principal point P_1, as shown in Fig. 3.3, is

$$N_1P_1 = F - F_1N_1 = F - F' \tag{3.18}$$

Using now Eq. (3.14), we find that

$$N_1P_1 = \left[1 - \frac{n'}{n}\right]F \tag{3.19}$$

and symmetrically, the distance from the nodal point N_2 to the principal plane P_2 is

$$N_2P_2 = \left[1 - \frac{n}{n'}\right]F' \tag{3.20}$$

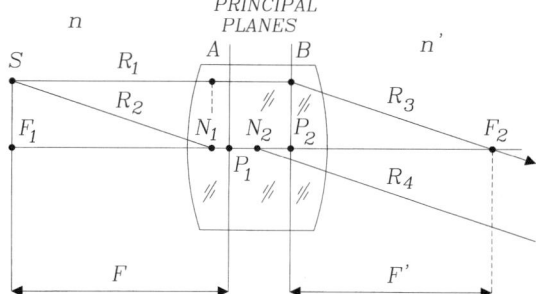

Figure 3.3 Cardinal points of a thick optical system.

The analogous thin lens expression for these relations is Eq. (3.18). If the object and image media are equal, the two nodal points coincide with the principal points. In this case the focal length may be measured by rotating the lens about a vertical axis until by trial and error the nodal point is found (Kingslake, 1932), as shown in Fig. 3.4. Nodal points, as well as principal points, receive the generic name of *cardinal points*.

An interesting consequence of the definition of nodal points is that they are images of each other with a unit lateral magnification. Thus, if the entrance pupil is located at the first nodal point, the exit pupil would be located at the second nodal point position and will have the same size as the entrance pupil.

Similarly to the nodal points, an optical system may also have in some cases (not always) a point with the property that an incident ray directed towards this point comes out of the system as emerging from the same point, as shown in Fig. 3.5(a) (Malacara, 1992). The direction of the incident ray

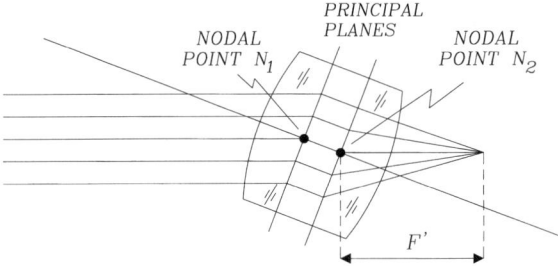

Figure 3.4 Measurement of the effective focal lens of a thick optical system.

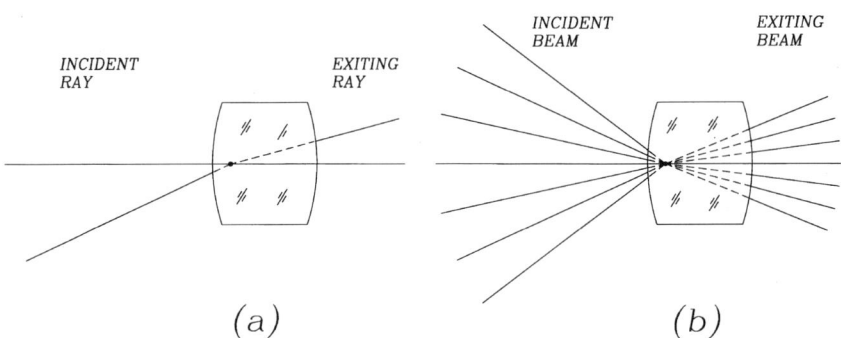

Figure 3.5 Points with incident and refracted rays intersecting at a common point on the optical axis.

Systems of Several Lenses and Thick Lenses

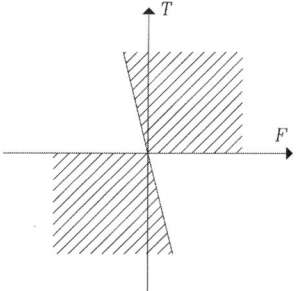

Figure 3.6 Regions for which the points illustrated in Fig. 3.5 exist.

and the emerging ray is not necessarily the same. Then, a converging beam focused on this point, as shown in Fig. 3.5(b), emerges from the system diverging from the same point, but not necessarily with the same angle of divergence as the entering beam.

The location of these points may be found by means of Eq. (3.17) (if the object and image media have the same refractive index), by imposing the condition:

$$L' + T = L \tag{3.21}$$

where T is the separation between the principal planes, so that this point is both the object and the image, at the same location. Thus, we may see that there are two of these points, at distances from the first principal point, given by

$$L = \frac{T \pm [T(T + 4F)]^{1/2}}{2} \tag{3.22}$$

Then, since the argument of the square root has to be positive, we see that these points exist only if

$$\frac{4F}{T} \geq -1 \tag{3.23}$$

as in the shaded regions shown in Fig. 3.6.

3.4 IMAGE FORMATION WITH A TILTED OR CURVED OBJECT

Let us consider a small plane object tilted with respect to the optical axis, in front of a convergent lens system as in Fig. 3.7. To find the inclination of the

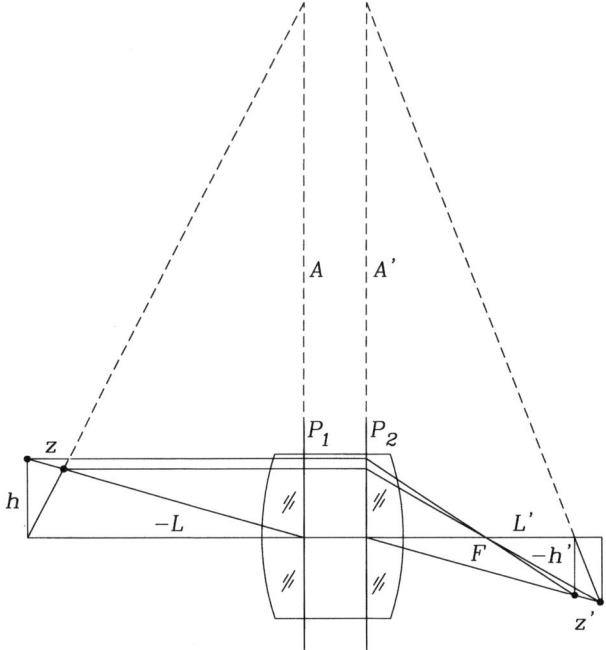

Figure 3.7 Image formation with a tilted object.

image plane let us begin by using Eqs. (1.66) and (1.68) to write

$$m^2 = \frac{z'}{z} \tag{3.24}$$

From this figure we can easily see that

$$\frac{z}{h} = -\frac{L}{A} \tag{3.25}$$

and

$$-\frac{z'}{h'} = \frac{L'}{A'} \tag{3.26}$$

where the distances A and A' are measured from the principal points to the lines of intersection of the principal planes with the plane inclined object and the plane inclined image, respectively. Using these three expressions and

Systems of Several Lenses and Thick Lenses 57

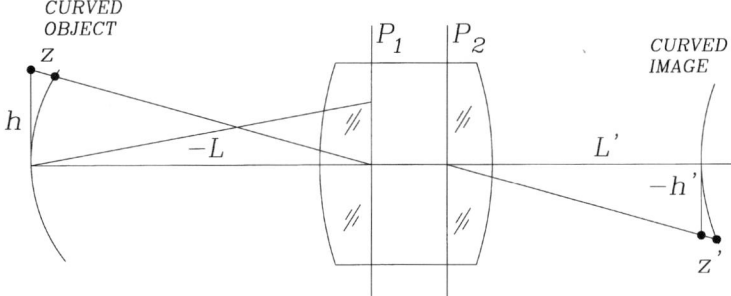

Figure 3.8 Image formation with a curved object.

Eq. (3.8) we can find that $A' = A$. When the imaging system is a thin lens, the object, the image, and the lens planes intersect at a common line.

It should be pointed out that when the object and the image planes are not parallel to each other a distortion of the image called keystone distortion appears. Then, the image of a square object is imaged as a trapezoid.

Let us now consider the case when the object is not flat but curved as in Fig. 3.8. Then, Eq. (3.24) for the sagittas z and z' of the curved object and image, respectively, remains valid. If the object and image radii of curvature are ρ and ρ', respectively, we can easily find that

$$m = -\frac{\rho}{\rho'} \tag{3.27}$$

3.5 THICK LENSES

Thick lenses have been widely studied in the literature (Herzberger, 1944, 1952). Let us consider a thick lens, as shown in Fig. 3.9, with thickness t and radii of curvature r_1 and r_2. To study this lens, let us first find the ratio of the meridional ray heights using Eq. (3.3). If we set $u_1 = 0$ we find the refracted angle u'_1 after the first surface:

$$u'_1 = -y_1 \left[\frac{n'_1 - n_1}{n'_1 r_1} \right] \tag{3.28}$$

on the other hand, this angle may be written as

$$u'_1 = -\frac{y_1 - y_2}{t} \tag{3.29}$$

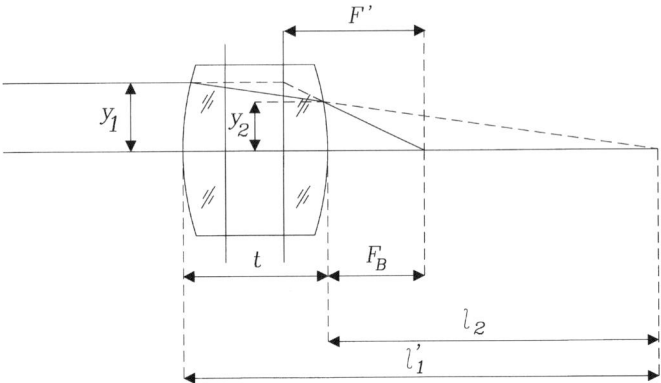

Figure 3.9 Light refraction in a thick lens.

hence, the ratio of the ray heights is

$$\frac{y_2}{y_1} = 1 - \left[\frac{n'_1 - n_1}{n'_1 r_1}\right] t \tag{3.30}$$

The effective focal length may be calculated with Eq. (3.4), obtaining

$$P = \frac{1}{F} = \left[\frac{n'_1 - n_1}{n'_2 r_1}\right] + \frac{y_2}{y_1}\left[\frac{n'_2 - n_2}{n'_2 r_2}\right] \tag{3.31}$$

but if we substitute here Eq. (3.30) and use the relation $n'_1 = n_2 = n$:

$$P = \frac{1}{F'} = \left[\frac{n - n_1}{n'_2 r_1}\right] - \left[\frac{n - n'_2}{n'_2 r_2}\right] + \frac{(n - n'_2)(n - n_1)}{n'_2 n r_1 r_2} t \tag{3.32}$$

This is a general expression, valid for any object and image medium. If this medium is air, $F = F'$ and the equation reduces to

$$\begin{aligned} P = \frac{1}{F} &= (n-1)\left[\frac{1}{r_1} - \frac{1}{r_2}\right] + \frac{(n-1)^2}{n r_1 r_2} \\ &= (n-1)\kappa + \frac{(n-1)^2}{n} c_1 c_2 t \end{aligned} \tag{3.33}$$

with $\kappa = c_1 - c_2$. Using the surface powers P_1 and P_2 defined in Chapter 2, we may write

$$P = P_1 + P_2 - P_1 P_2 \frac{t}{n} \tag{3.34}$$

An interesting particular case is that of a concentric lens, for which $r_1 = r_2 + t$. The effective focal length becomes

$$P = \frac{1}{F'} = -\frac{(n-1)^2 t}{n r_1 (r_1 - t)} \tag{3.35}$$

This lens has some interesting properties, as described by Rosin (1959), that makes it quite useful in many instruments.

Returning to the general case, however, the back focal length F_B of a thick lens may now be calculated if from Fig. 3.9 we observe that for the particular case of the lens in air:

$$F_B = \frac{y_2}{y_1} F' = \left[1 - \frac{(n-1)t}{n r_1} \right] F \tag{3.36}$$

or, alternatively, we may show that

$$\frac{1}{F_B} = (n-1) \left[\frac{1}{r_1 - t(n-1)/n} - \frac{1}{r_2} \right] \tag{3.37}$$

It is now easy to show from Eq. (3.32) that the second principal plane is at a distance from the second surface equal to

$$F_B - F = -\frac{(n-1)t}{n r_1} F = -P_1 F \frac{t}{n} \tag{3.38}$$

being positive if it is to the right of the last surface. We see that the position of the principal plane depends on the magnitude and sign of the first radius of curvature r_1.

In an analogous manner, the front focal length F_F is obtained by replacing r_1 with $-r_2$ and r_2 with $-r_1$. Thus,

$$\frac{1}{F_F} = (n-1) \left[\frac{1}{-r_2 - t(n-1)/n} + \frac{1}{r_1} \right] \tag{3.39}$$

Hence, the first principal plane is at a distance from the first surface equal to

$$F - F_F = -\frac{(n-1)t}{n r_2} F = P_2 F \frac{t}{n} \tag{3.40}$$

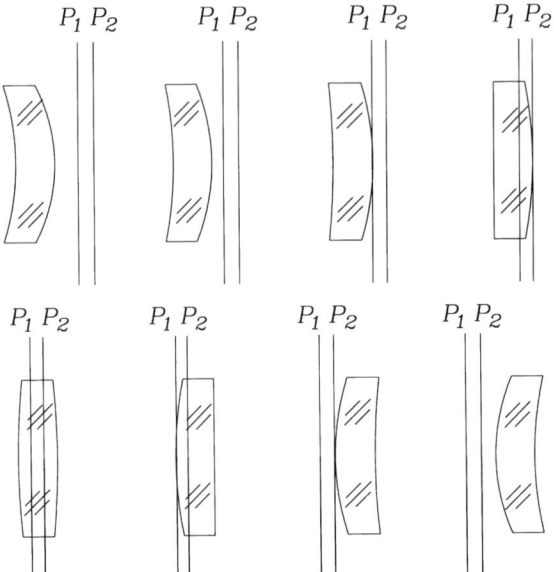

Figure 3.10 Positions of principal planes in a thick lens, for several bendings.

This distance is positive if the principal plane is to the right of the first surface.

We may easily prove that the separation T between the principal planes in any thick lens is equal to

$$T = \left[1 - \frac{F(P_1 + P_2)}{n}\right]t \sim (n-1)\frac{t}{n} \tag{3.41}$$

Thus, we see that the separation between the principal planes is almost constant, about one-third of the lens thickness, for any lens bending, if the lens is neither extremely thick, nor has a strong meniscus shape. The position of the principal planes for several lens shapes is illustrated in Fig. 3.10.

3.6 SYSTEMS OF THIN LENSES

Many optical devices may be designed using only thin lenses (Hopkins and Hanau, 1962). Let us consider the simplest case, of a system of two thin lenses separated by a finite distance. The effective focal length of a system of

Systems of Several Lenses and Thick Lenses

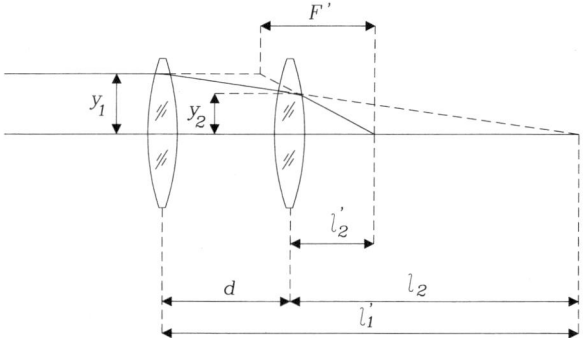

Figure 3.11 A system of two separated thin lenses.

two thin lenses separated by a distance d and with focal lengths f_1 and f_2, as shown in Fig. 3.11, may be found with Eq. (3.5):

$$P = \frac{1}{F} = \frac{1}{f_1} + \frac{y_2}{y_1}\left[\frac{1}{f_2}\right] \tag{3.42}$$

but from Fig. 3.11 we see that the ratio of the ray heights is

$$\frac{y_2}{y_1} = 1 - \frac{d}{f_1} \tag{3.43}$$

Hence, the effective focal length may be obtained as

$$P = \frac{1}{F} = \frac{1}{f_1} + \frac{1}{f_2} - \frac{d}{f_1 f_2} \tag{3.44}$$

or in terms of the power of the lenses:

$$P = P_1 + P_2 - P_1 P_2 d \tag{3.45}$$

Another common alternative expression is

$$F = \frac{f_1 f_2}{f_1 + f_2 - d} \tag{3.46}$$

If the two thin lenses are in contact with each other, this expression reduces to

$$\frac{1}{F} = \frac{1}{f_1} + \frac{1}{f_2} = P = P_1 + P_2 \tag{3.47}$$

The back focal length F_B of this system of two thin lenses is

$$F_B = \frac{y_2}{y_1} F = \left[1 - \frac{d}{f_1}\right] F \tag{3.48}$$

or

$$\frac{1}{F_B} = \frac{1}{f_1 - d} + \frac{1}{f_2} \tag{3.49}$$

We may now compute the distance from the second lens to the principal plane as

$$F_B - F = -\frac{d}{f_1} F = -P_1 f d \tag{3.50}$$

being positive if it is to the right of the second lens. Then, the position of the principal plane depends on the magnitude and sign of the focal length f_1.

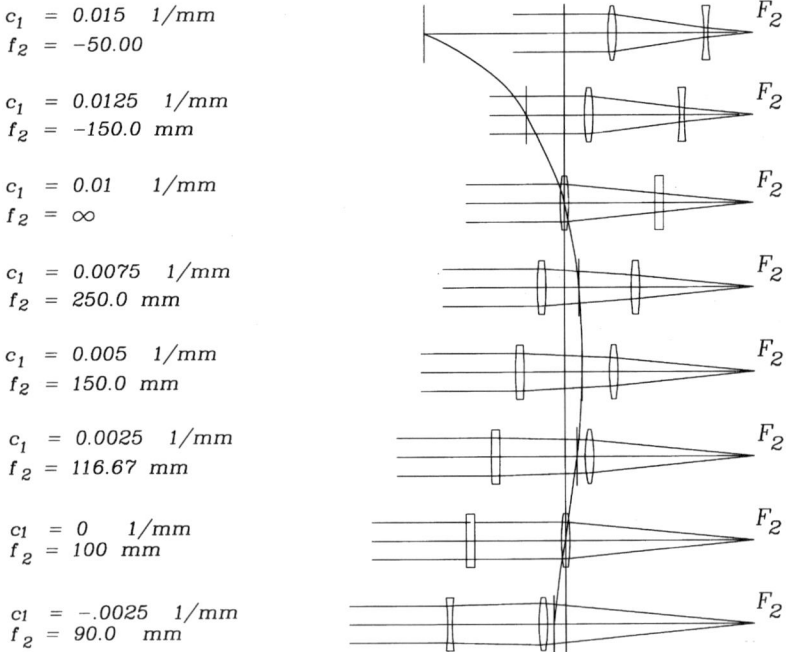

| c_1 | = | 0.015 | 1/mm |
| f_2 | = | −50.00 | |

| c_1 | = | 0.0125 | 1/mm |
| f_2 | = | −150.0 | mm |

| c_1 | = | 0.01 | 1/mm |
| f_2 | = | ∞ | |

| c_1 | = | 0.0075 | 1/mm |
| f_2 | = | 250.0 | mm |

| c_1 | = | 0.005 | 1/mm |
| f_2 | = | 150.0 | mm |

| c_1 | = | 0.0025 | 1/mm |
| f_2 | = | 116.67 | mm |

| c_1 | = | 0 | 1/mm |
| f_2 | = | 100 | mm |

| c_1 | = | −.0025 | 1/mm |
| f_2 | = | 90.0 | mm |

Figure 3.12 Position of the principal planes for a system of two separated thin lenses.

In a similar way, the front focal length may be written as

$$F_F - F = -\frac{d}{f_2}F = -P_2 f d \tag{3.51}$$

The separation T between the two principal planes is then given by

$$T = -[1 - F(P_1 + P_2)]d \tag{3.52}$$

The position of the principal planes for several lens combinations is illustrated in Fig. 3.12. Many interesting properties may be noticed by a close examination of this figure.

3.7 THE LAGRANGE INVARIANT IN A SYSTEM OF THIN LENSES

The Lagrange invariant in a system of thin lenses may adopt two special forms, one of them is in terms of the separation d and the heights of the meridional and principal rays. The other form is in terms of the power P of the lenses and the slopes of the meridional and principal rays. These two expressions are useful in the analysis of some of the first-order properties of systems of thin lenses. To find these expressions let us first write the already known Lagrange invariant in Eq. (1.63) in a lens forming part of a system of thin lenses. Assuming the lens to be in air ($n = 1$) we may write this invariant just after refraction on this lens as

$$\Lambda = \bar{y}u' - y\bar{u}' \tag{3.53}$$

The Lagrange invariant in the same space, but just before refraction at the next lens, has the same value, and may be written as

$$\Lambda = \bar{y}_{+1} u_{+1} - y_{+1} \bar{u}_{+1} \tag{3.54}$$

thus, using transfer relation (1.42) for both the meridional and the principal ray, we have

$$(\bar{y}_{+1} - \bar{y})u' = (y_{+1} - y)\bar{u}' \tag{3.55}$$

Substituting from this expression the value of the angles $\bar{u}'$ into Eq. (3.53) we have

$$\Lambda = \frac{\bar{y}(y - y_{+1}) - y(\bar{y} - \bar{y}_{+1})}{y - y_{+1}} u' \tag{3.56}$$

thus, using here Eqs. (1.47) and (1.42), we obtain

$$\Lambda = \frac{y\,\bar{y}_{+1} - y_{+1}\,\bar{y}}{l' - l_{+1}} \tag{3.57}$$

We may see that $l' - l_{+1}$ is the distance d between the two thin lenses; thus, we finally obtain the Lagrange invariant as

$$\Lambda = \frac{y\,\bar{y}_{+1} - y_{+1}\,\bar{y}}{d} \tag{3.58}$$

which relates the heights for the meridional and principal rays at one lens in the system with the corresponding ray heights at the next lens.

To find another form of the Lagrange invariant let us now use Eq. (3.6) to write

$$u'\bar{u}' = u'(\bar{u} + P\bar{y}) = (u + Py)\bar{u}' \tag{3.59}$$

and solving now for the power P and after some algebra we find the following expression for the Lagrange invariant:

$$\Lambda = \frac{u\bar{u}' - u'\bar{u}}{P} = \frac{u\,\bar{u}_{+1} - u_{+1}\,\bar{u}}{P} \tag{3.60}$$

3.8 EFFECT OF OBJECT OR STOP SHIFTING

In this section we will study the effect of shifting the object or the stop. There are some interesting relations between the meridional marginal ray and the principal ray that will be useful many times in the next chapters.

3.8.1 Shifting the Stop

Let us now use Lagrange's theorem to derive a useful relation describing the shifting of the stop to a new position, as shown in Fig. 3.13, along the optical axis, using paraxial approximations. A movement of the stop does not alter the meridional ray, since the object and image remain stationary. We assume that the stop diameter is changed (if necessary) while changing its position, so that the angle u remains constant. The object and image sizes also remain constant. Thus, the product hu does not change. Hence, if the

Systems of Several Lenses and Thick Lenses

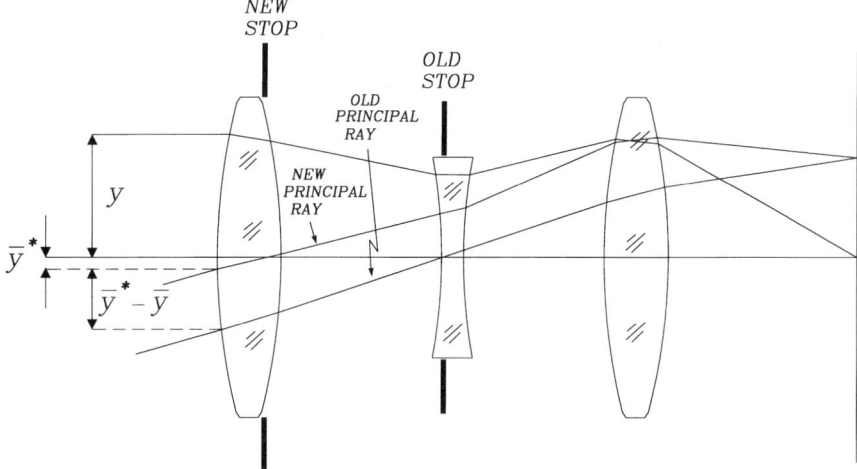

Figure 3.13 Stop shift in a thick system of lenses with ray heights at the first surface.

object and index media are the same, the Lagrange invariant remains constant with this stop shift. Then, we may write the Lagrange invariant as

$$\Lambda = n(y\bar{u} - \bar{y}u) = n(y\bar{u}^* - \bar{y}^* u) \tag{3.61}$$

and from this expression we find that

$$\frac{(\bar{y}^* - \bar{y})}{y} = \frac{(\bar{u}^* - \bar{u})}{u} \tag{3.62}$$

The left-hand side of this equation is invariant after refraction on the optical surface since the values of y and $\bar{y}$ do not change on refraction. In the same manner, the right-hand side is invariant after transferring from one surface to the next, since u and $\bar{u}$ do not change on this transfer. Thus, since both sides are equal, we conclude that both sides must be invariant for all system surfaces and equal to a constant $\bar{Q}$, given by

$$\bar{Q} = \frac{\bar{y}^* - \bar{y}}{y} \tag{3.63}$$

The direction of the principal ray in the object space must point to the entrance pupil, so that the principal ray passes through the center of the stop. In order to find this direction even if the stop is not at the first surface, we may use this relation. The principal ray may be traced through the system by first tracing a tentative principal ray with the stop at the first surface ($\bar{y}_1^* = 0$) and computing the value of $\bar{y}_s^*$ at the stop, selecting a new value for the desired height of the principal ray ($\bar{y}_s = 0$). Thus, a value of $\overline{Q}$ is obtained at the stop. Finally, the final principal ray is traced from the point object to a point on the first surface with coordinates (0, $\bar{y}_1$), given by

$$\bar{y}_1 = -\overline{Q} y_1 \tag{3.64}$$

3.8.2 Shifting Object and Image Planes

Using again Lagrange's theorem we may also derive a relation describing the effect of a shifting in the object position and its corresponding shifting in the image position. Let us assume that the object is displaced along the optical axis, but changing its size in such a way that the principal ray remains stationary, as shown in Fig. 3.14. The new image size is h'^*. The angle u also changes because its distance to the entrance pupil is modified. We may see that for the object medium:

$$hu = h^* u^* \tag{3.65}$$

Hence, if the object and image media are the same, the Lagrange invariant remains constant with this object and image displacements. We may then write

$$\Lambda = n(y\bar{u} - \bar{y}u) = n(y^*\bar{u} - \bar{y}u^*) \tag{3.66}$$

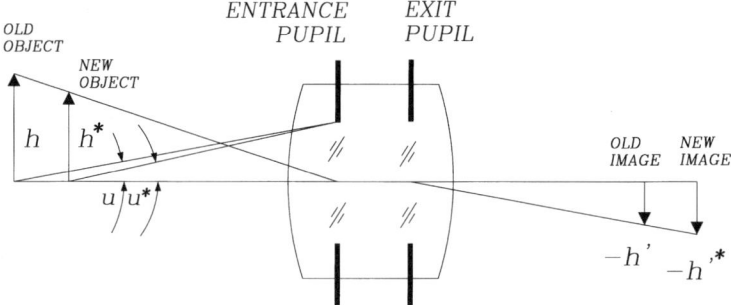

Figure 3.14 Object and image shifts in a thick system of lenses.

but from this expression we find that

$$\frac{(y^*-y)}{\bar{y}} = \frac{(u^*-u)}{\bar{u}} \tag{3.67}$$

where we may see, as in Section 3.8.1, that the left-hand side of this equation is invariant after refraction on the optical surface and the right-hand side is invariant after transferring from one surface to the next. Thus, we again conclude that both sides must be invariant for all system surfaces and equal to a constant Q. Hence,

$$Q = \frac{y^*-y}{\bar{y}} \tag{3.68}$$

Please notice that this Q is not the same as that used in Sections 1.3.2 and 1.4.2. However, there is no possibility of confusion because they will never be used together. This expression indicates that after the object and image shifts, the paraxial image height y changes at every surface with increments directly proportional to the principal ray height $\bar{y}$. One expression may be obtained from the other by interchanging the principal and the meridional rays.

At the object and image planes the value of y is zero. Thus, at these planes, after the object and image have been shifted, the value of the ratio $y/\bar{y}$ must be equal to Q. In other words, if two planes are conjugate to each other the value of $y/\bar{y}$ is equal at those planes.

3.9 THE DELANO y–$\bar{y}$ DIAGRAM

Delano (1963) proposed a diagram to analyze graphically an optical system with paraxial approximations. In this diagram the values of the meridional ray heights y and the values of the principal ray heights $\bar{y}$ at many different planes perpendicular to the optical axis are represented. No line can pass through the origin in this diagram, since it is impossible that the meridional ray height be zero at the stop or pupil. In other words, the image cannot be located at the pupil. Each point in this diagram corresponds to a plane perpendicular to the optical axis in the optical system. The straight line defined by corresponding to the object space is called the *object ray*. Similarly, the straight line corresponding to the image space is called the *image ray*.

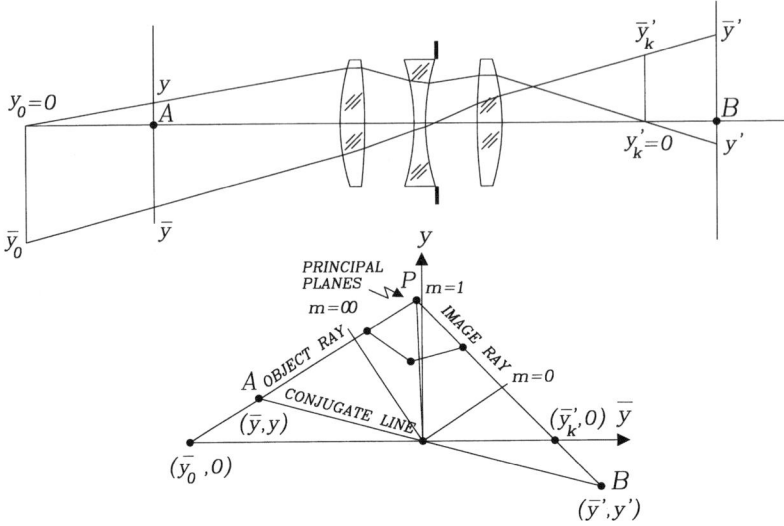

Figure 3.15 A $y-\bar{y}$ diagram with the object and image rays and some conjugate lines.

As illustrated in Fig. 3.15, any plane at the point A in the object space with coordinates $(\bar{y}, y)$ is on the object ray while its conjugate plane at the point B in the image space with coordinates $(\bar{y}', y')$ is on the image ray. The line joining these two points passes through the origin since, as pointed out in Section 3.8, $y/\bar{y}$ is equal to $y'/\bar{y}'$. This line is called a *conjugate line*. Any two conjugate points in the system can be joined by a conjugate line passing through the origin. Thus, the slope k of this line is given by

$$k = \frac{y}{\bar{y}} = \frac{y'}{\bar{y}'} \tag{3.69}$$

the magnification at the two conjugate planes under consideration is

$$m = \frac{y'}{y} = \frac{\bar{y}'}{\bar{y}} \tag{3.70}$$

and Shack (1973) points out that the slope k of the conjugate line is related to this magnification by

$$k = \frac{nu - mn'u'}{n\bar{u} - mn'\bar{u}'} \tag{3.71}$$

Systems of Several Lenses and Thick Lenses

Some special conjugate lines are interesting to examine:

1. One of these conjugate lines is the $\bar{y}$ axis, which joins the object and image planes. The magnification at the conjugate planes is that of the optical system with its object and image.
2. Another conjugate line is the y axis, which joins the entrance and exit pupils. The magnification is the magnification of the pupils.
3. The conjugate line from the origin to the point P joins the two principal planes with magnification equal to one.
4. The back focal plane is conjugate to the plane at an infinite distance in front of the system. The magnification for these planes is zero. It can be proved that the conjugate line joining them has a slope equal to the slope of the object ray. The back focal plane is located at the intersection of this line with the image ray.
5. The front focal plane is conjugate to the plane at an infinite distance after the system. The magnification for these planes is infinite. It can also be proved that the conjugate line joining these two planes has a slope equal to the slope of the image ray. The front focal plane is located at the intersection of this line with the object ray.

We may easily show in Fig. 3.17, with simple geometry, that the area of the triangle **FBC** is equal to

$$\text{Area} = \frac{y\,\bar{y}_{+1} - y_{+1}\,\bar{y}}{2} \tag{3.72}$$

Thus, using Eq. (3.58), the distance d between the two thin lenses represented at points **B** and **C** is given by

$$d = \frac{\text{Area}}{\Lambda} = \frac{y\,\bar{y}_{+1} - y_{+1}\,\bar{y}}{\Lambda} \tag{3.73}$$

Figure 3.16 shows some examples of this representation. The main properties of the y–$\bar{y}$ diagram are:

1. The polygon vertices are concave toward the origin for surfaces with positive power and vice versa.
2. The intersection between the first and the last segments (the object and the image rays) represents the two principal points of the system, located at a single point in this diagram, as shown in Fig. 3.17.
3. The area of the triangle formed by the origin and the two points representing two consecutive lenses in this diagram is directly proportional to the separation between these two lenses.

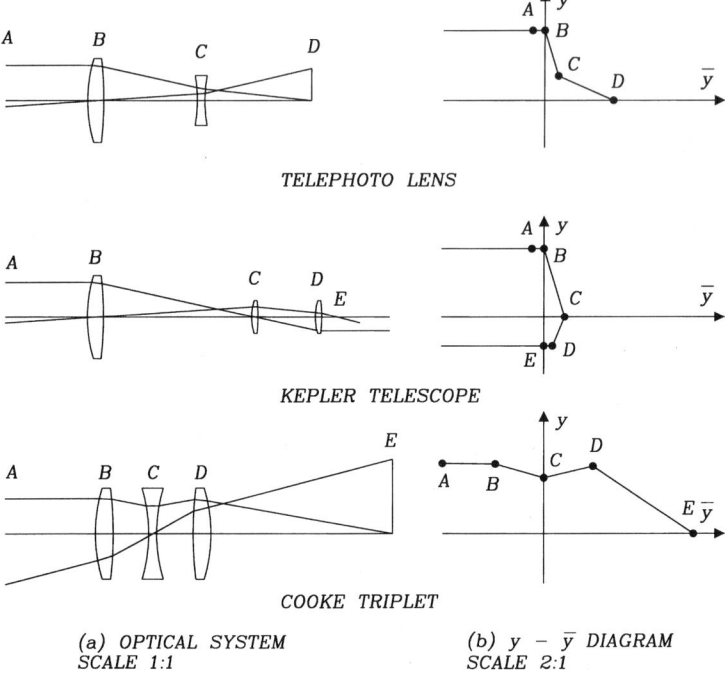

Figure 3.16 Three y–$\bar{y}$ diagrams. (a) Optical system (scale 1:1); (b) y–$\bar{y}$ diagram (scale 2:1).

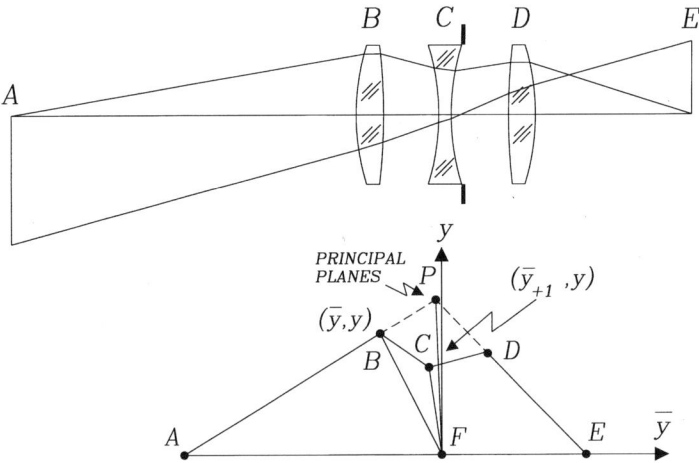

Figure 3.17 A y–$\bar{y}$ diagram of a triplet, used to find the area of triangle ABC.

Systems of Several Lenses and Thick Lenses

Many first-order operations commonly applied to optical systems can be performed in a simple manner using the y–$\bar{y}$ diagram. Next, we will describe two of them.

A Shift of the Stop

A shift of the stop must follow relation (3.56). So, a shift of the stop moves the vertices of this diagram along lines parallel to the $\bar{y}$ axis, with displacements directly proportional to the height y of the meridional ray. Then, the magnitude of this movement of the vertices, as shown in Fig. 3.18, must be such that the angle θ is the same for all points.

A Shift of the Object and Image

A shift of the object and image's positions must follow relation (3.68). Thus, this operation moves the vertices of the graph along lines parallel to the y axis, but with displacements proportional to the height $\bar{y}$ of the principal ray. As illustrated in Fig. 3.19, the angle θ is the same for all points.

Many other interesting first-order properties of optical systems may be derived from this diagram. First-order design of optical systems may be easily performed using this diagram, as pointed out with many examples by Shack (1973). Even complex systems like zoom lenses may be designed to first order with this tool (Besanmatter, 1980).

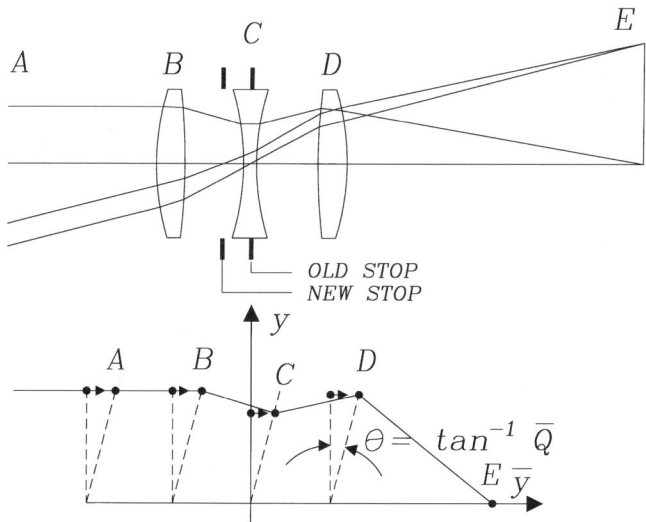

Figure 3.18 Stop shift in a y–$\bar{y}$ diagram.

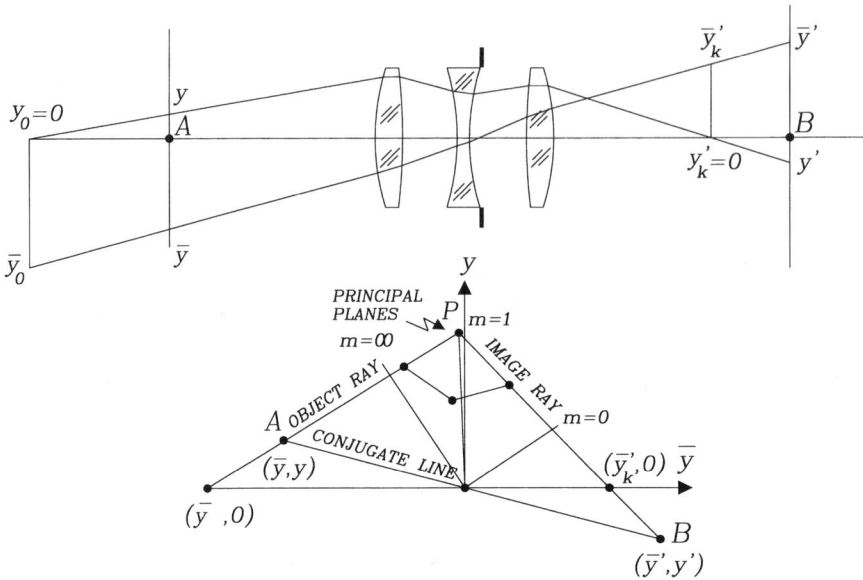

Figure 3.19 Object and image shifts in a y–$\bar{y}$ diagram.

REFERENCES

Besanmatter, W., "Designing Zoom Lenses Aided by the Delano Diagram," *Proc. SPIE*, **237**, 242 (1980).

Delano, E., "First-Order Design and the y–$\bar{y}$ Diagram," *Appl. Opt.*, **2**, 1251–1256 (1963).

Herzberger, M., "Replacing a Thin Lens by a Thick Lens," *J. Opt. Soc. Am.*, **34**, 114–115 (1944).

Herzberger, M., "Precalculation of Optical Systems," *J. Opt. Soc. Am.*, **42**, 637–640 (1952).

Hopkins, R. E. and Hanau, R., "Simple Thin Lens Optical Systems," in *Military Standardization Handbook: Optical Design, MIL-HDBK 141*, U.S. Defense Supply Agency, Washington, DC, 1962.

Kingslake, R., "A New Bench for Testing Photographic Objectives," *J. Opt. Soc. Am.*, **22**, 207–222 (1932).

Malacara, D., "A First Order Property of Some Thick Lenses and Systems," *Opt. Eng.*, **31**, 1546–1550 (1992).

Rosin, S., "Concentric Lens," *J. Opt. Soc. Am.*, **49**, 862–864 (1959).

Shack, R. V., "Analytic System Design with Pencil and Ruler—The Advantages of the y–$\bar{y}$ Diagram," *Proc. SPIE*, **39**, 127–140 (1973).

4
Spherical Aberration

4.1 SPHERICAL ABERRATION CALCULATION

Spherical aberration is the most important of all primary aberrations, because it affects the whole field of a lens, including the vicinity of the optical axis (Toraldo Di Francia, 1953). The name of this aberration comes from the fact that it is observed in most spherical surfaces, refracting or reflecting. The aberration is due to the different focus positions for marginal meridional and paraxial rays, as shown in Fig. 4.1. The value of this aberration may be calculated by means of many different methods.

Before calculating this aberration let us find some expressions for the values of the segments Q and Q', defined in Chap. 1, since they will be used several times in this section. From the definitions of the segments Q and Q' given in Eqs. (1.26) and (1.29) and some trigonometric work we may find that

$$\frac{Q}{r} = \sin I - \sin U = 2 \sin\left(\frac{I-U}{2}\right) \cos\left(\frac{I+U}{2}\right) \tag{4.1}$$

and

$$\frac{Q'}{r} = \sin I' - \sin U' = 2 \sin\left(\frac{I'-U'}{2}\right) \cos\left(\frac{I'+U'}{2}\right) \tag{4.2}$$

On the other hand, the value of the segment PA from the vertex to the intersection of the marginal ray with the optical surface, as shown in Fig. 4.2, is given by

$$PA = 2r \sin\left(\frac{I-U}{2}\right) = 2r \sin\left(\frac{I'-U'}{2}\right) \tag{4.3}$$

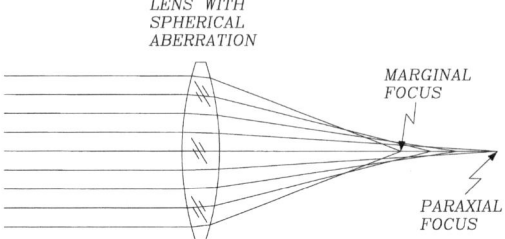

Figure 4.1 Spherical aberration in a lens.

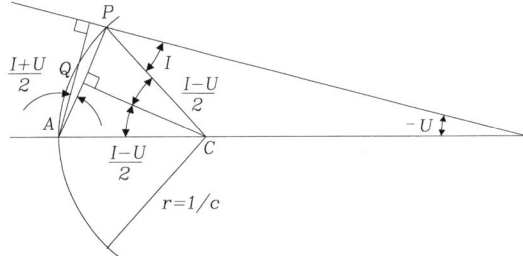

Figure 4.2 Refraction of a meridional ray at a spherical surface.

thus, the values of Q and Q' are equal to

$$Q = r(\sin I - \sin U) = PA \cos\left(\frac{I + U}{2}\right) \tag{4.4}$$

and

$$Q' = r(\sin I' - \sin U') = PA \cos\left(\frac{I' + U'}{2}\right) \tag{4.5}$$

Let us now proceed with the calculation of the spherical aberration, beginning with its formal definition, illustrated in Fig. 4.3. The longitudinal spherical aberration in the image is

$$SphL = L' - l' \tag{4.6}$$

and the longitudinal spherical aberration in the object is the aberration after the preceding surface, given by

$$SphL_{-1} = L - l \tag{4.7}$$

where the subscript -1 stands for this preceding surface.

Spherical Aberration

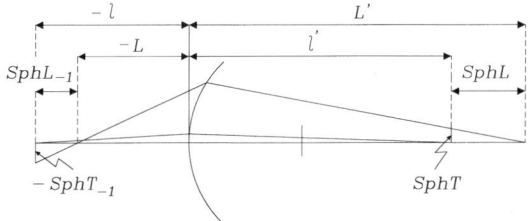

Figure 4.3 Definition of the longitudinal spherical aberration.

One method to find an expression for the spherical aberration has been described by Conrady (1957). The first step is to consider Eq. (1.36), which may be written as

$$\frac{n'}{L'} - \frac{n}{L} = \frac{n' - n}{r} + \frac{n \sin I}{r(\sin I - \sin U)} \left[1 - \frac{\sin I - \sin U}{\sin I' - \sin U'}\right] \quad (4.8)$$

but from Eq. (1.33) we have

$$\frac{\sin I}{\sin I - \sin U} = \frac{L - r}{L} \quad (4.9)$$

and using the values of Q and Q' in Eqs. (4.4) and (4.5) we obtain

$$\frac{n'}{L'} - \frac{n}{L} = \frac{n' - n}{r} + \frac{n(L - r)}{rL}\left[1 - \frac{Q}{Q'}\right] \quad (4.10)$$

This expression is exact for any meridional ray. For paraxial rays we have the Gauss equation:

$$\frac{n'}{l'} - \frac{n}{l} = \frac{n' - n}{r} \quad (4.11)$$

thus, subtracting one expression from the other we have

$$n'\frac{L' - l'}{L'l'} - n\frac{L - l}{Ll} = \frac{n(L - r)}{rL}\left[\frac{Q}{Q'} - 1\right] \quad (4.12)$$

obtaining

$$SphL = \left[\frac{nL'l'}{n'Ll}\right] SphL_{-1} + \frac{n(L - r)L'l'}{n'rL}\left[\frac{Q - Q'}{Q'}\right] \quad (4.13)$$

and, using Eqs. (4.4) and (4.5),

$$SphL = \left[\frac{nL'l'}{n'Ll}\right] SphL_{-1} + \frac{n(L-r)L'l'}{n'rL}$$
$$\times [(\cos(I+U)/2 - \cos(I'+U')/2)/\cos(I'+U')/2] \qquad (4.14)$$

which may be further developed with the trigonometric relation:

$$\cos\alpha - \cos\beta = 2\sin\left(\frac{\beta+\alpha}{2}\right)\sin\left(\frac{\beta-\alpha}{2}\right) \qquad (4.15)$$

and the value of $U = U' + I - I'$, to obtain the following relation in which the difference of cosines was replaced by a product of sines.

$$SphL = \left[\frac{nL'l'}{n'Ll}\right] SphL_{-1} + \frac{2n(L-r)L'l'}{n'rL}$$
$$\times [(\sin(I+U')/2 \times \sin(I'-I)/2)/\cos(I'+U')/2] \qquad (4.16)$$

This expression permits us to find in a simple manner the conditions for zero spherical aberration, as we will see later. The first term on the right-hand side of this expression represents the transferred longitudinal spherical aberration and the second term is the new spherical aberration introduced by this optical surface. The factor in front of the object's transferred spherical aberration $SphL_{-1}$ is the longitudinal magnification of the surface.

An elegant method to obtain an equivalent expression has been proposed by Delano (1952). We begin with Delano's expression in Eq. (1.56) for a paraxial and a marginal ray. We assume that the marginal and the paraxial rays originate at the same on-axis point in object space. From the longitudinal values for the spherical aberration defined in Eqs. (4.6) and (4.7), before and after refraction, we find (see Fig. 1.17)

$$SphL = -\frac{s'}{\sin U'}; \quad SphL_{-1} = -\frac{s}{\sin U} \qquad (4.17)$$

thus obtaining

$$SphL = \left[\frac{nu\sin U}{n'u'\sin U'}\right] SphL_{-1} - \frac{ni}{n'u'\sin U'}[Q'-Q] \qquad (4.18)$$

Spherical Aberration

Using now the same expressions for PA, Q, and Q' in Eqs. (4.3)–(4.5), we may obtain, using again Eq. (4.15),

$$SphL = \left[\frac{nu \sin U}{n'u' \sin U'}\right] SphL_{-1} + \frac{2niPA}{n'u'}\left[\frac{\sin(I+U')/2 \times \sin(I'-I)/2}{\sin U'}\right] \quad (4.19)$$

This expression, as well as Conrady's in Eq. (4.16), may be applied to an optical system formed by k centered surfaces along a common axis, with each relation being appropriate, depending on the circumstances. For small apertures PA approaches the meridional ray height y. Using the transfer relations $u_+ = u'$, $U_+ = U'$, $n_+ = n'$, we may obtain

$$SphL_k = \left[\frac{n_1 u_1 \sin U_1}{n'_k u'_k \sin U'_k}\right] SphL_0 + \sum_{j=1}^{k} \frac{2niPA}{n'_k u'_k}\left[\frac{\sin(I+U')/2 \times \sin(I'-I)/2}{\sin U'_k}\right] \quad (4.20)$$

where the subscript 0 is for the object, the subscript k for the last surface, and the subscript $k+1$ for the image's surface. All variables without subscript's are for surface j. The factor in front of $SphL_0$ is the longitudinal magnification [see Eq. (1.66)] of the whole optical system. The difference between the second term in Eq. (4.19) and the second term in Eq. (4.20) is that the longitudinal magnification of the part of the optical system after the surface under consideration has been added to Eq. (4.20) as a factor in the last term.

4.2 PRIMARY SPHERICAL ABERRATION

The primary spherical aberration is obtained if the aperture is large enough to deviate from the paraxial approximation to produce spherical aberration, but small enough to avoid high order terms. Thus, by using paraxial approximations in Eq. (4.20), with $y = PA$, the value of the primary spherical aberration is then easily found to be

$$SphL_k = \left[\frac{n_1 u_1^2}{n'_k u'^2_k}\right] SphL_0 + \sum_{j=1}^{k} \frac{yni(i+u')(i'-i)}{2n'_k u'^2_k} \quad (4.21)$$

where the factor in front of the spherical aberration of the object is as usual the longitudinal magnification of the optical system. This expression may also be written as

$$SphL_k = \left|\frac{n_1 u_1^2}{n'_k u'^2_k}\right| SphL_0 + \sum_{j=1}^{k} SphLC \qquad (4.22)$$

where the surface contribution $SphLC$ to the final longitudinal spherical aberration is given by

$$\begin{aligned} SphLC &= \frac{yni(i+u')(i'-i)}{2n'_k u'^2_k} \\ &= \frac{y(n/n')(n-n')(i+u')i^2}{2n'_k u'^2_k} \end{aligned} \qquad (4.23)$$

An alternative way of writing this expression, eliminating all angles is

$$SphLC = \frac{n y^4 F^2}{2 n'_k y_1^2} \left(\frac{n}{n'}-1\right)\left[\left(\frac{1}{r}-\frac{1}{l}\right)\frac{n}{n'}-\frac{1}{l}\right]\left(\frac{1}{r}-\frac{1}{l}\right)^2 \qquad (4.24)$$

where F is the effective focal length of the system to which the surface being considered belongs.

Replacing y by S, ($S^2 = x^2 + y^2$) due to the rotational symmetry of the optical system, the longitudinal spherical aberration may also be written as

$$SphL = a S^2 \qquad (4.25)$$

where a is the longitudinal spherical aberration coefficient.

We have derived the longitudinal spherical aberration because its expression will be frequently used when deriving the expressions for the primary off-axis aberrations, but the transverse spherical aberration is more useful and applied when actually calculating optical systems. Then, similarly to the longitudinal spherical aberration, the transverse spherical aberration may be written as

$$SphT = b S^3 \qquad (4.26)$$

where b is the transverse spherical aberration coefficient.

The value of the transverse spherical aberration is easily obtained from the value of the longitudinal spherical aberration, by using the next relation between the longitudinal and transverse aberrations (see Fig. 4.2):

$$u_1 = -\frac{SphT_0}{SphL_0} \quad u'_k = -\frac{SphT_k}{SphL_k} \qquad (4.27)$$

Spherical Aberration

which is valid not only for spherical aberration, but also for all other primary aberrations, as we will see later.

It is interesting to comment at this point that the longitudinal spherical aberration is always positive if the marginal focus is to the right of the paraxial focus, independently of the traveling direction for the light. Thus, a plane mirror in front of a lens, located between the lens and its focus, reverses the traveling direction of the light, as well as the relative positions of the marginal and paraxial foci. The sign of u'_k is also changed. Then, we may see that this flat mirror in front of the lens changes the sign of the longitudinal spherical aberration, but not the sign of the transverse spherical aberration.

Using Eq. (4.27) we may write the transverse spherical aberration as

$$SphT_k = \left[\frac{n_1 u_1}{n'_k u'_k}\right] SphT_0 + \sum_{j=1}^{k} SphTC \qquad (4.28)$$

where the transverse spherical aberration contribution is

$$SphTC = \frac{y(n/n')(n-n')(i+u')\,i^2}{2n'_k u'_k} \qquad (4.29)$$

which is frequently also written as

$$SphTC = \sigma\,i^2 \qquad (4.30)$$

where

$$\sigma = \frac{y(n/n')(n-n')(i+u')}{2n'_k u'_k} \qquad (4.31)$$

Let us now plot this transverse spherical aberration contribution as a function of the position l' for the image, using an object with a variable position on the optical axis, for a convex as well as for a concave optical surface, with a given index of refraction ($n_d = 1.5168$). The convergence angle for the incident beam is variable, so that the refracted beam has a constant angle of convergence equal to $u' = -0.3$ radian. We obtain the results in Fig. 4.4. We may see in this figure that there are three values of l' for which the spherical aberration becomes zero. These image positions are: (1) at the vertex of the surface ($l' = 0$), (2) at the center of curvature ($l' = r$), and (3) at a point on the optical axis such that $l' = r(n+1)/n$. We will examine this result with more detail later in this chapter. These graphs allow us to have a rough estimation of the

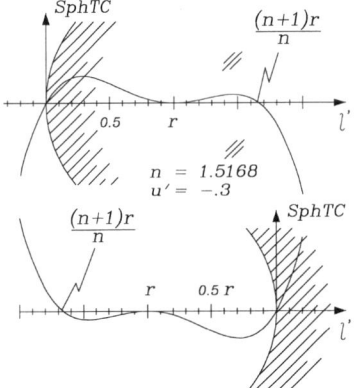

Figure 4.4 Transverse spherical aberration contribution at a spherical surface.

magnitude and sign of the transverse spherical aberration contributions of each surface in a lens, if we know the position of the image with respect to the center of curvature.

4.2.1 Spherical Aberration of a Thin Lens

The spherical aberration for the marginal ray of a lens, either thick or thin, may be computed from Eq. (4.24), by adding the contributions for the two surfaces. Thus, we may obtain

$$SphT = \frac{(n-1)l'_2}{2n^2 y_2} \left[\left(\frac{1}{r_1} - \frac{n+1}{l_1}\right) \left(\frac{1}{r_1} - \frac{1}{l_1}\right)^2 y_1^4 \right.$$

$$\left. - \left(\frac{1}{r_2} - \frac{n+1}{l'_2}\right) \left(\frac{1}{r_2} - \frac{1}{l'_2}\right)^2 y_2^4 \right] \quad (4.32)$$

If we particularize for the case of a thin lens we may obtain

$$SphT = \frac{(n-1)\kappa l'_2 y^3}{2n}$$

$$\times \left\{ [n^2 \kappa - c_1 + (n+1) v_1][n\kappa - 2(c_1 - v_1)] \right.$$

$$\left. + n(c_1 - v_1)^2 \right\} \quad (4.33)$$

Spherical Aberration

where

$$v_1 = \frac{1}{l_1}; \quad c_1 = \frac{1}{r_1}; \quad c_2 = \frac{1}{r_2}; \quad \kappa = c_1 - c_2 \tag{4.34}$$

We may see that the magnitude of the spherical aberration depends on the lens shape and also on the object and image positions. The following expression for the spherical aberration of thin lenses was computed by Conrady (1957) and many other authors, and may be found from the last expression and with the help of Eq. (2.13) as

$$SphT = -l'_k y^3 \big(G_1 \kappa^3 - G_2 \kappa^2 c_1 + G_3 \kappa^2 v_1 + G_4 \kappa c_1^2 \\ - G_5 \kappa c_1 v_1 + G_6 \kappa v_1^2 \big) \tag{4.35}$$

where $l'_k = f$ if the object is at infinity or $l'_k = [\kappa(n-1) + v_1]^{-1}$ if the object is at a finite distance, and the functions G are

$$G_1 = \frac{n^2(n-1)}{2} \quad G_2 = \frac{(2n+1)(n-1)}{2} \quad G_3 = \frac{(3n+1)(n-1)}{2}$$

$$G_4 = \frac{(n+2)(n-1)}{2n} \quad G_5 = \frac{2(n^2-1)}{n} \quad G_6 = \frac{(3n+2)(n-1)}{2n} \tag{4.36}$$

We see that the transverse spherical aberration increases with the cube of the aperture.

We have mentioned before that for a lens with a fixed focal length the magnitude of the spherical aberration depends on the lens shape and also on the object position. Given an object to lens distance, the lens shape for minimum spherical aberration may be obtained. Conversely, given a lens shape, the object position to minimize the spherical aberration may also be calculated.

The value of the curvature c_1 for a thin lens with the minimum value of the spherical aberration, as shown by Kingslake (1978), is given by

$$c_1 = \frac{n(n+1/2)\kappa + 2(n+1)v}{n+2}$$
$$= \frac{G_2 \kappa + G_5 v}{2 G_4} \tag{4.37}$$

where $v = 1/l$ is the inverse of the distance from the object to the lens. Thus, for the special case of an object at infinity:

$$c_1 = \frac{n(n+1/2)\kappa}{n+2} \tag{4.38}$$

The values of the transverse spherical aberration for a thin lens with a constant 100 mm focal length f is a function of the curvature c_1 of the front face. Differentiating Eq. (4.35) with respect to c_1, we find that there is a minimum value of the magnitude of this aberration for a certain value of the front curvature c_1, whose value depends on the object position.

The individual surface contributions when the object is at infinity (collimated incident light beam) as a function of the curvature c_1 of the front surface is shown in Fig. 4.5. We notice that the minimum value of the spherical aberration is obtained when the contributions $SphTC$ of each surface are almost equal. The optimum lens bending to obtain minimum spherical aberration is nearly when the angle of incidence for the incident ray is equal to the angle of refraction for the final refracted ray, but not exactly equal.

The transverse spherical aberration curve, as we see, does not pass through zero for any value of c_1, even if the object distance is changed, as shown in Fig. 4.6. However, if the thin lens is formed by two thin lenses in contact, one positive and one negative, made with glasses of different indices of refraction, the curve can be made to cross the c_1 axis. We may see in this figure that the magnitude of the minimum transverse spherical aberration decreases as the incident beam becomes more convergent. We also notice that the optimum lens bending for minimum transverse spherical aberration is a function of the object position.

It is important to notice that the value of the spherical aberration of a single lens whose bending has been optimized for minimum spherical aberration also depends on the refractive index. A higher refractive index is better.

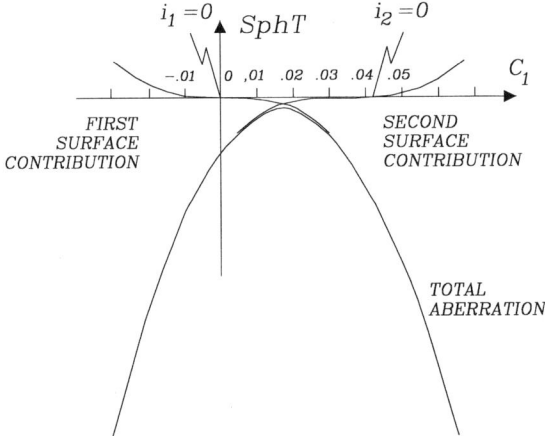

Figure 4.5 Contribution of each face of a lens to the total transverse spherical aberration of a thin lens.

Spherical Aberration

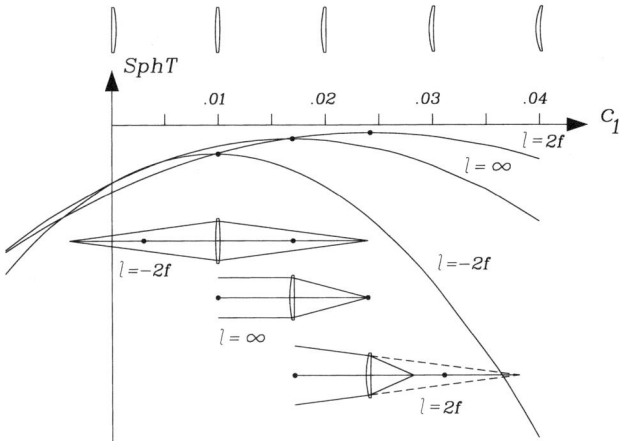

Figure 4.6 Transverse spherical aberration of a thin lens versus the front curvature for three different object positions.

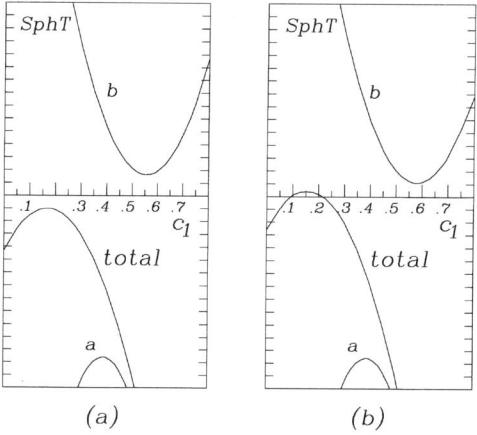

Figure 4.7 Contribution of each face and total transverse spherical aberration of a single thin lens and a thin doublet.

Figure 4.7 shows the transverse spherical aberration for a lens with a focal length equal to 100 mm and a diameter of 20 mm, made out of Schott's BK7 glass ($n_d = 1.5168$). This lens is made as a doublet, with two thin lenses in contact, with focal lengths equal to 38 mm and −61 mm. In Fig. 4.7(a) the two lenses are of BK7 glass. In Fig. 4.7(b) the positive lens is of BK7 glass, but the negative component is made out of Schott's F2 glass ($n_d = 1.6200$). The spherical aberration contributions of each component (1 and 2), as well as that of the complete doublet, are shown here. Only when the two glasses are different are there two solutions for zero spherical aberration. By selecting the right glass combination the paraboloidal graph may be made to just touch the c_1 axis. Then, only one solution exists for zero spherical aberration.

4.2.2 A System of Thin Lenses

When a single lens with its spherical aberration minimized is split into two elements so that the combination has the same power as the original lens, the spherical aberration is greatly reduced. To illustrate this fact, following Fischer and Mason (1987), let us consider a single lens with diameter D and focal length F which has been optimized for minimum spherical aberration. To split this lens into two we follow the next three steps:

1. We scale the lens by a factor of two, obtaining a lens with twice the diameter, twice the focal length, and twice the spherical aberration.
2. Now we reduce the diameter to the original value. The focal length does not change, but the transverse spherical aberration is reduced by a factor of 16, since the transverse spherical aberration grows with the fourth power of the aperture.
3. Finally, two identical lenses are placed in contact with each other. The combination duplicates the spherical aberration and reduces the focal length by half to the original value. The spherical aberration is now only one-eighth of the aberration in the initial lens.

Even further improvement can be achieved if the second lens is bent to its optimum shape with convergent light. The refractive index is a very important variable, since the spherical aberration decreases when the refractive index increases. Figure 4.8 shows the value of the marginal spherical aberration as a function of the refractive index for several thin lens systems. All systems have the same aperture ($D=0.333$) and focal length ($F=1$). We can see that with three lenses and a high index of refraction the spherical aberration becomes negative.

4.2.3 Spherical Aberration for a Plane-Parallel Plate in Converging Light

A plane-parallel plate may be shown to displace the image by an amount ΔL, depicted in Fig. 4.9, given by

$$\Delta L = \frac{t}{n}\left(n - \frac{\cos U}{\cos U'}\right) \tag{4.39}$$

Spherical Aberration

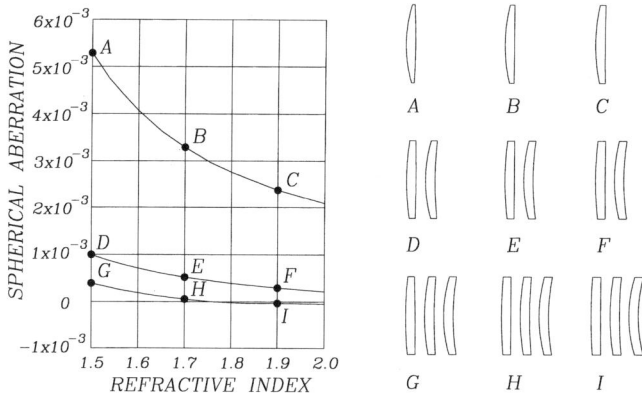

Figure 4.8 Spherical aberration in a single lens and in systems of two and three lenses.

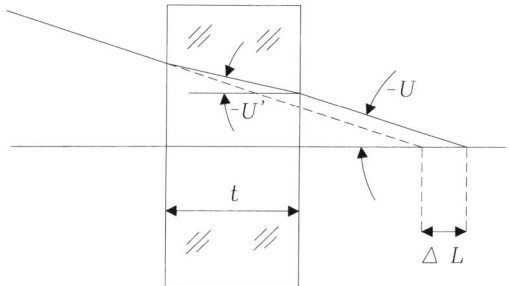

Figure 4.9 Spherical aberration of a thick plane-parallel glass plate.

for paraxial rays we have

$$\Delta l = \frac{t}{n}(n - 1) \qquad (4.40)$$

Hence, subtracting Eq. (4.39) from Eq. (4.40), the longitudinal spherical aberration *SphL* is equal to

$$SphL = \frac{t}{n}\left(1 - \frac{\cos U}{\cos U'}\right) \qquad (4.41)$$

with a paraxial approximation given by

$$SphL = \frac{t u^2}{2n} \qquad (4.42)$$

This expression is useful for calculating the spherical aberration introduced by the cover glass slip in microscopes. Although this cover glass is very thin, the angle of divergence is so large that a noticeable amount of spherical aberration is present.

4.3 ASPHERICAL SURFACES

Aspherical surfaces can have rotational symmetry or not, but the most common are of the first type. As described in more detail in Appendix 2, they can be mathematically represented by

$$Z = \frac{c S^2}{1 + \sqrt{1 - (K+1) c^2 S^2}} + A_1 S^4 + A_2 S^6 + A_3 S^8 + A_4 S^{10} \quad (4.43)$$

Where the first term represents a conic surface with rotational symmetry defined by its conic constant K, which is a function of the eccentricity of the conic surface. Alternatively this expression can be expanded as a spherical surface plus some aspheric deformation terms that include the effect of the conic shape. Then, we may find that

$$Z = \frac{c S^2}{1 + \sqrt{1 - c^2 S^2}} + B_1 S^4 + B_2 S^6 + B_3 S^8 + B_4 S^{10} \quad (4.44)$$

4.4 SPHERICAL ABERRATION OF ASPHERICAL SURFACES

An aspherical surface has an additional sagitta term given by the four aspheric terms in Eq. (4.44). As a first approximation, we may take only the first term. This term introduces a slope term in the surface, as shown in Fig. 4.10, given by

$$\frac{dZ}{dS} = 4 B_1 S^3 = \left(4 A_1 + \frac{K c^3}{2}\right) S^3 \quad (4.45)$$

This slope modification in the optical surface changes the slope of the refracted meridional ray by an amount:

$$dU' = \left(\frac{n - n'}{n'}\right) \frac{dZ}{dS} \quad (4.46)$$

Spherical Aberration 87

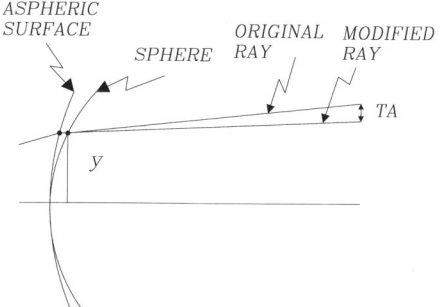

Figure 4.10 Spherical aberration of an aspheric surface.

then, this angle change introduces a transverse spherical aberration $SphT_\text{asph}$ term given by

$$dU' = \frac{SphT_\text{asph}}{l'} \tag{4.47}$$

thus, obtaining

$$SphT_\text{asph} = \frac{1}{2}\left(\frac{n-n'}{n'}\right)(8A_1 + Kc^3)l'\,S^3 \tag{4.48}$$

Thus, this aberration is propagated to the final image in the optical system with a factor given by the lateral magnification of the part of the optical system after the aspherical optical surface. Hence, by using Eq. (1.47) and the expression for the lateral magnification, the contribution of this surface to the final aberration is

$$SphTC_\text{asph} = -(8A_1 + Kc^3)\left(\frac{n-n'}{2}\right)\left(\frac{S^4}{n'_k u'_k}\right) \tag{4.49}$$

4.5 SURFACES WITHOUT SPHERICAL ABERRATION

A single optical surface may be completely free of spherical aberration under certain circumstances. These surfaces will be described in the following subsections.

4.5.1 Refractive Spherical Surfaces

The conditions for a single refractive optical surface to be free of spherical aberration may be found in several ways. From Eq. (4.12) we see that this

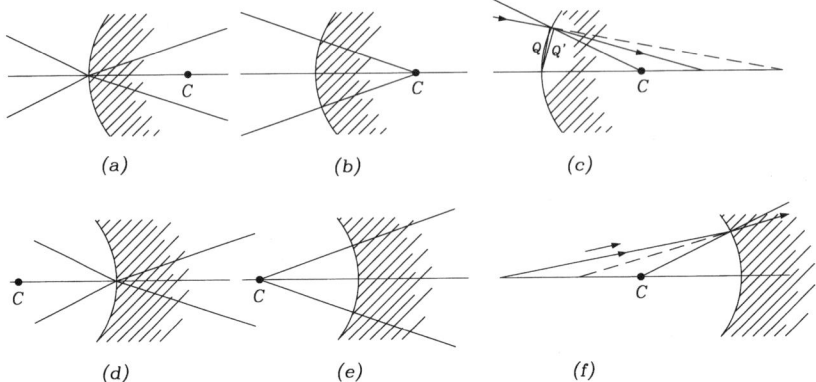

Figure 4.11 Three cases free of spherical aberration for a convex and a concave surface.

aberration is absent if $Q = Q'$. This is possible if (1) both Q and Q' are equal to zero, (2) when the object and the image coincide with the vertex of the surface, as in Fig. 4.11(a and d), (3) when the ray enters perpendicularly to the surface, with the object and the image at the center of curvature, as in Fig. 4.11(b and e), and (4) when the object and the image have certain positions such that the condition is satisfied, as in Fig. 4.11(c and f).

From Eq. (4.16) we may see the following conditions for zero spherical aberration: (1) $L' = 0$, (2) $L = r$ or equivalently $I = I'$, and (3) $I' = -U$. These are the same three conditions derived from Eq. (4.12). The first two cases are trivial and obvious, but the third case is the most interesting.

Applying the condition $I = -U'$ from Eq. (4.20) to Snell's law, we find that

$$\frac{n'}{n} \sin I' = -\sin U' \tag{4.50}$$

and using now Eq. (1.20) we find that

$$L' - r = \frac{n}{n'} r \tag{4.51}$$

and

$$L - r = \frac{n'}{n} r \tag{4.52}$$

These are the positions for the image and the object, respectively, in the configuration discovered by Abbe (1840–1905), director of the observatory at Jena and director of research of Zeiss. The object and image positions are called aplanatic Abbe points. There are many practical applications for these configurations free of spherical aberration.

4.5.2 Reflective Conic Surfaces

Reflecting conic surfaces with symmetry of revolution are free of spherical aberration if the object is placed at the proper position, depending on its conic constant value, as shown in Section A2.1.1. A paraboloid is free of spherical aberration if the object is at infinity, as shown in Fig. 4.12.

Ellipsoids and hyperboloids are also free of spherical aberration when the object, real or virtual, is placed at one of the foci. These are illustrated in Fig. 4.13, and may produce the four combinations of real and virtual objects and images.

4.5.3 Descartes' Ovoid

Rene Descartes found that refractive conics of revolution may also be free of spherical aberration. Such surfaces receive the name of cartesian ovoids. These ovoids are illustrated in Fig. 4.14, where n_1 is the internal index of refraction and n_2 is the external index of refraction. Let us consider two straight lines l_1 and l_2 from each focus to the ellipsoid surface, and one line l_3 traced from that point, parallel to the axis. A light ray may travel along l_1 and l_2 for any length of l_3 only if Fermat's law is satisfied as follows:

$$n_1 l_1 + n_2 l_3 = \text{constant} \tag{4.53}$$

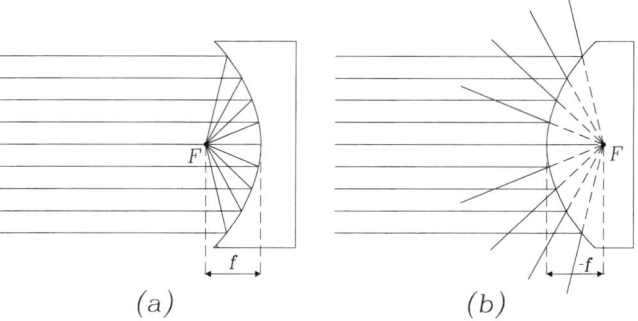

Figure 4.12 (a) Concave and (b) convex parabolic mirrors free of spherical aberration.

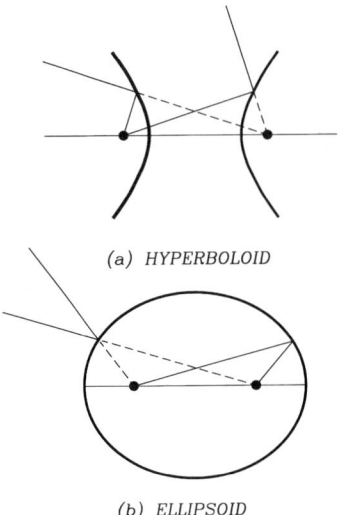

Figure 4.13 (a) Reflective hyperboloid and (b) reflective ellipsoid free of spherical aberration.

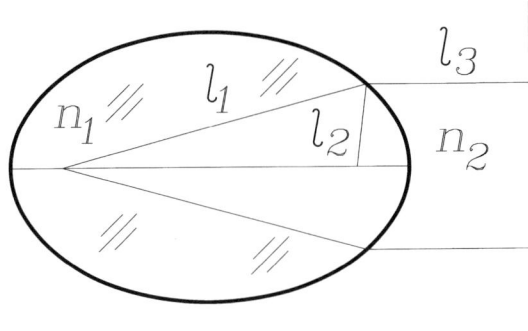

Figure 4.14 Descartes' ovoid.

On the other hand, from the properties of the ellipse, if the vertical straight line to the right is the directrix:

$$l_2 = e l_3 \tag{4.54}$$

where e is the eccentricity. Thus, we may find that

$$n_1 l_1 + n_2 \frac{l_2}{e} = \text{constant} \tag{4.55}$$

Spherical Aberration

Another ellipse property is

$$l_1 + l_2 = \text{constant} \tag{4.56}$$

but this is possible only if

$$n_1 = \frac{n_2}{e} \tag{4.57}$$

In conclusion, if the ratio of the internal and external refractive indices is chosen equal to a given ellipse eccentricity, there is no spherical aberration. Of course, we may also think that given the refractive indices, the eccentricity may be calculated. Figure 4.15 shows four possible configurations of the cartesian ovoid. In the two last cases the ovoid becomes a hyperboloid with rotational symmetry.

4.6 ABERRATION POLYNOMIAL FOR SPHERICAL ABERRATION

We found in Section 4.2 expressions for third order or primary spherical aberration. These results are valid for relatively small apertures, so that high order terms become negligible. If this is not the case, we have to consider

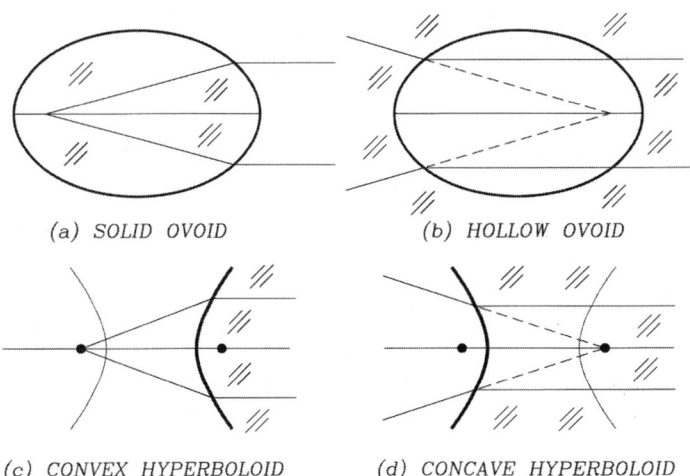

Figure 4.15 Ovoids and hyperboloids free of spherical aberration: (a) solid ovoid; (b) hollow ovoid; (c) convex hyperboloid; (d) concave hyperboloid.

terms with higher powers of the ray height. We will use the normalized distance:

$$\rho = \frac{S}{S_{\max}} \tag{4.58}$$

to represent the ray height at the exit pupil, where $S^2 = X^2 + Y^2$. Thus, we may write the exact (not primary) longitudinal spherical aberration LA_0 as

$$\begin{aligned} LA_0(\rho) &= a_0 + a_2\rho^2 + a_4\rho^4 + a_6\rho^6 + \cdots \\ &= \sum_{i=0}^{\infty} a_{2i}\rho^{2i} \end{aligned} \tag{4.59}$$

where the first constant term a_0 represents the longitudinal displacement of the reference paraxial focus; the second term $a_2\rho^2$ is the third order or primary longitudinal spherical aberration $SphL$; the third term $a_4\rho^4$ is the fifth order or secondary spherical aberration; and so on, for the rest of the terms.

It is more convenient to express the spherical aberration by the spherical transverse aberration TA_0 instead of the longitudinal aberration LA_0, related to each other by Eq. (4.27):

$$u'_k = -\frac{TA_0(\rho)}{LA_0(\rho)} = -\frac{\rho}{r_W} \tag{4.60}$$

where r_W is the *normalized radius of curvature* of the reference spherical wavefront with unit semidiameter. In general, we may write this radius of curvature as

$$r_W = \frac{l'_k - \bar{l}'_k}{S_{\max}} \tag{4.61}$$

thus, we may write

$$\begin{aligned} TA_0(\rho) &= b_1\rho + b_3\rho^3 + b_5\rho^5 + b_7\rho^7 + \cdots \\ &= \sum_{i=0}^{\infty} b_{2i+1}\rho^{2i+1} \end{aligned} \tag{4.62}$$

Spherical Aberration

where $b_1\rho$ is the transverse aberration for the paraxial rays at the observation plane, the second term $b_3\rho^3$ is the primary transverse spherical aberration *SphT*, and

$$b_{2i+1} = \frac{a_{2i}}{r_W} \tag{4.63}$$

Another useful way to write this aberration polynomial is by means of the wavefront aberration $W(\rho)$, defined as

$$W(\rho) = c_2\rho^2 + c_4\rho^4 + c_6\rho^6 + c_8\rho^8 + \cdots$$
$$= \sum_{i=0}^{\infty} c_{2i+2}\, \rho^{2i+2} \tag{4.64}$$

where the coefficients c_i may be easily found. These coefficients should not be mistaken for the curvature values, used in other parts of this book.

The focus displacement Δf of the observation plane with respect to the paraxial focus (positive if moved away from the lens) is given by

$$\Delta f = a_0 = b_1 r_W = -2\, c_2\, r_W^2 \tag{4.65}$$

4.6.1 Caustic

In the presence of spherical aberration the light rays follow the path illustrated in Fig. 4.16. The envelope of these rays is called the caustic (Cornejo and Malacara, 1978). We may see several interesting foci in this diagram: (1) the paraxial focus, (2) the position for zero wavefront deviation at the edge of the exit pupil, (3) the caustic waist position, (4) the marginal focus, and (5) the end of the caustic. The diagrams in Fig. 4.17 illustrate the transverse aberration curves and the wavefront deformations for these positions.

Assuming only the presence of primary spherical aberration and using Eqs. (4.63) and (4.65) in Eq. (4.59), the longitudinal spherical aberration measured at a plane with a focus shift with respect to the paraxial focus is given by

$$LA_0(\rho) = b_1 r_W + b_3 r_W \rho^2 = b_1 r_W + SphL(\rho) \tag{4.66}$$

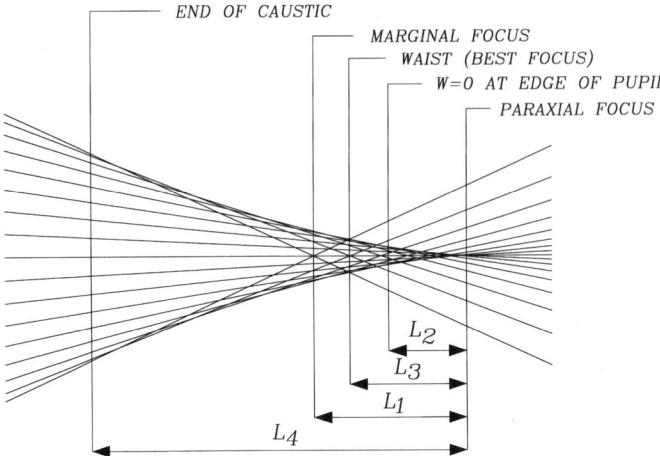

Figure 4.16 Primary spherical aberration caustic.

The transverse spherical aberration with the same focus shift is given by

$$TA_0(\rho) = b_1\,\rho + b_3\,\rho^3 = b_1\,\rho + SphT(\rho) \tag{4.67}$$

and the wavefront deviation measured at the exit pupil, with respect to a reference sphere with radius of curvature r_W, is

$$W(\rho) = -\frac{b_1\,\rho^2}{2r_W} - \frac{b_3\,\rho^4}{4r_W} \tag{4.68}$$

The *paraxial focus* is at the plane where there is no focus shift, i.e., when

$$b_1 = 0 \tag{4.69}$$

hence, from Eq. (4.66), the distance L_1 from the paraxial focus to the *marginal focus* is

$$L_1 = SphL(\rho_{\max}) = -b_3\,r_W\,\rho_{\max}^2 \tag{4.70}$$

which is by definition the marginal spherical aberration.

From Eq. 4.68, the axial position of the center of curvature of the reference sphere to obtain *zero wavefront deviation at the edge* of the exit

Spherical Aberration

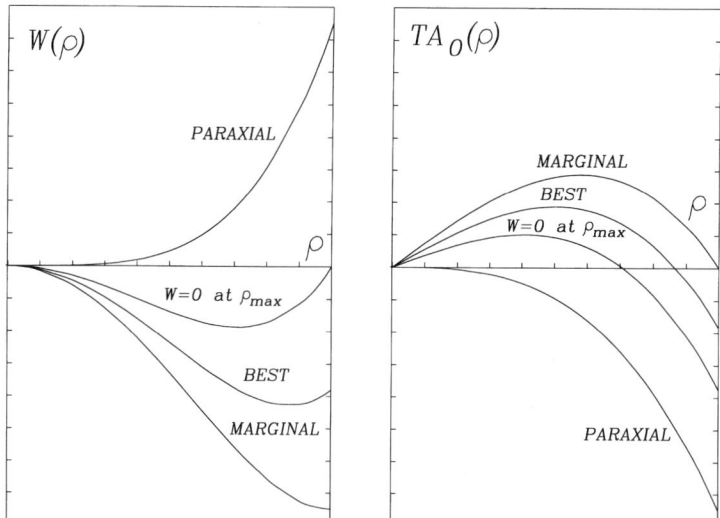

Figure 4.17 Wavefront deformations and transverse spherical aberration for several caustic planes.

pupil is obtained when

$$W(\rho_{\max}) = b_1 + \frac{b_3 \, \rho_{\max}^2}{2} = 0 \tag{4.71}$$

thus

$$b_1 = \frac{b_3 \, \rho_{\max}^2}{2} \tag{4.72}$$

hence, from Eq. 4.64, the distance L_2 from the paraxial focus to the position for zero wavefront deviation at the edge is

$$L_2 = -LA_0(\rho_{\max}) = \frac{b_3}{2} r_W \, \rho_{\max}^2 = \frac{L_1}{2} \tag{4.73}$$

The *caustic waist or best focus position* is obtained when the maximum zonal transverse spherical aberration is equal in magnitude but opposite in the sign to the marginal transverse spherical aberration. Thus, the value of $\rho = \rho_z$ for the maximum zonal aberration is found with the condition

$$\frac{d \, TA_0(\rho)}{d\rho} = b_1 + 3 \, b_3 \, \rho_z^2 = 0 \tag{4.74}$$

Then, to find the caustic waist we write

$$b_1 \, \rho_z + b_3 \, \rho_z^3 = -[b_1 \, \rho_{max} + b_3 \, \rho_{max}^3] \tag{4.75}$$

but substituting here the value of ρ_z, given by Eq. (4.73), we have

$$b_1 \, \rho_{max} + b_3 \, \rho_{max}^3 = -\frac{2 b_1}{3} \left[-\frac{b_1}{3 b_3} \right]^{1/2} \tag{4.76}$$

thus, obtaining for the distance from the paraxial focus to the caustic waist:

$$L_3 = -\frac{3 b_3}{4} r_W \rho_{max}^2 = \frac{3 L_1}{4} \tag{4.77}$$

Finally, to find the *end of the caustic* we impose the condition that the slope of the $TA(\rho)$ as a function of ρ plot is zero at the margin of the exit pupil. Thus, we write

$$\left[\frac{d \, TA_0(\rho)}{d\rho} \right]_{\rho = \rho_{max}} = b_1 + 3 b_3 \, \rho_{max}^2 = 0 \tag{4.78}$$

thus

$$b_1 = -3 b_3 \, \rho_{max}^2 \tag{4.79}$$

Hence, using again Eq. (4.66), the distance L_4 from the paraxial focus to the end of the caustic is

$$L_4 = -3 b_3 \, r_W \, \rho_{max}^2 = 3 L_1 \tag{4.80}$$

Figure 4.17 illustrates the wavefront deformation and the transverse aberrations for several focus positions.

4.6.2 Aberration Balancing

It has been shown in the last section how the image size may be optimized by introducing a focus shift that places the image plane at the waist of the caustic. This is all we can do if the lens has only primary spherical aberration. When designing a lens, however, high-order aberrations may be unavoidable. Then, it may be necessary to introduce some primary

Spherical Aberration

aberration in order to compensate (at least partially) the high-order aberrations. Let us consider an example of how third- and fifth-order spherical aberrations may be combined to improve the image. Let us assume that fifth-order aberration is introduced, so that the longitudinal and transverse spherical aberrations are zero for the marginal ray. Then, we may write

$$LA_0(\rho_{max}) = a_2\, \rho_{max}^2 + a_4\, \rho_{max}^4 = 0 \tag{4.81}$$

or

$$TA_0(\rho_{max}) = b_3\, \rho_{max}^3 + b_5\, \rho_{max}^5 = 0 \tag{4.82}$$

With this condition, as depicted in Fig. 4.18, it may be shown that the maximum values of these longitudinal and transverse aberrations, respectively, occur for the following values of ρ:

$$\rho_{maxLA} = \frac{\rho_{max}}{\sqrt{2}} = 0.707\, \rho_{max} \tag{4.83}$$

and

$$\rho_{maxTA} = \sqrt{\frac{3}{5}}\, \rho_{max} = 0.775\, \rho_{max} \tag{4.84}$$

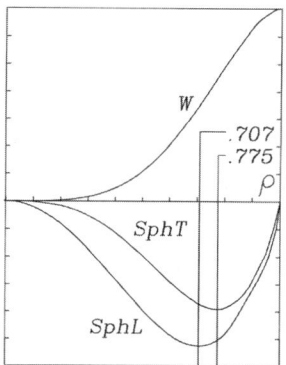

Figure 4.18 Transverse and longitudinal spherical aberration maxima when these aberrations are zero at the edge. The wavefront shape is also shown.

4.7 HIGH-ORDER SPHERICAL ABERRATION

The high-order spherical aberration term may appear for two reasons:
1. Because it is produced by the refractive surface at the same time as the primary aberrations. This high-order aberration may be computed by subtracting the primary aberration in Eq. (4.21) from the exact aberration in Eq. (4.20).
2. Because any traveling wavefront is continuously changing its shape along its trajectory. If we neglect diffraction effects, only the spherical or plane wavefronts remain spherical or plane, respectively, as they travel in space. A wavefront with only primary spherical aberration develops high-order spherical aberration as it travels.

Let us now consider in some detail this mechanism for the appearance of high-order spherical aberration. A wavefront with only primary spherical aberration is depicted in Fig. 4.19. It may be shown that if we define

$$\eta = \frac{z}{r_W} \tag{4.85}$$

the ray heights y_1 and y_2 on the planes $\mathbf{P}_1$ and $\mathbf{P}_2$, respectively, are related by

$$y_2 = y_1(1 - \eta) + TA\eta \tag{4.86}$$

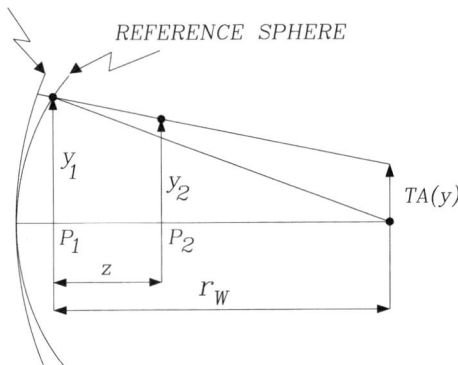

Figure 4.19 Wavefront shape change along traveling path.

Spherical Aberration

where *TA* is the transverse aberration. If this transverse aberration is due only to primary spherical aberration at the plane $\mathbf{P}_1$, we may write the transverse aberration as

$$TA = a_3 y_1^3 \tag{4.87}$$

Then, from Eqs. (4.85) and (4.86) we may approximately obtain

$$y_1 = \frac{y_2}{(1-\eta)} - \frac{\eta a_3}{(1-\eta)^3} y_2^3 \tag{4.88}$$

hence, it can be proved that the transverse aberration as a function of the ray height on the plane $\mathbf{P}_2$ is approximately given by

$$TA = \frac{a_3}{(1-\eta)^3} y_2^3 - \frac{3\eta a_3^2}{(1-\eta)^5} y_2^5 \tag{4.89}$$

The constant $1/(1-\eta)^3$ in the first term is just a scaling factor due to the smaller diameter of the wavefront. The second term is new and represents the fifth-order spherical aberration due to the propagation of the wavefront from plane $\mathbf{P}_1$ to plane $\mathbf{P}_2$.

4.8 SPHERICAL ABERRATION CORRECTION WITH GRADIENT INDEX

We have shown in this chapter that if a single lens is made with homogeneous glass the spherical aberration cannot be corrected. As shown in Fig. 4.6, even with the optimum shape, a residual aberration remains. If the object is at infinity, the optimum shape is nearly a plano convex lens with its flat surface on the back. From Eq. (4.35) the residual spherical aberration is given by

$$\begin{aligned} SphT &= -\frac{y^3}{f^2(n-1)^3}(G_1 - G_2 + G_4) \\ &= \frac{y^3}{2f^2(n-1)^2}\left(n^2 - 2n + \frac{2}{n}\right) \end{aligned} \tag{4.90}$$

whose maximum value, at the edge of a lens with diameter *D* is

$$SphT_{\max} = \frac{D^3}{16f^2(n-1)^2}\left(n^2 - 2n + \frac{2}{n}\right) \tag{4.91}$$

This aberration can be corrected with only spherical surfaces using an axial gradient index lens as described by Moore (1977). The solution is to introduce the axial gradient with the refractive index along the optical axis as represented by

$$n(z) = N_{00} + N_{01} z + N_{02} z^2 + \cdots \tag{4.92}$$

where N_{00} is the refractive index at the vertex A of the first lens surface (see Fig. 4.20). In this lens the paraxial focal length is larger than the marginal focal length. So, to correct the spherical aberration the refractive index has to be higher at the center of the lens than at the edge. Figure 4.20 shows a lens with an axial gradient. A good correction is obtained with a linear approximation of expression (4.90). The gradient index on the convex surface has a depth equal to the sagitta z_0 of this surface. Thus, using Eq. (4.92) we have

$$n(z_0) = n = N_{00} + N_{01} z_0 \tag{4.93}$$

With this gradient the refractive index for a ray passing through the edge of the lens is a constant equal to n while for the ray along the optical axis it decreases linearly from the point A to the point B, until it reaches the value n. The net effect is that the optical path for the paraxial rays becomes larger with the presence of the gradient index, but the marginal rays are not affected. The optical path difference OPD introduced by the gradient index can be shown to be

$$OPD(y) = -\frac{N_1 z}{2} = -\frac{N_1 y^4}{8 f^2 (n-1)} \tag{4.94}$$

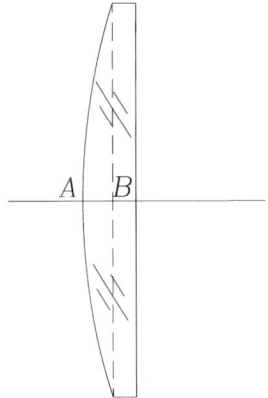

Figure 4.20 Axial gradient lens free of spherical aberration.

where z is the sagitta at the height y on the lens. Thus, the spherical aberration $SphT$ is

$$SphT = -f\frac{dOPD}{dy} = \frac{N_1 y^3}{2f(n-1)^2} \qquad (4.95)$$

whose maximum value is, at the edge of a lens with diameter D,

$$SphT_{max} = \frac{N_1 D^3}{16f(n-1)^2} \qquad (4.96)$$

With the proper gradient index magnitude the spherical aberration can be corrected. Then, equating expressions (4.91) and (4.96) and assuming a value $n(z_0) = 1.5$, the change Δn from the point B to the point C along the optical axis can be shown to be given by

$$\Delta n = N_{01} z_0 = \frac{(n^2 - 2n + 2/n)}{8(n-1)} \frac{1}{(f/D)^2} = \frac{0.15}{(f/D)^2} \qquad (4.97)$$

where (f/D) is the f number of the lens. Thus, the necessary Δn for an $f/4$ lens is only 0.0094 while for an $f/1$ lens it is 0.15.

REFERENCES

Conrady, A. E., *Applied Optics and Optical Design*, Dover Publications, New York, 1957.
Cornejo, A. and Malacara, D., "Caustic Coordinates in Platzeck–Gaviola Test of Conic Mirrors," *Appl. Opt.*, **17**, 18–19 (1978).
Delano, E., "A General Contribution Formula for Tangential Rays," *J. Opt. Soc. Am.*, **42**, 631–633 (1952).
Fischer, R. E. and Mason, K. L., "Spherical Aberration—Some Fascinating Observations," *Proc. SPIE*, **766**, 53–60 (1987).
Kingslake, R., *Lens Design Fundamentals*, Academic Press, New York, 1978.
Moore, D. T., "Design of a Single Element Gradient-Index Collimator," *J. Opt. Soc. Am.*, **67**, 1137 (1977).
Toraldo Di Francia, G., "On the Image Sharpness in the Central Field of a System Presenting Third- and Fifth-Order Spherical Aberration," *J. Opt. Soc. Am.*, **43**, 827–835 (1953).

5
Monochromatic Off-Axis Aberrations

5.1 OBLIQUE RAYS

Spherical aberration affects the image over the whole field of view of a lens, including the optical axis and its vicinity. There are some other image defects or aberrations that affect only the image points off the optical axis. The farther this point is from the optical axis, the larger the aberration is. In this chapter we will describe these aberrations, first studied during the 1850s by L. von Seidel. For this reason the primary or third-order aberrations are known as Seidel aberrations.

The number of important contributors to this field is enormous. It would be impossible here just to mention them. Probably the most complete study of the primary optical aberrations is that of A. E. Conrady, during the 1930s. The reader of this book is strongly advised to examine the classical references on this subject by Conrady (1957, 1960) and Kingslake (1965, 1978). The more advanced students are referred to the important work of Buchdahl (1948, 1954, 1956, 1958a,b,c, 1959, 1960a,b,c, 1961, 1962a,b, 1965, 1970) on the high-order aberration theory. Important work on this subject has also been performed by many other authors, like Cruickshank and Hills (1960), Focke (1965), Herzberger and Marchand (1952, 1954), Hopkins and Hanau (1962), Hopkins et al. (1955), and many others, as we may see in the list of references at the end of this chapter.

In this chapter we will study the off-axis aberrations of centered systems. These are systems that have a common optical axis, where all centers of curvature of the optical surfaces lie on this optical axis. However, there are some optical systems where the optical surfaces may be either tilted or laterally shifted with respect to the optical axis, producing some aberrations, as described by Epstein (1949) and Ruben (1964).

The study of this chapter should be complemented with an examination of Appendix 1 at the end of the book, where most of the notation used in this chapter is defined. There are four off-axis monochromatic aberrations, namely, Petzval curvature, coma, astigmatism, and distortion. These aberrations will be studied in this chapter in some detail. To begin this study, we must first introduce some basic concepts.

Oblique rays, also frequently called *skew rays* are those emitted by an off-axis object point not contained in the meridional plane. A particular important case of oblique rays are the sagittal rays. To obtain some understanding of how an oblique sagittal ray propagates let us consider Fig. 5.1.

The principal ray passes through the center of the entrance pupil with semidiameter y and is then refracted at the optical surface. By definition, the auxiliary axis **AC** passes through the off-axis object and the center of curvature of the surface. Since the optical surface is a sphere, the system has rotational symmetry around this auxiliary axis. If a meridional paraxial ray is traced from the object, almost parallel to the auxiliary axis, its image is also on the auxiliary optical axis, at the point **P**, as shown in Fig. 5.2. The exit pupil is not in contact with the surface, but on its left side, so that the principal ray $\mathbf{A_0 U}$ enters the surface at the point $\mathbf{A_0}$. Tangential rays, $\mathbf{T_1 T}$ through $\mathbf{T_1}$, and $\mathbf{T_2 T}$ through $\mathbf{T_2}$, on the entrance pupil, cross the auxiliary optical axis, but not at the point **P**, but $\mathbf{B_1}$ and $\mathbf{B_2}$ due to the spherical aberration. The ray from $\mathbf{T_1}$ has more spherical aberration than the ray from $\mathbf{T_2}$, so they cross each other at the point **T**. This is the tangential focus.

The two sagittal rays, $\mathbf{S_1 S}$ through $\mathbf{S_1}$, and $\mathbf{S_2 S}$ through $\mathbf{S_2}$, on the entrance pupil, are at equal distances $\mathbf{AS_1}$ and $\mathbf{AS_2}$ from the auxiliary optical axis so, after refraction, they cross each other at a point **S** on the auxiliary optical axis.

In Figs. 5.1 and 5.2 we may identify the following reference points:

W: Paraxial axial image (gaussian)

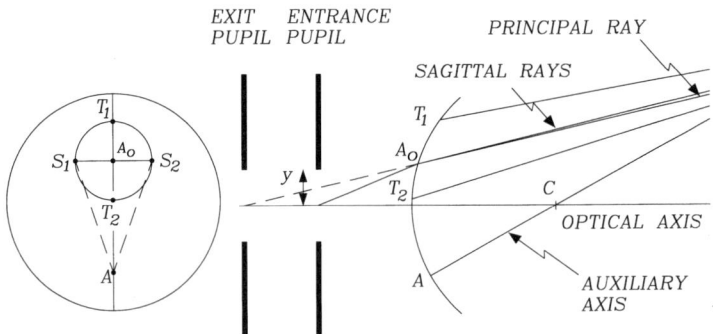

Figure 5.1   Tangential and sagittal rays refracted on a spherical surface.

Monochromatic Off-Axis Aberrations

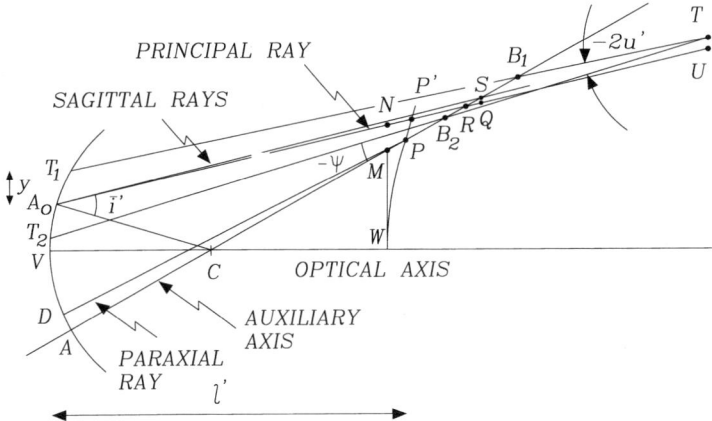

Figure 5.2 Some rays and points to illustrate the formation of the primary aberrations.

- **M:** Point on auxiliary axis above point **W** (gaussian image if the field is flat)
- **P:** Petzval focus
- **P′:** Point on intersection of principal ray and Petzval surface
- **N:** Point on principal ray above point **M**
- **S:** Marginal sagittal image, on intersection of marginal sagittal rays with auxiliary axis
- **R:** Intersection of principal ray with auxiliary axis
- **Q:** Point on principal ray, below point **S**
- **B$_1$:** Intersection of upper tangential ray with auxiliary axis
- **B$_2$:** Intersection of lower tangential ray with auxiliary axis
- **T:** Marginal tangential focus at intersection of marginal tangential rays
- **U:** Point on principal ray below point **T**
- **V:** Vertex of optical surface

The point **W** is the paraxial image of an on-axis object (not shown in the figure). The Petzval focus **P** is defined as the focus for the paraxial rays **DP**, close to the auxiliary optical axis, originating from an off-axis point, in the same plane as the on-axis object producing image **W**. Let us assume that the point **A$_0$**, where the light beam coming from the off-axis object point enters the surface, shifts down over the optical surface to the point **A** on the auxiliary axis. Then, the sagittal and the tangential foci move until they fuse together in one point at the Petzval focus. In other words, the astigmatism disappears.

For the principal ray, the angle ψ in Fig. 5.2 may be related to the angle $\bar{i}'$, by means of Eq. (1.39) and taking $\bar{u}'$ equal to ψ, by

$$-\frac{\psi}{r} = \frac{\bar{i}'}{l'-r} \tag{5.1}$$

and for the meridional ray, the angle u' is related to the angle i', using the same expression (1.39), by

$$-\frac{u'}{r} = \frac{i'}{l'-r} \tag{5.2}$$

Thus, obtaining

$$\psi = \left(\frac{\bar{i}'}{i'}\right)u' = \left(\frac{\bar{i}}{i}\right)u' \tag{5.3}$$

where the last term is obtained using Snell's law. On the other hand, it is easy to see that in a first approximation we may write

$$\psi = -\frac{AA_0}{l'} = \frac{AA_0}{y}u' \tag{5.4}$$

thus

$$AA_0 = \psi\frac{y}{u'} = \left(\frac{\bar{i}}{i}\right)y \tag{5.5}$$

This is a useful relation that we will use later.

All of the rays we have considered in Fig. 5.2 originate at an off-axis point object and all of them cross the auxiliary optical axis at some point, as shown in Fig. 5.3. The location of these intersections may be calculated with the help of Eq. (4.25) for the spherical aberration, with reference to point **P**. Thus, for each of the rays in Figs. 5.1 and 5.2 we may write:

1. For the principal ray:

$$PR = a(AA_0)^2 \tag{5.6}$$

but using Eq. (5.5):

$$PR = ay^2\left(\frac{\bar{i}}{i}\right)^2 = SphL\left(\frac{\bar{i}}{i}\right)^2 \tag{5.7}$$

Monochromatic Off-Axis Aberrations

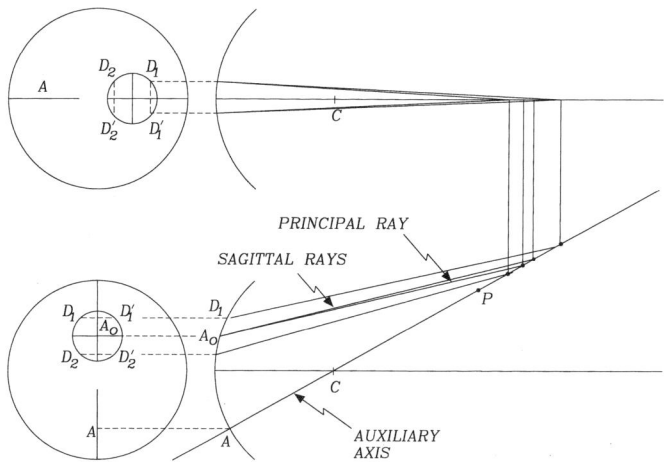

Figure 5.3 Upper and side view of some rays refracted on a spherical surface.

2. For the marginal sagittal rays passing through S_1 and S_2:

$$PS = a(AS_2)^2 = a(AA_0)^2 + a(A_0 S_2)^2 = PR + RS \qquad (5.8)$$

thus, we may see that

$$RS = ay^2 = SphL \qquad (5.9)$$

Since RS decreases with the value of y, we may see that for the paraxial sagittal rays passing through S'_1 and S'_2 the value of RS approaches zero. Thus, the paraxial sagittal focus coincides with the point **R**.

3. For the upper marginal tangential ray, passing through T_1:

$$\begin{aligned} PB_1 &= a(AT_1)^2 = a(AA_0 + A_0 T_1)^2 \\ &= a(AA_0)^2 + a(A_0 T_1)^2 + 2a(AA_0)y \\ &= PR + RS + SB_1 \end{aligned} \qquad (5.10)$$

4. For the lower marginal tangential ray, passing through T_2:

$$\begin{aligned} PB_2 &= a(AT_2)^2 = a(AA_0 - A_0 T_2)^2 \\ &= a(AA_0)^2 + a(A_0 T_2)^2 - 2a(AA_0)y \\ &= PR + RS - SB_2 \end{aligned} \qquad (5.11)$$

thus,

$$SB_1 = SB_2 = 2a(AA_0)y \tag{5.12}$$

and with Eq. (5.9):

$$B_1B_2 = 4a(AA_0)y = \frac{4SphL(AA_0)}{y} \tag{5.13}$$

This result shows that the length of the segment $\mathbf{B}_1\mathbf{B}_2$ is directly proportional to y, indicating that the position of the tangential image $\mathbf{T}$ in a first approximation does not change with the value of y. In other words, the marginal tangential focus and the paraxial tangential focus, formed by the rays through T'_1 and T'_2 are at the same point $\mathbf{T}$. Now, using Eq. (5.5) we write

$$B_1B_2 = 4SphL\left(\frac{\bar{i}}{i}\right) \tag{5.14}$$

With this background, we may now define the primary aberrations as follows:

Primary spherical aberration $= SphL = \mathbf{RS}$—From Eq. (5.9) this distance is equal to the on-axis spherical aberration for a marginal ray with height y.

Sagittal coma $= Coma_S = \mathbf{SQ}$—This is the transverse distance from the marginal sagittal image to the principal ray. This distance is zero for paraxial rays. Then, this aberration is defined only for marginal rays.

Tangential coma $= Coma_T = \mathbf{TU}$—This is the transverse distance from the marginal tangential image to the principal ray.

Sagittal longitudinal astigmatism $= AstL_S = \mathbf{P'R}$—If the pupil semi-diameter y is reduced, the sagittal focus S approaches the point $\mathbf{R}$, as shown by Eq. (5.9). Thus, the paraxial sagittal focus is the point $\mathbf{R}$. The sagittal longitudinal astigmatism is defined as the longitudinal distance from the paraxial sagittal focus to the Petzval surface.

Tangential longitudinal astigmatism $= AstL_T = \mathbf{P'T}$—From Eq. (5.13) we see that the distance B_1B_2 increases linearly with y. Thus, we may see that in a first approximation the marginal tangential focus $\mathbf{T}$ is at the same position as the paraxial tangential focus. Then, the tangential longitudinal astigmatism is defined as the longitudinal distance from the paraxial tangential focus T to the Petzval surface.

Petzval curvature = *Ptz* = **MP**—This is the longitudinal distance from the Petzval focus to the ideal image point **M** of the off-axis object located in the same plane as the on-axis object producing image **W**.

Distortion = *Dist* = **MN**—This is the transverse distance from the ideal off-axis image **M** to the principal ray.

We see that all primary aberrations arise due to the existence of the spherical aberration. With these results we are ready to study these aberrations, first described by Seidel.

5.2 PETZVAL CURVATURE

To find the Petzval curvature we have to use an optical configuration without astigmatism. We have seen before that this condition is fulfilled if the light beam from the object is narrow and travels along the auxiliary axis. Let us consider a spherical refracting surface as in Fig. 5.4 where we have a spherical refractive surface with an auxiliary axis. With a very small aperture, so that all the other primary aberrations may be neglected, the image of point **W** on the optical axis is at **W'**. Circles C_1 and C_2 are concentric with the optical surface. The incident beam is rotated about the center of curvature to move the beam off-axis and preserve the spherical symmetry. Then, due to this spherical symmetry about **C**, the image of point **A** in circle C_1 is **A'** in circle C_2. The image of **B** is at point **P**. The distances AB and $A'P$ are related by the longitudinal magnification of the system, as

$$A'P = \overline{m}AB \tag{5.15}$$

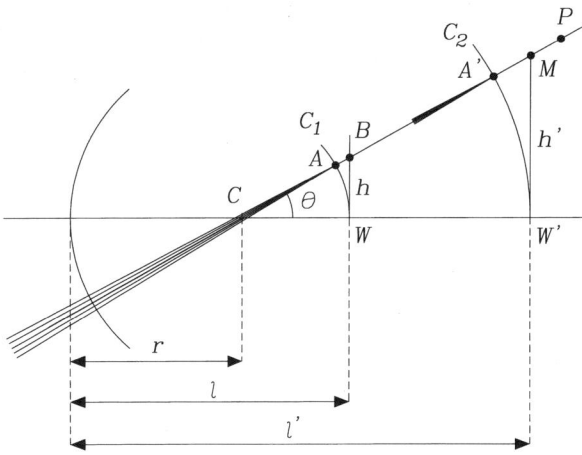

Figure 5.4 Petzval theorem demonstration.

Assuming now that the field is small (angle θ small) we approximate the sagitta AB by

$$AB = \frac{h^2}{2(l-r)} \tag{5.16}$$

and similarly

$$A'M = \frac{h'^2}{2(l'-r)} \tag{5.17}$$

From Fig. 5.4, the distance MP is

$$MP = A'P - A'M \tag{5.18}$$

thus, using Eqs. (5.15)–(5.17) we obtain

$$MP = \frac{h^2}{2(l-r)}\overline{m} - \frac{h'^2}{2(l'-r)} \tag{5.19}$$

but since the longitudinal magnification is given by

$$\overline{m} = \frac{n' h'^2}{n h^2} \tag{5.20}$$

we may write

$$MP = \frac{n' h'^2}{2n(l-r)} - \frac{h'^2}{2(l'-r)} \tag{5.21}$$

On the other hand, from Eqs. (1.38)–(1.41) we may find that

$$\frac{n'-n}{r} = \frac{n}{l'-r} - \frac{n'}{l-r} \tag{5.22}$$

obtaining

$$Ptz = MP = -\frac{h'^2}{2}\left(\frac{n'-n}{nr}\right) \tag{5.23}$$

This result gives the Petzval curvature, i.e., the field curvature in the absence of astigmatism, for only one optical surface. For a centered system of k surfaces we have to add the contributions of all surfaces. As when studying the spherical aberration, the contribution of surface j to the final longitudinal displacement of the image is the displacement MP multiplied by the longitudinal magnification $\overline{m}_j$ of the part of the optical system after the surface being considered. This longitudinal magnification is

$$\overline{m} = \frac{n'_k h'^2_k}{n' h'^2} \tag{5.24}$$

Thus, the Petzval field contribution is given by

$$PtzC = \overline{m} MP \tag{5.25}$$

hence

$$PtzC = -\frac{h'^2_k n'_k}{2} \left(\frac{n' - n}{nn' r} \right) \tag{5.26}$$

This is the sagitta of the focal surface. Then, the radius of curvature is

$$\frac{1}{r_{ptz}} = -n'_k \sum_{j=1}^{k} \left(\frac{n' - n}{nn' r} \right) \tag{5.27}$$

This is the *Petzval theorem* and the focal surface that it defines is called the *Petzval surface*.

For the particular case of a thin lens, using Eq. (2.8), we may find that

$$\frac{1}{r_{ptz}} = -\frac{1}{nf} = -\frac{P}{n} \tag{5.28}$$

and for a system of thin lenses, with any separation between them:

$$\frac{1}{r_{ptz}} = -\sum_{i=1}^{k} \frac{1}{nf} = -\sum_{i=1}^{k} \frac{P_i}{n_i} \tag{5.29}$$

Then, in a third-order approximation, the Petzval surface is spherical and has a curvature that is directly proportional to the sum of the powers of the thin lenses forming the system, divided by the refractive index on the glass. If a flat surface is used to examine the image with the on-axis image being focused, the off-axis images will be defocused, with a degree of defocusing (image size) growing with the square of the image height. It is interesting to notice that the Petzval curvature does not depend on the position of the object along the optical axis. In other words, the Petzval curvature is the same for any pair of object–image conjugates. The control of Petzval curvature has been described by several researchers, e.g., Wallin (1951).

5.3 COMA

We will now see how the coma aberration for a complete system, as well as the surface contributions, may be calculated.

5.3.1 Offense Against the Sine Condition

Thus, let us consider an optical system, as shown in Fig. 5.5. The marginal sagittal image is at the point **S** above the principal ray. The paraxial sagittal image is on the principal ray and its height h'_k is calculated with the Lagrange theorem. The marginal sagittal image H'_{Sk} is calculated with the optical sine theorem. The sagittal coma arises because the image lateral magnification is different for the paraxial and for the marginal sagittal rays.

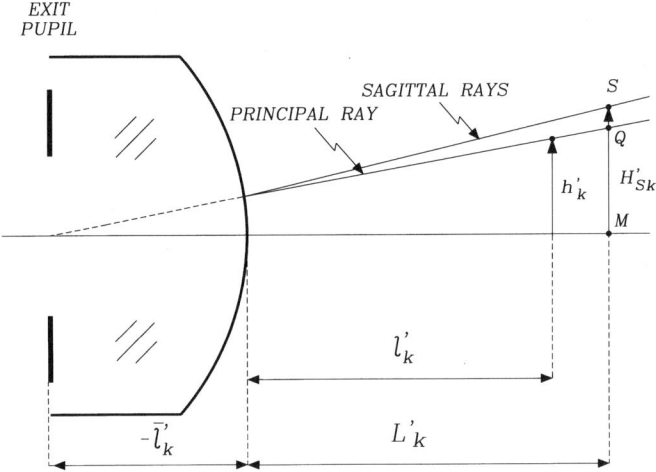

Figure 5.5 Formation of the coma aberration.

Besides, due to the spherical aberration, these two images are at different planes. The primary sagittal coma, represented by $Coma_S$, is defined by the lateral distance SQ from the marginal sagittal focus to the principal ray. Thus,

$$Coma_S = SQ = H'_{Sk} - QM \qquad (5.30)$$

The value of $Coma_S$ is positive when the sagittal focus is above the principal ray. To evaluate the sagittal coma let us first define a coefficient OSC (*Offense against the sine condition*) as follows:

$$OSC = \frac{H'_{Sk}}{QM} - 1 \qquad (5.31)$$

Thus, the primary sagittal coma is (assuming $QM \sim h'_k$)

$$Coma_S = OSC h'_k \qquad (5.32)$$

Since OSC is a constant for a given system, an important conclusion is that the aberration of $Coma_S$ increases linearly with the image height.

Now, the magnitude of QM is related to the paraxial image height h'_k on the paraxial focus plane by

$$QM = \frac{L'_k - \bar{l}'_k}{l'_k - \bar{l}'_k} h'_k \qquad (5.33)$$

hence the OSC is given by

$$OSC = \left[\frac{l'_k - \bar{l}'_k}{L'_k - \bar{l}'_k}\right] \frac{H'_{Sk}}{h'_k} - 1 \qquad (5.34)$$

If there is no coma in the object, we have $h_{S1} = h_1$, then, using the Lagrange and optical sine theorems we may find that

$$OSC = \left[\frac{l'_k - \bar{l}'_k}{L'_k - \bar{l}'_k}\right] \frac{h_{S1} u'_k \sin U_1}{h_1 u_1 \sin U'_k} - 1 \qquad (5.35)$$

An alternative expression is

$$OSC = \left[\frac{l'_k - \bar{l}'_k}{L'_k - \bar{l}'_k}\right] \frac{M}{m} - 1 = \left[\frac{l'_k - \bar{l}'_k}{L'_k - \bar{l}'_k}\right] \frac{u'_k \sin U_1}{u_1 \sin U'_k} - 1 \qquad (5.36)$$

where m and M are the paraxial and marginal image magnifications, respectively. In terms of the spherical aberration the OSC may be written as

$$OSC = \left[1 - \frac{SphL}{L'_k - \bar{l}'_k}\right]\frac{M}{m} - 1 \tag{5.37}$$

It is easy to see that when there is no spherical aberration the coma is absent if the principal surface is not a plane but a sphere centered at the focus. For infinite object distances the OSC may be calculated by

$$OSC = \left[\frac{l'_k - \bar{l}'_k}{L'_k - \bar{l}'_k}\right]\frac{F_M}{F} - 1 = \left[\frac{l'_k - \bar{l}'_k}{L'_k - \bar{l}'_k}\right]\frac{Yu'_k}{y \sin U'_k} - 1 \tag{5.38}$$

where F_M and F are the marginal (with incident ray height Y) and paraxial (with incident ray height y) focal lengths, measured as in Fig. 5.6, along the refracted rays.

5.3.2 Coma Contribution of Each Surface

We have derived an expression that allows us to compute the final primary coma in an optical system. However, we do not have any information about the coma contribution of each surface in the system. There are several possible ways of calculating these contributions. One is by using the results for a complete system, but taking the system as only one surface. Once the aberration for a single surface is obtained, the contribution of

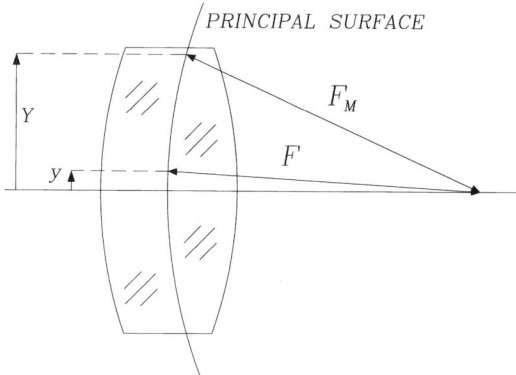

Figure 5.6 Principal surface in a system free of spherical aberration and of coma.

Monochromatic Off-Axis Aberrations

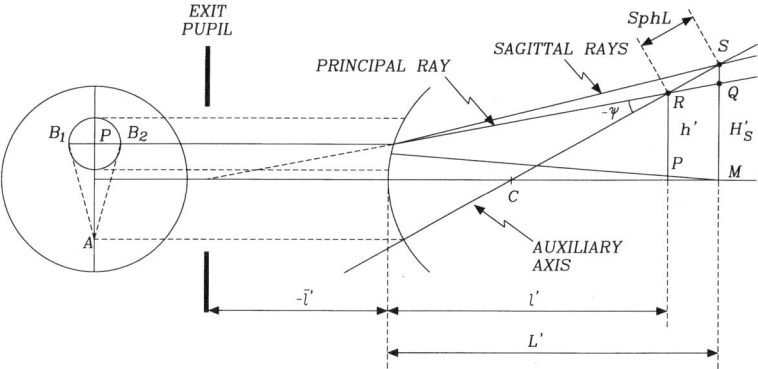

Figure 5.7 Some rays to illustrate the formation of the coma aberration.

this surface in a system of several surfaces is obtained by multiplication by the magnification of the rest of the system, after the surface. The transverse magnification is used for the transverse aberrations and the longitudinal magnification for the longitudinal aberrations.

In Fig. 5.7 the sagittal coma for a single surface has been defined by the distance SQ, which may be written as

$$Coma_S = SQ = -RS\psi \tag{5.39}$$

and by using Eq. (5.3) and the definition of the spherical aberration as $SphL = RS$:

$$Coma_S = -SphL\left(\frac{\bar{i}}{i}\right)u' = SphT\left(\frac{\bar{i}}{i}\right) \tag{5.40}$$

or, multiplying both sides of this expression by the transverse magnification of the part the optical system after this surface, we have the following contribution of the optical surface to the total coma in the system:

$$Coma_S C = SphTC\left(\frac{\bar{i}}{i}\right) \tag{5.41}$$

or, using Eq. (5.30):

$$Coma_S C = \sigma i \bar{i} \tag{5.42}$$

It is possible to show after some geometric considerations, as described by Conrady (1957), from Fig. 5.2, that

$$Coma_T = 3SphT\left(\frac{\bar{i}}{i}\right) \tag{5.43}$$

or

$$Coma_T = 3Coma_S \tag{5.44}$$

5.3.3 Coma in a Single Thin Lens

The primary or third-order coma aberration for any object position and lens bending, but with the stop at the lens, may be calculated with the following expression given by Conrady (1957):

$$Coma_S = h'_k y^2 \left(\frac{1}{4}G_5 \kappa c_1 - G_7 \kappa v_1 - G_8 c^2\right) \tag{5.45}$$

where the function G_5 has been defined in Chap. 4 and the functions G_7 and G_8 are defined by

$$G_7 = \frac{(2n+1)(n-1)}{2n}; G_8 = \frac{n(n-1)}{2} \tag{5.46}$$

This aberration as a function of the curvature c_1 is shown in Fig. 5.8. The coma can be made equal to zero with almost the same bending that produces the minimum spherical aberration, when the stop is in contact with the lens. If the lens has a large spherical aberration, the coma may be corrected only with the stop shifted with respect to the lens.

5.4 ASTIGMATISM

The primary or third-order longitudinal sagittal astigmatism, as shown in Fig. 5.2, is the distance from the Petzval surface to the sagittal (paraxial) surface. The longitudinal tangential astigmatism is the distance from the Petzval surface to the tangential surface. The astigmatism is positive if the sagittal focus is farther away from the optical surface than the Petzval focus. Thus, the sagittal longitudinal astigmatism on a single spherical surface is given by

$$AstL_S = P'R \simeq PR \tag{5.47}$$

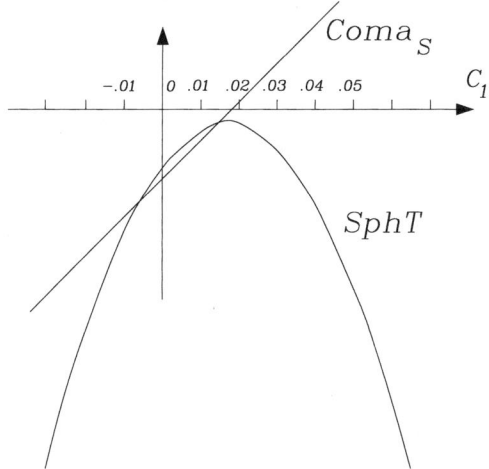

Figure 5.8 Variation of the transverse spherical aberration and coma versus the curvature of the front face in a thin lens.

but using Eqs. (5.7) and (5.9):

$$PR = RS\left(\frac{\bar{i}}{i}\right)^2 \tag{5.48}$$

Thus, the longitudinal sagittal astigmatism becomes

$$AstL_S = SphL\left(\frac{\bar{i}}{i}\right)^2 \tag{5.49}$$

If we examine Eq. (5.3), we may easily conclude that the astigmatism increases with the square of the angle ψ shown in Fig. 5.7. On the other hand, the magnitude of this angle depends on the position of the exit pupil and also on the image height. Given an optical system, with its stop fixed, the angle ψ grows approximately linearly with the image height for small values of ψ. This means that the astigmatism increases with the square of the image height.

Now, multiplying by u'_k both sides of Eq. (5.49) we obtain

$$AstT_S = SphT\left(\frac{\bar{i}}{i}\right)^2 \tag{5.50}$$

This is the sagittal astigmatism for a single spherical surface. The contribution of this surface to the final astigmatism in a complete system with several surfaces is this astigmatism multiplied by the longitudinal magnification of the part of the optical system after this surface. Then, multiplying both sides of Eq. (5.49) by this magnification, we have

$$AstL_S\,C = SphLC\left(\frac{\bar{i}}{i}\right)^2 \tag{5.51}$$

and in an analogous manner, the contribution to the transverse sagittal astigmatism is

$$AstT_S\,C = SphTC\left(\frac{\bar{i}}{i}\right)^2 = \sigma\bar{i}^2 \tag{5.52}$$

The tangential astigmatism may also be found from Fig. 5.2. In a first approximation it easy to see that

$$ST = \frac{B_1\,B_2}{2u'}\psi \tag{5.53}$$

then, using here Eqs. (5.3) and (5.14) we find that

$$ST = 2SphL\left(\frac{\bar{i}}{i}\right)^2 \tag{5.54}$$

The primary tangential astigmatism $P'T$ is then easily shown to be

$$AstL_T = 3SphL\left(\frac{\bar{i}}{i}\right)^2 \tag{5.55}$$

In conclusion, as in the case of the coma, the primary sagittal astigmatism and the primary tangential astigmatism are related by a factor of three to the coma, as follows:

$$AstT_T = 3\,AstT_S \tag{5.56}$$

If the object is moved on the object plane, the sagittal image moves on the sagittal surface and the tangential image on the tangential surface. Since the astigmatism grows with the square of the image height if the field is relatively small, the shape of the sagittal and tangential surfaces is almost spherical. If by any method, e.g., moving the stop position, the magnitude of the astigmatism is changed, then the sagittal and tangential surfaces become closer or separate more, but keeping constant the 1:3 relation. They join together in a single surface, which by definition is the Petzval surface.

When the beam propagates, different image shapes are generated for different observation planes. As shown in Fig. 5.19, inside and outside of focus the image is elliptical, but with different orientations.

5.4.1 Coddington Equations

Henry Coddington in 1829 in London derived two equations to find the positions of the sagittal and the tangential images in a single refractive surface. These two equations are similar to the Gauss equation and may be considered a generalization of it. There are several possible ways to derive these equations, but here we will present the method described by Kingslake (1978).

Tangential Image

Let us consider Fig. 5.9 with a spherical refractive surface and two very close meridional rays originating at object **B** and then refracted near the point **P**. These two rays are differentially refracted and converge to the tangential image B_T. To find the position of this point of convergence we first define the central angle $\theta = \overline{U} + \overline{I}$ and then differentiate it as follows:

$$d\theta = d\overline{U} + d\overline{I} \tag{5.57}$$

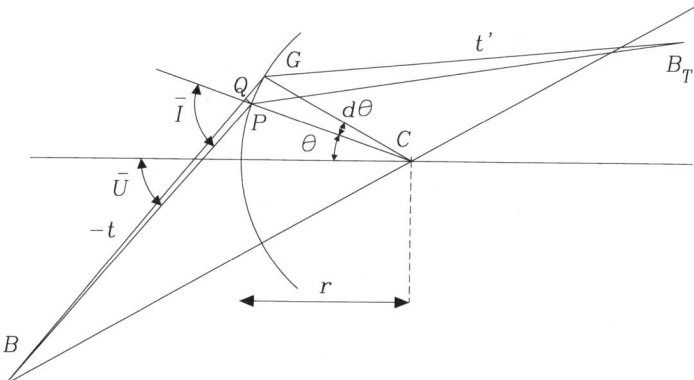

Figure 5.9 Derivation of Coddington equation for tangential rays.

The distance from the object **B** to the point **P** is t. Then, the small segment PQ is

$$PQ = t\mathrm{d}\overline{U} = PG\cos\overline{I} = r\cos\overline{I}\mathrm{d}\theta$$
$$= t(\mathrm{d}\theta - \mathrm{d}\overline{I}) = r\cos\overline{I}\mathrm{d}\theta \quad (5.58)$$

hence, we may write

$$\mathrm{d}\overline{I} = \left(1 - \frac{r\cos\overline{I}}{t}\right)\mathrm{d}\theta \quad (5.59)$$

and in a similar manner for the refracted ray:

$$\mathrm{d}\overline{I}' = \left(1 - \frac{r\cos\overline{I}'}{t}\right)\mathrm{d}\theta \quad (5.60)$$

Differentiating the expression for Suell's law of refraction we find

$$n\cos\overline{I}\mathrm{d}\overline{I} = n'\cos\overline{I}'\mathrm{d}\overline{I}' \quad (5.61)$$

finally obtaining

$$\frac{n'\cos\overline{I}'^2}{t'} - \frac{n\cos\overline{I}^2}{t} = \frac{n'\cos\overline{I}' - n\cos\overline{I}}{r} \quad (5.62)$$

This expression becomes the Gauss equation when the object height is zero, so that $\overline{I}$ and $\overline{I}'$ are zero.

Sagittal Image

The corresponding equation for the sagittal image may now be found with the help of Fig. 5.10. First, we have to remember that the sagittal image is on the auxiliary axis. Thus, we only have to find the intersection of the refracted principal ray with the auxiliary axis. Since the area of the triangle **BPB**$_S$ is the area of the triangle **BPC** plus the area of the triangle **PCB**$_S$, we may write these areas as

$$\frac{1}{2}ss'\sin(180° - \overline{I} + \overline{I}') = -\frac{1}{2}sr\sin(180° - \overline{I}) + \frac{1}{2}s'r\sin\overline{I}' \quad (5.63)$$

hence

$$-ss'\sin(\overline{I} - \overline{I}') = -sr\sin\overline{I} + s'r\sin\overline{I}' \quad (5.64)$$

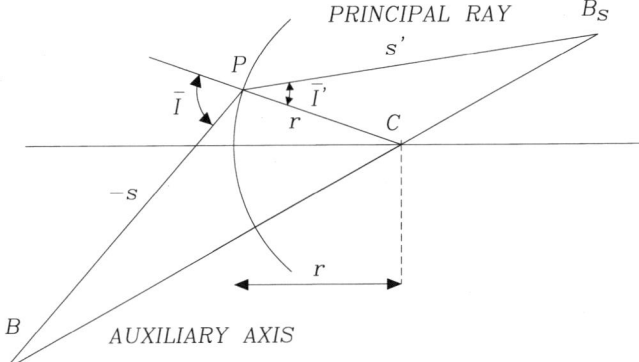

Figure 5.10 Derivation of Coddington equation for sagittal rays.

Then, after some algebraic steps and using the law of refraction we finally have

$$\frac{n'}{s'} - \frac{n}{s} = \frac{n'\cos\bar{I}' - n\cos\bar{I}}{r} \tag{5.65}$$

Again, this expression becomes the Gauss equation when $\bar{I}$ and $\bar{I}'$ are zero. These equations are frequently used to evaluate the astigmatism of optical systems.

5.4.2 Relations Between Petzval Curvature and Astigmatism

With a third-order approximation, if the field is relatively small (semifield smaller than about 10°) the sagittal (S), tangential (T), and Petzval (P) surfaces may be represented by spherical surfaces, as shown in Fig. 5.11, where these aberrations are positive. The separation between these surfaces follows relation (5.56). The sagittas for sagittal and tangential surfaces are equal to $AstL_S + Ptz$ and $AstL_t + Ptz$, respectively. The surface of best definition is between the sagittal and tangential surfaces. Thus, the sagitta for the surface of best definition is

$$Best = Ptz + \frac{AstT_s + AstT_t}{2} = Ptz + 2\,AstT_s \tag{5.66}$$

If the surface of best definition has to be flat, this sagitta has to be equal to zero, otherwise the radius of curvature for this surface is

$$r_{best} = \frac{h'^2}{2(Ptz + 2\,AstT_s)} \tag{5.67}$$

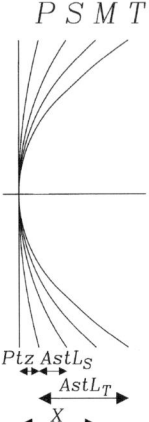

Figure 5.11 Astigmatic surfaces and definitions of sagittal and tangential astigmatism.

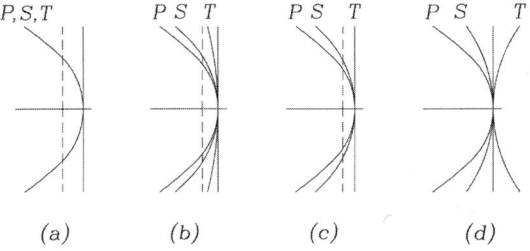

Figure 5.12 Astigmatic curves for different amounts of astigmatism with a constant Petzval curvature.

We have proved that the Petzval curvature depends only of the total power of the lenses forming the optical system and not on the lens shapes (bending) nor on the stop position. Thus, by bending and selecting the stop's position we may only change the astigmatism. On the other hand, the tangential and sagittal astigmatisms are always in a 3:1 relation. So, these focal surfaces may have different curvatures, as shown in Fig. 5.12.

To obtain the best overall image definition we may position the observation plane (screen, photographic film, or detector) at the places indicated with a dotted line. The four figures show the same Petzval surface with increasing amount of the magnitude of astigmatism. If there is no astigmatism, but there is a negative Petzval curvature, as in most cases, as shown in Fig. 5.12(a), the image is perfect and well defined over the whole Petzval surface. In this case the observing screen may be curved as in some astronomical instruments, or a field flattener may be used. As stated by

Conrady (1957), the astigmatism in Fig. 5.12(b) is a better choice for astronomical photography, where the field is not very large. Then, there is some astigmatism, but the optimum focal plane, between the sagittal and tangential surfaces, is flatter. If elongated images are not satisfactory, as in the case of photographic cameras, where the field is wide, the large astigmatism in Fig. 5.12(d) is a compromise, where the best-definition surface is flat. The price is a large astigmatism with the size of the image growing toward the edge. The best choice for most practical purposes is to reduce the astigmatism a little bit with respect to that in Fig. 5.12(d), by choosing a flat tangential field, as in Fig. 5.12(c).

An important practical conclusion that should always be in mind is that in a system with a negative Petzval sum, which occurs most of the time, the best overall image is obtained only if positive astigmatism is present.

For semifields larger than about 20° significant amounts of high-order astigmatism may appear, making the sagittal and tangential surfaces to deviate strongly from the spherical shape. In this case, high-order aberrations should be used to balance the primary aberrations, as shown in Fig. 5.13. The two astigmatic surfaces should cross near the edge of the field.

5.4.3 Comatic and Astigmatic Images

We have seen that coma and astigmatism are two different aberrations, but they are not independent in a single optical surface. Both are present and closely interrelated through the spherical aberration. Figure 5.14 shows how the aberrations of astigmatism and coma change for different values of the ratio $\bar{i}/i$. We may observe that for values of this ratio smaller than one the coma dominates, but for values greater than one the situation is reversed. Only in a complex system, with several centered spherical optical surfaces,

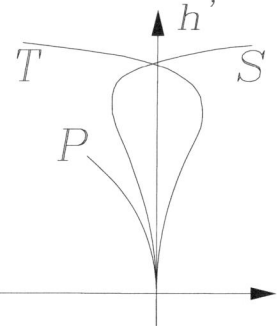

Figure 5.13 Astigmatic curves with high-order aberrations for a large field.

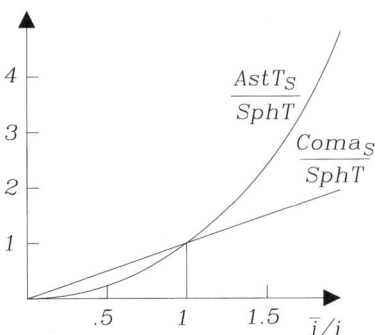

Figure 5.14 Variations in the ratios $AstT_S/SphT$ and $Coma_S/SphT$ versus the ratio $\bar{i}/i$.

may a single aberration, coma, or astigmatism be present. The image structure for each of these aberrations will now be described.

To understand how the coma image appears, let us divide the exit pupil into concentric rings as in Fig 5.15. The paths for some rays from one of the rings in the pupil are illustrated in this figure. Here, we may appreciate the following facts:

1. Rays symmetrical with respect to the meridional plane, $\mathbf{D}_i$ and $\mathbf{D}'_i$, cross each other at a point on the meridional plane, $\mathbf{P}_i$.
2. All the points $\mathbf{P}_i$ lie on a straight line, parallel to the principal ray. This line is called the *characteristic focal line*. A *diapoint* is defined as the point intersection of an oblique ray and the meridional plane. Thus, the focal characteristic line may also be defined as the *locus of diapoints* for the rays passing through a ring on the exit pupil.
3. Each circle on the exit pupil generates a characteristic focal line, parallel to the principal ray. The smaller the circle on the exit pupil is, the closer the characteristic focal line gets to the principal ray.
4. The tangential rays from $\mathbf{D}_0$ and $\mathbf{D}_4$ cross at the tangential focus, on the focal plane.
5. The sagittal rays from $\mathbf{D}_2$ and $\mathbf{D}'_2$ cross at the sagittal focus, on the focal plane.
6. Each ring also on the exit pupil also becomes a small ring on the focal plane. However, one turn around on the exit pupil becomes two turns on the image.
7. The complete comatic image is formed with all the rings, becoming smaller as they shift along the meridional plane (y axis).

A stereo pair of images showing the ray paths passing through a ring on the entrance pupil in the presence of coma is illustrated in Fig. 5.16. The

Monochromatic Off-Axis Aberrations

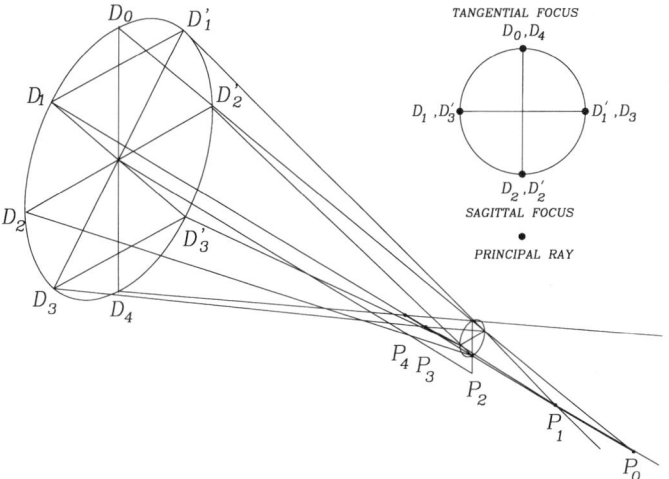

Figure 5.15 Rays around a ring on the pupil in the presence of coma.

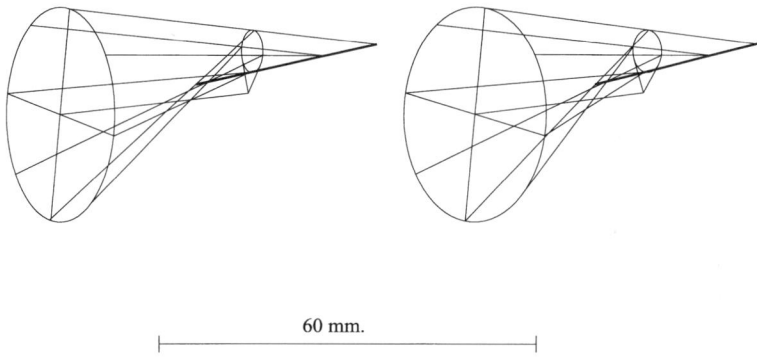

60 mm.

Figure 5.16 Stereo pair of images showing the ray paths from a ring on the pupil in the presence of coma.

final structure of a comatic image with positive coma is illustrated in Fig. 5.17. In this figure we see graphical definitions of the transverse sagittal and tangential coma.

In an optical system with pure astigmatism (without coma), the rays from a ring on the exit pupil travel as in Fig. 5.18. Here, we may appreciate the following:

1. Rays symmetrical with respect to the meridional plane, $\mathbf{D}_i$ and $\mathbf{D}'_i$, cross each other at a point on the meridional plane, $\mathbf{P}'_i$. The letter i stands for 0, 1, 2, 3, or 4 in Fig. 5.18.

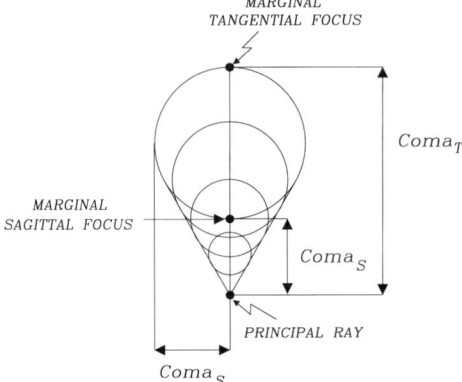

Figure 5.17 Formation of the comatic image.

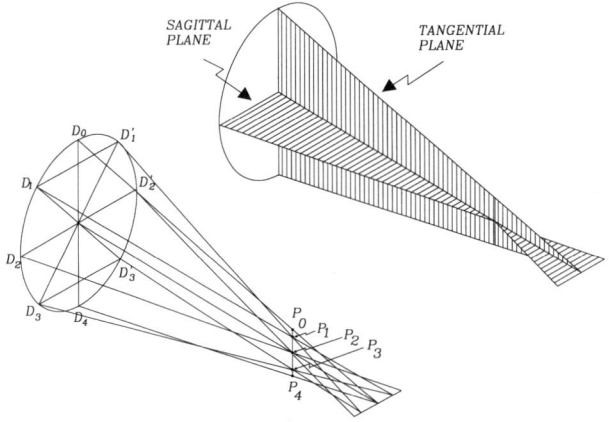

Figure 5.18 Rays forming the image in a system with astigmatism.

2. All the points P_i lie on a straight line, perpendicular to the principal ray. This line is the *characteristic focal line*. In this case this is also the sagittal focus.
3. Each circle on the exit pupil generates a characteristic focal line; all are placed on the corresponding sagittal focus. Thus, in a single optical surface all characteristic focal lines are parallel to each other and perpendicular to the principal ray. In an optical system without spherical aberration all characteristic focal lines collapse in a single line.
4. At an intermediate plane between the sagittal and the tangential focus, the image is a small circle.

Monochromatic Off-Axis Aberrations

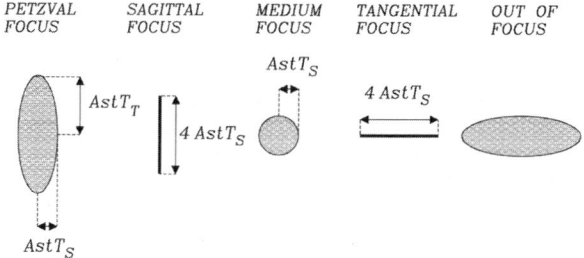

Figure 5.19 Astigmatic images in different focal surfaces.

The astigmatic images at several focal planes are illustrated in Fig. 5.19. The magnitudes of the sagittal and tangential transverse astigmatisms are shown here. Now, let us make some general considerations about the characteristic focal line:

1. The focal characteristic line for a single refracting surface is on the auxiliary axis, as shown in Fig. 5.3.
2. In general, there is a focal characteristic line for each concentric ring on the exit pupil, and all lines are parallel to each other.
3. In a system with pure coma the focal characteristic lines are parallel to the principal ray and in a system with pure astigmatism they are perpendicular to the principal ray. Thus, it is clear that in a single surface we cannot isolate astigmatism and coma, because the auxiliary optical axis is always inclined with respect to the principal ray.
4. The center of the characteristic focal line is the sagittal focus. The extremes are defined by the marginal tangential rays.
5. In a complete optical system the characteristic focal line is in general inclined to a certain angle θ with respect to the principal ray. Then, this inclination is given by the relative amounts of these two aberrations as follows:

$$\tan \theta = -\frac{AstT_S}{Coma_S} u'_k \tag{5.68}$$

A little more insight and understanding about the structure of the astigmatic and comatic images may be obtained with a detailed examination of Fig. 5.20. This figure plots the locus of the intersections on the focal plane of the light rays passing through a circular ring on the entrance pupil. These plots are taken at different equidistant focal plane positions, but different

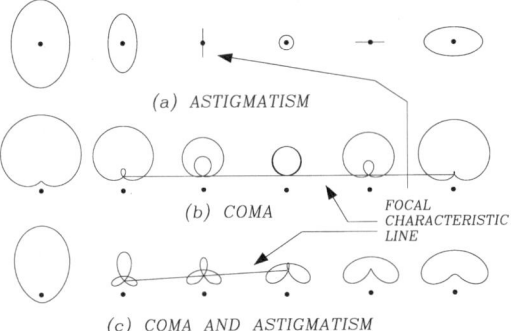

Figure 5.20 Images in different focal surfaces in the presence of coma and astigmatism: (a) astigmatism; (b) coma; (c) coma and astigmation.

for the three cases shown. The large dot represents the principal ray intersection and a line joining them would be the principal ray. The focal characteristic line (or diapoint locus) is graphically represented in these three cases.

5.4.4 Aplanatic Refractive Surfaces

From Eqs. (5.40) and (5.49) we see that both, the coma and astigmatism contributions of a spherical optical surface are zero if (1) the contribution to the spherical aberration is zero or (2) the principal ray is perpendicular to the surface ($\bar{i}=0$). However, it is important to be careful with the cases in which the meridional ray is perpendicular to the surface ($i'=0$), because a division by zero might occur in the factor ($\bar{i}/i'$). If we see that Eq. (5.23) for the spherical aberration contribution may be written as

$$SphLC = \frac{y(n/n')(n-n')(i+u')i^2}{2n'_k u'^2_k} \tag{5.69}$$

we then see that the astigmatism may be different from zero when the spherical aberration is zero and the meridional ray is normal to the surface.

It is interesting to consider the three cases of spherical aberration zero and coma zero in a single surface, described in Chap. 4. We may see that these cases also have zero coma and zero astigmatism. An exception, as we proved, is the case of astigmatism, when the object and the image are both at the center of curvature.

Another method to prove that the coma aberration is zero in the three cases of spherical aberration zero is by calculating the *OSC*. The *OSC* for any optical system, with zero spherical aberration ($L'_k = l'_k$), from Eq. (5.37), is

$$OSC = \frac{u'_k \sin U}{u \sin U'_k} - 1 \tag{5.70}$$

However, for a single reflecting surface, from Eqs. (1.19)–(1.21) we find that

$$\frac{M}{m} = \frac{u' \sin U}{u \sin U'} = \left(\frac{L'-r}{L-r}\right)\left(\frac{l-r}{l'-r}\right) \tag{5.71}$$

Since the surface is assumed to be free of spherical aberration ($L' = l'$), and the object is assumed to be a point ($L = l$), we see that both sides of this expression are one. Then, it is easy to see that OSC and hence the sagittal coma are zero in the three cases in which the spherical aberration is zero.

Abbe called these surfaces *aplanatic* due to their simultaneous absence of spherical aberration and coma. In general, we say that an optical system is aplanatic if it is simultaneously free from spherical aberration and coma. According to a more recent definition, an optical system is said to be aplanatic if the image is perfect (aberration free) and can be moved to any point inside a small region centered on the optical axis, without introducing any aberrations. This aberration-free region is called the aplanatic region.

Similarly, a system is isoplanatic if the image can be moved to any point inside of a small region near the optical axis, without altering the image structure, i.e., without changing the aberrations. This zone near the optical axis is the isoplanatic region.

5.5 DISTORTION

When all aberrations are absent, the image of a point object is located at a point named the gaussian image. If the object height is h, the gaussian image height is exactly equal to mh', where m is the first-order lateral magnification, obtained from the Lagrange theorem. With reference to Fig. 5.2, we may see that if the coma aberration is zero, both the tangential image **T** and the sagittal image **S** are on the principal ray, at points **U** and **Q**, respectively. If there is no spherical aberration, the marginal sagittal focus **S** coincides with the paraxial sagittal focus **R**, so the point **Q** and the point **R** become the same. If there is no astigmatism, the paraxial sagittal focus **R** as well as the tangential focus **T** coincide at a single point **P′** on the Petzval surface.

Let us further assume that there is no Petzval curvature. Then, the image point would be at the point **N** on the principal ray. In conclusion, even if all aberrations we have studied are absent from the optical system, the image might still be laterally deviated with respect to the gaussian image point **M**.

The distortion aberration is due to a deviation in the actual image height, determined by the principal ray, with respect to the gaussian image height h'. Thus, if we plot the actual image height, defined by the intersection of the principal ray with the gaussian focal plane, as a function of the object height h' the result is not a line, as shown in Fig. 5.21. The distortion may be negative or positive. If a square is imaged with a lens having distortion, the result may be as shown in Fig. 5.22(a) or (b). Positive distortion is also called pincushion distortion, and negative distortion is also called barrel distortion, due to the aspect of the image of a square.

Positive distortion occurs when the principal ray is above the ideal gaussian image. Let us consider again Fig. 5.2, with a single refractive surface with the principal ray and the auxiliary axis. The ideal or gaussian image is then at the point **M** on the auxiliary axis. Then, the value of the distortion is the distance from the gaussian image **M** to the intersection **N** of the principal ray with the focal plane containing **M**. Thus, by observing the figure, we see that the value of the distortion is

$$Dist = -MR\psi = -(PR + MP)\psi \tag{5.72}$$

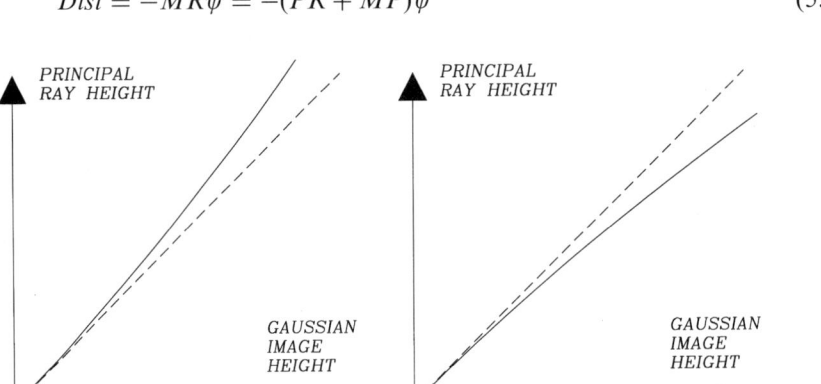

(a) POSITIVE DISTORTION   (b) NEGATIVE DISTORTION

Figure 5.21 Principal ray height versus the gaussian image height in the presence of distortion: (a) positive distortion; (b) negative distortion.

Monochromatic Off-Axis Aberrations

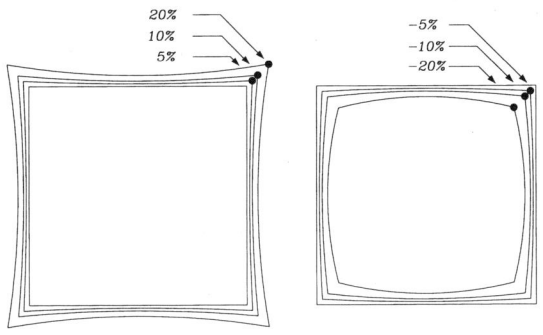

(a) POSITIVE DISTORTION (PINCUSHION)

(b) NEGATIVE DISTORTION (BARREL)

Figure 5.22 Images of a square with both possible signs of distortion. The distortion percentages at the corners of the square are indicated: (a) positive distortion (pincushion); (b) negative distortion (barrel).

but from using Eqs. (5.7) for the value of PR and since MP is the value of Ptz, given in Eq. (5.21), the value of the distortion is

$$Dist = -\left[SphL\left(\frac{\bar{i}}{i}\right)^2 + Ptz\right]\psi \tag{5.73}$$

and using now the value of ψ in Eq. (5.3):

$$Dist = SphT\left(\frac{\bar{i}}{i}\right)^3 - Ptz\left(\frac{\bar{i}}{i}\right)u' \tag{5.74}$$

The distortion for this surface may also be written as

$$Dist = Coma_S\left(\frac{\bar{i}}{i}\right)^2 - Ptz\left(\frac{\bar{i}}{i}\right)u' \tag{5.75}$$

This is the distortion produced by only one surface. As usual, the aberration contribution of this surface to the final astigmatism in a complete system is obtained by multiplying this result by the longitudinal magnification of the optical system after this surface. Then, we have

$$DistC = Coma_S\, C\left(\frac{\bar{i}}{i}\right)^2 - PtzC\left(\frac{\bar{i}}{i}\right)u'_k \tag{5.76}$$

It is possible to show that an alternative expression, better adapted for numerical calculation, is

$$DistC = \bar{\sigma}\, i\bar{i} + \frac{h'_k}{2}(\bar{u}'^2 - \bar{u}^2) \tag{5.77}$$

where

$$\bar{\sigma} = \frac{\bar{y}(n/n')(n-n')(\bar{i}+\bar{u}')}{2n'_k u'_k} \tag{5.78}$$

The aberration of distortion may be exactly computed by tracing the principal ray and subtracting the image height obtained with this ray from the gaussian image height.

This distortion increases with the image height and may be represented by a polynomial with odd powers of this the image height. Thus, the primary distortion term grows with the cube of the image height. This is the reason for the appearance of the image of a square in the presence of this aberration.

Frequently, the distortion is expressed as a percentage of the gaussian image height, as shown in Fig. 5.22. If the object is a square, the value of the distortion at the corners is exactly twice the value of the distortion at the middle of the sides.

5.6 OFF-AXIS ABERRATIONS IN ASPHERICAL SURFACES

As shown in Chap. 4, an aspheric surface has a deformation with respect to the sphere that introduces an additional spherical aberration term given by

$$SphTC_{\text{asph}} = Dy^3 \tag{5.79}$$

where S has been replaced by the symbol y and the constant D is defined as

$$D = -(8A_1 + Kc^3)\left(\frac{n-n'}{2n'_k u'_k}\right)y \tag{5.80}$$

The ratio y/u'_k is approximately constant.

This term changes the slope of the refracted light rays, and thus their transverse aberrations by an amount equal to this transverse spherical aberration. The value of y is the distance from the optical axis (vertex of the surface) to the intersection of the light ray with the optical surface. Then, for

Monochromatic Off-Axis Aberrations

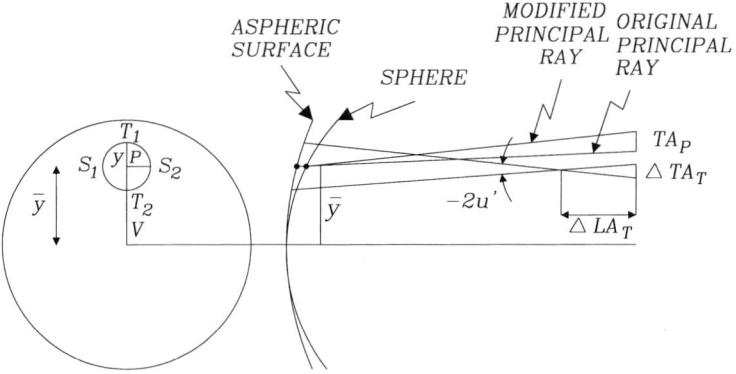

Figure 5.23 Aberrations introduced by aspherizing an optical surface.

the rays shown in Fig. 5.23, these additional transverse aberrations are as follows:

1. For the principal ray, passing through **P**:

$$TA_P = D\bar{y}^3 \tag{5.81}$$

2. For the upper tangential ray, passing through **T**$_1$:

$$TA_{T1} = D(\bar{y} + y)^3 = D(\bar{y}^3 + y^3 + 3y\bar{y}^2 + 3y^2\bar{y}) \tag{5.82}$$

3. For the lower tangential ray, passing through **T**$_2$:

$$TA_{T2} = D(\bar{y} - y)^3 = D(\bar{y}^3 - y^3 - 3y\bar{y}^2 + 3y^2\bar{y}) \tag{5.83}$$

4. For the sagittal rays, passing through **S**$_1$ and **S**$_2$:

$$TA_S = D(\bar{y}^2 + y^2)^{3/2} \simeq D(\bar{y}^3 + \bar{y}y^2) \tag{5.84}$$

With these results we may now find the additional aberration terms due to the aspheric deformation. The additional term to be added to the coma contribution $Coma_S C$ will be represented by $Coma_S C_{\text{asph}}$ and is given by

$$Coma_S\, C_{\text{asph}} = TA_S - TA_P = SphTC_{\text{asph}}\left(\frac{\bar{y}}{y}\right) \tag{5.85}$$

since both the principal ray and the sagittal rays are modified.

The vertical shift of the marginal tangential focus is obtained by taking the average of the transverse aberration for these rays. Since the distance from this tangential focus to the principal ray is the tangential coma, the additional term to be added to $Coma_T$ is given by

$$Coma_{T\,\mathrm{asph}} = \frac{TA_{T1} + TA_{T2}}{2} - TA_P \tag{5.86}$$

thus, we may find the tangential coma contribution:

$$Coma_T\,C_{\mathrm{asph}} = 3\,SphTC_{\mathrm{asph}}\left(\frac{\bar{y}}{y}\right) \tag{5.87}$$

As we see, the 1:3 relation between the sagittal and the tangential coma is preserved.

The sagittal rays are not contained in the meridional plane. The sagittal transverse aberration TA_S due to the aspheric deformation may easily seen to be in the direction VS_2 shown in Fig. 5.23.

The horizontal component (x direction) of the transverse aberration for the sagittal rays is in opposite directions for the two rays and has a value equal to

$$TA_{Sx} = TA_S\left(\frac{y}{\bar{y}}\right) = D(\bar{y}^3 + \bar{y}\,y^2)\left(\frac{y}{\bar{y}}\right) \tag{5.88}$$

Neglecting the second term to introduce a paraxial approximation, the paraxial sagittal focus is longitudinally displaced by an amount equal to the contribution to the longitudinal sagittal astigmatism, as follows:

$$AstL_{S\,\mathrm{asph}} = -\frac{TA_{Sx}}{u'} = -\frac{D\bar{y}^2 y}{u'} \tag{5.89}$$

then, we obtain using Eqs. (5.79) and (4.27), the aspheric contribution to the longitudinal sagittal astigmatism as

$$AstL_S\,C_{\mathrm{asph}} = SphLC_{\mathrm{asph}}\left(\frac{\bar{y}}{y}\right)^2 \tag{5.90}$$

As shown in Fig. 5.23, the difference in the transverse aberrations in the vertical direction for the two marginal tangential rays introduces a longitudinal displacement ΔLA_T of the tangential focus, given by

$$\Delta LA_T = -\frac{\Delta TA_T}{2u'} = -\frac{TA_{T1} - TA_{T2}}{2u'}$$

$$= -\frac{D(y^3 + 3y\,\bar{y}^2)}{u'} \tag{5.91}$$

Monochromatic Off-Axis Aberrations

thus, the longitudinal shift of the paraxial tangential focus is given by

$$\Delta LA_T = -\frac{3Dy\bar{y}^2}{u'} \tag{5.92}$$

where the cubic term in y was neglected to introduce the paraxial approximation. It is interesting to notice that, as opposed to the case of spherical surfaces, where the tangential focus is fixed for paraxial and marginal rays, here, we have a spherical aberration effect. Thus, we have a paraxial tangential focus and a marginal tangential focus at a different position. This focus shift is the astigmatism contribution introduced by the aspheric surface, which may be written

$$AstL_T C_{\text{asph}} = -3\, SphLC_{\text{asph}} \left(\frac{\bar{y}}{y}\right)^2 \tag{5.93}$$

We see that the 1:3 relation is also preserved for the astigmatism contributions due to the aspheric deformation.

The transverse aberration for the principal ray is equal to the contribution to the distortion introduced by the aspheric deformation, as follows:

$$DistC_{\text{asph}} = SphTC_{\text{asph}} \left(\frac{\bar{y}}{y}\right)^3 \tag{5.94}$$

5.7 ABERRATIONS AND WAVEFRONT DEFORMATIONS

According to the Fermat principle, the optical path from any point object to its image must be a constant for all ray paths, if the image is perfect. However, if the image has aberrations the wavefront exiting the optical system is not spherical, but has some deformations (Miyamoto, 1964). These aberrations may then be computed if the real optical path from the object to the image is calculated.

Let us consider Fig. 5.24, where the optical path difference for the ray refracted at the vertex of the surface and the ray refracted at the point **P** is

$$\begin{aligned} OPD &= [nBP + n'PB'] - [nAB + n'AB'] \\ &= n'[PB' - AB'] + n[BP - AB] \end{aligned} \tag{5.95}$$

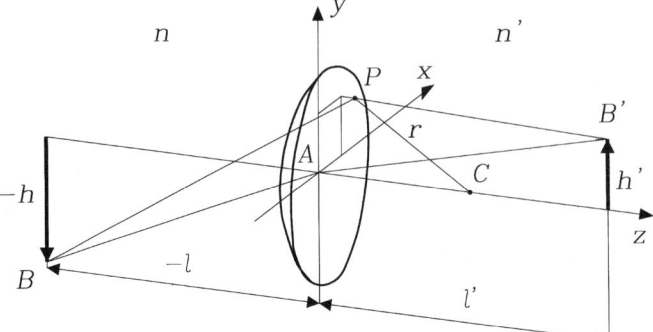

Figure 5.24 Optical paths of rays and their associated wavefronts.

The sagitta z at this surface is given by

$$z = \frac{x^2 + y^2 + z^2}{2r} = \frac{s^2}{2r} = r - \sqrt{r^2 - (x^2 + y^2)} \tag{5.96}$$

where $s^2 = x^2 + y^2 + z^2 = S^2 + z^2$. Thus, we may find that

$$AB = l\sqrt{1 + \left(\frac{h}{l}\right)^2} \tag{5.97}$$

and

$$BP^2 = (-l + z)^2 + (-h + y)^2 + x^2 \tag{5.98}$$

thus,

$$BP = l\sqrt{1 + \left[\left(\frac{h}{l}\right)^2 - \frac{s^2}{l}\left(\frac{1}{r} - \frac{1}{l}\right) - 2\frac{hy}{l^2}\right]} \tag{5.99}$$

Now, expanding in a Taylor series, assuming that $a \ll 1$:

$$\sqrt{1 + a} = 1 + \frac{1}{2}a - \frac{1}{8}a^2 + \frac{1}{16}a^3 - \frac{5}{128}a^4 + \cdots \tag{5.100}$$

Hence, we may find that, eliminating all terms higher than h^3 and S^4,

$$BP - AB = -\frac{s^2}{2}\left(\frac{1}{r} - \frac{1}{l}\right) - \frac{hy}{l} - \frac{s^4}{8l}\left(\frac{1}{r} - \frac{1}{l}\right)^2$$

$$-\frac{h^2 y^2}{2 l^3} + \frac{s^2 h^2}{4 l^2}\left(\frac{1}{r} - \frac{1}{l}\right)$$

$$-\frac{s^2 hy}{2 l^2}\left(\frac{1}{r} - \frac{1}{l}\right) + \frac{h^3 y}{2 l^3} + \cdots \qquad (5.101)$$

and similarly for $PB' - AB'$. Then, the final result for the optical path difference is

$$OPD = -\frac{s^2}{2}\left[n\left(\frac{1}{r} - \frac{1}{l}\right) - n'\left(\frac{1}{r} - \frac{1}{l'}\right)\right] - y\left[\frac{nh}{l} - \frac{n'h'}{l'}\right]$$

$$-\frac{s^4}{8}\left[\frac{n}{l}\left(\frac{1}{r} - \frac{1}{l}\right)^2 - \frac{n'}{l'}\left(\frac{1}{r} - \frac{1}{l'}\right)^2\right]$$

$$-\frac{s^2 y}{2}\left[\frac{nh}{l^2}\left(\frac{1}{r} - \frac{1}{l}\right) - \frac{n'h'}{l'^2}\left(\frac{1}{r} - \frac{1}{l'}\right)\right]$$

$$+\frac{s^2}{4}\left[\frac{nh^2}{l^2}\left(\frac{1}{r} - \frac{1}{l}\right) - \frac{n'h'^2}{l'^2}\left(\frac{1}{r} - \frac{1}{l'}\right)\right]$$

$$-\frac{y^2}{2}\left[\frac{nh^2}{l^3} - \frac{n'h'^2}{l'^3}\right] + \frac{y}{2}\left[\frac{nh^3}{l^3} - \frac{n'h'^3}{l'^3}\right] \cdots \qquad (5.102)$$

The first term represents a defocusing term or a change in the spherical reference wavefront. If we make this term equal to zero the points **B** and **B'** are in planes conjugate to each other and we obtain the Gauss equation.

The second term is a transverse displacement of the image or a tilt of the reference spherical wavefront in the y direction. If this term is made equal to zero, the Lagrange theorem is obtained. Then, the point **B'** is the conjugate (image) of point **B**.

The third term is the primary spherical aberration of a single spherical surface, expressed as a wavefront deformation. The ray transverse spherical may be obtained from this expression by derivation with respect to S, as shown in Section 1.9.

The fourth term represents the primary coma on a single surface, again, as a wavefront deformation. The sagittal coma may be obtained by derivation with respect to x and the tangential coma by derivation with respect to y.

The fifth and sixth terms combined represent the Petzval curvature and the primary astigmatism. The last term is a tilt of the wavefront in the y direction, produced by the primary distortion.

It must be pointed out that these aberrations assume that the stop is in contact with the optical surface. However, these expressions may be generalized to include any stop shift.

5.8 SYMMETRICAL PRINCIPLE

A system is fully symmetrical when one half of the system is identical to the other half, including object and image. Then, the stop is always at the center of the system and the magnification is -1, as shown in Fig. 5.25.

In this system the symmetrical wavefront aberrations $[W(y) = W(-y)]$ are doubled, but the antisymmetrical $[W(y) = -W(-y)]$ wavefront aberrations are canceled out. Then, the coma, distortion, and magnification chromatic aberrations are automatically made zero.

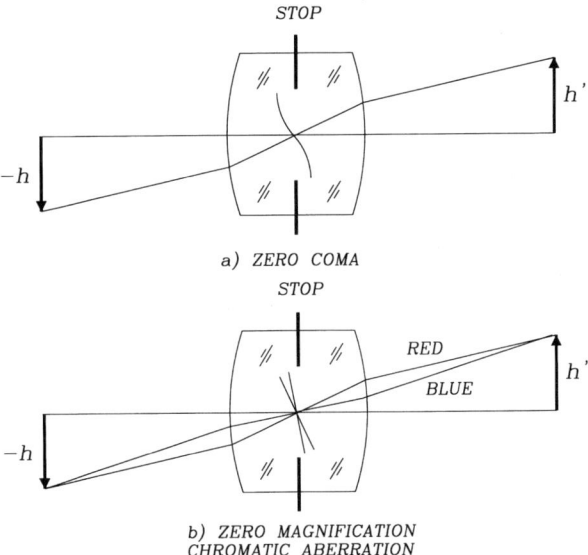

Figure 5.25 Symmetrical optical system: (a) zero coma; (b) zero magnification chromatic aberration.

If the system is symmetrical, but the conjugates are not equal, still the antisymmetrical aberrations are small. Fortunately, full symmetry is not necessary to obtain zero antisymmetric aberrations. To understand how this may be done, we have to remember some first-order properties of the principal planes. The nodal points coincide with the principal points when the object medium and the image medium are both air. If the stop is at one of the principal planes, the entrance pupil coincides with one of the principal planes and the exit pupil with the other. Then, the principal ray will pass through the nodal points. We may then say that, if the principal planes are fixed for all obliquity angles, the system is free of distortion. More generally, we may say that, if the entrance and exit pupils are fixed for all obliquity angles, the system is free of distortion. Similarly, as we will see in Chapter 6, if the entrance and exit pupils are fixed for all colors, the system is free of magnification chromatic aberration. These are the Bow–Sutton conditions.

5.9 STOP SHIFT EQUATIONS

When the stop is shifted to a new position, the principal ray height changes at every surface. However, we have shown in Eq. (3.63) that the ratio of the change in the principal ray height to the meridional ray height is a constant for all surfaces in the system, as follows:

$$\overline{Q} = \frac{\overline{y}^* - \overline{y}}{y} \tag{5.103}$$

where we represent the modified value when the stop is moved to a new position, with an asterisk.

Before we derive the expressions for the change in the aberrations when the stop is shifted, let us find some useful relations. We see from Fig. 5.26 that when we shift the principal ray:

$$\overline{u}'^*_k - \overline{u}'_k = \left[\frac{1}{l'_k - \overline{l}'^*_k} - \frac{1}{l'_k - \overline{l}'_k} \right] h'_k$$

$$= \frac{\overline{y}^*_k - \overline{y}_k}{l'_k} = \frac{(\overline{y}^*_k - \overline{y}_k) u'_k}{y_k} \tag{5.104}$$

hence, the value of the constant parameter $\overline{Q}$ is given by

$$\overline{Q} = \left[\frac{1}{l'_k - \overline{l}'^*_k} - \frac{1}{l'_k - \overline{l}'_k} \right] \frac{h'_k}{u'_k} \tag{5.105}$$

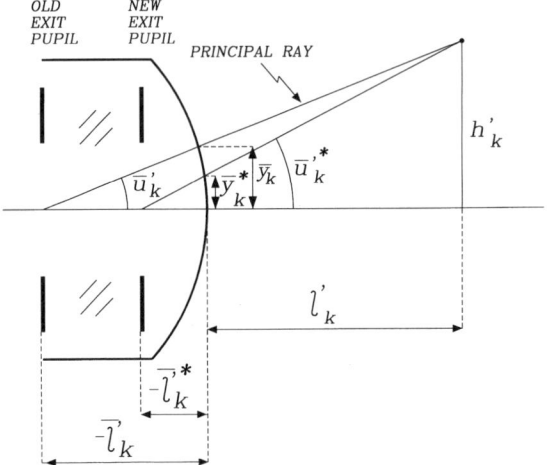

Figure 5.26 Change in the off-axis aberrations by a shift in the stop.

Another useful relation may be obtained from Eq. (5.5) as follows (see Fig. 5.2):

$$\bar{i}' = A A_0 \frac{i'}{y} = (AV + \bar{y})\frac{i'}{y} \tag{5.106}$$

where the distance AV remains constant with a stop shift. Similarly, for the shifted stop we may write

$$\bar{i}'^* = (AV + \bar{y}^*)\frac{i'}{y} \tag{5.107}$$

and subtracting from Eq. (5.105):

$$\bar{i}'^* - \bar{i}' = (\bar{y}^* - \bar{y})\frac{i'}{y} \tag{5.108}$$

Thus, using the definition of $\overline{Q}$, we may find that

$$\overline{Q} = \frac{\bar{i}'^* - \bar{i}'}{i'} = \frac{\bar{i}^* - \bar{i}}{i} \tag{5.109}$$

From this equation we may also see that

$$\bar{i}'^{*2} - \bar{i}'^2 = \overline{Q}(\bar{i}'^* + \bar{i}')i' \tag{5.110}$$

Monochromatic Off-Axis Aberrations

which may be transformed with the help of Eq. (5.109) into

$$\left(\frac{\bar{i}'^*}{i'}\right)^2 = \left(\frac{\bar{i}'}{i'}\right)^2 + 2\overline{Q}\left(\frac{\bar{i}'}{i'}\right) + \overline{Q}^2 \qquad (5.111)$$

These expressions are all the tools we need to find the change in the primary aberrations with a shift in the stop. The value of the spherical aberration does not depend on the position of the stop because the image is on the axis. Thus, for the case of spherical aberration we may write

$$SphT^* = SphT \qquad (5.112)$$

It is easy to see from Eqs. (5.40) and (5.109) that for the case of $Coma_S$ we have

$$Coma_S^* = SphT^*\left(\frac{\bar{i}'^*}{i'}\right)$$
$$= Coma_S + SphT\,\overline{Q} \qquad (5.113)$$

This equation shows that the coma aberration does not change its value by shifting the stop if the spherical aberration of the system is zero. If we want to correct the coma by selecting a position for the stop, the system must have spherical aberration.

For the astigmatism, using Eqs. (5.50) and (5.111), we may write

$$AstT_S^* = SphT^*\left(\frac{\bar{i}'^*}{i'}\right)^2$$
$$= AstT_S + 2\,Coma_S\,\overline{Q} + SphT\,\overline{Q}^2 \qquad (5.114)$$

The Petzval curvature does not change with a shift in the position of the stop; thus, we write

$$Ptz^* = Ptz \qquad (5.115)$$

In a similar manner, we may find the stop shift equation for distortion, from Eq. (5.75), as follows:

$$Dist^* = SphT^*\left(\frac{\bar{i}^*}{i}\right)^3 - Ptz^*\left(\frac{\bar{i}^*}{i}\right)u'$$
$$= Dist + (3\,AstT_S + PtzT)\overline{Q} + 3\,Coma_S\,\overline{Q}^2 + SphT\,\overline{Q}^3 \qquad (5.116)$$

An interesting consequence of these stop shift equations is that if a system has all primary aberrations corrected, a shift in the stop does not change the state of correction.

REFERENCES

Buchdahl, H. A., "Algebraic Theory of the Primary Aberrations of the Symmetrical Optical System," *J. Opt. Soc. Am.*, **38**, 14–19 (1948).
Buchdahl, H. A., *Optical Aberration Coefficients*, Oxford University Press, London, 1954. (Reprinted with Buchdahl's articles by Dover Publications, New York, 1968.)
Buchdahl, H. A., "Optical Aberration Coefficients: I. The Coefficient of Tertiary Spherical Aberration," *J. Opt. Soc. Am.*, **46**, 941–943 (1956).
Buchdahl, H. A., "Optical Aberration Coefficients: II. The Tertiary Intrinsic Coefficients," *J. Opt. Soc. Am.*, **48**, 563–567 (1958a).
Buchdahl, H. A., "Optical Aberration Coefficients: III. The Computation of the Tertiary Coefficients," *J. Opt. Soc. Am.*, **48**, 747–756 (1958b).
Buchdahl, H. A., "Optical Aberration Coefficients: IV. The Coefficient of Quaternary Spherical Aberration," *J. Opt. Soc. Am.*, **48**, 757–759 (1958c).
Buchdahl, H. A., "Optical Aberration Coefficients: V. On the Quality of Predicted Displacements," *J. Opt. Soc. Am.*, **49**, 1113–1121 (1959).
Buchdahl, H. A., "Optical Aberration Coefficients: VI. On the Computations Involving Coordinates Lying Partly in Image Space," *J. Opt. Soc. Am.*, **50**, 534–539 (1960a).
Buchdahl, H. A., "Optical Aberration Coefficients: VII. The Primary, Secondary and Tertiary Deformations and Retardation of the Wavefront," *J. Opt. Soc. Am.*, **50**, 539–544 (1960b).
Buchdahl, H. A., "Optical Aberration Coefficients: VIII. Coefficient of Spherical Aberration of Order Eleven," *J. Opt. Soc. Am.*, **50**, 678–683 (1960c).
Buchdahl, H. A., "Optical Aberration Coefficients: IX. Theory of Reversible Optical Systems," *J. Opt. Soc. Am.*, **51**, 608–616 (1961).
Buchdahl, H. A., "Optical Aberration Coefficients: X. Theory of Concentric Optical Systems," *J. Opt. Soc. Am.*, **52**, 1361–1367 (1962a).
Buchdahl, H. A., "Optical Aberration Coefficients: XI. Theory of a Concentric Corrector," *J. Opt. Soc. Am.*, **52**, 1367–1372 (1962b).
Buchdahl, H. A., "Optical Aberration Coefficients: XII. Remarks Relating to Aberrations of Any Order," *J. Opt. Soc. Am.*, **55**, 641–649 (1965).
Buchdahl, H. A., *Introduction to Hamiltonian Optics*, Cambridge University Press, Cambridge, UK, 1970.
Conrady, A. E., *Applied Optics and Optical Design, Part I*, Dover Publications, New York, 1957.
Conrady, A. E., *Applied Optics and Optical Design, Part II*, Dover Publications, New York, 1960.
Cruickshank, F. D. and Hills, G. A., "Use of Optical Aberration Coefficients in Optical Design," *J. Opt. Soc. Am.*, **50**, 379–387 (1960).

Epstein, L. I, "The Aberrations of Slightly Decentered Optical Systems," *J. Opt. Soc. Am.*, **39**, 847–853 (1949).

Focke, J., "Higher Order Aberration Theory," in *Progress in Optics*, E. Wolf, ed., Vol. IV, Chap. I, North Holland, Amsterdam, 1965.

Herzberger, M. and Marchand, E., "Image Error Theory for Finite Aperture and Field," *J. Opt. Soc. Am.*, **42**, 306–321 (1952).

Herzberger, M. and Marchand, E., "Tracing a Normal Congruence through an Optical System," *J. Opt. Soc. Am.*, **44**, 146–154 (1954).

Hopkins, R. E. and Hanau, R., "Aberration Analysis and Third Order Theory," in *Military Standardization Handbook*: *Optical Design*, *MIL-HDBK 141*, U.S. Defense Supply Agency, Washington, DC, 1962.

Hopkins, R. E., McCarthy, C. A., and Walters, R., "Automatic Correction of Third Order Aberrations," *J. Opt. Soc. Am.*, **45**, 363–365 (1955).

Kingslake, R., "Lens Design," in *Applied Optics and Optical Engineering*, R. Kingslake, ed., Vol. III, Chap. 1, Academic Press, San Diego, CA, 1965.

Kingslake, R., *Lens Design Fundamentals*, Academic Press, San Diego, CA, 1978.

Miyamoto, K., "Wave Optics and Geometrical Optics in Optical Design," in *Progress in Optics*, Vol. 1, E. Wolf, ed., North Holland, Amsterdam, 1964.

Ruben, P., "Aberrations Arising from Decentrations and Tilts," *J. Opt. Soc. Am.*, **54**, 4552 (1964).

Wallin, W., "The Control of Petzval Curvature," *J. Opt. Soc. Am.*, **41**, 1029–1032 (1951).

6
Chromatic Aberrations

6.1 INTRODUCTION

The value of the refractive index of any transparent material is a function of the wavelength (color) of the light. In general, in the visible spectrum the index of refraction increases with the frequency. In other words, it is higher for violet light than for red light. Figure 1.3 shows some curves displaying how the index of refraction changes with the wavelength for two typical glasses.

The chromatic dispersion of glasses in the spectral range not including absorption frequencies can be represented by several approximate expressions. The simplest one was proposed by Cauchy in Prague:

$$n = A_0 + \frac{A_1}{\lambda^2} + \frac{A_2}{\lambda^4} \tag{6.1}$$

This formula is accurate only to the third or fourth decimal place in some cases. An empirically improved formula was proposed by Conrady (1960) as follows:

$$n = A_0 + \frac{A_1}{\lambda^2} + \frac{A_2}{\lambda^{3.5}} \tag{6.2}$$

within an accuracy of one unit in the fifth decimal place. Better formulas have been proposed by several authors, e.g., by Herzberger (1942, 1959).

From a series expansion of a theoretical dispersion formula, a more accurate expression was used by Schott for many years. Recently, However, Schott has adopted a more accurate expression called the Sellmeier formula, derived from classical dispersion theory. This formula permits the interpolation of refractive indices for the entire visual range, from infrared to ultraviolet, with a precision better than 1×10^{-5}, and it is written as

$$n^2 = \frac{B_1\lambda^2}{\lambda^2 - C_1} + \frac{B_2\lambda^2}{\lambda^2 - C_2} + \frac{B_3\lambda^2}{\lambda^2 - C_3} \tag{6.3}$$

The coefficients are computed by glass manufacturers using the refractive indices values for several melt samples. The values for these coefficients for each type of glass are supplied by the glass manufacturers.

Chromatic aberration has been widely described in the literature by many authors, among others, by Cruickshank (1946), Herzberger and Salzberg (1962), and Herzberger and Jenkins (1949). This aberration may be obtained with strictly paraxial rays, i.e., with only first-order (gaussian) theory.

6.2 AXIAL CHROMATIC ABERRATION

Primary approximation expressions for chromatic aberrations may be found using only first-order theory (gaussian optics). To find an expression for the axial chromatic aberration, illustrated in Fig. 6.1, let us represent the refractive index for red light by n_C and the refractive index for blue light by n_F; we may write the Gauss law for these two colors by

$$\frac{n'_C}{l'_C} - \frac{n_C}{l_C} = \frac{n'_C - n_C}{r} \tag{6.4}$$

and

$$\frac{n'_F}{l'_F} - \frac{n_F}{l_F} = \frac{n'_F - n_F}{r} \tag{6.5}$$

and subtracting the second expression from the first we have

$$\frac{n'_C}{l'_C} - \frac{n'_F}{l'_F} - \frac{n_C}{l_C} + \frac{n_F}{l_F} = \frac{(n'_C - n'_F) - (n_C - n_F)}{r} \tag{6.6}$$

if we now define:

$$\Delta n = n_F - n_C \tag{6.7}$$

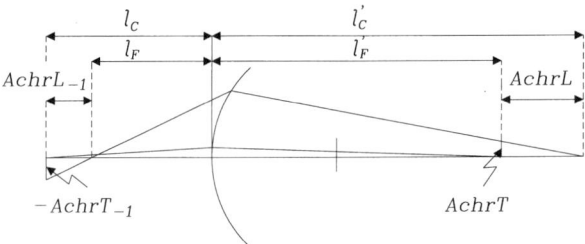

Figure 6.1 Definition of the object and image chromatic aberration.

Chromatic Aberrations

and

$$\Delta n' = n'_F - n'_C \tag{6.8}$$

we obtain

$$\frac{n'_C(l'_F - l'_C)}{l'_C l'_F} - \frac{n_C(l_F - l_C)}{l_C l_F} = \Delta n \left[\frac{1}{r} - \frac{1}{l_F}\right] - \Delta n' \left[\frac{1}{r} - \frac{1}{l'_F}\right] \tag{6.9}$$

or approximately

$$\frac{n'_C(l'_F - l'_C)}{l'^2} - \frac{n_C(l_F - l_C)}{l^2} = \Delta n \left[\frac{1}{r} - \frac{1}{l}\right] - \Delta n' \left[\frac{1}{r} - \frac{1}{l'}\right] \tag{6.10}$$

Now, using the relations $u = -y/l$, $u' = -y/l'$, and $(1/r - 1/l) = i/y$, which may be obtained from Eq. (1.51), we find after defining $AchrL = l'_F - l'_C$ and $AchrL_{-1} = l_F - l_C$, as in Fig. 6.1,

$$n'u'^2 AchrL - nu^2 AchrL_{-1} = yni \left[\frac{\Delta n}{n} - \frac{\Delta n'}{n'}\right] \tag{6.11}$$

Next, as in the spherical aberration in Section 4.1, we write this expression for every surface in the optical system and add. After some algebra, using the transfer relations, we obtain

$$AchrL_k = AchrL_0 \left[\frac{n_1 u_1^2}{n'_k u'^2_k}\right] + \sum_{i=0}^{k} \frac{yni}{n'_k u'^2_k} \left[\frac{\Delta n}{n} - \frac{\Delta n'}{n'}\right] \tag{6.12}$$

where the subscript 0 is for the object, or

$$AchrL_k = AchrL_0 \left[\frac{n_1 u_1^2}{n'_k u'^2_k}\right] + \sum_{i=0}^{k} AchrLC \tag{6.13}$$

where the quantity in square brackets on the right-hand side of this equation is the longitudinal magnification of the part of the optical system after the surface being considered, and $AchrLC$ is the longitudinal primary chromatic aberration contribution, given by

$$AchrLC = \frac{yni}{n'_k u'^2_k} \left[\frac{\Delta n}{n} - \frac{\Delta n'}{n'}\right] \tag{6.14}$$

As in the case of spherical aberration, we may also write an expression for the axial transverse primary chromatic aberration as follows:

$$AchrT_k = AchrT_0 \left[\frac{n_1 u_1}{n'_k u'_k}\right] + \sum_{i=0}^{k} AchrTC \qquad (6.15)$$

where

$$AchrTC = \frac{yni}{n'_k u'_k} \left[\frac{\Delta n}{n} - \frac{\Delta n'}{n'}\right] \qquad (6.16)$$

6.2.1 Axial Chromatic Aberration of a Thin Lens

The axial chromatic aberration contribution of a thin lens may be found by using the expressions in the preceding section. However, it is simpler to obtain it directly from the thin-lens relation, Eq. (3.8). So, we may write

$$\frac{1}{l'_C} - \frac{1}{l_C} = (n_C - 1)\kappa \qquad (6.17)$$

where κ is the total lens curvature ($\kappa = c_1 - c_2$), and

$$\frac{1}{l'_F} - \frac{1}{l_F} = (n_F - 1)\kappa \qquad (6.18)$$

Thus, subtracting the second expression from the first:

$$\frac{l'_F - l'_C}{l'_F l'_C} - \frac{l_F - l_C}{l_F l_C} = (n_C - n_F)\kappa \qquad (6.19)$$

Again, using the definitions for $AchrL$ and Δn, this expression may be approximated by

$$\frac{AchrL}{l'^2} = \frac{AchrL_0}{l^2} - \Delta n \kappa \qquad (6.20)$$

where the aberration with subscript 0 is on the object medium. Substituting the value for the curvature κ from Eq. (3.8) and using the relations $u = -y/l$ and $u' = -y/l'$, we find that

$$AchrL = AchrL_0 \frac{u^2}{u'^2} - \frac{y^2}{u'^2 f(n-1)} \Delta n \qquad (6.21)$$

Chromatic Aberrations

The Abbe number is a glass characteristic and is defined in Appendix 3 as $V = (n-1)/\Delta n = (n_D - 1)/(n_F - n_C)$; thus, we may write

$$AchrL = AchrL_0 \frac{u^2}{u'^2} - \frac{y^2}{u'^2 fV} \tag{6.22}$$

We use now the well known procedure of writing this expression for a centered system with k thin lenses, by adding in the corresponding longitudinal magnifications. After some algebraic steps, using the transfer equations, we may write

$$AchrL = AchrL_0 \frac{u_1^2}{u_k'^2} - \frac{1}{u_k'^2} \sum_{i=0}^{k} \frac{y_i^2 P_i}{V_i} \tag{6.23}$$

where P_i is the power of each lens. Thus, the transverse axial chromatic aberration, multiplying by u_k', is given by

$$AchrT = AchrT_0 \frac{u_1^2}{u_k'} - \frac{1}{u_k'} \sum_{i=0}^{k} \frac{y_i^2 P_i}{V_i} \tag{6.24}$$

6.2.2 Achromatic Doublet

From Eq. (6.24) we see that an achromatic doublet formed by two thin lenses in contact is obtained with the condition:

$$f_1 V_1 = -f_2 V_2 \tag{6.25}$$

where the effective focal length F of the combination is given by Eq. (3.44). Then, we may find that the focal length of the first element has to be

$$f_1 = F\left[1 - \frac{V_2}{V_1}\right] \tag{6.26}$$

and the focal length of the second element is

$$f_2 = F\left[1 - \frac{V_1}{V_2}\right] \tag{6.27}$$

As we should expect, the absolute value of the power for the positive lens is greater than the absolute value of the power for the negative lens, since the total power is positive. One conclusion is that an achromatic thin-lens system can be made only with two glasses with different Abbe numbers. Newton has been sometimes criticized because he said that an achromatic lens could not be constructed. The reason is that during his time most known glasses had the same Abbe number.

The power of each component is inversely proportional to the difference between the Abbe numbers. Hence, in order to have thin lenses with low powers, the Abbe numbers must be as different as possible. Unfortunately, this condition is incompatible with the condition for almost equal partial dispersion ratios, in order to have low secondary color, as required by Eq. (6.30).

Once the focal length of each component is calculated, we have the bending of both lenses as degrees of freedom to correct the spherical aberration and coma. We have seen that the solution for zero coma in a lens is very close to a point for maximum spherical aberration (minimum absolute value).

If we design an achromatic lens, as shown in Fig. 6.2, with two different types of glass, the two selected wavelengths (frequently C and F) will have the same focal length, but it will be different for all other colors. The focal length as a function of the wavelength for an achromatic lens and a single lens are compared in Fig. 6.3.

An alternative manner of deriving the conditions for an achromatic doublet is by writing the power for a doublet from Eq. (2.8) as follows:

$$P = \frac{1}{F} = \frac{1}{f_1} + \frac{1}{f_2} = (n_1 - 1)\kappa_1 + (n_2 - 1)\kappa_2 \tag{6.28}$$

Thus, the doublet is achromatic if

$$\frac{dP}{d\lambda} = \kappa_1 \frac{dn_1}{d\lambda} + \kappa_2 \frac{dn_2}{d\lambda} = 0 \tag{6.29}$$

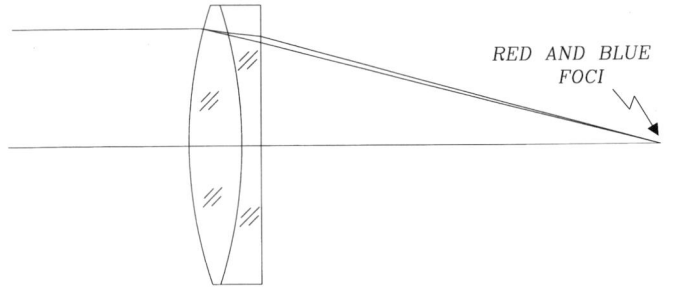

Figure 6.2 A doublet without chromatic aberration.

Chromatic Aberrations

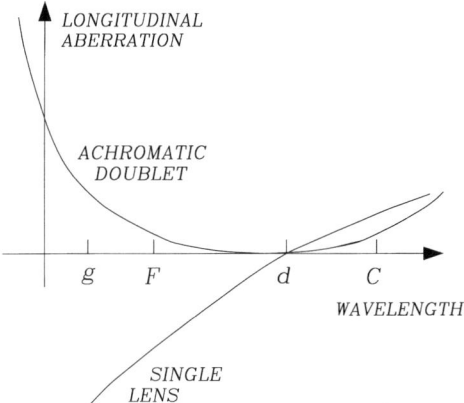

Figure 6.3 Change of the longitudinal chromatic aberration with the wavelength for an achromatic doublet and for a single lens.

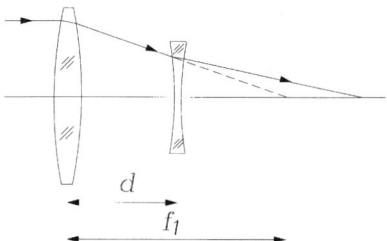

Figure 6.4 Achromatic system with two spaced elements.

is satisfied for a wavelength near the middle of the range of wavelengths between C and F. It can be shown that this expression is equivalent to Eq. (6.25).

6.2.3 Achromatic Doublet with Separated Elements

An achromatic system formed by two separated elements, as shown in Fig. 6.4, is called a *dialyte*. Again, from Eq. (6.24) we see that an achromatic doublet formed by two thin lenses separated by a finite distance d is obtained with the condition:

$$\frac{f_1 V_1}{y_1} = -\frac{f_2 V_2}{y_2} \tag{6.30}$$

On the other hand, from Eq. (3.43) and defining the ratio $k = d/f_1$, we find that

$$\frac{y_2}{y_1} = 1 - \frac{d}{f_1} = 1 - k \tag{6.31}$$

Since the effective focal length F of the combination is given by Eq. (3.42), we may find that the focal length of the first element is

$$f_1 = F\left[1 - \frac{V_2}{V_1(1-k)}\right] \tag{6.32}$$

and the focal length of the second element is

$$f_2 = F(1-k)\left[1 - \frac{V_1(1-k)}{V_2}\right] \tag{6.33}$$

We may see that, as the lenses are separated (k increased), the absolute value of the power of both elements increases. However, the power of the negative element increases faster. When $k = 0.225$ the absolute values of the power of both lenses are equal.

6.2.4 Axial Chromatic Aberration Correction with One Glass

The axial chromatic aberration may also be corrected with only one type of glass if either two separated lenses or a thick lens is made. Let us consider first the case of two separated thin lenses, as shown in Fig. 6.5.

The system is corrected for axial chromatic aberration if the back focal length is constant for all wavelengths. This is possible for a short range of wavelengths if the derivative of the back focal length (or the back power) with respect to the refractive index is made equal to zero. Thus, from Eqs. (2.8) and (3.49) we may write

$$P_B = \frac{1}{F_B} = \frac{1}{1/(n-1)\kappa_1 - d} + (n-1)\kappa_2 \tag{6.34}$$

and taking the derivative of this power:

$$\frac{dP_B}{dn} = \frac{\kappa_1}{[1-(n-1)\kappa_1 d]^2} + \kappa_2 = 0 \tag{6.35}$$

Chromatic Aberrations

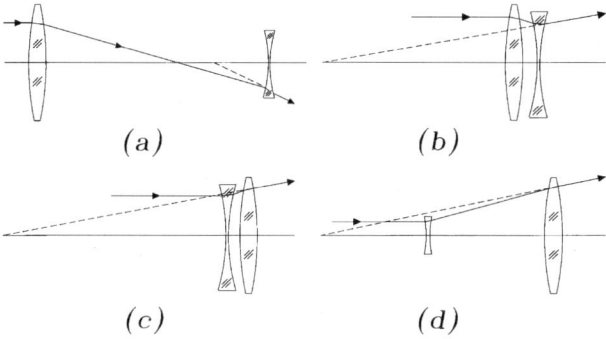

Figure 6.5 System with two separated elements without axial chromatic aberration.

we obtain

$$f_2 = -f_1 \left[1 - \frac{d}{f_1}\right]^2 = -f_1[1-k]^2 \tag{6.36}$$

where k has been defined as $k = d/f_1$. If we use the expression for the effective focal length, Eq. (3.46), we find that

$$f_1 = \frac{k}{(k-1)} F \tag{6.37}$$

and

$$f_2 = -k(k-1)F \tag{6.38}$$

As Kingslake (1978) points out, we see that for a positive system the value of k has to be greater than one, making a very long system. Since the separation is greater than the focal length of the first system, the focus is inside the system, as shown in Fig. 6.5(a). This is called a Schupmann lens.

In the case of a negative system (divergent), a long or compact system may be obtained. The system is very compact with the positive lens in the front, as in Fig. 6.5(b) or, with the negative system in the front, as in Fig. 6.5(c). A long system may also be obtained, as shown in Fig. 6.5(d).

Let us now consider the case of a thick lens with zero axial chromatic aberration. The procedure to design this lens is the same as for the system of two separated thin lenses. The first step is to obtain the derivative of the back power of the thick lens in Eq. (3.37) and to make it equal to zero. Thus, we may obtain

$$\frac{dP_B}{dn} = \frac{[r_1 - t(n-1)^2/n^2]}{[r_1 - t(n-1)/n]^2} - \frac{1}{r_2} \qquad (6.39)$$

hence, the radii of curvature of the thick lens must satisfy the relation:

$$\frac{r_2}{r_1} = \frac{[1 - k(n-1)/n]^2}{[1 - k(n-1)^2/n^2]} \qquad (6.40)$$

where k has been defined as $k = t/r_1$. Figure 6.6 shows this ratio of the radii of curvature as a function of k in the interval from -1 to $+1$.

It may be shown that this ratio is positive for values of k less than $n^2/(n-1)^2 \sim 9$, which is an extremely thick lens. Using relation (6.38) and Eq. (3.33), we may see that the effective focal length of a thick achromatic lens is given by

$$F = -\frac{[1 - k(n-1)/n]}{[k(n-1)^2/n^2]} r_1 \qquad (6.41)$$

and then, it is possible to show that this focal length is positive only if

$$k = \frac{t}{r_1} \geq \frac{n}{n-1} \sim 3 \qquad (6.42)$$

assuming that the thickness t is always positive. Hence, the lens is corrected for axial chromatic aberration and has a positive effective focal length only

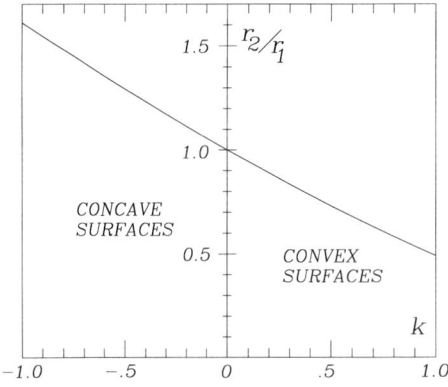

Figure 6.6 Ratio r_2/r_1 for different values of k in a thick lens corrected for axial chromatic aberration.

Chromatic Aberrations

if it is very thick. Then, the focus is inside of the lens body and the refracted beam will be divergent anyway, after exiting the lens. Figure 6.7 shows some possible configurations for this lens.

6.2.5 Spherochromatism

The magnitude of the spherical aberration is a function of the refractive index, so it is reasonable to expect a variation in the spherical aberration with the wavelength. The magnitude of the spherochromatism has been defined as follows:

$$\begin{aligned}\text{Spherochromatism} &= SphL_F - SphL_C = (L' - l')_F - (L' - l')_C \\ &= (L'_F - L'_C) - (l'_F - l'_C) \\ &= AchrL_{\text{marginal}} - AchrL_{\text{paraxial}}\end{aligned} \quad (6.43)$$

The transverse spherical aberration curves for three different wavelengths are shown in Fig. 6.8. We may see as expected, that the

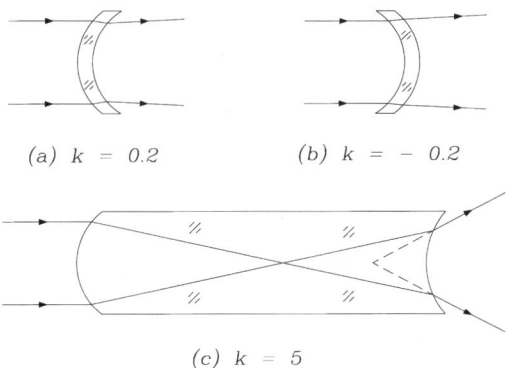

Figure 6.7 Thick lenses corrected for axial chromatic aberration: (a) $k=0.2$; (b) $k=-0.2$; (c) $k=5.0$.

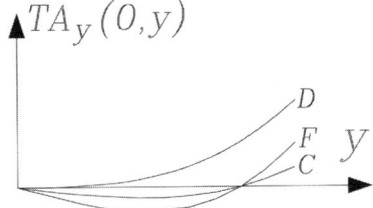

Figure 6.8 Transverse aberration for three colors in a system with spherochromatism.

curves show a good achromatic correction only for a pair of wavelengths, and that the spherical aberration correction is good only for one wavelength. The effect of this aberration in catadioptric (a system formed by lenses and mirrors) systems has been studied by Stephens (1948).

6.2.6 Conrady's *D–d* Method of Achromatization

Spherochromatism, as explained in Section 6.2.5, is the change in the values of the axial chromatic aberration with the height y of the ray. Thus, correcting the axial chromatic aberration for paraxial rays does not mean that the axial chromatic aberration for marginal rays is also zero. The best choice is then to correct the axial chromatic aberration for the rays in the zone at 0.7 of the semidiameter of the entrance pupil.

An equivalent way of looking at the same condition is as shown in Fig. 6.9. A white light spherical wavefront entering the lens system, from an on-axis object point is refracted passing through the lens. The blue (F) and red (C) colors produce two different wavefronts going out of the optical system. Let us now assume that the two wavefronts touch each other at the center and at the edge in order to minimize the axial chromatic aberration for the whole aperture. This obviously means that the two wavefronts are parallel to each other at about the 0.7 zone. Since the two wavefronts are parallel, the two rays from this zone are traveling along the same path and thus cross the optical axis at about the same point.

Summarizing, the optimum condition for the correction of the axial chromatic aberration is obtained when the ray aberration for the rays from the 0.7 zone is zero or, equivalently, when the wavefronts touch at the center and at the edge.

Based on this result, Conrady (1960) suggested the *D–d* method of achromatization. Using Fermat's principle, the two wavefronts touch at the

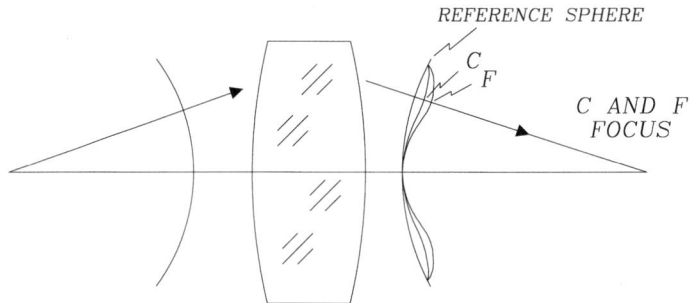

Figure 6.9 Conrady's *D–d* method to correct the achromatic aberration.

center when the optical path is the same for the two paraxial rays with colors C and F. Thus,

$$\sum_{i=0}^{k} dn_C = \sum_{i=0}^{k} dn_F \qquad (6.44)$$

and the wavefronts touch at the edge when the optical path is the same for two marginal rays with these colors C and F:

$$\sum_{i=0}^{k} Dn_C = \sum_{i=0}^{k} Dn_F \qquad (6.45)$$

Now, we have that

$$\sum_{i=0}^{k} dn_C = \sum_{i=0}^{k} Dn_C \qquad (6.46)$$

only if the two wavefronts touch each other at the edge, over the reference sphere, which is not necessarily the case. Thus, subtracting Eq. (6.43) from Eq. (6.44), we obtain

$$\sum_{i=0}^{k} (D-d)n_C = \sum_{i=0}^{k} (D-d)n_F \qquad (6.47)$$

or

$$\sum_{i=0}^{k} (D-d)(n_F - n_C) = 0 \qquad (6.48)$$

In the practical computation of this expression we can make the following approximations. The traveled distances are not exactly the same for both colors because different colors follow slightly different paths, but they may be considered equal. This approximation is quite accurate. Also, to save some computation time, $(D-d)$ does not need to be calculated in air, since $(n_F - n_C) = 0$ in a nondispersive medium like air or a vacuum.

As shown by Feder (1954), the relation between the axial transverse aberration and the $(D-d)$ sum is given by

$$\frac{AchrT}{F} = -\frac{d\left(\sum_{i=0}^{k}(D-d)(n_F - n_C)\right)}{dy} \qquad (6.49)$$

The similarity between this expression and Eq. (1.76) is evident.

6.3 SECONDARY COLOR ABERRATION

The focal length for the wavelengths between the two colors being selected for achromatization is different from that for these colors. This is the secondary color. The effect of the secondary color is so important that it has been widely studied and described in the literature, e.g., by Cartwright (1939), Christen (1981), Herzberger and McClure (1963), Smith (1959), Stephens (1957, 1959, 1960), Willey (1962), and Wynne (1977, 1978). Since the axial achromatic contribution of a thin lens for colors F and C, from Eq. (6.21) and taking the longitudinal magnification factor into account,

$$AchrLC = -\frac{y^2}{u'^2 f} \frac{n_F - n_C}{(n-1)} = -\frac{y^2}{u'^2 fV} \quad (6.50)$$

by analogy, the axial achromatic contribution for colors λ and F is

$$AchrLC_{\lambda F} = -\frac{y^2}{u'^2 f} \frac{n_\lambda - n_F}{(n-1)} \quad (6.51)$$

Thus, by taking the ratio of the two expressions we find that

$$AchrLC_{\lambda F} = AchrLC \left[\frac{n_\lambda - n_F}{n_F - n_C}\right] \quad (6.52)$$

The quantity in square brackets is defined as the partial dispersion ratio from λ to F, written as $P_{\lambda F}$. Thus,

$$P_{\lambda F} = \frac{n_\lambda - n_F}{n_F - n_C} \quad (6.53)$$

Thus, by using Eq. (6.48), for an axially centered system of k surfaces we have

$$AchrLC_{\lambda Fk} = -\frac{1}{u'^2} \sum_{i=0}^{k} \frac{P_{\lambda F} y^2}{fV} \quad (6.54)$$

For an achromatic doublet with two thin lenses in contact, this expression for the magnitude of the secondary axial color aberration becomes

$$AchrLC_{\lambda F} = -F\left[\frac{P_{\lambda F1} - P_{\lambda F2}}{V_1 - V_2}\right] \quad (6.55)$$

Chromatic Aberrations

This means that the doublet is apochromatic if the two partial dispersions $P_{\lambda F1}$ and $P_{\lambda F2}$ for the two glasses are equal. Figure A3.3 shows a plot with many commercial optical glasses showing their value of the partial dispersion versus the Abbe number. If an achromatic doublet is made with two glasses represented in this graph, the axial secondary color aberration is directly proportional to the slope of the straight line joining these two points. Unfortunately, this slope is almost the same for any pair of glasses, with the exception of some fluor–crown glasses and fluorite.

Another procedure to obtain the condition for apochromatism of a doublet can be obtained from Eq. (6.29), which is satisfied for a wavelength approximately between the C and F lines. If this expression is satisfied for all wavelengths in this range, we can show that

$$\frac{d}{d\lambda}\left(\frac{dn_1/d\lambda}{dn_2/d\lambda}\right) = 0 \tag{6.56}$$

This condition (Perrin, 1938) is equivalent to the condition that the two partial dispersions of the glasses should be equal. The problem of selecting the glasses for apochromatism has also been studied by Lessing (1957, 1958).

6.3.1 Apochromatic Triplet

Another method to correct the secondary color is by means of the use of three glasses forming a triplet. In order to have a focal length F for the system, we write

$$\frac{1}{F} = \frac{1}{f_1} + \frac{1}{f_2} + \frac{1}{f_3} \tag{6.57}$$

From Eq. (6.25), in order to correct for red (C) and blue light (F), for a system of three lenses we may write

$$\frac{1}{f_1 V_1} + \frac{1}{f_2 V_2} + \frac{1}{f_3 V_3} = 0 \tag{6.58}$$

and, from Eq. (6.54), in order to have the yellow (D) color at a common focus with the red and blue light:

$$\frac{P_1}{f_1 V_1} + \frac{P_2}{f_2 V_2} + \frac{P_3}{f_3 V_3} = 0 \tag{6.59}$$

This system of three equations may be solved to obtain the following focal lengths for the three lenses:

$$\frac{1}{f_1} = \frac{1}{F}\frac{V_1[P_3 - P_2]}{\Delta} \tag{6.60}$$

$$\frac{1}{f_2} = \frac{1}{F}\frac{V_2[P_1 - P_3]}{\Delta} \tag{6.61}$$

and

$$\frac{1}{f_3} = \frac{1}{F}\frac{V_3[P_2 - P_1]}{\Delta} \tag{6.62}$$

where P_i are the partial dispersions and Δ is the determinant:

$$\Delta = \begin{vmatrix} P_1 & V_1 & 1 \\ P_2 & V_2 & 1 \\ P_3 & V_3 & 1 \end{vmatrix} \tag{6.63}$$

We may see that the value of this determinant is the area of the triangle connecting the points representing the three glasses in a diagram of the partial dispersion P as a function of the Abbe number V. Thus, if this system of equations is to have a solution, this triangle must not have a zero area.

6.4 MAGNIFICATION CHROMATIC ABERRATION

The magnification chromatic aberration, also frequently called lateral chromatic aberration or lateral color, appears when the images produced by different colors have different sizes on the observing plane. The effect of this aberration is a blurring of the image detail for off-axis points. The farther away from the axis, the greater the aberration (O'Connell, 1957).

To find an expression for the magnification chromatic aberration, let us consider an optical system, as shown in Fig. 6.10. The paraxial sagittal image for red light is at the point **S** on the red principal ray. The paraxial sagittal image for blue light is at the point **R** on the blue principal ray. The heights h'_F and h'_C may both be calculated with the Lagrange theorem. As pointed out before, the magnification chromatic aberration arises because the image magnification is different for the red and blue paraxial sagittal rays. Due to the axial chromatic aberration, these two images are at different planes. The magnification chromatic aberration, represented by

Chromatic Aberrations

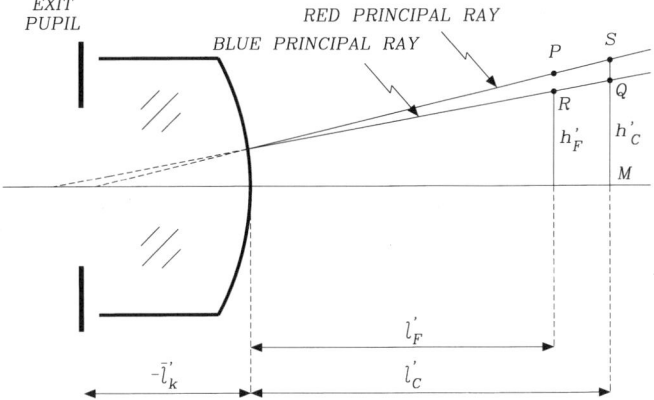

Figure 6.10 Principal rays in a system with axial and magnification chromatic aberration.

Mchr, is defined by the lateral distance SQ from the blue paraxial sagittal focus to the red principal ray (or vice versa, by the distance $PR \sim SQ$).

Now, let us consider again Fig. 6.10 and define a quantity CDM (*chromatic difference of magnification*), analogous to OSC. We assume in this figure that the exit pupil position is at the same position for colors C and F. This is not strictly true but it may be considered so in a first approximation, if $(l'_F - \bar{l}'_k)$ is large compared with the usually small distance between the pupils for the two colors C and F. Thus, we may write

$$QM = \left[\frac{l'_F - \bar{l}'_k}{l'_C - \bar{l}'_k}\right] h'_C \tag{6.64}$$

Then, the quantity CDM is defined as

$$\begin{aligned} CDM &= \frac{QS}{MQ} = \frac{MS - MQ}{MQ} = \frac{MS}{MQ} - 1 \\ &= \frac{h'_F}{MQ} - 1 \end{aligned} \tag{6.65}$$

but substituting the value of MQ:

$$CDM = \left[\frac{l'_C - \bar{l}'_k}{l'_F - \bar{l}'_k}\right] \frac{h'_F}{h'_C} - 1 \tag{6.66}$$

or equivalently,

$$CDM = \left[\frac{l'_C - \bar{l}'_k}{l'_F - \bar{l}'_k}\right]\frac{u'_F}{u'_C} - 1 \tag{6.67}$$

Then, the transverse magnification or lateral chromatic aberration $Mchr$ is given by

$$Mchr' = \left[\left[\frac{l'_C - \bar{l}'_k}{l'_F - \bar{l}'_k}\right]\frac{u'_F}{u'_C} - 1\right]h' \tag{6.68}$$

which is valid for any object distance. For an object at infinity this expression may be rewritten after some algebra and approximating l'_C by $l'_F = l'_K$:

$$Mchr' = \left[\frac{F_F - F_C}{F} - \frac{AchrL}{l'_k - \bar{l}'_k}\right]h' \tag{6.69}$$

where F_C and F_F are the effective focal lengths for red and blue light, respectively. We see that this chromatic aberration depends both on the axial chromatic difference and on the change in the magnitude of the effective focal length with the color.

Let us now find the surface contribution to this aberration. From Fig. 6.11 we may see that the magnification chromatic aberration is given by

$$MchrC = SQ = -RS\psi \tag{6.70}$$

Thus, since RS is the axial elements aberration, by using Eq. (5.3), we may obtain the magnification chromatic aberration as

$$Mchr = -AchrLu'_k\left(\frac{\bar{i}}{i}\right) = AchrT\left(\frac{\bar{i}}{i}\right) \tag{6.71}$$

As explained in Section 5.4, the angle ratio of the angles of refraction for the principal and meridional rays increases linearly with the image height for small fields. Thus, the primary (first order) magnification chromatic aberration, as the primary (third order) coma, increases linearly with the image height.

Chromatic Aberrations

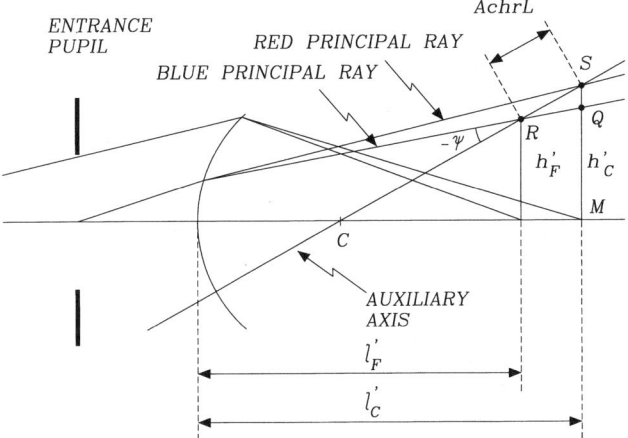

Figure 6.11 Calculation of the magnification chromatic aberration.

The contribution to the final magnification chromatic aberration in the whole optical system as

$$MchrC = -AchrLC\,u'_k\left(\frac{\bar{i}}{i}\right) = AchrTC\left(\frac{\bar{i}}{i}\right) \tag{6.72}$$

6.4.1 Stop Shift Equation

As in the case of the spherical aberration and the Petzval curvature, the axial achromatic aberration remains unchanged with a shift in the stop:

$$AchrT^* = AchrT \tag{6.73}$$

In a similar way to the procedure used for the case of coma, using Eq. (5.108), we may find for the magnification chromatic aberration:

$$MchrT^* = MchrT + AchrTQ \tag{6.74}$$

6.4.2 Correction of the Magnification Chromatic Aberration

The magnification chromatic and the axial chromatic aberrations are closely interrelated. They may appear in many different combinations, for example:

1. When both axial and magnification chromatic aberrations are corrected, as illustrated in Fig. 6.12(a), the red and blue images are in the

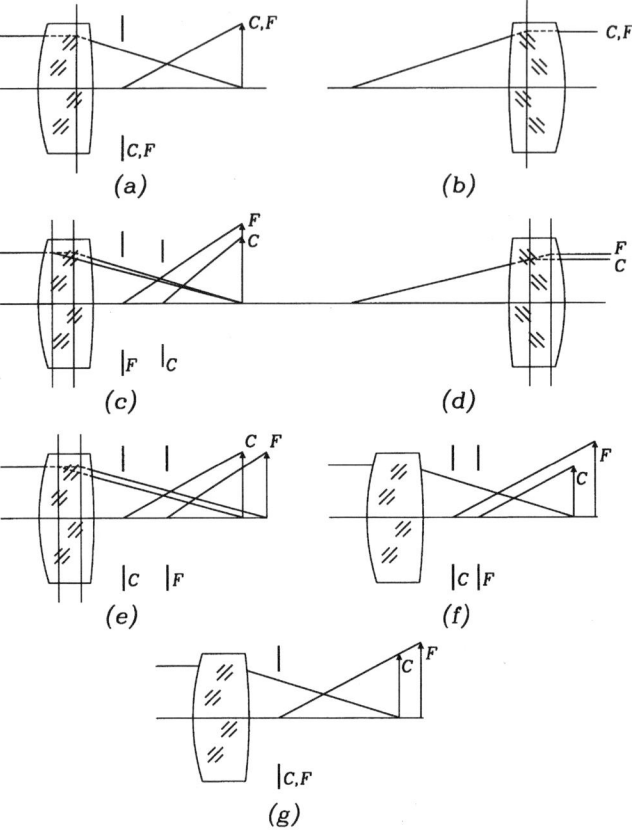

Figure 6.12 Red and blue meridional and principal rays in optical systems with different amounts of axial and magnification chromatic aberrations.

same plane and have the same size. It can be shown that in this case the two meridional rays as well as the two principal rays for the two colors (C and F) follow the same paths. Also, the two exit pupils are at the same position and have the same diameters. For the particular case when the object is at an infinite distance, if this system is reversed as in Fig. 6.12(b), after refraction, the red and blue rays follow the same path, parallel to the optical axis.

2. If the axial chromatic aberration is corrected but not the magnification chromatic aberration, as in Fig. 6.12(c), the red and blue images are in the same plane but have different size. The two meridional rays as well as the two principal rays follow different paths. Either the positions for the exit pupils or their sizes or both are different. Considering

again the particular case when the object is at an infinite distance, if the system is reversed as in Fig. 6.12(d), the exiting meridional rays will follow different but parallel trajectories. The effective focal lengths for the red and blue colors are equal.

3. Figure 6.12(e) shows the case when the lateral magnification is the same for colors C and F, i.e., the two images have the same size but they have different positions. The two refracted meridional rays have the same slope. In this case the two chromatic aberrations are present.

4. Let us now consider the case when the angular magnification M is the same for red and blue colors as illustrated in Fig. 6.12(f), but neither the axial chromatic nor the magnification chromatic aberrations are corrected. The exiting principal rays for the two colors follow different but parallel paths. The two exit pupils have the same size but different positions. Their separation is frequently called the *axial chromatic aberration of the pupil*. When their size is different the difference in their semidiameters is called the *magnification chromatic aberration of the pupil*.

5. Finally, the system in Fig. 6.12(g) has not been corrected for axial chromatic aberration but the magnification chromatic aberration is fully corrected.

If the object is at an infinite distance and the red and blue images have the same size, the effective focal lengths for blue and red light are equal ($F_C = F_F$). Thus, from Eq. (6.66), we may see that if $F_C = F_F$ the magnification chromatic aberration is zero only if one of the following conditions is satisfied:

1. The axial chromatic aberration is corrected.
2. The power of the system is zero (infinite effective focal length).
3. The exit pupil of the system is at infinity ($l'_k - l_k$), or in other words, that the principal ray is parallel to the optical axis.

The effective focal length and the back focal length are equal in a thin lens. Thus, in a thin achromatic lens both the axial achromatic and the magnification chromatic aberrations are corrected. Another interesting conclusion is that a system of two separated lenses has both chromatic aberrations corrected only if the two components are individually corrected for axial chromatic aberration.

6.4.3 Magnification Chromatic Aberration Correction with One Glass

The magnification chromatic aberration may also be corrected, as the axial chromatic aberration, with only one kind of glass, provided that the system is not thin. If the exit pupil is at infinity (back telecentric), or at a long distance from the system compared with its focal length, the only necessary condition is that the effective focal length of the system for the blue and red colors be equal. Then, the system produces an image as illustrated in Fig. 6.12(d). Let us consider two cases: (1) a system of two separated thin lenses and (2) a thick lens.

From Eqs. (3.8) and (4.37), the effective focal length of a system of two thin lenses separated by a distance d is given by

$$P = \frac{1}{F} = (n-1)\kappa_1 + (n-1)\kappa_2 - d(n-1)^2 \kappa_1 \kappa_2 \qquad (6.75)$$

thus, differentiating with respect to n, we find that

$$\frac{dP}{dn} = \kappa_1 + \kappa_2 - 2d(n-1)\kappa_1\kappa_2 = 0 \qquad (6.76)$$

obtaining the condition:

$$d = \frac{f_1 + f_2}{2} \qquad (6.77)$$

Thus, the system is corrected for the magnification chromatic aberration if the average of their focal lengths is equal to their separation and the exit pupil is at infinity, as shown in Fig. 6.13.

Let us consider now the case of a single thick lens with thickness t, corrected for magnification chromatic aberration, as shown in Fig. 6.14. From Eq. (3.33) we find that

$$P = \frac{1}{F} = (n-1)\kappa + \frac{(n-1)^2}{n} tc_1 c_2 \qquad (6.78)$$

thus, differentiating with respect to n, we find that

$$\frac{dP}{dn} = \kappa + \frac{(n^2 - 1)}{n^2} tc_1 c_2 = 0 \qquad (6.79)$$

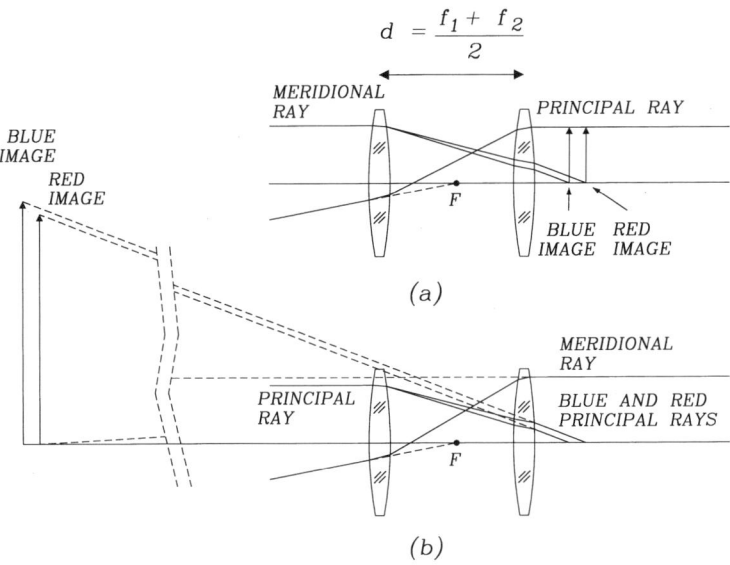

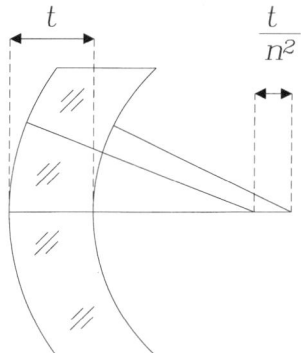

Figure 6.13 System with two elements with the same glass, corrected for magnification chromatic aberration. (a) The object is located at infinity and (b) the intrance pupil in located at infinity.

Figure 6.14 A thick lens corrected for magnification chromatic aberration.

obtaining the condition:

$$t = \frac{(r_1 - r_2)n^2}{(n^2 - 1)} \tag{6.80}$$

We may see that the separation between the centers of curvature of the two lens surfaces has to be equal to t/n^2, as shown in Fig. 6.14. Finally, we should remember that the magnification chromatic aberration in this thick lens, is corrected with relation (6.77), and the system of two thin lenses, is corrected using relation (6.74), only if the principal ray on the image space is parallel to the optical axis, or in other words, if the exit pupil is at infinity.

REFERENCES

Cartwright, C. H., "Lithium-Fluoride Quartz Apochromat," *J. Opt. Soc. Am.*, **29**, 350–351 (1939).

Christen, R., "An Apochromatic Triplet Objective," *Sky and Telescope*, 376–381 (Oct. 1981).

Conrady, A. E., *Applied Optics and Optical Design, Part 2*, Dover Publications, New York, 1960.

Cruickshank, F. D., "On the Primary Chromatic Coefficients of a Lens System," *J. Opt. Soc. Am.*, **36**, 103–107 (1946).

Feder, D. P., "Conrady's Chromatic Condition," *J. Res. Natl Bur. Std.*, **53**, 47 (1954).

Herzberger, M., "The Dispersion of Optical Glass," *J. Opt. Soc. Am.*, **32**, 70–77 (1942).

Herzberger, M., "Colour Correction in Optical Systems and a New Dispersion Formula," *Opt. Acta*, **6**, 197 (1959).

Herzberger, M. and Jenkins, F. A., "Color Correction in Optical Systems and Types of Glass," *J. Opt. Soc. Am.*, **39**, 984–989 (1949).

Herzberger, M. and McClure, N. R., "The Design of Superachromatic Lenses," *Appl. Opt.*, **2**, 553–560 (1963).

Herzberger, M. and Salzberg, C. D., "Refractive Indices of Infrared Optical Materials and Color Correction of Infrared Lenses," *J. Opt. Soc. Am.*, **52**, 420–427 (1962).

Kingslake, R., *Lens Design Fundamentals*, Academic Press, New York, 1978.

Lessing, N. V. D. W., "Selection of Optical Glasses in Apochromats," *J. Opt. Soc. Am.*, **47**, 955–958 (1957).

Lessing, N. V. D. W., "Further Considerations on the Selection of Optical Glasses in Apochromats," *J. Opt. Soc. Am.*, **48**, 269–273 (1958).

O'Connell, J. M., "Variation of Photographic Resolving Power with Lateral Chromatic Aberration," *J. Opt. Soc. Am.*, **47**, 1018–1020 (1957).

Perrin, F. H., "A Study of Harting's Criterion for Complete Achromatism," *J. Opt. Soc. Am.*, **28**, 86–93 (1938).

Smith, W. J., "Thin Lens Analysis of Secondary Spectrum," *J. Opt. Soc. Am.*, **49**, 640–641 (1959).

Stephens, R. E., "Reduction of Sphero-Chromatic Aberration in Catadioptric Systems," *J. Opt. Soc. Am.*, **38**, 733–735 (1948).

Stephens, R. E., "Secondary Chromatic Aberration," *J. Opt. Soc. Am.*, **47**, 1135 (1957).

Stephens, R. E., "Selection of Glasses for Three Color Achromats," *J. Opt. Soc. Am.*, **49**, 398–401 (1959).

Stephens, R. E., "Four Color Achromats and Superchromats," *J. Opt. Soc. Am.*, **50**, 1016–1019 (1960).

Willey, R. R., "Machine-Aided Selection of Optical Glasses for Two-Element, Three Color Achromats," *Appl. Opt.*, **1**, 368–369 (1962).

Wynne, C. G., "Secondary Spectrum with Normal Glasses," *Opt. Commun.*, **21**, 419 (1977).

Wynne, C. G., "A Comprehensive First-Order Theory of Chromatic Aberration. Secondary Spectrum Correction without Special Glasses," *Opt. Acta*, **25**, 627–636 (1978).

7
The Aberration Polynomial

7.1 WAVE ABERRATION POLYNOMIAL

In a general manner, without assuming any symmetries, the shape of a wavefront may be represented by the polynomial:

$$W(x,y) = \sum_{i=0}^{k} \sum_{j=0}^{i} c_{ij} x^j y^{i-j} \tag{7.1}$$

including high-order aberration terms, where k is the degree of this polynomial. In polar coordinates we define

$$x = S \sin\theta \tag{7.2}$$

and

$$y = S \cos\theta \tag{7.3}$$

where the angle θ is measured with respect to the y axis, as shown in Fig. 7.1. Then, the wavefront shape may be written as

$$W(S,\theta) = \sum_{n=0}^{k} \sum_{l=0}^{n} S^n (a_{nl} \cos^l\theta + b_{nl} \sin^l\theta) \tag{7.4}$$

where the $\cos\theta$ and $\sin\theta$ terms describe the symmetrical and antisymmetrical components of the wavefront, respectively. However, not all possible values of n and l are permitted. To have a single valued function we must satisfy the condition:

$$W(S,\theta) = W(-S, \theta + \pi) \tag{7.5}$$

Then, it is easy to see that n and l must both be odd or both even. If this expression for the wavefront is converted into cartesian coordinates $W(x, y)$,

171

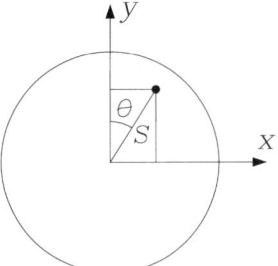

Figure 7.1 Polar coordinates for the ray on the entrance pupil of an optical system.

it becomes an infinite series, unless $l \leq n$. Thus, if we want Eq. (7.4) to be equivalent to the finite series in Eq. (7.1), we impose this condition, that is almost always satisfied, except in some very rare cases related to rotational shearing interferograms, as pointed out by Malacara and DeVore (1992).

Now, if we restrict ourselves to the case of a wavefront produced by an axially symmetric optical system, with a point object displaced along the y axis, the wavefront is symmetric about the tangential or meridional plane, obtaining

$$W(S,\theta) = \sum_{n=0}^{k} \sum_{l=0}^{n} S^n a_{nl} \cos^l \theta \qquad (7.6)$$

As shown by Hopkins (1950), if we include in this expression the image height h and impose the conditions:

$$W(S,\theta,h) = W(S,-\theta,h) \qquad (7.7)$$

because of the symmetry about the meridional plane, and

$$W(S,\theta,h) = W(S,\theta+\pi,-h) \qquad (7.8)$$

because of the rotational symmetry of the lens system about the optical axis, the wavefront expression may be shown to have only terms of the form:

$$S^2, \; hS\cos\theta, \; h^2 \qquad (7.9)$$

and their products. An interesting consequence is that the sum of the powers of S and h is always an even number. The greater this number, the higher the aberration order. The wavefront may be represented by a linear combination of these aberrations, with terms $_kW_{nl} S^n h^k \cos^l\theta$, where k is the power of the image height h, n is the power of the aperture S, and l is the power of $\cos\theta$, with $l \leq n$. So, we obtain the wavefront $W(S, \theta, h)$ as a sum of one or more of the following terms:

Constant term $(n+k=0)$
    $_0w_{00}$ Piston term (constant OPD)

First-order terms $(n+k=2)$
    $_2w_{00}\ h^2$ Parabolic field phase term
    $_1w_{11}\ S\ h\cos\theta$ Tilt about x axis (image y displacement)
                                with magnification change
    $_0w_{20}\ S^2$ Defocusing

Third-order or primary aberrations $(n+k=4)$
    $_3w_{11}\ S\ h^3\cos\theta$ Distortion
    $_2w_{20}\ S^2\ h^2$ Petzval curvature
    $_2w_{22}\ S^2\ h^2\cos^2\theta$ Primary astigmatism
    $_1w_{31}\ S^3\ h\cos\theta$ Primary (circular) coma
    $_0w_{40}\ S^4$ Primary spherical aberration

Fifth-order aberrations $(n+k=6)$
    $_2w_{40}\ S^4\ h^3$ Secondary field phase term
    $_1w_{51}\ S^5\ h\cos\theta$ Linear fifth-order coma
    $_0w_{60}\ S^6$,   Fifth-order spherical aberration
    etc.   (7.10)

Each of these terms has a name, but not all are higher order terms. Hopkins (1950) has proposed the following general names:

1. Spherical aberrations: terms independent of θ ($k=0$).
2. Comatic aberrations: terms with odd powers of $\cos\theta$ (k odd).
3. Astigmatic aberrations: terms with even powers of $\cos\theta$ (k even).

As an example, the fifth-order aberration $_3w_{33}\ S^3\ h^3\cos^3\theta$ is named an elliptical coma because when added to the primary (circular) coma, illustrated in Fig. 5.17, it transforms the circles into ellipses. This aberration isolated (without the primary coma) is also sometimes called triangular astigmatism by telescope makers, because it appears on-axis, due to a mirror deformation and not as a result of an off-axis displacement of the image.

In the particular case of a centered optical system having only primary aberrations, if the image height dependence is not considered, and the field phase terms are neglected, the wavefront aberration polynomial may be written in a more compact manner as described by Kingslake (1925–1926):

$$W(S,\theta) = F + ES\cos\theta + DS^2 + CS^2(1 + 2\cos^2\theta)$$
$$+ BS^3\cos\theta + AS^4 \qquad (7.11)$$

where $S^2 = x^2 + y^2$. In cartesian coordinates we may write this polynomial as

$$W(x,y) = F + Ey + D(x^2 + y^2) + C(x^2 + 3y^2)$$
$$+ By(x^2 + y^2) + A(x^2 + y^2)^2 \qquad (7.12)$$

where:

$A = {_0}w_{40}$	Spherical aberration coefficient
$B = {_1}w_{31}h$	Coma coefficient
$C = 0.5 {_2}w_{22} h^2$	Astigmatism coefficient
$D = {_0}w_{20} - 0.5 {_2}w_{22}h^2$	Defocusing coefficient
$E = {_1}w_{11}h + {_3}w_{11}$	Tilt about the x axis (image displacement along the y axis)
$F = {_1}w_{00}$	Constant or piston term (OPD)

The defocusing and the astigmatism coefficients have a different definition, in order to have a 1:3 relation between the sagittal and the tangential curvatures in the expression for the astigmatism. It is important to notice that a positive transverse or longitudinal ray aberration means a negative wavefront aberration and vice versa.

These wavefront aberration coefficients may be related to the Seidel or primary aberrations by differentiation of the aberration polynomial in Eq. (7.12) and using Eqs. (1.75) and (1.76), in order to obtain the transverse aberration values, as follows:

$$A = -\frac{SphT}{4r_W y^3} = -\frac{SphL}{4y^4} = -\frac{a}{4y^2} \qquad (7.13)$$

$$B = -\frac{Coma_S}{r_W x^2} = -\frac{Coma_T}{3r_W y^2} \qquad (7.14)$$

$$C = -\frac{\Delta f_S}{2r_W^2} = -\frac{\Delta f_T}{6r_W^2} \qquad (7.15)$$

$$D = -\frac{\Delta f_A}{2r_W^2} \qquad (7.16)$$

and

$$E = -\frac{\Delta h}{r_W} \qquad (7.17)$$

where r_W is the radius of curvature of the reference sphere (distance from the exit pupil to the gaussian image), Δh is the transverse image displacement along the y axis, measured with respect to the yellow gaussian image, and

$\Delta f_A =$ axial focus displacement
$\Delta f_S =$ sagittal focus displacement
$\Delta f_T =$ tangential focus displacement

These image displacements are related to the sagittal and tangential astigmatisms and to the Petzval curvature by

$$\Delta f_S = AstL_S + Ptz - z \tag{7.18}$$

and

$$\Delta f_T = AstL_T + Ptz - z \tag{7.19}$$

where z is the sagitta of the focal surface (if curved). We must remember that the value of Ptz is referred to an ideally flat focal plane, and the focus displacement is measured with respect to the actual curved focal surface.

7.2 ZERNIKE POLYNOMIALS

The actual wavefront deformations may be represented by means of many types of analytical functions. However, the analytical function may not exactly describe the actual wavefront. The fit error is the difference between the actual wavefront W' and the analytical wavefront W. We may then define a quantity called the *fit variance* σ_f^2 to characterize the quality of the fit as follows:

$$\sigma_f^2 = \frac{\int_0^1 \int_0^{2\pi} (W' - W)^2 \rho \, d\rho \, d\theta}{\int_0^1 \int_0^{2\pi} \rho \, d\rho \, d\theta} = \frac{1}{\pi} \int_0^1 \int_0^{2\pi} (W' - W)^2 \rho \, d\rho \, d\theta \tag{7.20}$$

We may notice that the normalizing factor in front of the integral is $1/\pi$. When the fit variance is zero, the analytic function is an exact representation of the real wavefront.

The mean wavefront deformation W_{av} including the normalizing factor is defined by

$$W_{av} = \frac{1}{\pi} \int_0^1 \int_0^{2\pi} W(\rho,\theta) \rho \, d\rho \, d\theta \tag{7.21}$$

All wavefront deformations are measured with respect to a spherical reference. The center of curvature of this spherical wavefront is near the gaussian image. Any displacement with respect to the gaussian image appears as a wavefront tilt and any longitudinal displacement appears as a defocusing term. However, the position of the center of curvature is not enough to define completely the spherical reference, since the radius of curvature is also needed. Any change in this radius of curvature introduces a modification in the constant (piston) term. This last term is the only one that does not affect the position of the image, which is the position of the center of curvature, nor the image structure.

The wavefront variance σ_w^2 is defined as

$$\sigma_w^2 = \frac{1}{\pi} \int_0^1 \int_0^{2\pi} (W(\rho,\theta) - W_{av})^2 \rho \, d\rho \, d\theta$$

$$= \frac{1}{\pi} \int_0^1 \int_0^{2\pi} W^2(\rho,\theta) \rho \, d\rho \, d\theta - W_{av}^2 \qquad (7.22)$$

which represents the root mean squared (*rms*) value of the wavefront deformations, with respect to the reference spherical wavefront. As we have explained before, the reference spherical wavefront may be defined with any value of the radius of curvature (piston term) without modifying the position of the center of curvature or the image structure. Nevertheless, the value of the wavefront variance may be affected by this selection. A convenient way to eliminate this problem is to define the reference sphere in the definition of the wavefront variance as one with the same position as the mean wavefront deformation. This is the reason for subtracting W_{av} in Eq. (7.22).

The most commonly used functions to represent analytically the wavefront deformations are the Zernike polynomials, due to their unique and desirable properties. We will now briefly describe this polynomial representation (Malacara and De Vore, 1992; Wyant and Creath, 1992), without restricting ourselves to the case of a wavefront with symmetry about the y axis, as in the case of a wavefront produced by a centered system. Zernike polynomials $U(\rho, \theta)$ are written in polar coordinates and are orthogonal in the unit circle (exit pupil with radius one), with the orthogonality condition:

$$\int_0^1 \int_0^{2\pi} U_n^l U_{n'}^{l'} \rho \, d\rho \, d\theta = \frac{\pi}{2(n+1)} \delta_{nn'} \delta_{ll'} \qquad (7.23)$$

where $\rho = S/S_{max}$. The Zernike polynomials are represented with two indices n and l, since they are dependent on two coordinates. The index n is the degree of the radial polynomial and l is the angular dependence index. The numbers n and l are both even or both odd, making $n - l$ always even, as shown in Section 7.1, in order to satisfy Eq. (7.5). There are $(1/2)(n+1)(n+2)$ linearly independent polynomials Z_n^l of degree $\leq n$, one for each pair of numbers n and l. Thus, these polynomials can be separated into two functions, one depending only on the radius ρ and the other being dependent only on the angle θ, as follows:

$$U_n^l = R_n^l \begin{bmatrix} \sin \\ \cos \end{bmatrix} l\theta = U_n^{n-2m} = R_n^{n-2m} \begin{bmatrix} \sin \\ \cos \end{bmatrix} (n - 2m)\theta \qquad (7.24)$$

where the sine function is used when $n - 2m > 0$ (antisymmetric functions) and the cosine function is used when $n - 2m \leq 0$ (symmetric functions). Thus, in a centered optical system all terms with the sine function are zero and only the cosine terms remain. The degree of the radial polynomial $R_n^l(\rho)$ is n and $0 \leq m \leq n$. It may be shown that $|l|$ is the minimum exponent of these polynomials R_n^l. The radial polynomial is given by

$$R_n^{n-2m}(\rho) = R_n^{-(n-2m)}(\rho)$$
$$= \sum_{s=0}^{m} (-1)^s \frac{(n-s)!}{s!(m-s)!(n-m-s)!} \rho^{n-2s} \qquad (7.25)$$

All Zernike polynomials $U_n(\rho)$ may be ordered with a single index r, defined by

$$r = \frac{n(n+1)}{2} + m + 1 \qquad (7.26)$$

The first 15 Zernike polynomials are shown in Table 7.1. Kim and Shannon (1987) have shown isometric plots for the first 37 Zernike polynomials. Figure 7.2 shows isometric plots for some of these polynomials.

The triangular and "ashtray" astigmatisms may be visualized as the shape that a flexible disk adopts when supported on top of three or four supports equally distributed around the edge. However, according to Hopkins' notation, the triangular astigmatism is really a comatic term (elliptical coma). It should be pointed out that these polynomials are

Table 7.1 First 15 Zernike Polynomials

n	m	r	Zernike polynomial	Meaning
0	0	1	1	Piston term
1	0	2	$\rho \sin\theta$	Tilt about x axis
1	1	3	$\rho \cos\theta$	Tilt about y axis
2	0	4	$\rho^2 \sin 2\theta$	Astigmatism with axis at $\pm 45°$
2	1	5	$2\rho^2 - 1$	Defocusing
2	2	6	$\rho^2 \cos 2\theta$	Astigmatism, axis at $0°$ or $90°$
3	0	7	$\rho^3 \sin 3\theta$	Triangular astigmatism, based on x axis
3	1	8	$(3\rho^3 - 2\rho) \sin\theta$	Primary coma along x axis
3	2	9	$(3\rho^3 - 2\rho) \cos\theta$	Primary coma along y axis
3	3	10	$\rho^3 \cos 3\theta$	Triangular astigmatism, based on y axis
4	0	11	$\rho^4 \sin 4\theta$	Ashtray astigmatism, nodes on axes
4	1	12	$(4\rho^4 - 3\rho^2) \sin 2\theta$	
4	2	13	$64\rho^4 - 6\rho^2 + 1$	Primary spherical aberration
4	3	14	$(4\rho^4 - 3\rho^2) \cos 2\theta$	
4	4	15	$\rho^4 \cos 4\theta$	Ashtray astigmatism, crests on axis

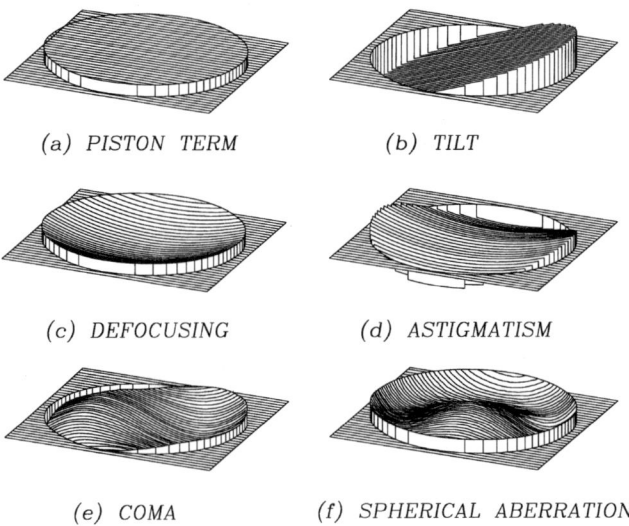

(a) PISTON TERM (b) TILT

(c) DEFOCUSING (d) ASTIGMATISM

(e) COMA (f) SPHERICAL ABERRATION

Figure 7.2 Wavefront shapes for some aberrations: (a) piston term; (b) tilt; (c) defocusing; (d) astigmatism; (e) coma; (f) spherical aberration.

Aberration Polynomial

orthogonal by the definition in Eq. (7.23), only if the pupil is circular, without any central obscurations.

Any continuous wavefront shape $W(x,y)$ may be represented by a linear combination of these Zernike polynomials, as follows:

$$W(x,y) = \sum_{n=0}^{k} \sum_{m=0}^{n} A_{nm} U_{nm} = \sum_{r=0}^{L} A_r U_r \qquad (7.27)$$

Given a selected power L, the coefficients A_r are found by many of several possible procedures, so that the fit variance defined in Eq. (7.20) is minimized.

The advantage of expressing the wavefront by a linear combination of orthogonal polynomials is that the wavefront deviation represented by each term is a best fit (minimum fit variance) with respect to the actual wavefront. Then, any combination of these terms must also be a best fit. Each Zernike polynomial is obtained by adding to each type of aberration, the proper amount of piston, tilt, and defocusing, so that the rms value σ_w^2, for each Zernike polynomial, represented by Eq. (7.24) is minimized. As an example, let us consider the term for spherical aberration, where we may see that a term $+1$ (piston term) and a term $-6\rho^2$ (defocusing) has been added to the spherical aberration term $6\rho^4$. This term minimizes the deviation of this polynomial with respect to a flat wavefront.

The practical consequence of the orthogonality of the Zernike polynomials is that any aberration terms, like defocusing, tilt, or any other, may be added or subtracted from the wavefront function $W(x,y)$ without losing the best fit.

Using the orthogonality condition in Eq. (7.23), the mean wavefront deformation of each Zernike polynomial may be shown to be

$$W_{av} = \frac{1}{\pi} \int_0^1 \int_0^{2\pi} U_r(\rho,\theta) \rho \, d\rho \, d\theta$$

$$= \frac{1}{2}; \text{ if}: r = 1$$

$$= 0; \text{ if}: r > 1 \qquad (7.28)$$

in other words, the mean wavefront deformation is zero for all Zernike polynomials, with the exception of the piston term. Thus, the wavefront variance, defined in Eq. (7.22) is given by

$$\sigma_W^2 = \frac{1}{2} \sum_{r=1}^{L} \frac{A_r^2}{n+1} - W_{av}^2 = \frac{1}{2} \sum_{r=2}^{L} \frac{A_r^2}{n+1} \qquad (7.29)$$

where, from Eq. (7.25), n is related to r by

$$n = \text{next integer greater than } \frac{-3 + [1 + 8r]^{1/2}}{2} \qquad (7.30)$$

The aberrations of a centered optical system are symmetrical with respect to the meridional plane, but this is not the general case, e.g., when decentered or tilted surfaces are present. We may see that with Zernike polynomials not only aberrations symmetrical with respect to the meridional plane may be represented. For example, a coma or astigmatism aberration with any orientation in the x–y plane may also be represented. A coma aberration with a 10^0 inclination with respect to the x axis may be written as a combination of coma along the y axis ($r = 9$) and coma along the x axis ($r = 8$). As shown by Malacara (1983), these two terms may be combined in a single term, where the orientation angle is a parameter and the magnitude of the combined aberration is another parameter.

The wavefront deformation may be obtained by integration of the transverse aberration values or by direct computation of the optical path difference, as we will see in Chap. 9. Once some values of the wavefront are determined, the analytic wavefront expression in terms of Zernike polynomials may be obtained by a two-dimensional least squares fit as shown by Malacara and DeVore (1992) and by Malacara et al. (1990).

7.3 WAVEFRONT REPRESENTATION BY AN ARRAY OF GAUSSIANS

Frequently, a wavefront is measured or calculated only at some sampling points, e.g., in phase-shifting interferometry and when calculating the wavefront at an array of points, as in a spot diagram. The need for an analytical representation of the wavefront may arise. In this case Zernike polynomials may be employed. However, the description of a wavefront shape can be inaccurate with a polynomial representation if sharp local deformations are present. The largest errors in the analytical representation occur at these deformations and near the edge of the pupil. In this case an analytical representation by an array of gaussians may give better results, as shown by Montoya-Hernandez et al. (1999). Let us assume that we have a two-dimensional array of $(2M + 1) \times (2N + 1)$ gaussians with a separation d as shown in Fig. 7.3. The height w_{nm} of each gaussian is adjusted to obtain the desired wavefront shape $W(x, y)$ as follows:

$$W(x,y) = \sum_{m=-M}^{M} \sum_{n=N}^{N} w_{nm} e^{-((x-md)^2 + (y-nd)^2)/\rho^2} \qquad (7.31)$$

Aberration Polynomial

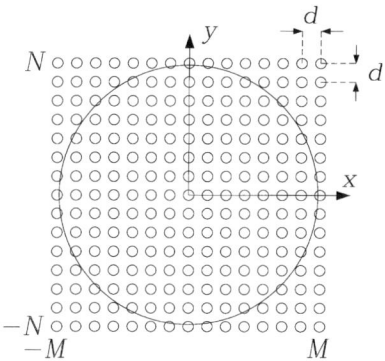

Figure 7.3 Sampling of a wavefront shape with a two-dimensional array of gaussians.

The Fourier transform $F\{W(x, y)\}$ of the function $W(x, y)$ represents the spatial frequency content of this wavefront and it is given by

$$F\{W(x,y)\} = \pi\rho^2 e^{[-\pi^2\rho^2(f_x-f_y)]} \sum_{m=-M}^{M} \sum_{n=N}^{N} w_{nm} e^{-i2\pi d(mf_x - nf_y)} \qquad (7.32)$$

The separation d and the width ρ of the gaussians are two important parameters to be selected. To understand how these values are found let us consider a one-dimensional function $g(x)$, which is sampled by a comb function $h(x)$ as shown in Fig. 7.4(a). We assume that the function $g(x)$ is band limited, with a maximum spatial frequency f_{max}. According to the sampling theorem the comb sampling frequency should be less than half this frequency f_{max} so that the function $g(x)$ can be fully reconstructed.

The Fourier transform of the product of two functions is equal to the convolution of the Fourier transforms of the two functions, as follows:

$$F\{g(x)h(x)\} = G(f) * H(f) \qquad (7.33)$$

where the symbol $*$ represents the convolution operation.

We see in Fig. 7.4(b) that in the Fourier or frequency space an array of lobes represents each one the Fourier transforms of the sampled function. If the sampling frequency is higher than $2f_{max}$ the lobes are separated without any overlapping. Ideally, they should just touch each other. The function $g(x)$ is well represented only if all lobes in the Fourier space are filtered out with the only exception of the central lobe.

To perform the necessary spatial filtering the comb function is now replaced by an array of gaussians as in Fig. 7.5(a). In the Fourier space the

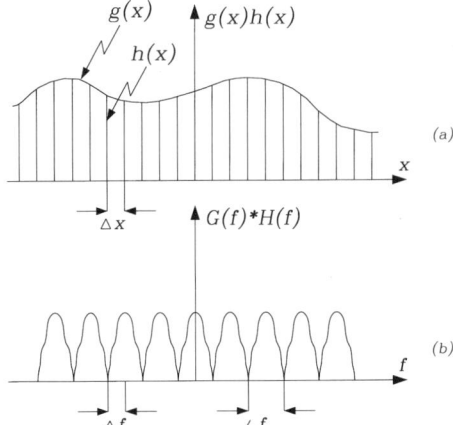

Figure 7.4 Sampling of a one-dimensional function with a comb function.

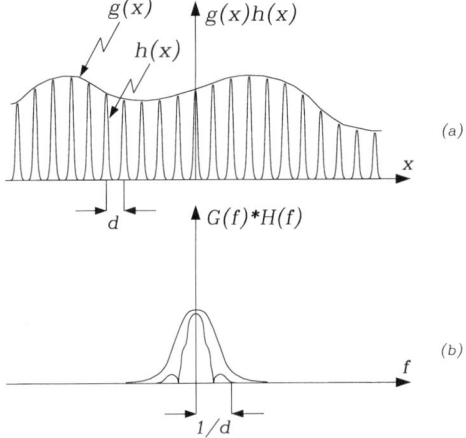

Figure 7.5 Sampling a one-dimensional function with an array of gaussians.

Fourier transform of these gaussians appears as a modulating envelope that filters out the undesired lobes as in Fig. 7.5(b). To obtain a good filtering the gaussians should have a width ρ approximately equal to the array separation d.

The remaining parameter to be determined is the gaussian height w_{nm}. This can be done by an iterative procedure. To obtain the wavefront deformation at a given point it is not necessary to evaluate all the gaussian heights, since the contribution of the gaussians decay very fast with their distance to that point. The height of each gaussian is adjusted until the

Aberration Polynomial 183

function $g(x)$ has the desired value at that point. A few iterations are sufficient.

7.4 TRANSVERSE ABERRATION POLYNOMIALS

In a lens design or ray tracing program the transverse aberrations are more easily obtained and analyzed than the wavefront aberrations. Given the wavefront aberration, the transverse aberrations may be found, as we described before, by differentiation of the aberration polynomial in Eq. (7.12) and using Eqs. (1.75) and (1.76). Thus, we may find the transverse aberrations along the x and y axes as

$$TA_x(x,y) = -[2(D+C)x + 2Bxy + 4A(x^2+y^2)x]r_W \quad (7.34)$$

and

$$TA_y(x,y) = -[E + 2(D+3C)y + B(3y^2+x^2) + 4A(x^2+y^2)y]r_W \quad (7.35)$$

where r_W is the radius of curvature of the wavefront, as defined before. Let us now study in more detail these transverse aberration functions.

7.4.1 Axial, Meridional, and Sagittal Plots

To analyze an optical system design, a fan of meridional rays and a fan of sagittal rays is traced through the system, as shown in Fig. 7.6. The rays are equally spaced on the entrance pupil, on the x and y axes. Off-axis as well as axial fans of rays are traced. These plots are extremely important evaluation tools in modern design. Any lens designer must understand them very well.

Axial plots—An axial fan of rays from an on-axis point object is traced through the optical system. The heights of the rays on the entrance pupil are selected at equal y intervals. The function describing the transverse aberration $TA_y(0,y)$ versus the ray height y on the entrance pupil is antisymmetric, due to the symmetry of the optical system about the meridional plane, so only the light rays on the positive side of the y axis are necessary. Thus, from Eq. 7.35 we may write

$$\begin{aligned}TA_y(0,y) &= a_1 y + a_3 y^3 + a_5 y^5 \\ &= -[2Dy + 4Ay^3]r_W + SphT_5 \\ &= \Delta f_A \frac{y}{r_W} + SphT + SphT_5\end{aligned} \quad (7.36)$$

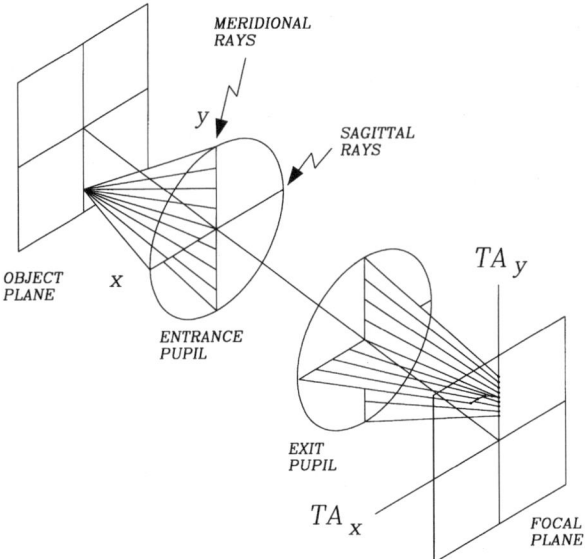

Figure 7.6 Meridional and sagittal fans of rays traced through an optical system.

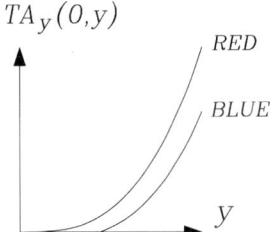

Figure 7.7 Axial plot for axial chromatic aberration.

where $SphT_5$ is the fifth-order transverse spherical aberration. Only the primary spherical aberration with a cubic term, the fifth-order spherical aberration with a fifth power term, and the focus shift with a linear term are present in this plot. Figure 7.7 shows an axial plot for a lens with spherical and chromatic aberrations.

To analyze an axial plot a straight line tangent to the graph on the y axis crossing is drawn, as shown in Fig. 7.8. The distance from a point P on the straight line to the curve is the magnitude of the transverse spherical aberration. The slope of the straight line is equal to $\Delta f_A/r_W$.

Aberration Polynomial

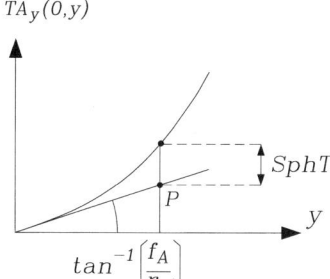

Figure 7.8 Axial plot for spherical aberration and defocusing.

If two axial plots are obtained, one for blue light (F) and one for red light (C), the axial chromatic aberration is given by

$$AchrL = (\Delta f_A)_F - (\Delta f_A)_C \tag{7.37}$$

and the spherochromatism is given by

$$Spherochromatism = SphL_F - SphL_C \tag{7.38}$$

Meridional plots—A meridional plot is obtained when the object is off-axis and the rays are on the meridional plane. Then, the coordinate x on the entrance pupil is zero, so $TA_x(x, y)$ becomes zero and $TA_y(x, y)$ is given by

$$\begin{aligned} TA_y(0,y) &= a_0 + a_1 y + a_2 y^2 + a_3 y^3 + a_5 y^5 \\ &= -[E + 2(D + 3C)y + 3By^2 + 4Ay^3]r_W + SphT_5 \\ &= \Delta h + (\Delta f_A + \Delta f_T)\frac{y}{r_W} + Coma_T + SphT + SphT_5 \end{aligned} \tag{7.39}$$

We see that both the axial focus shift and the tangential surface curvature produce a linear term in this plot. As in the axial plot, if a straight line, tangent to the curve on the intersection of this plot with the y axis is drawn, the slope is equal to $(\Delta f_A + \Delta f_T)/r_W$. Since Δf_A may be independently found from an axial trace, the sagitta Δf_T for the tangential focal surface may be determined. If the axial trace is made without any focus displacement (at the gaussian plane), the slope of the meridional plot directly gives the sagitta of the tangential focal surface, as follows:

$$\Delta f_T = r_W \frac{d TA_y(0,y)}{dy} = 6 r_W C \tag{7.40}$$

If the Petzval curvature is known, the tangential astigmatism may be calculated with this value of the tangential focus displacement, using Eq. (7.19).

On the meridional plot, the tangential coma produces a parabolic term, the primary spherical aberration produces a cubic term, and the fifth-order aberration a fifth degree term. We see that symmetrical as well as antisymmetrical transverse ray aberrations appear and hence the plot does not in general have any symmetry. Figure 7.9 shows meridional plots for some aberrations.

The height of the graph at the point it crosses the y axis is equal to the image displacement Δh. If the rays are traced in yellow light, this image height is the distortion. If two meridional plots are obtained, one with red (C) light and another with blue light (F), the magnification chromatic aberration is given by

$$Mchr = \Delta h_F - \Delta h_C \tag{7.41}$$

Assuming that no high-order aberrations are present, from the meridional plots we may obtain the magnitudes of the spherical aberration

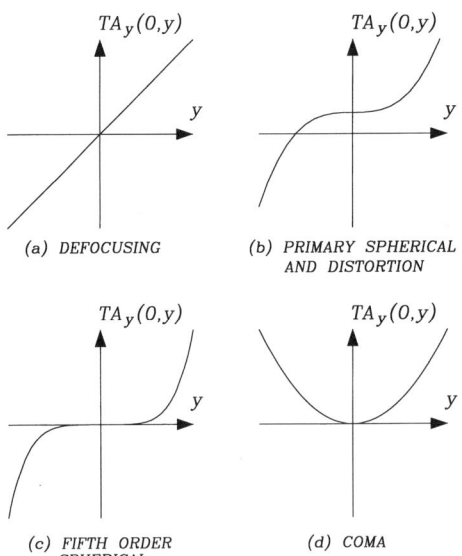

Figure 7.9 Axial plots for some aberrations: (a) defocusing; (b) primary spherical and distortion; (c) fifth-order spherical; (d) coma.

Aberration Polynomial

and coma. Subtracting two symmetrically placed points on this plot we may obtain

$$\frac{TA_y(0,y) - TA_y(0,y)}{2} = a_1 y + a_3 y^3 + a_5 y^5$$
$$= -[2(D + 3C)y + 4Ay^3]r_W + SphT_5$$
$$= (\Delta f_A + \Delta f_r)\frac{y}{r_W} + SphT + SphT_5 \qquad (7.42)$$

and adding them we find that

$$\frac{TA_y(0,y) + TA_y(0,-y)}{2} = a_0 + a_2 y^2$$
$$= -[E + 3By^2]r_W$$
$$= \Delta h + Coma_T \qquad (7.43)$$

Figure 7.10 shows meridional plots for some combination of aberrations.

In a more quantitative manner, to avoid a graphic estimation of the slope of the plot, we may calculate the values of the transverse aberration at six points, e.g., at the edge of the pupil (y_m), at one-half of the radius ($y_m/2$),

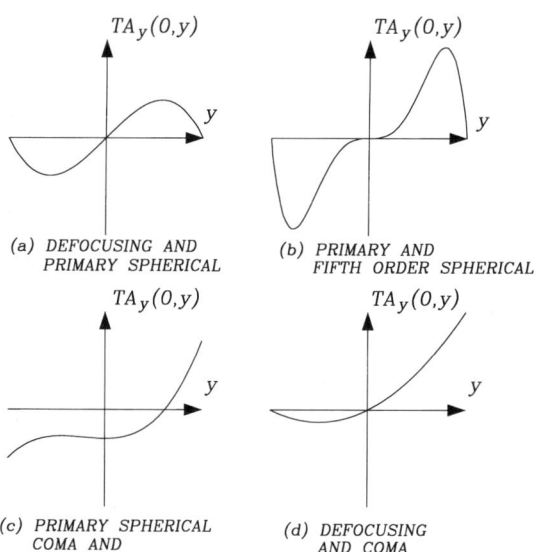

Figure 7.10 Meridional plots for some aberrations: (a) defocusing and primary spherical; (b) primary and fifth-order spherical; (c) primary spherical coma and distortion; (d) defocusing and coma.

and at one-tenth of the radius ($y_m/10$) as follows:

$$TA_y(0,y_m) = a_1 y_m + a_3 y_m^3 + a_5 y_m^5$$
$$TA_y(0,y_m/2) = \frac{a_1}{2} y_m + \frac{a_3}{8} y_m^3 + \frac{a_5}{16} y_m^5$$
$$TA_y(0,y_m/10) = \frac{a_1}{10} y_m + \frac{a_3}{100} y_m^3 + \frac{a_5}{1000} y_m^5 \qquad (7.44)$$

and then the coefficients a_1, a_3, a_5 may be obtained from

$$a_1 = \frac{5[TA(0,y_m/10) - TA(0,-y_m/10)]}{y_m} - \frac{a_3}{10} y_m^2 - \frac{a_5}{1000} y_m^4$$
$$a_3 = \frac{4[TA(0,y_m/2) - TA(0,-y_m/2)]}{y_m^3} - \frac{4a_1}{y_m^2} - \frac{a_5}{8} y_m^2$$
$$a_5 = \frac{[TA(0,y_m) - TA(0,y_m)]}{2} - \frac{a_1}{y_m^4} - \frac{a_3}{y_m^2} \qquad (7.45)$$

in an iterative manner, in no more than two or three passes, taking a_3 and a_5 equal to zero in the first equation and a_5 equal to zero in the second equation, in the first pass. In the same manner, we obtain for the even power terms:

$$a_0 = TA(0,0)$$

$$a_2 = \frac{TA(0,y_m) + TA(0,-y_m)}{2y_m^3} - \frac{a_0}{y_m^2} \qquad (7.46)$$

Sagittal plots—These plots are obtained when the y coordinate on the entrance pupil is equal to zero. The first plot is

$$TA_x(x,0) = -[2(D+C)x + 4Ax^3]r_W + SphT_5$$
$$= (\Delta f_A + \Delta f_S)\frac{x}{r_W} + SphT + SphT_5 \qquad (7.47)$$

which detects and measures only antisymmetric transverse aberrations, like spherical aberration and defocusing. The second sagittal plot is

$$TA_y(x,0) = -[E + Bx^2]r_W$$
$$= \Delta h + Coma_S \qquad (7.48)$$

showing only symmetric transverse aberrations like $Coma_S$. In the first term of the first sagittal plot we have the astigmatism and the focus

Aberration Polynomial

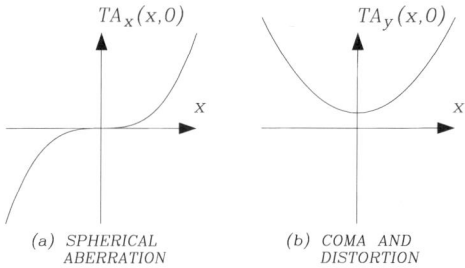

Figure 7.11 Sagittal plots for some aberrations: (a) spherical aberration; (b) coma and distortion.

shift represented. We may see that the amount of astigmatism may be obtained by subtracting the linear terms (slope differences) of the meridional $TA_y(0, y)$ and the sagittal $TA_x(x, 0)$ plots. In the second sagittal plot $TA_y(x, 0)$ only the coma aberration is present. Figure 7.11 shows the sagittal plots for an optical system with spherical aberration, coma, and distortion.

Subtracting the meridional from the sagittal plots and making $x = y$, we may also obtain

$$TA_y(0,y) - TA_x(x,0) = -[E + 2Cy]r_W$$
$$= \Delta h + \Delta f_S \frac{y}{r_W} \qquad (7.49)$$

Thus, with these plots we may obtain the magnitudes of all primary aberrations. However, a more common approach in practice is to estimate them from the primary aberration coefficients. The great advantage of these plots over the primary aberration coefficients is that information about high-order aberrations is also obtained.

REFERENCES

Hopkins, H. H., *Wave Theory of Aberrations*, Clarendon Press, Oxford, 1950.

Kim, C.-J. and Shannon, R., "Catalog of Zernike Polynomials," in *Applied Optics and Optical Engineering*, Vol. X, Chap. 4, R. Shannon and J. C. Wyant, eds., Academic Press, New York, 1987.

Kingslake, R., "The Interferometer Patterns due to the Primary Aberrations," *Trans. Opt. Soc.*, **27**, 94 (1925–1926).

Malacara, D., "Set of Orthogonal Primary Aberration Coefficients," *Appl. Opt.*, **22**, 1273–1274 (1983).

Malacara, D. and DeVore, S. L., *Optical Shop Testing*, 2nd ed., D. Malacara, ed., Chap. 13, John Wiley, New York, 1992.

Malacara, D., Carpio, J. M. and Sánchez, J. J., "Wavefront Fitting with Discrete Orthogonal Polynomials in a Unit Radius Circle," *Opt. Eng.*, **29**, 672–675 (1990).

Montoya-Hernández, M., Servin, M., Malacara-Hernández, D. and Paez, G., "Wavefront Fitting Using Gaussian Functions," *Opt. Commun.*, **163**, 259–269 (1999).

Wyant, J. C. and Creath, K., "Basic Wavefront Aberration Theory for Optical Metrology," in *Applied Optics and Optical Engineering*, Vol. XI, Chap. 1, R. Shannon and J. C. Wyant, eds., Academic Press, Boston, MA, 1992.

8
Diffraction in Optical Systems

8.1 HUYGENS–FRESNEL THEORY

As we pointed out in Chap. 7, light may be considered in a first approximation as a bundle of rays, but the real nature is that of a wave. Frequently, the geometrical optics approximation is not accurate enough to describe and explain some phenomena or image structures. Then, it is necessary to use diffraction theory. In this chapter we will briefly describe this theory with a special emphasis on its applications to the study of images. This subject has been treated in many books and journal publications; see e.g., Born and Wolf (1964) and Malacara (1988).

There are many theories that explain diffraction phenomena, but the simplest one is the Huygens–Fresnel theory, which is surprisingly accurate in most cases. This theory assumes that a wavefront may be considered to emit secondary wavelets as passing through an aperture as shown in Fig. 8.1. This secondary Huygens wavelets were postulated by Christian Huygens in 1678 in Holland, but this theory was not enough to explain diffraction effects quantitatively. Many years later, in 1815 in France, Agoustin Arago Fresnel considered that Huygens wavelets must interfere with their corresponding phase when arriving at the observing screen. This theory is sufficient to explain all diffraction effects appearing in optical systems, with the exception of the value of the resulting phase. However, the calculated irradiance values for a point light source (plane wavefront) are extremely accurate. Many other theories have been postulated to improve the results of the Huygens–Fresnel model in some particular cases, but we do not need them for our purposes.

In any diffraction experiment the important elements are the light source, the diffracting aperture, and the observing screen. If any of the two distances, the distance from the light source to the diffracting aperture or the distance from the diffracting aperture to the observing screen, or both, are finite, we have the so called Fresnel diffraction theory. If both distances are infinite, then we have a Fraunhofer diffraction configuration. Let us now

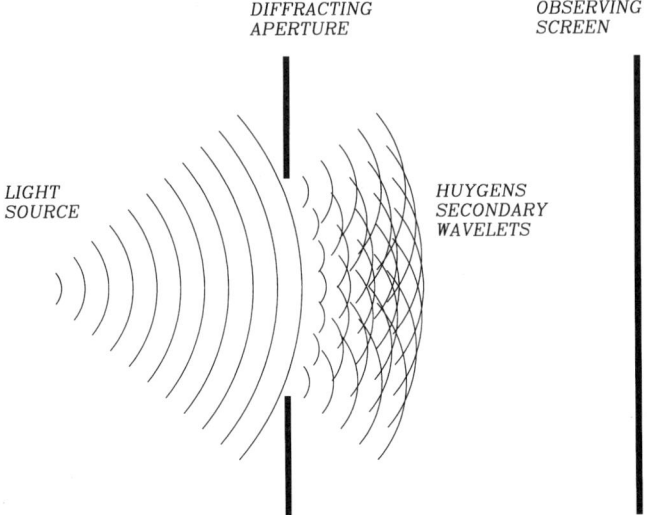

Figure 8.1 Diffraction of a wavefront of light passing through an aperture.

consider in the following sections some examples of Fresnel and Fraunhofer diffraction phenomena.

8.2 FRESNEL DIFFRACTION

An interesting case of Fresnel diffraction is that of the diffraction by a circular aperture, with the geometry depicted in Fig. 8.2. To add and consider the interference of the Huygens wavelets with their proper phase at the observing screen it is simple only at the observing point **P** on the optical axis, due to the circular symmetry of the aperture. As we may see, all Huygens wavelets emanating from a point on an imaginary thin ring centered on the aperture would have the same phase on the observing point **P**. Thus, the difference in phase between the light that passes the aperture through its center (optical axis) and the light passing through this ring with radius S is given by

$$\delta = KS^2 \tag{8.1}$$

with

$$K = \frac{\pi(a+b)}{ab\lambda} \tag{8.2}$$

Diffraction in Optical Systems

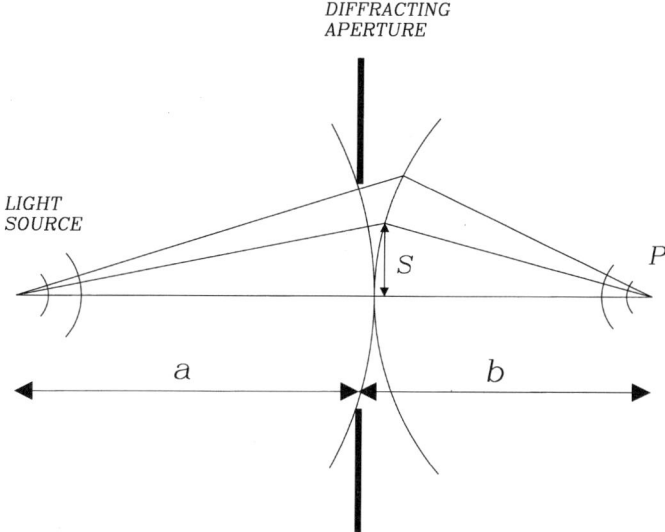

Figure 8.2 Geometry for the diffraction of a spherical wavefront on-axis passing through a circular aperture.

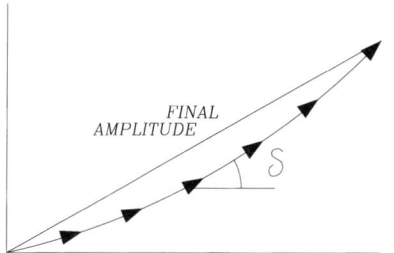

Figure 8.3 Vector addition of amplitude contributions on diffraction.

where a and b are the distances from the point light source to the center of the diffracting aperture and from this diffracting aperture to the observing point, respectively. The contribution dI to the irradiance, at the point of observation, of the light from the ring on the aperture is directly proportional to its area. This area is directly proportional to the ring width dS and its radius S. If we add these contributions to the irradiance, like vectors, with an angle between them equal to their phase difference, we find the curve in Fig. 8.3, represented mathematically by

$$dx = A\, s\, dS\, \cos\delta \qquad (8.3)$$

and

$$dy = A\,s\,dS\sin\delta \tag{8.4}$$

Then, using Eq. (8.1) and integrating we may find that

$$x^2 = \left[y - \frac{a}{2K}\right]^2 = \left[\frac{A}{2K}\right]^2 \tag{8.5}$$

This expression represents a circle with its center on the y axis and tangent to the x axis. This means that if the aperture diameter is increased continuously the irradiance at the point of observation on the optical axis is going to oscillate, passing through values of maximum amplitude and values with zero amplitude. Obviously, it is difficult to observe these oscillations experimentally for a very large aperture, because the oscillations will be quite rapid and decrease in amplitude as the aperture diameter becomes larger, as shown in Fig. 8.4. Then, the curve in Eq. (8.5) instead of being a circle, is a spiral, as shown in Fig. 8.5.

As an application of this theory, let us design a pinhole camera to image the Sun. The optimum size for the pinhole is the minimum diameter that produces the maximum irradiance at the point of observation. From Fig. 8.5 we see that this diameter is such that the phase difference between the ray traveling along the optical axis and the ray diffracted from the edge of the aperture is equal to $\pi/2$. Thus, the radius S_0 of this aperture is given by

$$S_0 = \sqrt{\frac{ab\lambda}{2(a+b)}} \tag{8.6}$$

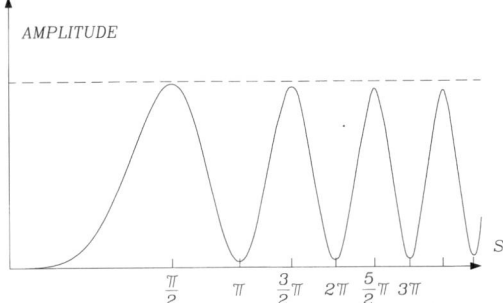

Figure 8.4 Amplitude variations along the radius of the circular aperture illuminated with a spherical wavefront.

Diffraction in Optical Systems 195

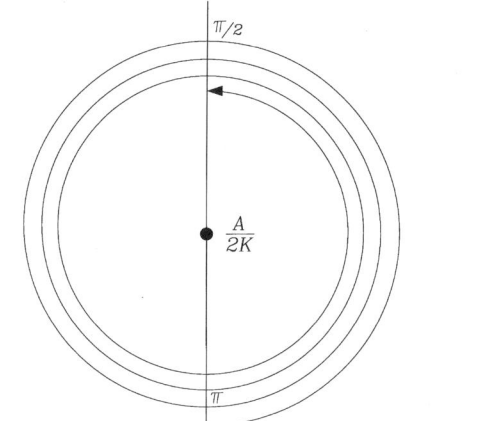

Figure 8.5 Spiral representing the sum of the amplitude contributions on the observing point for a spherical wavefront on-axis illuminating a circular aperture.

Since the distance a from the Sun to the aperture is infinity, this expression becomes

$$S_0 = \sqrt{\frac{b\lambda}{2}} \tag{8.7}$$

The distance b determines the size and irradiance of the image of the Sun. The size d of the image increases linearly with this distance b, while the irradiance decreases inversely with this distance. Thus, given b, if the angular diameter of the Sun is θ in radians, the linear diameter d of the image is

$$d = \theta b \tag{8.8}$$

8.3 FRAUNHOFER DIFFRACTION

The most interesting type of diffraction, from the point of view of lens designers is when both the distance from the light source to the aperture and the distance from the aperture to the observation plane are infinity. This is known as *Fraunhofer diffraction*.

In the case of lenses the diffracting aperture is the finite size of the lens or, to be more precise, the finite diameter of the entrance pupil. The distance from the light source to the aperture may not be infinite, but generally it is long enough to be considered so. The distance from the aperture to the image is not physically infinite, but is optically infinite, because the lens focuses in a common point all rays in an incident beam of parallel (collimated) rays, as shown in Fig. 8.6. Then, the observing plane may be considered optically placed at infinity from the diffracting aperture. As a typical and interesting example of Fraunhofer diffraction let us consider a diffracting slit. The slit has a width $2a$, it is in the plane x–y and centered on the y axis as shown in Fig. 8.7.

From a direct application of the Huygens–Fresnel model of diffraction, as illustrated in Fig. 8.7, the amplitude $U(\theta)$ on a point in the direction θ over the observation screen is given by

$$U(\theta) = A \int_{-a}^{a} e^{iky\sin\theta} \, dy \tag{8.9}$$

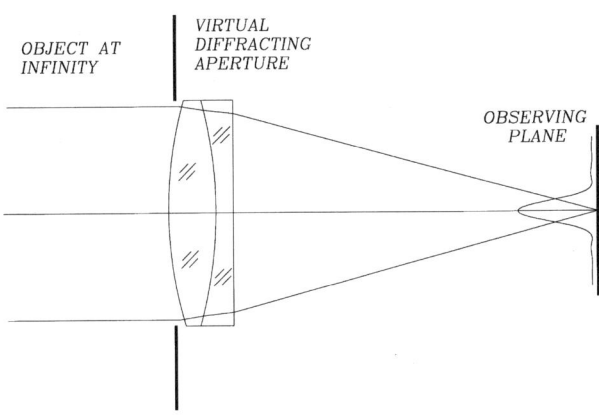

Figure 8.6 Observing the diffraction image of a point object.

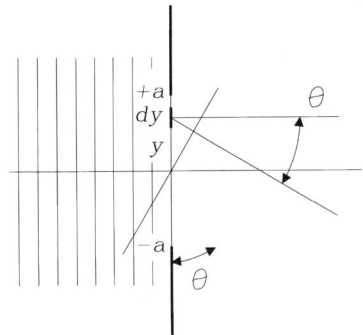

Figure 8.7 Geometry to calculate the Fraunhofer diffraction pattern.

Diffraction in Optical Systems

where A is a constant, and k is given by

$$k = \frac{2\pi}{\lambda} \tag{8.10}$$

The exponential term is the phase of the ray passing through the point y with respect to the ray passing through the origin. Integrating, we obtain

$$\begin{aligned} U(\theta) &= Aa\left[\frac{e^{iky\sin\theta} - e^{iky\sin\theta}}{2ika\sin\theta}\right] \\ &= U_0\left[\frac{\sin(ka\sin\theta)}{ka\sin\theta}\right] \\ &= U_0\,\mathrm{sinc}\,(ka\sin\theta) \end{aligned} \tag{8.11}$$

where U_0 is a constant. The irradiance distribution $I(\theta)$ is obtained by taking the square conjugate of the amplitude function $U(\theta)$. Since this amplitude distribution is real, the irradiance distribution is

$$I(\theta) = I_0 \mathrm{sinc}^2(ka\sin\theta) \tag{8.12}$$

The function amplitude $U(\theta)$ and its corresponding irradiance pattern $I(\theta)$ is plotted in Fig. 8.8. We may see that the first minimum (zero value) of the irradiance, or dark fringe, occurs for an angle θ given by

$$\sin\theta = \frac{\lambda}{2a} \tag{8.13}$$

As is expected, this angle decreases for wider slits.

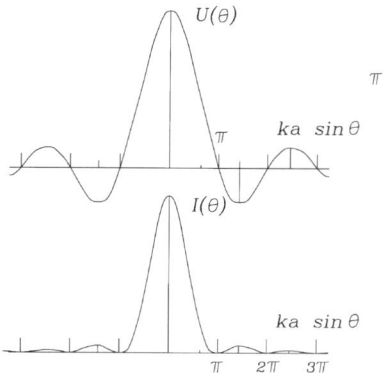

Figure 8.8 Amplitude and irradiance patterns for Fraunhofer diffraction of a slit.

8.3.1 Circular Aperture

The most important aperture in Fraunhofer diffraction is the circular aperture, since most lenses have this shape. In an analogous manner to the slit, generalizing Eq. (8.9) for two dimensions, the Fraunhofer irradiance distribution is given by

$$U(\theta_x,\theta_y) = \frac{A}{\pi} \int\!\!\int_\sigma e^{ik(x\sin\theta_x + y\sin\theta_y)}\, dx\, dy \tag{8.14}$$

where the factor $1/\pi$ has been placed in front of the integral, so that the amplitude $U(0,0)$ at the center of the image becomes one, when the amplitude A of the incident light beam, as well as the radius of the exit pupil, have also a unitary value. If we use polar coordinates, as shown in Fig. 8.1 and Eqs. (8.2) and (8.3), we may write for the polar coordinates of the diffracted ray:

$$\sin\theta_x = \sin\theta_\rho \cos\phi \tag{8.15}$$

and

$$\sin\theta_x = \sin\theta_\rho \sin\phi \tag{8.16}$$

where $\sin\theta_\rho$ is the radial angular distance and ϕ is the angular coordinate. Then, Eq. (8.14) may be written

$$U(\theta_\rho,\phi) = \frac{A}{\pi} \int_0^a\!\!\int_0^{2\pi} e^{ikS\sin\theta_\rho \cos(\theta-\phi)}\, S\, dS \tag{8.17}$$

Then, applying this expression to the circular aperture we find that

$$U(\theta_\rho) = 2U_0 \left[\frac{J_1(ka\sin\theta_\rho)}{ka\sin\theta_\rho}\right] \tag{8.18}$$

where a is the semidiameter of the circular aperture and $J_1(x)$ is the first-order Bessel function. This irradiance distribution, called the Airy function, shown in Fig. 8.9, is given by

$$I(\theta_\rho) = 4I_0 \left[\frac{J_1(ka\sin\theta_\rho)}{ka\sin\theta_\rho}\right]^2 \tag{8.19}$$

Diffraction in Optical Systems

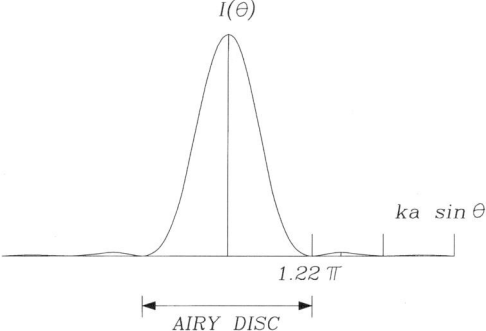

Figure 8.9 Fraunhofer diffraction pattern produced by a circular aperture.

The angular semidiameter θ_ρ of the first dark ring, also called the Airy disk, is given by

$$\sin \theta_\rho = 1.22 \frac{\lambda}{2a} = 1.22 \frac{\lambda}{D} \tag{8.20}$$

Thus, we may see that the angular resolution of a perfect lens depends only on the diameter of the lens. The Airy disk has a semidiameter $d/2$ given by

$$\frac{d}{2} = 1.22 \frac{F\lambda}{D} \tag{8.21}$$

where D is the aperture diameter and F is the effective focal length. If we assume a wavelength λ equal to 500 nm the Airy disk diameter becomes

$$d = 1.22 \frac{F}{D} = 1.22 \, FN \, \mu m \tag{8.22}$$

thus, we may say that the Airy disk linear diameter is approximately equal to the *f*-number FN in micrometers. The structure and characteristics of this diffraction image has been studied in detail by Stoltzmann (1980) and Taylor and Thompson (1958). Most of the light energy (about 84%) is in the central nucleus (Airy disk), as shown in the radial distribution of energy in Fig. 8.10.

8.3.2 Annular Aperture

Many instruments, like the Cassegrain telescope, have a central opaque disk at the center of the entrance pupil. Then, the effective pupil is not a clear

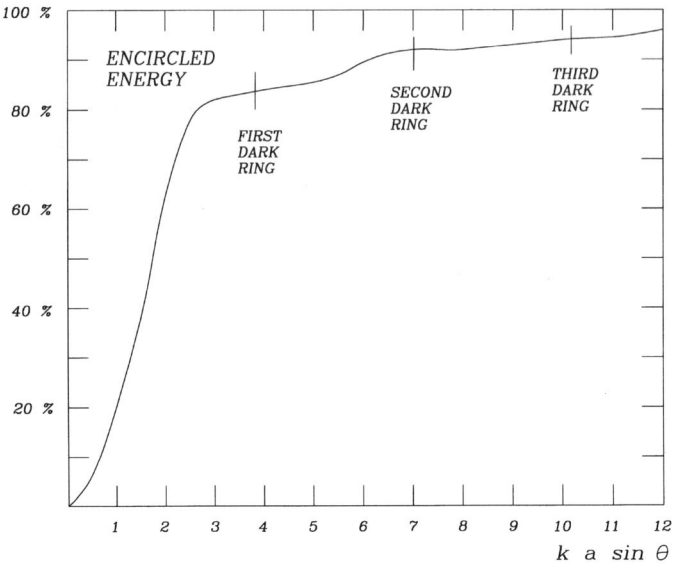

Figure 8.10 Encircled energy in the Fraunhofer diffraction pattern of a circular aperture.

disk, but an annular aperture. In this case the amplitude on the diffraction image would be given by

$$U(\theta_p) = 2U_0 a_1^2 \left[\frac{J_1(ka_1 \sin \theta_p)}{ka_1 \sin \theta_p} \right] - 2U_0 a_2^2 \left[\frac{J_1(ka_2 \sin \theta_p)}{ka_2 \sin \theta_p} \right] \quad (8.23)$$

where a_1 and a_2 are the semidiameter of the aperture and the semidiameter of the central disk, respectively. The obscuration ratio is defined as $\eta = a_2/a_1$. As the obscuration ratio increases, the diffraction image also increases its size. Figure 8.11 shows the radial distribution of energy for different obscuration ratios. The effect of an annular aperture has been studied by Taylor and Thompson (1958) and Welford (1960).

Since the aperture shape and size have an influence on the image structure, many authors have investigated many different ways of modifying the entrance pupil to improve the image. This procedure is called apodization. (Barakat, 1962a; Barakat and Levin, 1963a).

8.4 DIFFRACTION IMAGES WITH ABERRATIONS

In the presence of aberrations the image of the point object is not the Airy function. In this case the image may be found by integration of the

Diffraction in Optical Systems

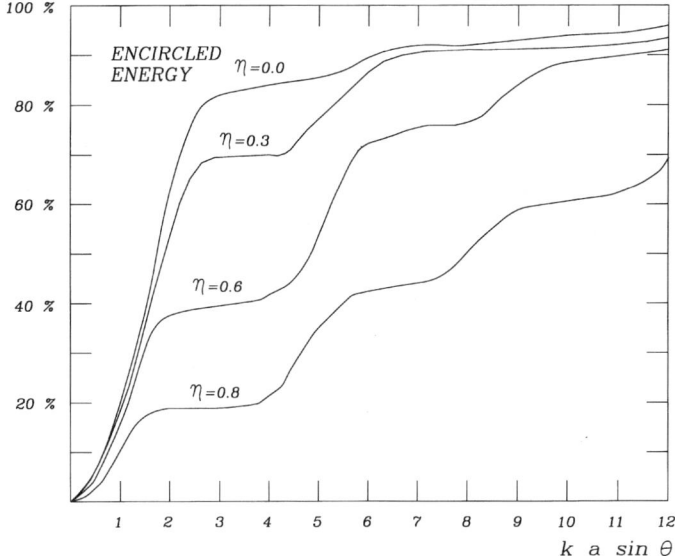

Figure 8.11 Encircled energy in the Fraunhofer diffraction pattern of a circular aperture with an annular aperture.

two-dimensional diffraction integral given by Eq. (8.17). However, in this case amplitude A over the exit pupil is not constant and has to be inside the integral sign. The amplitude $A(S, \theta)$ including the phase distribution, may be expressed as

$$A(S,\theta) = E(S,\theta)e^{ikW(S,\theta)} \tag{8.24}$$

where $E(S, \theta)$ is the amplitude distribution on the entrance pupil, which is in general constant but not always. $W(S,\theta)$ is the wavefront shape in the presence of the aberrations in the optical system. Using polar coordinates, this diffraction image in the presence of aberrations may be written as

$$U(\theta_x,\theta_y) = \frac{1}{\pi} \iint_\sigma E(x,y) e^{ik[x\sin\theta_x + y\sin\theta_y + W(x,y)]} \, dx \, dy \tag{8.25}$$

or, in polar coordinates, as

$$U(\theta_\rho,\phi) = \frac{1}{\pi} \int_0^a \int_0^{2\pi} E(S,\theta) e^{ik[S\sin\theta_\rho \cos(\theta-\phi) + W(S,\theta)]} \, S \, dS \, d\theta \tag{8.26}$$

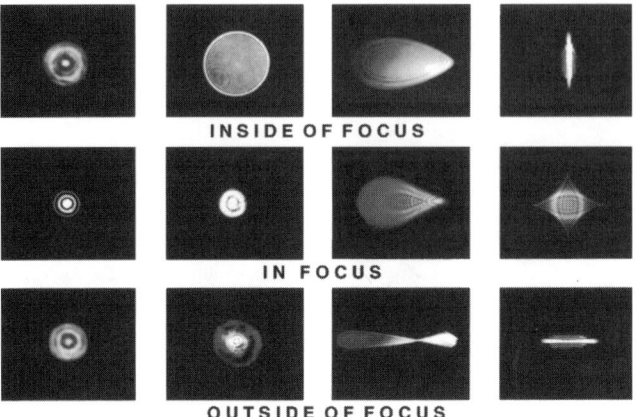

Figure 8.12 Fraunhofer diffraction images in the presence of aberrations.

The aberrated images of a point source in the presence of aberrations have a typical characteristic structure. Beautiful photographs may be found in the book by Cagnet *et al.* (1962). The influence of the aberrations on the images, taking into account diffraction, have been studied by several authors (Barakat, 1961; Barakat and Houston, 1964a; Maréchal, 1947). Figure 8.12 shows examples of diffraction images in the presence of primary aberrations.

8.5 STREHL RATIO

When the aberrations are large the image size is larger than the Airy disk. Since the amount of light forming the image is the same with and without aberrations, the irradiance at the center of the images has to decrease when the image size increases. From Eq. (8.26), assuming a constant amplitude illumination over the entrance pupil, the amplitude at the center of the image (optical axis) is

$$U(\theta_\rho,\phi) = \frac{1}{\pi} \int_0^1 \int_0^{2\pi} e^{ikW(\rho,\theta)} \rho \, d\rho \, d\theta \qquad (8.27)$$

where, as explained before, the quantity $1/\pi$ in front of the integral is a normalization factor to make the amplitude equal to one when there are no aberrations, and the entrance pupil has a semidiameter equal to one.

The Strehl ratio is defined as the ratio of the irradiance at the center of the aberrated diffraction image to that of a perfect image and it is given by

$$\text{Strehl ratio} = \left| \frac{1}{\pi} \int_0^1 \int_0^{2\pi} e^{ikW(\rho,\theta)} \rho \, d\rho \, d\theta \right|^2$$

$$= \left| \frac{1}{\pi} \int_0^1 \int_0^{2\pi} \cos kW(\rho,\theta) \rho \, d\rho \, d\theta \right.$$

$$\left. + \frac{i}{\pi} \int_0^1 \int_0^{2\pi} \sin kW(\rho,\theta) \rho \, d\rho \, d\theta \right| \quad (8.28)$$

which, by assuming that $W(\rho,\theta)$ is small compared with λ, may also be written as

$$\text{Strehl ratio} = \left| \frac{1}{\pi} \int_0^1 \int_0^{2\pi} [1 - \frac{1}{2}(kW)^2 + ikW] \rho \, d\rho \, d\theta \right|^2 \quad (8.29)$$

and transformed by separating each term into a different integral, to

$$\text{Strehl ratio} = \left| 1 - \frac{k^2}{2\pi} \int_0^1 \int_0^{2\pi} W^2 \rho \, d\rho \, d\theta + \frac{ik}{\pi} \int_0^1 \int_0^{2\pi} W \rho \, d\rho \, d\theta \right|^2$$

$$\simeq \left[1 - \frac{k^2}{2\pi} \int_0^1 \int_0^{2\pi} W^2 \rho \, d\rho \, d\theta \right]^2 + \frac{2k}{\pi} \left[\int_0^1 \int_0^{2\pi} W \rho \, d\rho \, d\theta \right]^2$$

$$\simeq 1 - \frac{k^2}{\pi} \int_0^1 \int_0^{2\pi} W^2 \rho \, d\rho \, d\theta + k^2 W_{av}^2 \quad (8.30)$$

However, from the definition of the wavefront variance σ_W^2 in Eq. (8.22), we obtain

$$\text{Strehl ratio} \simeq 1 - k^2 \sigma_W^2 \quad (8.31)$$

We see that the Strehl ratio is a function only of the wavefront variance, or the square of the rms wavefront deviation. This expression is valid for Strehl ratios as low as 0.5.

8.6 OPTICAL TRANSFER FUNCTION

Another method to specify the resolving power of an optical imaging system is by means of the *optical transfer function* (OTF), described and studied by many authors (Baker, 1992; Barakat, 1962b, 1964; Barakat and Houston, 1963a,b, 1965; Barakat and Levin, 1963a,b; Barnes, 1971; Hopkins, 1957; Linfoot, 1955, 1956, 1964; Smith, 1963; Wolf, 1952). This function is defined as the contrast in the image of a sinusoidal grating with a given spatial frequency, defined by

$$\omega = \frac{2\pi}{L} \tag{8.32}$$

Let us assume that we form the image of an object containing a wide spectrum of spatial frequencies and then analyze the frequency content in the image of this object. Then, the OTF is the ratio of the amplitude of a given spatial frequency in the image to the amplitude of the component with the same spatial frequency in the object. If the object contains all spatial frequencies with a constant amplitude, the OTF becomes the Fourier transform of the image. Such an object is a point object and its image is *point spread function* (PSF). Hence, the OTF is simply the Fourier transform of the point spread function. If $T(x,y)$ is the amplitude (and phase) distribution on the exit pupil of the optical system, called the *pupil function* (PF), from Eq. (8.14), the amplitude distribution on the image of a point object, called the *amplitude point spread function* (APSF), is given by

$$A(x_F, y_F) = \frac{1}{\pi} \int\!\!\int_\sigma T(x,y) e^{ik(xx_F + yy_F)/r_W}\, dx dy \tag{8.33}$$

where (x_F, y_F) are the coordinates in the focal plane and r_W is the radius of curvature of the wavefront at the exit pupil. The integration is made over the exit pupil area σ. The pupil function $T(x,y)$ may be written as

$$T(x,y) = E(x,y) e^{ikW(x,y)} \tag{8.34}$$

where $E(x,y)$ represents the amplitude distribution (without any phase information) over the exit pupil and $W(x,y)$ is the wavefront deformation on this pupil. The *PSF* is then given by the complex square of the amplitude in the image:

$$S(x_F, y_F) = A(x_F, y_F) A^*(x_F, y_F) \tag{8.35}$$

and substituting here the value of $A(x, y)$ we obtain

$$S(x_F,y_F) = \frac{1}{\pi^2} \iint T(x,y) e^{ik(xx_F+yy_F)/r_W} \, dx \, dy$$
$$\times \iint T^*(x,y) e^{-ik(xx_F+yy_F)/r_W} \, dx \, dy \tag{8.36}$$

The variables x and y in the second integral are replaced by x' and y', in order to move everything under the four integral signs, obtaining

$$S(x_F,y_F) = \frac{1}{\pi^2} \iiiint T(x,y)$$
$$\times T^*(x',y') e^{-ik(x_F(x-x')+y_F(y-y'))/r_W} \, dx \, dy \, dx' \, dy' \tag{8.37}$$

Once the point spread function is computed by means of this expression, or by any other procedure, the optical transfer function $F(\omega_x, \omega_y)$ may be obtained from the Fourier transform of the point spread function $S(x,y)$ as follows:

$$F(\omega_x,\omega_y) = \iint_\sigma S(x_F,y_F) e^{i(\omega_x x_F, \omega_y y_F)} \, dx_F \, dy \tag{8.38}$$

We see that in general this OTF is complex and, thus it has a real and an imaginary term. The modulus of the OTF is called the *modulation transfer function* (MTF) and represents the contrast in the image of a sinusoidal periodic structure. The imaginary term receives the name of *phase transfer function* (PTF) and gives information about the spatial phase shifting or any contrast reversal (when the phase shift is 180°) in the image. Since the OTF is the Fourier transform of a real function, it is hermitian. This means that the real part is symmetrical and the imaginary part is antisymmetrical. In other words,

$$F(\omega_x,\omega_y) = F^*(-\omega_x,\omega_y); \; F(\omega_x,\omega_y) = F^*(\omega_x, -\omega_y) \tag{8.39}$$

Due to the symmetry of the optical system about the meridional plane, the point spread function satisfies the condition:

$$S(x_F,y_F) = S(-x_F,y_F) \tag{8.40}$$

thus, we may show that $F(\omega_x,0)$ is real. If the PSF has the additional property that

$$S(x_F, y_F) = S(x_F, -y_F) \tag{8.41}$$

as in any rotationally symmetric aberration like defocusing, spherical aberration, or astigmatism, the optical transfer function $F(\omega_x,\omega_y)$ is real.

There is an alternative method to obtain the OTF if the pupil function $T(x,y)$ is known. To show this method let us assume that we know the optical transfer function $F(\omega_x,\omega_y)$; then, the PSF would be the inverse Fourier transform as follows:

$$S(x_F, y_F) = \frac{1}{4\pi^2} \iint_\sigma F(\omega_x, \omega_y) e^{i(\omega_x x_F + \omega_y y_F)} \, d\omega_x \, d\omega_y \tag{8.42}$$

Now, except for a constant, Eqs. (8.37) and (8.42) are identical if we set

$$\omega_x = \frac{k}{r_W}(x - x') \tag{8.43}$$

$$\omega_y = \frac{k}{r_W}(y - y') \tag{8.44}$$

and

$$F(\omega_x, \omega_y) = \frac{4r_W^2}{k^2} \iint_\sigma T(x,y) T^*\left(x - \frac{r_W}{k}\omega_x, y - \frac{r_W}{k}\omega_y\right) dx\, dy \tag{8.45}$$

We may now use this expression to compute the OTF by the convolution of the pupil function. If the entrance pupil is illuminated with a constant amplitude light beam we have $E(x,y) = 1$. Then, we may write

$$F(\omega_x, \omega_y) = \frac{4r_W^2}{k^2} \iint_\sigma \exp ik\left(W(x,y) - W\left(x - \frac{r_W}{k}\omega_x, y - \frac{r_W}{k}\omega_y\right)\right) dx\, dy$$

$$\tag{8.46}$$

Figure 8.13 shows schematically the mutual relations between the main functions described in this section. The constant in front of Eqs. (8.45) and (8.46) is ignored and substituted by another such that the OTF at the origin ($\omega_x = \omega_y = 0$) is real and equal to one. This is equivalent to setting the total energy in the PSF equal to one. This is the normalized OTF.

Diffraction in Optical Systems

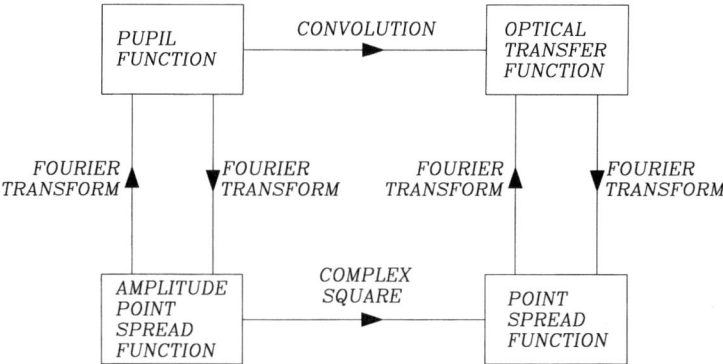

Figure 8.13 Transformation relations for some mathematical operations.

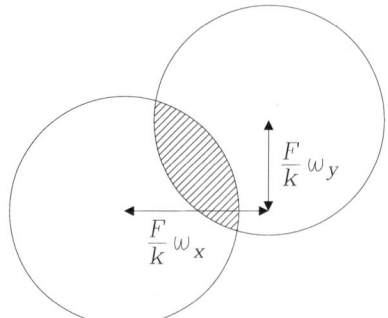

Figure 8.14 Two identical mutually displaced pupils for calculating the optical transfer function.

An interesting case to study is when the optical system is free of aberrations. Then, $W(x,y) = 0$, inside the exit pupil clear parts, and the OTF is the common area of two images of the exit pupil laterally displaced with respect to each other, as shown in Fig. 8.14. We may also see that the maximum spatial frequency of this function is

$$\omega_x = \frac{kD}{r_W} \tag{8.47}$$

but using now the definition of spatial frequency in Eq. (8.32), we see that the linear resolving power is given by

$$L \geq \frac{\lambda r_W}{D} \tag{8.48}$$

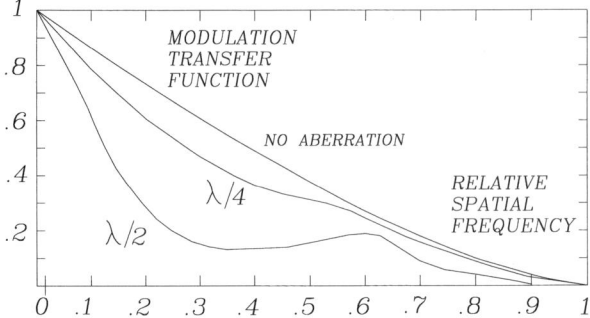

Figure 8.15 Modulation transfer function for a perfect system and for systems with a small amount of spherical aberration.

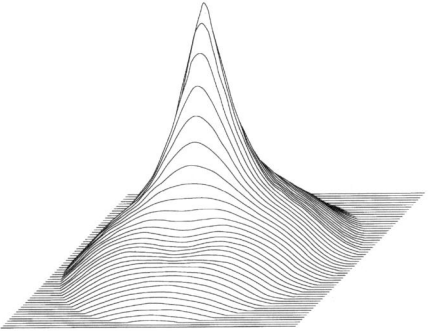

Figure 8.16 Isometric representation of a modulation transfer function for an off-axis image.

where D is the diameter of the exit pupil. This expression is almost the same as Eq. (8.21).

Figure 8.15 shows some the normalized MTF for a perfect lens with several amounts of defocusing. The OTF for an annular aperture has been studied by O'Neill (1956).

The MTF off-axis does not necessarily have rotational symmetry. Then, the normalized MTF has to be represented by a surface, as shown in Fig. 8.16.

8.6.1 Optical Transfer Function and Strehl Ratio

From Eq. (8.42), the Strehl ratio is given by

$$\text{Strehl ratio} = S(0,0) = \frac{1}{4\pi^2} \iint_\sigma F(\omega_x, \omega_y) \, d\omega_x \, d\omega_y \qquad (8.49)$$

Diffraction in Optical Systems

and using now the property that the OTF is hermitian, and hence the real part is symmetrical and the imaginary part is antisymmetrical, we may prove that

$$\text{Strehl ratio} = \frac{1}{4\pi^2} \iint_\sigma \Re[F(\omega_x,\omega_y)] \, d\omega_x \, d\omega \tag{8.50}$$

where the symbol $\Re$ represents the real part of a function.

The Strehl ratio and the wavefront variance are directly related to each other by Eq. (8.31). Thus, we may conclude that the wavefront variance determines the area under the surface representing the real part of the OTF. On the other hand, since this MTF at high spatial frequencies increases with this volume, the response of the system to such frequencies is determined only by the wavefront variance.

8.7 RESOLUTION CRITERIA

There are many resolution criteria that can be used to specify the quality of an optical system or to specify construction tolerances. Next, we will describe a few of these criteria. An important variable when defining the resolution of an optical system is the image detector being used. It is not the same to detect the image with a photographic plate as with the eye.

If we have two close point objects, their images will also be close to each other and their diffraction images may overlap. This overlapping may be so large that only one image is observed. The problem then is to define how close these images may be and still detect two separate images. There are several different criteria, applicable to different conditions. For example, if the two interfering images are completely coherent to each other, as in the case of two stars in a telescope, the irradiances of the two images will add. On the other extreme, if the object being imaged is illuminated with coherent light, e.g., with a laser, the amplitudes will add to their corresponding phase, producing a different result. There may also be intermediate situations, as in a microscope, where the object is illuminated with partially coherent light. One more variable that should be taken into account is the ratio of the intensities of the two images.

One of these resolution criteria is the Rayleigh criterion (Barakat, 1965; Murty, 1945), which applies to incoherent images with equal irradiances. It says that two images are just resolved when separated by their Airy disk radius, as shown in Fig. 8.17. It should be pointed out here that strictly speaking this Rayleigh criterion assumes that two neighboring points in an image are incoherent to each other, so that their intensities and

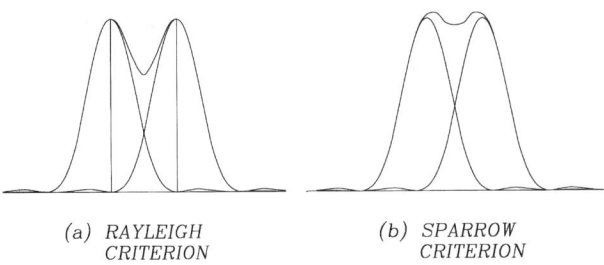

Figure 8.17 Rayleigh and Sparrow resolution criteria: (a) Rayleigh criterion; (b) Sparrow criterion.

not their amplitudes have to be added. This means that the luminous object has to be spatially incoherent or if it is opaque it has to be illuminated with a spatially incoherent light beam. This hypothesis is clearly valid in the case of telescopes and photographic cameras, but not completely in the case of microscopes.

There is another resolution limit frequently used by astronomers, called the Sparrow criterion (Sparrow, 1916). According to the astronomer Dawes, two stars with the same intensity may be separated when observing with the naked eye in a telescope, when they are actually closer than the Rayleigh criterion suggests. This separation is about 0.84 the radius of the Airy disk.

Ideally, the wavefront in an optical system forming an image of a point object should be spherical. Even if this wavefront is spherical the image is not a point but a diffraction image with some rings around, known as the Airy disk, as we have studied before. This diffraction image finite size limits the resolution of perfect optical systems. Lord Rayleigh in 1878 pointed out that the diffraction image remains almost unchanged if the converging wavefront deviates from a perfect sphere by less than about one-quarter of a wavelength. This is the Rayleigh limit, which is widely used by lens designers as an aid in setting optical tolerances.

It has been found that wavefront deviations of up to twice the Rayleigh limit in the central disk in the diffraction image do not substantially increase the image in diameter. However, the image contrast may decrease due to the presence of a halo around the central image.

The local wavefront deformations can be smaller than one-quarter of the wavelength, but the transverse aberrations (wavefront slopes) can be very high. This is why halos are produced. This quarter of a wave criterion is quite useful, but with very complicated wavefront deformations a better image analysis may be necessary. These effects have been studied by Maréchal and Françon (1960). The Maréchal criterion (1947) establishes that the image degradation due to the presence of aberrations is not

Diffraction in Optical Systems

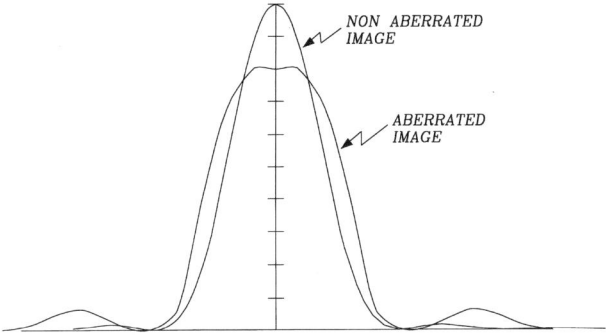

Figure 8.18 Strehl resolution criterion

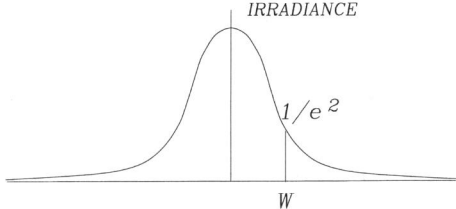

Figure 8.19 Gaussian irradiance distribution across a gaussian beam.

noticeable when the Strehl value is greater than 0.8. (See Fig. 8.18). This Strehl value corresponds to an rms wavefront deformation of about $\lambda/14$. The Maréchal criterion coincides with the Rayleigh criterion for the case of spherical aberration.

When the transverse aberrations are large compared with the Airy diffraction image, the effects of diffraction may be neglected. In this case, tolerances are frequently based on the OTF.

8.8 GAUSSIAN BEAMS

A gaussian beam is one in which its amplitude and irradiance have a distribution with rotational symmetry and decreases from the optical axis to the edge with a gaussian shape, as shown in Fig. 8.19. Then, the amplitude would be represented by

$$A(\rho) = E_0 e^{-\rho^2/w^2} \tag{8.51}$$

and the irradiance by

$$I(\rho) = I_0 e^{-2\rho^2/w^2} \tag{8.52}$$

where ρ is the distance from the point being considered to the optical axis and, w is the value of ρ when the irradiance is $1/e^2$ of its axial value. These beams appear in the light beams emited by gas lasers, and have very interesting and important properties that have been studied by Kogelnik (1959, 1979). A spherical convergent gaussian wavefront becomes flat and gaussian at the focus. This is easy to understand if we remember that the Fourier transform of a gaussian function is also a gaussian function. After going through this focus the wavefront diverges again with an spherical shape and a gaussian distribution of amplitudes. As shown in Fig. 8.20, the beam is perfectly symmetrical, with the center of symmetry at the focus. This focus or minimum diameter of the beam is called a waist. The semidiameter w_0 of the waist is related to the angle of convergence θ by

$$\theta = \frac{\lambda}{\pi w_0} \tag{8.53}$$

Far from the waist or focus, the center of curvature of the wavefront is at the center of this waist, but as the wavefront gets closer, it becomes flatter. Finally, at the waist, it is perfectly flat. If we define the Rayleigh range as

$$z_R = \frac{w_0}{\theta} \tag{8.54}$$

then, at a distance z from the waist the radius of curvature R of the wavefront is

$$R = z + \frac{\pi^2 w_0^4}{\lambda^2 z} = z + \frac{z_R^2}{z} \tag{8.55}$$

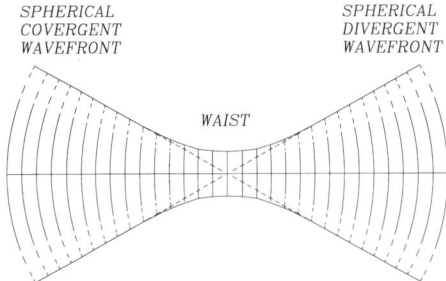

Figure 8.20 Propagation of a gaussian beam.

Diffraction in Optical Systems 213

We then see that the center of curvature of the wavefront is at a distance from the waist given by the last term in this expression. This center of curvature is at the waist only when

$$z \gg z_R = \frac{w_0}{\theta} \tag{8.56}$$

The semidiameter w at a distance z from the waist is given by

$$w = \left[w_0^2 + \frac{\lambda^2 z^2}{\pi^2 w_0^2} \right]^{1/2} = w_0 \left[1 + \frac{z^2}{z_R^2} \right]^{1/2} \tag{8.57}$$

8.8.1 Focusing and Collimating a Gaussian Beam

Gaussian beams with a large angle of convergence are focused in a small spot, and beams with a small angle of convergence are focused in a large spot. The minimum spot size is at the gaussian waist, not at the focus of the lens. The focus of the lens is at the center of curvature of the wavefront, at a distance R from this wavefront.

Due to diffraction a flat wavefront with finite extension cannot keep its flatness along its traveling path, due to diffraction effects. A wavefront with a round shape and constant amplitude will diffract, producing a divergent wavefront with a complicated shape. A flat gaussian beam is also affected by diffraction, producing a divergent beam. In this case, however, the wavefront is always spherical, with the center of curvature getting closer to the center of the initial flat wavefront (waist) as it travels.

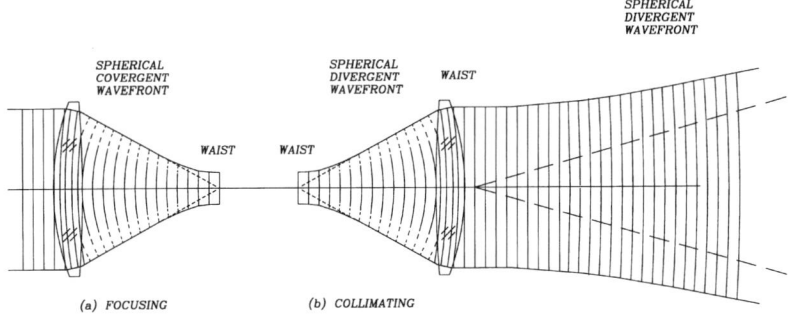

Figure 8.21 Focusing and collimation of a gaussian beam: (a) Focusing; (b) collimating.

The divergence angle will also approach the value given by Eq. (8.33) at distances far from the waist. Thus, the divergence is larger for smaller, initially flat, wavefronts. If a gaussian beam is collimated with a lens, another gaussian beam with its waist at the exit pupil of the lens is produced. The center of curvature of the entering wavefront must be at the focus of the lens, not at the waist, as shown in Fig. 8.21. To reduce the divergence angle, the diameter of the waist must be increased by means of an inverted telescope.

REFERENCES

Baker, L., ed., *Selected Papers on Optical Transfer Function: Foundation and Theory*, SPIE Optical Engineering Press, Bellingham, WA, 1992.

Barakat, R., "Total Illumination in a Diffraction Image Containing Spherical Aberration," *J. Opt. Soc. Am.*, **51**, 152–157 (1961).

Barakat, R., "Application of Apodization to Increase Two-Point Resolution by the Sparrow Criterion: I. Coherent Illumination," *J. Opt. Soc. Am.*, **52**, 276–283 (1962a).

Barakat, R., "Computation of the Transfer Function of an Optical System from the Design Data for Rotationally Symmetric Aberrations: I. Theory," *J. Opt. Soc. Am.*, **52**, 985–991 (1962b).

Barakat, R., "Numerical Results Concerning the Transfer Functions and Total Illuminance for Optimum Balanced Fifth-Order Spherical Aberration," *J. Opt. Soc. Am.*, **54**, 38–44 (1964).

Barakat, R., "Rayleigh Wavefront Criterion," *J. Opt. Soc. Am.*, **55**, 641–649 (1965).

Barakat, R. and Houston, A., "Reciprocity Relations Between the Transfer Function and Total Illuminance 1," *J. Opt. Soc. Am.*, **53**, 1244–1249 (1963a).

Barakat, R. and Houston, A., "Modulation of Square Wave Objects in Incoherent Light," *J. Opt. Soc. Am.*, **53**, 1371–1376 (1963b).

Barakat, R. and Houston, A., "Diffraction Effects of Coma," *J. Opt. Soc. Am.*, **54**, 1084–1088 (1964).

Barakat, R. and Houston, A., "Transfer Function of an Annular Aperture in the Presence of Spherical Aberration," *J. Opt. Soc. Am.*, **55**, 538–541 (1965).

Barakat, R. and Levin, E., "Application of Apodization to Increase Two-Point Resolution by the Sparrow Criterion: II. Incoherent Illumination," *J. Opt. Soc. Am.*, **53**, 274–282 (1963a).

Barakat, R. and Levin, E., "Transfer Functions and Total Illuminance of High Numerical Aperture Systems Obeying the Sine Condition," *J. Opt. Soc. Am.*, **53**, 324–332 (1963b).

Barnes, K. R., *The Optical Transfer Function*, American Elsevier, New York, 1971.

Born, M. and Wolf, E., *Principles of Optics*, Pergamon Press, New York, 1964.

Cagnet, M., Françon, M., and Thrierr, J., *Atlas of Optical Phenomena*, Prentice Hall, Englewood Cliffs, NJ, 1962.

Hopkins, H. H., "The Numerical Evaluation of the Frequency Response of Optical Systems," *Proc. Phys. Soc. (London)*, **B70**, 1002–1005 (1957).
Kogelnik, H. W., "On the Propagation of Gaussian Beams of Light Through Lenslike Media Including those with a Loss or Gain Variation," *Appl. Opt.*, **4**, 1562–1569 (1959).
Kogelnik, H. W., "Propagation of Laser Beams," in *Applied Optics and Optical Engineering*, Vol. VII, Chap. 6, R. R. Shannon and J. C. Wyant, eds., Academic Press, San Diego, CA, 1979.
Linfoot, E. H., "Information Theory and Optical Images," *J. Opt. Soc. Am.*, **45**, 808–819 (1955).
Linfoot, E. H., "Transmission Factors and Optical Design," *J. Opt. Soc. Am.*, **46**, 740–752 (1956).
Linfoot, E. H., *Fourier Methods in Optical Design*, Focal, New York, 1964.
Malacara, D., "Diffraction and Scattering," Chap. IV in *Methods of Experimental Physics, Physical Optics and Light Measurements*, Vol. 26, D. Malacara, ed., Academic Press, San Diego, CA, 1988.
Maréchal, A., "Etude des Effets Combine de la Diffraction et des Aberration Geometriques sur l'Image d'un Point Lumineux," *Rev. Opt.*, **9**, 257–297 (1947).
Maréchal, A. and Françon, M., *Diffraction. Structure Des Images*, Editions de la Revue D'Optique Theorique et Instrumentale, Paris, 1960.
Murty, M. V. R. K., "On the Theoretical Limit of Resolution," *J. Opt. Soc. Am.* **47**, 667–668 (1945).
O'Neill, E. L., "Transfer Function for an Annular Aperture," *J. Opt. Soc. Am.*, **46**, 258–288 (1956).
Smith, W. F., "Optical Image Evaluation and the Transfer Function," *Appl. Opt.*, **2**, 335–350 (1963).
Sparrow, C. M. "On Spectroscopy Resolving Power," *Astrophys. J.*, **44**, 76–86 (1916).
Stoltzmann, D. E., "The Perfect Point Spread Function," in *Applied Optics and Optical Engineering*, Vol. IX, R. R. Shannon and J. C. Wyant, eds., Academic Press, San Diego, CA, 1980.
Taylor, C. A. and Thompson, B. J., "Attempt to Investigate Experimentally the Intensity Distribution near the Focus in the Error-Free Diffraction Pattern of Circular and Annular Apertures," *J. Opt. Soc. Am.*, **48**, 844–850 (1958).
Welford, W. T., "Use of Annular Apertures to Increase Focal Depth," *J. Opt. Soc. Am.*, **50**, 749–753 (1960).
Wolf, E., "In a New Aberration Function of Optical Instruments," *J. Opt. Soc. Am.*, **42**, 547–552 (1952).

9
Computer Evaluation of Optical Systems

9.1 MERIDIONAL RAY TRACING AND STOP POSITION ANALYSIS

In general, the full real aperture of a lens system has to be greater than the diameter of the stop, as shown in Fig. 9.1, in order to allow light beams from off-axis object points to enter the lens. On-axis, the effective clear aperture is smaller than the lens. If we obtain the meridional plot for the full lens aperture with an enlarged stop in contact with the first surface of the lens, or at any other selected place, we obtain a graph like that in Fig. 9.1. After placing the stop with the correct diameter in its final place, only one region of this graph will be used. This kind of enlarged meridional plot is very useful in many systems to determine the best position for the final stop and, by analyzing it, we may obtain the following information:

1. The height TA_0 is the lateral image displacement due to distortion.
2. The difference in heights TA_0 for several colors is the magnification chromatic aberration, if plots for different wavelengths are obtained.
3. The slope of the plot in the region selected is an indication of a local off-axis meridional defocusing or, in other words, the tangential field curvature.
4. The curvature of the plot is an indication of the magnitude of the tangential coma.
5. An **S**-shaped plot (cubic component) represents the spherical aberration.
6. At the minimum M_1 or at the maximum M_2 of this plot the tangential field is flat (zero slope), but there is coma (curvature of the plot).
7. The maximum and minimum regions may have different amounts of distortion if their plot heights are different.

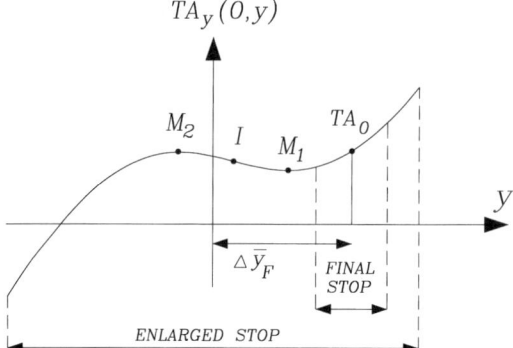

Figure 9.1 Meridional plot of an optical system using an enlarged stop.

8. At the inflection point **I** there is no coma (curvature zero), but there is tangential field curvature (slope different from zero).
9. If the inflection point has zero slope the tangential field is flat and there is no coma. This condition may be frequently obtained by bending of the lenses.

Once the final principal ray height has been selected from examination of this plot, the final position for the stop has to be calculated by any desired procedure. One method is by tracing a meridional ray and the preliminary principal ray with the enlarged stop and then using the stop shift relation in Eq. (3.63). A position for the stop has to be found such that the principal ray at the preliminary enlarged stop moves to the desired height. The y–$\bar{y}$ diagram described in Chap. 3 may also be used to perform the stop shift. For the particular case of a single thin or thick lens with an object at infinity, the preliminary enlarged stop may be placed at the first surface, as shown in Fig. 9.2. Then, the final stop position may be found from the selected final principal ray height. If this ray height is positive, the final stop is in front of the lens, at a distance from the front surface, given by

$$\bar{l}_F = -F \frac{\bar{y}_F}{h'} \tag{9.1}$$

or, if the ray height is negative, the pupil is in the back of the lens, at a distance from the last surface, given by

$$\bar{l}_B = F_B - \frac{2F - F_F}{[1 - (\bar{y}_B/h')]} \tag{9.2}$$

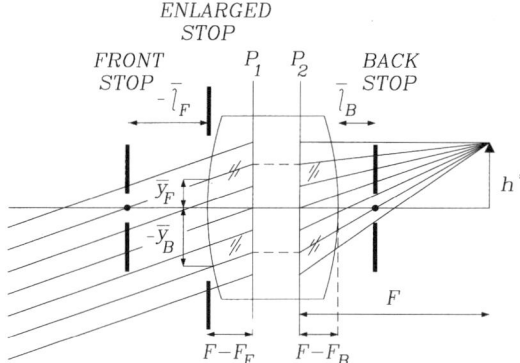

Figure 9.2 Calculation of the stop position after using the enlarged stop.

where F is the effective focal length, F_F is the front focal length, F_B is the back focal length and h' is the image height.

9.2 SPOT DIAGRAM

The prediction of the image quality of optical systems can be performed with several different procedures, as will be explained in this chapter (Wetherell, 1980). If the transverse aberrations are much larger than the Airy disk diameter, so that we may neglect diffraction effects, the geometrical or ray evaluation is enough to predict the actual performance of the lens system. One of several geometric methods is the spot diagram. A rectangular or polar array of rays is traced through the entrance pupil of the optical system, as shown in Fig. 9.3. Then, the intersection of these rays with the focal plane is plotted. This plot, called the spot diagram (Herzberger, 1947, 1957; Linfoot, 1955; Lucy, 1956; Miyamoto, 1963; Stavroudis and Feder, 1954), represents the values of the transverse aberrations TA_x and TA_y for each ray. Due to the symmetry of the system the spot diagram is also symmetrical about the y axis.

The spot diagrams give a visual representation of the energy distribution in the image of a point object. Figure 9.4 shows some spot diagrams traced with a rectangular array of rays. If several spot diagrams are obtained, for different colors, the chromatic aberration may also be evaluated.

9.2.1 Geometrical Spot Size

The spot diagram data may be used to obtain useful information regarding the quality of the image, as the geometrical spot size of the image and the

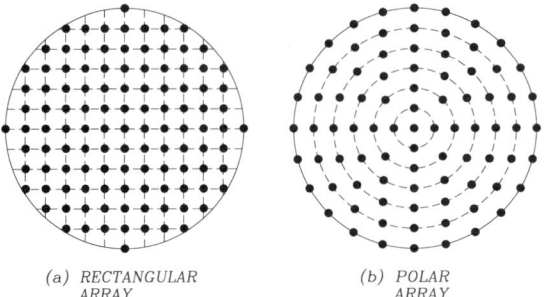

(a) RECTANGULAR ARRAY   *(b) POLAR ARRAY*

Figure 9.3 Rectangular and polar arrays of rays on the entrance pupil, to obtain the spot diagram: (a) rectangular array; (b) polar array.

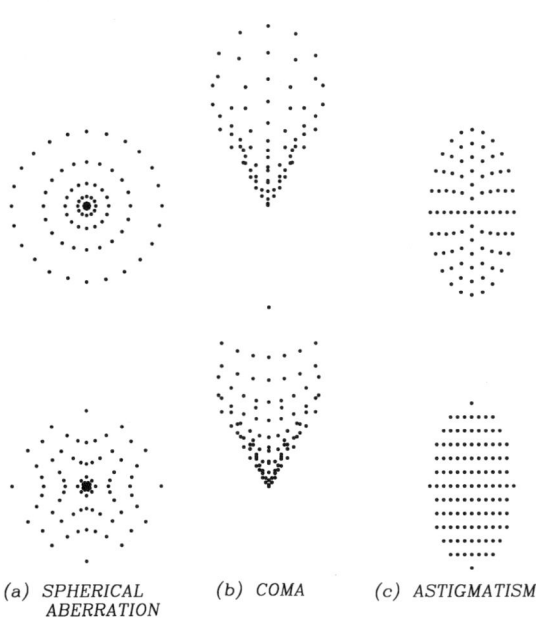

(a) SPHERICAL ABERRATION   *(b) COMA*   *(c) ASTIGMATISM*

Figure 9.4 Spot diagrams for some primary aberrations, using polar and rectangular arrays: (a) spherical aberration; (b) coma; (c) astigmatism.

radial energy distribution, as we will now describe. With the definition of the radial transverse aberration as

$$TA_\rho^2 = TA_x^2 + TA_y^2 \tag{9.3}$$

where we must remember that these ray transverse aberrations TA_x and TA_y are measured with respect to the gaussian image, for the yellow line (d or e). The average transverse aberration is the height of the centroid of the aberrated image, also measured with respect to the gaussian image. Taking into account the symmetry about the meridional plane, it may be written as

$$\overline{TA} = \frac{\sum_{i=1}^{N} TA_y}{N} \qquad (9.4)$$

where the sum is performed over all the rays in the spot diagram. The geometrical spot size is easily defined in terms of the transverse aberrations, by the variance of the transverse aberrations, which is the square of the root mean square spot size TA_{rms}, as follows:

$$TA_{rms}^2 = \frac{\sum_{i=1}^{N}[TA_x^2 + (TA_y - \overline{TA})^2]}{N} = \frac{\sum_{i=1}^{N} TA_\rho^2}{N} - \overline{TA}^2 \qquad (9.5)$$

In this expression, the reference for the calculation of the image size is its centroid, which, in general, is not at the gaussian image. If the reference is the gaussian image, the last term must be made equal to zero. The accuracy of expressions (9.4) and (9.5) is critically dependent on two factors, as shown by Forbes (1988). The first one is the type of array being used to trace the rays for the spot diagram, and the second is the number of rays. The exact result is obviously obtained only when the number of rays tends to infinity.

If we have an infinite number of rays, so that the transverse aberration $TA_\rho(\rho,\theta)$ is a continuous function, Eqs. (9.4) and (9.5) become

$$\overline{TA} = \frac{1}{\pi} \int_0^1 \int_0^{2\pi} TA_y(\rho,\theta)\, \rho\, d\rho\, d\theta \qquad (9.6)$$

and

$$TA_{rms}^2 = \frac{1}{\pi} \int_0^1 \int_0^{2\pi} [TA_x^2 + (TA_y - \overline{TA})^2] \rho\, d\rho\, d\theta$$

$$= \frac{1}{\pi} \int_0^1 \int_0^{2\pi} TA_\rho^2(\rho,\theta)\, \rho\, d\rho\, d\theta - \overline{TA}^2 \qquad (9.7)$$

This expression may be accurately evaluated using gaussian quadrature integration with a procedure by Forbes (1988), to be described here.

Now, let us place the rays on the entrance pupil in rings, all with the same number of rays, with a uniform distribution given by

$$\theta_k = \frac{\pi(k - 1/2)}{N_\theta} \tag{9.8}$$

where N_θ is the number of points in one-half of a circle. It is easy to see that the angular dependence of the transverse aberration function $TA_\rho(\rho,\theta)$ is very smooth. This integral may be accurately represented by

$$TA_{rms}^2 = \frac{2}{N_\theta} \sum_{k=1}^{N_\theta} \int_0^1 TA_\rho^2(\rho,\theta_k)\, \rho\, d\rho - \overline{TA}^2 \tag{9.9}$$

On the other hand, for a centered system the transverse aberration function $TA_\rho(\rho,\theta)$ is symmetrical about the meridional plane; hence, in an analogous manner to the proof in Section 7.1 for the wavefront aberrations, we may also prove that this transverse aberration function contains only terms with ρ^2 and with $r \cos \theta$. Also, because of this symmetry, all odd powers of $\cos \theta$ add to zero in the angular sum. This means that we have to consider in this integral only even powers of ρ. If we define for convenience a new variable $\sigma = \rho^2$, we obtain

$$TA_{rms}^2 = \sum_{k=1}^{N_\theta} \int_0^1 \frac{1}{N_\theta} TA_\rho^2(\sigma^{1/2},\theta_k) d\sigma - \overline{TA}^2 \tag{9.10}$$

The gaussian method of integration permits us to evaluate a definite integral as follows:

$$\int_0^1 f(x) dx = \sum_{j=1}^N w_j f(x_j) \tag{9.11}$$

where $f(x)$ is a polynomial whose value has been sampled at N points. The gaussian sampling positions, x_i and w_i, are the gaussian weights selected to make the integral exact when the degree of the polynomial is less than or equal to $2N - 1$. Thus, using this method to evaluate the radial integral in Eq. (9.10), using again the normalized radius ρ instead of $\sigma^{1/2}$, we find that

$$TA_{rms}^2 = \sum_{k=1}^{N_\theta} \sum_{j=1}^{N_\rho} w_j(\rho) TA_\rho^2(\rho_j,\theta_k) - \overline{TA}^2 \tag{9.12}$$

with N_r being the number of rings where the gaussian sampling points are located. If we trace the spot diagram with only nine rays, with $N_r = 3$ and $N_\theta = 3$, as shown in Fig. 9.5(a) and suggested by Forbes (1988), we obtain an accuracy of 1%. The gaussian sampling points and the weights for this ray configuration are presented in the Table 9.1. The weights $w_j(\rho)$ have been normalized so that $N_\theta[w_1(\rho) + w_2(\rho) + w_3(\rho)] = 1$.

This distribution of the rays on the entrance pupil just described is not totally satisfactory, because the central ray in the pupil or principal ray is not included. Forbes has described an alternative scheme called Radau integration that solves this problem. The distribution of rays in the pupil for the use of the Radau integration is shown in Fig. 9.5(b). The Radau constants to be used in Eq. (9.12) are listed in Table 9.2, where the weights $w_j(\rho)$ have been normalized so that $N_\theta[w_1(\rho) + w_2(\rho) + w_3(\rho) + w_4(\rho)] = 1$.

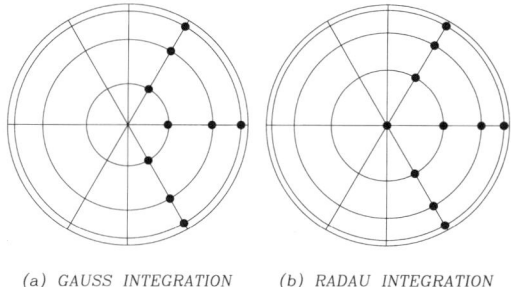

(a) GAUSS INTEGRATION   (b) RADAU INTEGRATION

Figure 9.5 Distribution of rays in the pupil for Radau integration: (a) Gauss integration; (b) Radau integration.

Table 9.1 Gaussian Integration Parameters for $N_r = 3$ and $N_\theta = 3$

j	$w_j(\rho)$	ρ_j
1	0.09259259	0.33571069
2	0.14814815	0.70710678
3	0.09259259	0.94196515

Table 9.2 Radau Integration Parameters for $N_r = 4$ and $N_\theta = 3$

j	$w_j(\rho)$	ρ_j
1	0.02083333	0.00000000
2	0.10961477	0.46080423
3	0.12939782	0.76846154
4	0.73487407	0.95467902

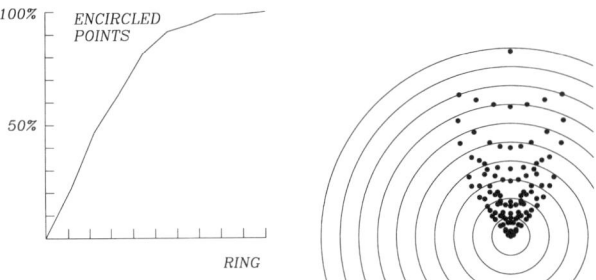

Figure 9.6 Computing encircled energy from spot diagrams.

9.2.2 Radial Energy Distribution

The radial (or encircled) energy distribution may easily be obtained from the spot diagram by counting the number of points in the diagram, inside of circles with different diameters (Barakat and Morello, 1964), as illustrated in Fig. 9.6. This is a one-dimensional result, but very valuable for estimating the resolving power of the lens system.

9.3 WAVEFRONT DEFORMATION

The wavefront shape may be obtained in many ways, as will be described in the following sections.

9.3.1 Calculation from Transverse Aberrations Data

One method to calculate the wavefront shape is by numerical integration of the transverse aberrations in the spot diagram, using relations in Section 5.7 and the trapezoidal rule, as in the Hartmann test (Ghozeil, 1992).

The integration with the trapezoidal rule is exact only if the only aberrations present are tilt, defocusing, and astigmatism. If there is spherical aberration, coma, and high-order aberrations the results may have a large error. In the Hartmann test this problem is greatly reduced by measuring the transverse aberration to be integrated, with respect to the ideal aberrated position. In lens design evaluation the analogous procedure consists of the following steps:

1. The aberration coefficients are calculated from meridional and sagittal plots using data contained in the spot diagram, as described in Chap. 8. These coefficients give us the wavefront shape assuming that only primary aberrations plus the fifth-order

spherical aberration are present. To improve these results the following steps are then performed.
2. The spot diagram corresponding to these aberration coefficients is calculated, by differentiation, in order to obtain the transverse aberrations.
3. The transverse aberrations on the actual spot diagram are subtracted from the transverse aberrations on the calculated spot diagram.
4. The transverse aberration differences are then integrated to obtain the high-order wavefront distortions. For an integration along a line parallel to the x axis we may write

$$W_{nm} = \frac{\sigma}{2r_W} \sum_{i=i_1}^{i=n} [TA(x_{i-1}, y_m) - TA(x_i, y_m)] \qquad (9.13)$$

and for an integration along a line parallel to the y axis.
5. These high-order wavefront distortions are added to the preliminary wavefront calculated in the first step. This is the desired wavefront shape.

9.3.2 Direct Calculation of the Optical Path

The wavefront deformation in an optical system can be calculated by tracing rays through the optical system (Plight, 1980). In this method the optical path traveled by the light rays through the optical system is directly obtained when doing the ray tracing. By Fermat's principle the optical path traveled by the light rays from the point object to the point image is a constant if the image is perfect. Since the image is not perfect, not all rays go to the ideal point image and the refracted wavefront is not spherical, but has some deformations as in the example in Fig. 9.7.

To calculate the wavefront deformation we first define the position and radius of curvature of a reference sphere. The natural selection is a sphere tangent to the exit pupil, with center of curvature at the gaussian

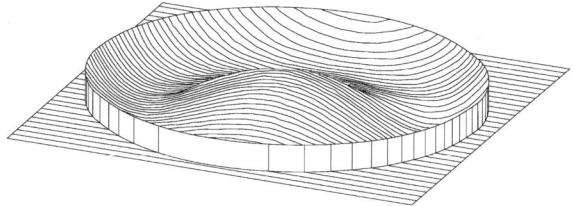

Figure 9.7 Wavefront from an optical system.

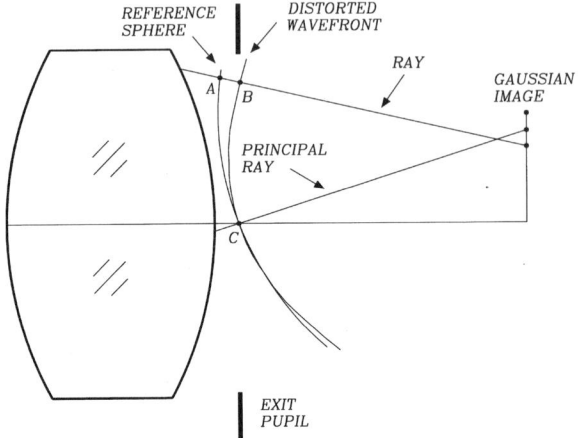

Figure 9.8 Computing the wavefront shape from optical paths.

image, as shown in Fig. 9.8. Then, by ray tracing, the rays are traced from the object point (not shown in the figure) to point **A** on the reference sphere, and the added optical path is calculated. By Fermat's principle, the optical path from the point object to point **B** on the wavefront is equal to the optical path from the point object to point **C** on the intersection of the principal ray with the wavefront. Thus, the wavefront deformation is given by the optical path along the principal ray to the point **C** on the wavefront, minus the optical path along the traced ray, to the point **A** on the reference sphere.

This method has been described by Welford (1986) and used by Marchand and Phillips (1963). The total optical path OP_{total} through the system from an off-axis point object to the reference wavefront is

$$OP_{\text{total}} = \sum_{j=0}^{k} OP_j \qquad (9.14)$$

where the first surface is number zero and the reference wavefront is surface number k. The optical path OP_j between surface j and surface $j+1$, as shown in Fig. 9.9, using Eq. A4.6, is given by

$$OP_j = n_j^2 \frac{Z_{j+1} - Z_j + t_j}{M_j} \qquad (9.15)$$

where, as defined in Section A4.1.1, M_j is the third cosine director multiplied by the refractive index, and Z_j and Z_{j+1} are the sagittas for the first and second surfaces.

Computer Evaluation of Optical Systems

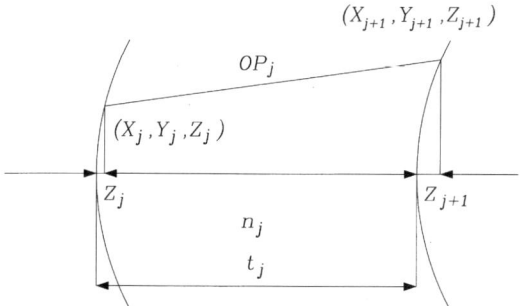

Figure 9.9 Computing optical paths.

The wavefront shape is then found by subtracting from this optical path the optical path along the principal ray as follows:

$$W(x,y) = OP_{\text{total}}(x,y) - \overline{OP}_{\text{total}}(x,y)$$
$$= \sum_{j=0}^{k} n_j^2 \left[\frac{Z_{j+1} - Z_j + t_j}{M_j} - \frac{\overline{Z}_{j+1} - \overline{Z}_j + t_j}{\overline{M}_j} \right] \quad (9.16)$$

where as usual, the bar indicates that the variable is for the principal ray.

In general, the magnitudes of these two numbers to be subtracted are as large as t_j and must be computed with an accuracy of a small fraction of a wavelength. This implies computing numbers with a precision of at least eight to nine digits, which is not easy. As suggested by Welford (1986) to solve this problem we may write

$$\frac{1}{M} - \frac{1}{\overline{M}} = \frac{\overline{M} - M}{\overline{M} M} = \frac{\overline{M}^2 - M^2}{\overline{M} M (\overline{M} + M)}$$
$$= \frac{K^2 + L^2 - \overline{K}^2 - \overline{L}^2}{\overline{M} M (\overline{M} + M)} \quad (9.17)$$

where we have used the relation $K^2 + L^2 + M^2 = n^2$. Then, substituting this result into Eq. (9.16) we finally obtain

$$W(x,y) = \sum_{j=0}^{k} n_j^2 \left[t_j \frac{K_j^2 + L_j^2 - \overline{K}_j^2 - \overline{L}_j^2}{\overline{M}_j M_j (\overline{M}_j + M_j)} + \frac{Z_{j+1} - Z_j}{M_j} - \frac{\overline{Z}_{j+1} - \overline{Z}_j}{\overline{M}_j} \right]$$

(9.18)

In this manner, the accuracy is greatly increased to an acceptable magnitude, because now the difference between very large numbers is not taken. Instead, operations with relatively small numbers are involved.

9.3.3 Conrady's Method to Compute Wavefront Deformation

Conrady (1960) proposed another method to compute the wavefront deformation, based on results from ray tracing of an axial beam of rays. His method is capable of great accuracy, but has two problems, namely, that it applies only on-axis and to centered systems with rotational symmetry and that it requires a lot computation.

To describe this method, let us consider Fig. 9.10, where a meridional ray has been refracted in a spherical surface with vertex **A**, at point **P**. Let us assume that the incident wavefront is perfectly spherical, with center of curvature at **B**. Then, a point **O** is on the intersection of the incident ray and this wavefront. Thus, the optical path difference (*OPD*) introduced by this surface is

$$OPD = n'AB' - nOP - n'PB' \qquad (9.19)$$

We trace a circle passing through **P**, and a point **E**, with center at **B'** (this circle represents the refracted wavefront only if there is no *OPD* introduced). Then, the optical path difference may also be expressed by

$$OPD = n'AE - nOP \qquad (9.20)$$

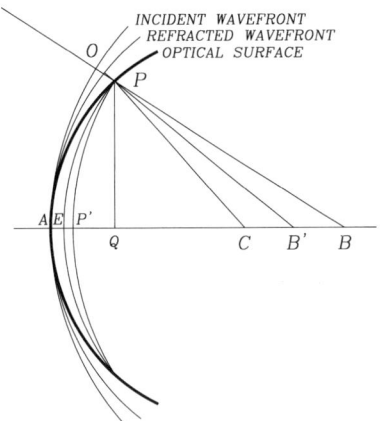

Figure 9.10 Conrady's method to compute the wavefront on-axis for a centered optical system.

Now, we trace a circle through **P** and a point **P′**, with center at **B**. Then, since OP is equal to AP' we may see, observing Fig. 9.10, that

$$AE = AQ - EQ; \qquad OP = AQ - P'Q \qquad (9.21)$$

Also, we may see that the distance or sagitta AQ is given by

$$Z = AQ = Y\tan\frac{U+I}{2} = Y\tan\frac{U'+I'}{2} \qquad (9.22)$$

and the distance EQ by

$$EQ = Y\tan\frac{U}{2} \qquad (9.23)$$

Then, using these expressions and an analogous relation for $P'Q$, we find the OPD to be

$$OPD = n'Y\left[\tan\frac{U'+I'}{2} - \tan\frac{U'}{2}\right] - nY\left[\tan\frac{U+I}{2} - \tan\frac{U}{2}\right] \qquad (9.24)$$

With some trigonometrical work this expression becomes

$$OPD = n'Y\frac{\sin I'/2}{\cos(U'+I')/2\cos U'/2} - nY\frac{\sin I/2}{\cos(U+I)/2\cos U/2} \qquad (9.25)$$

The next step is to multiply numerator and denominator of the first term by $2\cos I'/2$ and similarly the second term by $2\cos I/2$, and using sine law, we find that

$$OPD = \frac{n'Y\sin I'}{2\cos(U+I/2)}\left[\frac{1}{\cos(I'/2)\cos(U'/2)} - \frac{1}{\cos(I/2)\cos(U/2)}\right] \qquad (9.26)$$

After some more algebraic steps, the final expression for the OPD is

$$OPD = \frac{n'Y\sin(U-U')/2 \times \sin(I-U')/2}{2\cos(U/2)\cos(I/2)\cos(U'/2)\cos(I'/2)\cos(U+I/2)} \qquad (9.27)$$

This is the OPD for a single surface, but for the whole system it is just the sum of the surface contributions.

9.4 POINT AND LINE SPREAD FUNCTION

The point and line spread functions are two different calculated functions that permit the evaluation of the quality of the image produced by an optical system (Barakat and Houston, 1964; Jones, 1958; Malacara, 1990; Marchand, 1964). The point spread function is the irradiance in the image of a point source in an optical system. This function may be obtained in a number of ways. One way is with the Fourier transform of the pupil function as shown in Chap. 8, where the amplitude on the image was given by

$$A(x_F, y_F) = \iint_\sigma T(x,y) e^{ik(xx_F + yy_F)/F} \, dx \, dy \tag{9.28}$$

where (xF, yF) are the coordinates in the focal plane and the integration is made over the entrance pupil area σ. The amplitude $T(x,y)$ on the entrance pupil, or pupil function, is given by

$$T(x,y) = E(x,y) e^{ikW(x,y)} \tag{9.29}$$

where $E(x, y)$ is the amplitude distribution over the exit pupil and $W(x, y)$ is the wavefront deformation on this pupil. The point spread function is then given by the complex square of the amplitude in the image:

$$S(x_F, y_F) = A(x_F, y_F) A^*(x_F, y_F) \tag{9.30}$$

If the entrance pupil has a constant illumination [$E(x, y) = $ constant)], Eq. (9.28) may be written as

$$A(x_F, y_F) = \iint_\sigma e^{ik[(xx_F + yy_F)/F + W(x,y)]} \, dx \, dy \tag{9.31}$$

This integral may now be evaluated by dividing the aperture (region of integration) into small squares, as shown in Fig. 9.11. Then, if the center of each square has coordinates (x_0, y_0), the wavefront on this small square may be written as

$$W(x,y) = W_0 + \frac{TA_x}{F}(x - x_0) + \frac{TA_y}{F}(y - y_0) \tag{9.32}$$

where W_0 is the wavefront deviation at the center of the square, found with any of the methods described in Section 9.3. Then, by integrating, we may

Computer Evaluation of Optical Systems

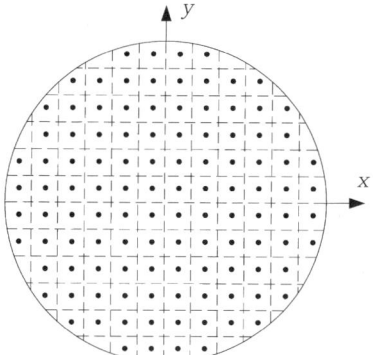

Figure 9.11 Ray distribution on entrance pupil to compute the diffraction image.

find that

$$A(x_F, y_F) = \sum \text{sinc}\left[\frac{kA}{2F}(TA_{xj} + x_F)\right] \text{sinc}\left[\frac{kA}{2F}(TA_{yj} + y_F)\right]$$
$$\times \exp ik\left(W_{0j} + \frac{x_F x_{0j} + y_F y_{0j}}{F}\right) \quad (9.33)$$

where A is the length of one side of the small square and the function sinc φ is equal to $(\sin \varphi)/\varphi$. This is the superposition of many Fraunhofer diffraction patterns produced by each of the small squares on the entrance pupil. These patterns are added with a phase factor due to their relative position on the pupil. The centers of these diffraction patterns correspond to the ray intersections with the focal plane (spot diagram points). These patterns must overlap, in order to produce a continuous spread function. They should not be separated as in Fig. 9.12. A safe condition is that the maximum value of the transverse aberration should be smaller than the diffraction pattern size. Thus, we may write this condition as

$$TA_{x\max} = \frac{\lambda F}{A} \quad (9.34)$$

If the aberrations are large, we may compute the point spread function by evaluating the density of points in the spot diagram. To have good accuracy, the number of points must be as large as possible.

The image of a line object is called the line spread function, and it is used many times compared to the point spread function. To have information in several directions, line spread functions would have to be computed for several object orientations.

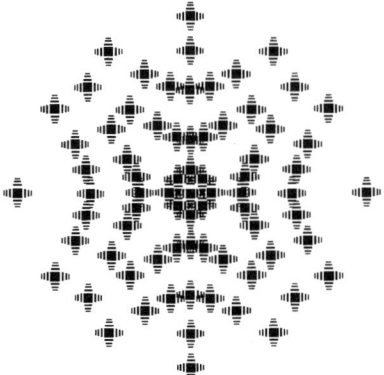

Figure 9.12 Nonoverlapping diffraction images from each square in Fig. 9.11.

9.5 OPTICAL TRANSFER FUNCTION

The optical transfer function, already studied in some detail in Chap. 8, can be calculated for an optical system by several different methods (Barakat and Morello, 1962; Heshmaty-Manesh and Tam, 1982; Kidger, 1978; Plight, 1978). Here, we describe some practical procedures for its computation from lens data.

Hopkins (1957) suggested a method to compute the modulation transfer function by integration of Eq. (9.28). The two laterally sheared apertures are divided into small squares as in Fig. 9.11 and the wavefront on each square is represented by Eq. (9.32). Then, we may obtain the following expression for the diffraction optical transfer function:

$$F(\omega_x,\omega_y) = \sum \mathrm{sinc}\left[\frac{kA}{2F}(TA_{xj} - TA_{xjs})\right]\mathrm{sinc}\left[\frac{kA}{2F}(TA_{yj} - TA_{yjs})\right]$$
$$\times \exp ik(W_{0j} - W_{0js}) \qquad (9.35)$$

where the subscript s stands for the sheared exit pupil by an amount ($F\omega_x/k$, $F\omega_y/k$).

9.5.1 Geometrical Optical Transfer Function

When the wavefront aberration is very large compared with the wavelength, we may approximate the optical transfer function in Eq. (8.46) by

$$F(\omega_x,\omega_y) = \int\int \exp iF\left(\frac{\partial W(x,y)}{\partial x}\omega_x + \frac{\partial W(x,y)}{\partial y}\omega_y\right) dx\,dy \qquad (9.36)$$

Computer Evaluation of Optical Systems

or by using relations (1.75) and (1.76):

$$F(\omega_x,\omega_y) = \iint \exp i[TA_x(x,y)\omega_x + TA_y(x,y)\omega_y]\,dx\,dy \qquad (9.37)$$

The region of integration is the common area between the two laterally sheared pupils, as in Fig. 8.14, but since the aberration is large, the cut-off spatial frequency is reached before the shear is large. So, the region of integration may be considered to be the whole circular pupil.

If the OTF is computed from ray tracing data (spot diagram), by dividing the aperture into small squares, the geometrical transfer function may be written as

$$F(\omega_x,\omega_y) = \sum_{i=1}^{N} \cos(TA_x\omega_x + TA_y\omega_y)$$

$$+ i\sum_{i=1}^{N} \sin(TA_x\omega_x + TA_y\omega_y) \qquad (9.38)$$

where the sum is performed for all rays on a spot diagram. This geometrical approximation is surprisingly accurate. An aberration of a few wavelengths is large enough to produce the same result with the exact OTF and with the geometrical approximation.

For very low spatial frequencies, assuming that the transverse aberration is small compared with the spatial period, we may approximately write

$$F(\omega_x,\omega_y) = \sum_{i=1}^{N}\left[1 - \frac{(TA_x\omega_x + TA_y\omega_y)^2}{2}\right]$$

$$+ i\sum_{i=1}^{N}[TA_x\omega_x + TA_y\omega_y] \qquad (9.39)$$

This is the optical transfer function for low spatial frequencies and its square is the modulation transfer function:

$$F(\omega_x,\omega_y)F^*(\omega_x,\omega_y) = N - \sum_{i=1}^{N}(TA_x\omega_x + TA_y\omega_y)^2$$

$$+ \left[\sum_{i=1}^{N}(TA_x\omega_x + TA_y\omega_y)\right]^2 \qquad (9.40)$$

From Eq. (9.5) we may see that, for an optical system with a rotationally symmetric wavefront in which the value of ω is the same in all directions, we may write

$$F(\omega_x,\omega_y)F^*(\omega_x,\omega_y) = 1 - \omega^2 T A_{\text{rms}}^2 \qquad (9.41)$$

This means that the modulation transfer function for low spatial frequencies is determined by the root mean square size of the geometrical image. Conversely, if a system is optimized for the lowest value of the geometrical image, the optimization on the optical transfer function is for the lower range of spatial frequencies.

9.6 TOLERANCE TO ABERRATIONS

It is not enough to evaluate the aberration of an optical system. They can never be made perfect, so we must know when to stop trying to improve it. The maximum allowed aberration in a given optical system depends on its intended use. We will now study a few different cases.

    1. *Interferometric quality*—A lens or optical system to be used in an interferometer may for some particular applications require a wavefront as good as possible. Then, primary as well as high-order aberration terms must be highly corrected. Once the best lens is obtained, the wavefront deviations are minimized with the proper amount of tilts and defocusing. A wavefront deformation less than $\lambda/100$ is some times required.

    2. *Diffraction limited*—An image-forming lens with small aberrations, so that the highest possible theoretical resolution is obtained, is said to be diffraction limited. The image size is the Airy diffraction image. According to the Rayleigh criterion, an optical system produces a diffraction-limited image if the wavefront deformation has a maximum absolute value less than one-quarter of the wavelength. It must be pointed out, however, that this principle is strictly valid only for pure primary spherical aberration. If the wavefront is smooth enough as in a lens with only primary spherical aberration, the ray transverse aberrations are less than the Airy disk radius.

    3. *Visual quality*—A lens to be used for visual observations may sometimes require diffraction-limited resolution. This is the case of high magnification lenses, like microscope objectives. In many other low magnification systems, like eyepieces or low-power telescopes, the limiting factor is the resolution of the eye. In this case, the geometric image (transverse aberrations) must be smaller than the angular resolution of the

eye, which is about one minute of arc. The accomodation capacity of the eye must be taken into account when evaluating visual systems.

4. *Atmospheric seeing limited*—This is the case of the terrestrial astronomical telescope, in which the angular resolution is limited by the atmospheric turbulence or "seeing." Then, the transverse aberrations must be smaller than the size of the seeing image. This image size is between 1 and 1/10 arcsec, depending on the atmospheric conditions and the quality of the telescope site.

9.6.1 Curvature and Thickness Tolerances

The tolerance in the deviation of the construction value of the curvature or thickness of a lens with respect to the ideal design value depends on many factors. One of them is the allowed deviation in the performance of the system, as described in the preceding section. The tolerance specification has to take into account the cost, which is not linear with the magnitude of the tolerance. The manufacturing cost grows almost exponentially with the tightness of the tolerance. Thus, the tolerance should be as tight as the previously set allowance in the cost of production permits it and not higher. Another important factor is the technical capability of the optical shop in charge of the manufacturing process. Given an optical system not all optical surfaces are equally sensitive to figure or curvature deviations. The system designer has to evaluate the sensitivity of each design parameter before establishing the tolerances.

An idea of the magnitudes of the tolerances for different qualities of optical instrument was given by Shannon (1995) (see Table 9.3).

It is quite important to keep in mind that not only will the deviation in one parameter contribute to the image degradation, but a combination of all deviations also will. The final error can be estimated by the rms value of all tolerances, Thus, if there are N parameters and the tolerance in the

Table 9.3 Typical Manufacturing Tolerances in a Lens System

Parameter	Commercial	Precision	High *precision*
Wavefront deformation	0.25 wave rms	0.1 wave rms	0.05 wave rms
Radius of curvature	1.0%	0.1%	0.01%
Thickness	± 0.2 mm	± 0.5 mm	± 0.01 mm
Decentration	0.1 mm	0.01 mm	0.001 mm
Tilt	1 arcmin	10 arcsec	1 arcsec

Source: Shannon (1995).

parameter i produces an error W_i, the final estimated rms error W_{rms} is given by

$$W_{\text{rms}} = \sqrt{\sum_{i=0}^{N} W_i^2} \tag{9.42}$$

REFERENCES

Barakat, R. and Houston, A., "Line Spread Function and Cumulative Line Spread Function for Systems with Rotational Symmetry," *J. Opt. Soc. Am.*, **54**, 768–773 (1964).

Barakat, R. and Morello, M. V., "Computation of the Transfer Function of an Optical System from the Design Data for Rotationally Symmetric Aberrations: II. Programming and Numerical Results," *J. Opt. Soc. Am.*, **52**, 992–997 (1962).

Barakat, R. and Morello, M. V., "Computation of the Total Illuminance (Encircled Energy) of an Optical System from the Design Data for Rotationally Symmetric Aberrations," *J. Opt. Soc. Am.*, **54**, 235–240 (1964).

Conrady, A. E., *Applied Optics and Optical Design, Part Two*, Dover Publications, New York, 1960.

Forbes, G. W., "Optical System Assessment for Design: Numerical Ray Tracing in the Gaussian Pupil," *J. Opt. Soc. Am. A*, **5**, 1943 (1988).

Ghozeil, I., "Hartmann and Other Screen Tests," in *Optical Shop Testing*, D. Malacara, ed., John Wiley, New York, 1992.

Herzberger, M., "Light Distribution in the Optical Image," *J. Opt. Soc. Am.*, **37**, 485–493 (1947).

Herzberger, M., "Analysis of Spot Diagrams," *J. Opt. Soc. Am.*, **47**, 584–594 (1957).

Heshmaty-Manesh, D. and Tam, S. C., "Optical Transfer Function Calculation by Winograd's Fast Fourier Transform," *Appl. Opt.*, **21**, 3273–3277 (1982).

Hopkins, H. H., "Geometrical–Optical Treatment of Frequency Response," *Proc. Phys. Soc. B*, **70**, 1162–1172 (1957).

Jones, R. C., "On the Point and Line Spread Functions of Photographic Images," *J. Opt. Soc. Am.*, **48**, 934–937 (1958).

Kidger, M. J., "The Calculation of the Optical Transfer Function Using Gaussian Quadrature," *Opt. Acta*, **25**, 665–680 (1978).

Linfoot, E. H., "Plate Diagram Analysis," in *Recent Advances in Optics*, Chap. 4, Clarendon Press, Oxford, UK, 1955.

Lucy, F. A., "Image Quality Criteria Derived from Skew Traces," *J. Opt. Soc. Am.*, **46**, 699–706 (1956).

Malacara, D., "Diffraction Performance Calculations in Lens Design," *Proceedings of the International Lens Design Conference*, Monterey, CA, June 1990.

Marchand, E. W., "Derivation of the Point Spread Function from the Line Spread Function," *J. Opt. Soc. Am.*, **54**, 915–919 (1964).

Marchand, E. W. and Phillips, R., "Calculation of the Optical Transfer Function from Lens-Design Data," *Appl. Opt.*, **2**, 359–364 (1963).

Miyamoto, K., "Image Evaluation by Spot Diagram Using a Computer," *Appl. Opt.*, **2**, 1247–1250 (1963).

Plight, A. M., "The Rapid Calculation of the Optical Transfer Function for On-Axis Systems Using the Orthogonal Properties of Tchebycheff Polynomials," *Opt. Acta*, **25**, 849–860 (1978).

Plight, A. M., "The Calculation of the Wavefront Aberration Polynomial," *Opt. Acta*, **27**, 717–721 (1980).

Shannon, R. E., "Tolerancing Techniques," in *Handbook of Optics*, Vol. I, M. Bass, ed., MacGraw-Hill, New York, 1995.

Stavroudis, O. and Feder, D. P., "Automatic Computation of Spot Diagrams," *J. Opt. Soc. Am.*, **44**, 163–170 (1954).

Welford, W. T., *Aberrations of Optical Systems*, Adam Hilger, Bristol, UK, 1986.

Wetherell, W. B., "The Calculation of Image Quality," in *Applied Optics and Optical Engineering*, Vol. VIII, Chap. 6, R. R. Shannon and J. C. Wyant, eds., Academic Press, San Diego, CA, 1980.

10
Prisms

10.1 TUNNEL DIAGRAM

Prisms and mirror systems are important parts in optical systems and have been studied by many authors, e.g., by Hopkins (1962, 1965). The presence of a prism in an optical system has many effects that must be taken into account when designing such a system. Among these effects we can mention the following:

1. A change in the direction of propagation of the light.
2. A transformation on the image orientation.
3. An image displacement along the optical axis.
4. The limited sizes of their faces may act as stops, limiting the lateral extension of the light beam.
5. Some aberration contributions are added, mainly spherical and axial chromatic aberrations, even with flat faces.

It is easily proved that the longitudinal displacement d of the image, introduced by a glass plate with thickness L, is given by

$$d = \frac{(n-1)}{n} L \tag{10.1}$$

The primary spherical aberration introduced by the prism may be calculated as shown in Section 4.2.3.

All these effects may easily be taken into account while designing an optical system by unfolding the prism in every reflection to find the equivalent flat parallel glass block. Then, we obtain what is called a tunnel diagram for the prism, as in the example in Fig. 10.1.

10.2 DEFLECTING A LIGHT BEAM

Let us consider the reflection of a light beam in a system of two reflecting faces with one of these faces rotated at an angle θ relative to the other, as

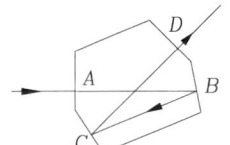

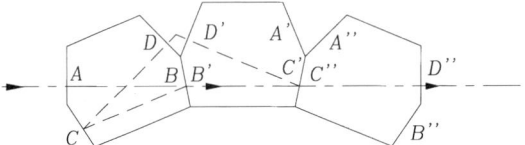

Figure 10.1 Tunnel diagram for a prism.

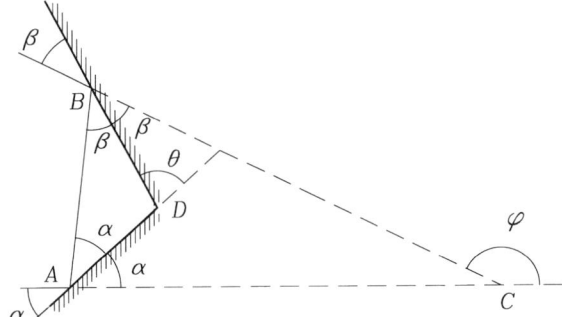

Figure 10.2 Reflection in a system of two flat mirrors.

shown in Fig. 10.2. We will prove that the direction of propagation of the light beam will be changed by an angle 2θ, independently of the direction of incidence with respect to the system, as long as the incident ray is in a common plane with the normals to the two reflecting surfaces.

In the triangle **ABC** we see that

$$\phi = 2\alpha + 2\beta \tag{10.2}$$

and in the triangle **ABD**:

$$\theta = \alpha + \beta \tag{10.3}$$

then, we see that

$$\phi = 2\theta \tag{10.4}$$

Prisms

In conclusion, if the angle between the two mirrors is θ, the light ray will deviate by an angle φ, independently of the direction of incidence of the light ray. There are several prisms that use this property.

By means of three reflections in three mutually perpendicular surfaces a beam of light may also be deflected by an angle of 180°, reflecting it back along a trajectory in a parallel direction to the incident light beam. To show this let us use the vectorial reflection law, which may be obtained from the vectorial refraction law in Eqs. (1.16) and (1.17), by setting $n' = -n$, as

$$\mathbf{S}' = \mathbf{S} + (2\cos I)\mathbf{p}$$
$$\mathbf{S} + 2(\mathbf{S} \cdot \mathbf{p})\mathbf{p} \tag{10.5}$$

where $\mathbf{S}$ and $\mathbf{S}'$ are vectors along the incident and the reflected ray, respectively. If we have three reflecting surfaces with their three normals not coplanar, we may write for the first reflecting surface:

$$\mathbf{S}'_1 = \mathbf{S}_1 + 2(\mathbf{S}_1 \cdot \mathbf{p}_1)\mathbf{p}_1 \tag{10.6}$$

for the second surface:

$$\mathbf{S}'_2 = \mathbf{S}_2 + 2(\mathbf{S}_2 \cdot \mathbf{p}_2)\mathbf{p}_2 \tag{10.7}$$

and for the third surface:

$$\mathbf{S}'_3 = \mathbf{S}_3 + 2(\mathbf{S}_3 \cdot \mathbf{p}_3)\mathbf{p}_3 \tag{10.8}$$

We may now assume that the ray is first reflected on the surface number one and last on the surface number three. Then,

$$\mathbf{S}_2 = \mathbf{S}'_1 \tag{10.9}$$

and

$$\mathbf{S}_3 = \mathbf{S}'_2 \tag{10.10}$$

Hence, the final ray direction is then given by

$$\mathbf{S}'_3 = \mathbf{S}_1 + 2(\mathbf{S}_1 \cdot \mathbf{p}_1)p_1 + 2(\mathbf{S}_2 \cdot \mathbf{p}_2)\mathbf{p}_2 + 2(\mathbf{S}_3 \cdot \mathbf{p}_3)\mathbf{p}_3 \tag{10.11}$$

Since the three reflecting surfaces are mutually perpendicular we have that $\mathbf{p}_1 \cdot \mathbf{p}_2 = \mathbf{p}_2 \cdot \mathbf{p}_3 = \mathbf{p}_1 \cdot \mathbf{p}_3$. Then, using these relations we may find that

$$\mathbf{S}'_3 = \mathbf{S}_1 + 2(\mathbf{S}_1 \cdot \mathbf{p}_1)\mathbf{p}_1 + 2(\mathbf{S}_1 \cdot \mathbf{p}_2)p_2 + 2(\mathbf{S}_1 \cdot \mathbf{p}_3)\mathbf{p}_3 \qquad (10.12)$$

but since the vectors $\mathbf{p}_1$, $\mathbf{p}_2$, and $\mathbf{p}_3$ form an orthogonal base, we may show that

$$\mathbf{S}'_3 = \mathbf{S}_1 - 2\mathbf{S}_1 = -\mathbf{S}_1 \qquad (10.13)$$

proving that a system of three mutually perpendicular reflectors is a retroreflecting system. There are many uses of this result, as we will see later.

10.3 TRANSFORMING AN IMAGE

In this chapter we will describe some prisms made out of isotropic materials, such as glass, and with flat faces. We will consider prisms that change the direction of propagation of the light without any chromatic dispersion. Thus, this light deviation is produced by internal reflection if the internal angle of incidence is greater than the critical angle, or by coating the surface with a reflective coating. Besides changing the light direction, these prisms also produce a change in the image orientation, which may be described by some basic image transformations, illustrated in Fig. 10.3 and defined as follows:

1. An *inversion* is a geometric reflection about a horizontal axis.
2. A *reversion* is a geometric reflection about a vertical axis.
3. A *reflection on an inclined axis*, at an angle θ.
4. A *rotation* by an angle θ.

Any mirror reflection (including reflections on spherical mirrors) produces a reflection transformation. The axis for this operation is perpendicular to both the incident and the reflected beams. Obviously, the

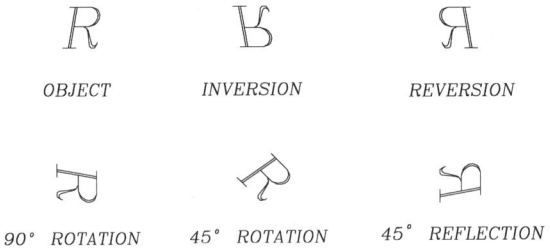

Figure 10.3 Image transformations.

operations of inversion and reversion are particular cases of the general reflection transformation. Two succesive reflections may be easily shown to be equivalent to a rotation, as follows:

Reflection at α_1 + *reflection at* α_2 = *rotation by* $2(\alpha_2 - \alpha_1)$

An image is said to be readable, if its original orientation may be recovered with a rotation. Thus, a rotation does not change the readability of an image. It is easy to prove the important conclusion that an even number of reflections produces a readable image. On the other hand, an odd number of reflections always gives a nonreadable image. Thus, a rotation can be produced only with an even number of reflections.

Two transformations may be combined to produce another transformation, as in the following examples that may be considered as particular cases of the general relation just described:

Inversion + *reversion* = *rotation by* $180°$
Inversion + *rotation by* $180°$ = *reversion*
Reversion + *rotation by* $90°$ = *reflection at* $45°$

We may also show that if the axis of a reflection transformation rotates, the resulting image also rotates, in the same direction and with twice the angular speed. Thus, a practical consequence is that all inverting systems may be converted into reversing systems by rotation by an angle of $90°$.

A system of plane mirrors with arbitrary orientations have two distinct effects: (1) the beam direction is changed and (2) the image orientation is also modified. Both of these effects may be studied using matrices. The problem of the optical axis deflection and the problem of the image orientation has been treated by many authors, e.g., Pegis and Rao (1963), Walles and Hopkins (1964), Walther (1964) and Berkowitz (1965). The mirror system is described using an orthogonal system of coordinates x_0, y_0, z_0 in the object space, with z_0 being along the optical axis and pointing in the traveling direction of the light. Then, for a single mirror we have the following linear transformation with a symmetrical matrix:

$$\begin{bmatrix} l \\ m \\ n \end{bmatrix} = \begin{bmatrix} (1-2L^2) & (-2LM) & (-2LN) \\ (-2LM) & (1-2M^2) & (-2MN) \\ (-2LN) & (-2MN) & (1-2N^2) \end{bmatrix} \begin{bmatrix} l_o \\ m_o \\ n_o \end{bmatrix} \quad (10.14)$$

where (l, m, n) and (l_o, m_o, n_o) are the direction cosines of the reflected and incident rays, respectively. The quantities (L, M, N) are the direction cosines of the normals to the mirror.

To find the final direction of the beam, the reflection matrices for each mirror are multiplied in the order opposite to that in which the light rays strike the mirrors. On the other hand, to find the image orientation, the matrices are multiplied in the same order that the light strikes the mirrors.

10.4 DEFLECTING AND TRANSFORMING PRISMS

These prisms, besides transforming the image orientation, bend the optical axis, changing the direction of propagation of the light. There are many prisms of this kind. Here, we will just describe a few examples.

10.4.1 Deflecting Prisms

To describe all the deflecting prisms would be impossible, so, we only describe some of the main types, which are:

1. Right angle prism
2. Amici prism
3. Pentaprism
4. Wollaston prism

The *right angle* prism is the simplest of all prisms and in most of the cases, it can be replaced by a flat mirror. The image produced by this prism is not readable, since there is only one reflection, as shown by Fig. 10.4(a). This prism can be modified to produce a readable image. This is accomplished by substituting the hypotenuse side by a couple of mutually perpendicular faces, forming a roof, to obtain an *Amici prism* as shown in Fig. 10.4(b).

Both rectangular and Amici prisms can be modified to deflect a beam of light 45° instead of 90° as in the prisms shown in Fig. 10.5.

In the prisms previously described, the deflecting angle depends on the angle of incidence. It is possible to design a prism in which the deflecting angle is independent of the incidence angle. This is accomplished with two reflecting surfaces instead of just one. By using the property described in Section 10.2, the deflection angle is twice the angle between the two mirrors or reflecting surfaces.

This property is used in the *Wollaston* prism, shown in Fig. 10.6, and in the *pentaprism*, shown in Fig. 10.7. In the Wollaston prism both reflecting surfaces form a 45° angle and the deflecting angle is 90°. In the pentaprism both surfaces form an angle of 135° and the deflection angle is 270°.

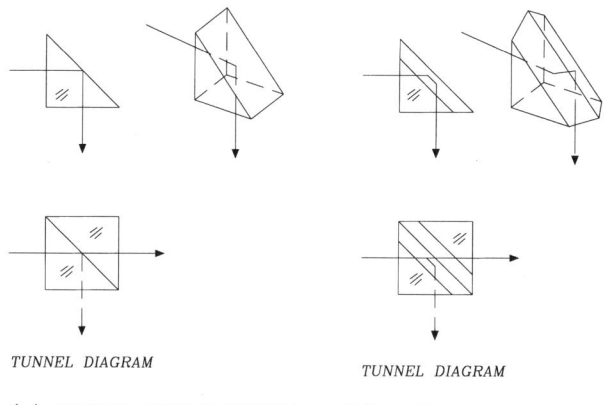

Figure 10.4 (a) Right angle and (b) Amici prisms.

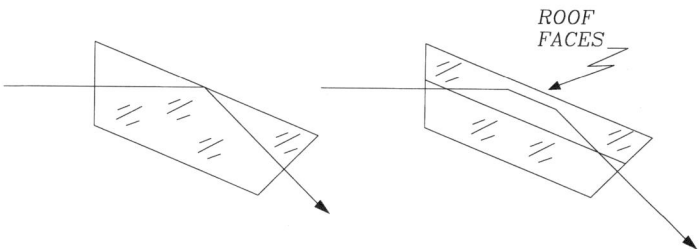

Figure 10.5 Deflecting dove prism.

In both of the previously described prisms, the image is readable, since there are two reflections. The pentaprism is more compact and simpler to build; hence, it is more commonly used.

Although both prisms can be modified to obtain a 45° deflection, it results in an impractical and a complicated shape. To obtain a 45° deflection independent of the incidence angle, the prism shown in Fig. 10.8 is preferred. These prisms are used in microscopes, to obtain a more comfortable observing position.

Another 45° deflecting prism, similar to the pentaprism, is shown in Fig. 10.1.

10.4.2 Retroreflecting Systems

A *retroreflecting prism* is a particular case of a constant deviation prism, in which the deflecting angle is 180°.

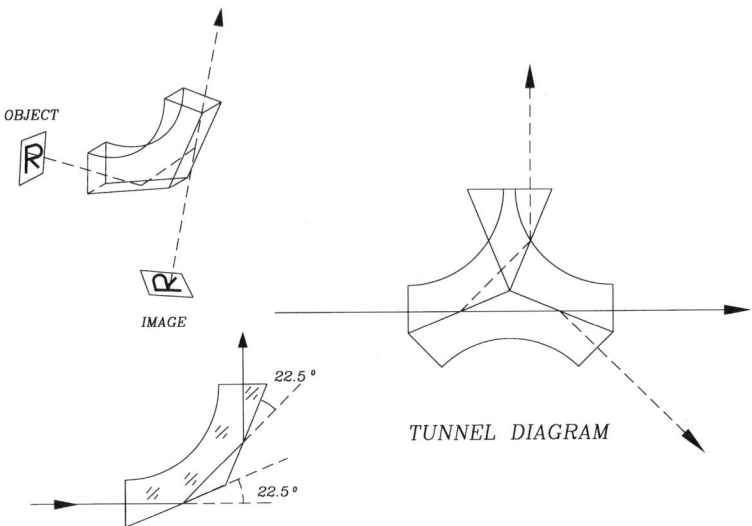

Figure 10.6 Wollaston prism.

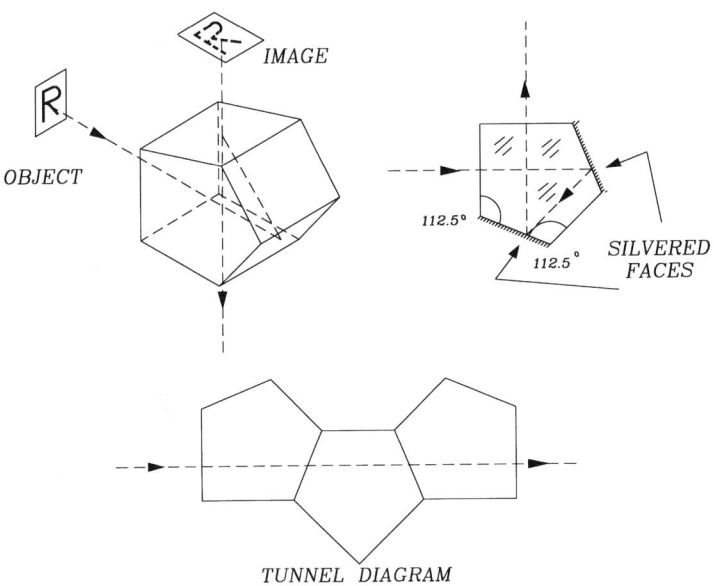

Figure 10.7 Pentaprism.

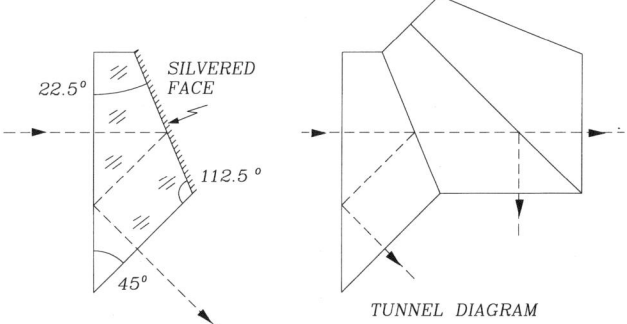

Figure 10.8 Forty-five degrees deflecting prism.

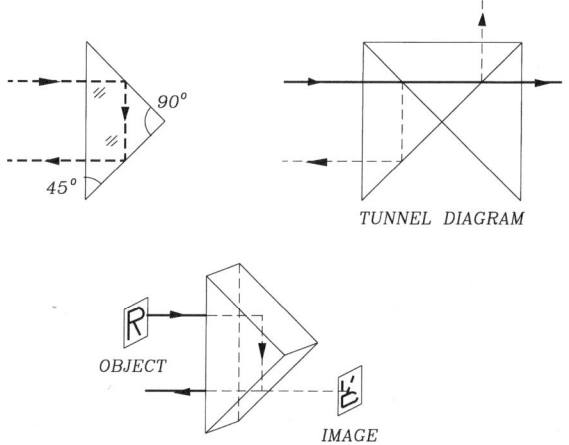

Figure 10.9 Rectangular retroreflecting prism.

A rectangle prism can be used as a retroreflecting prism with the shape shown in Fig. 10.9. In such a case, it is called a *Porro prism*. The Porro prism is a perfect retroreflector, assuming that the incident ray is coplanar with the normals to the surfaces.

A perfect retroreflecting prism without the previous constraint is made with three mutually perpendicular reflecting surfaces. This prism, shown in Fig. 10.10, is called a *cube corner prism*.

Cube corner prisms are very useful in optical experiments where a 180° reflection is needed. Uses for the cube corner retroreflector are found in applications where the prism can wobble or jitter or is difficult to align because it is far from the light source. Applications for this prism range from the common ones like reflectors in a car's red back light to the highly

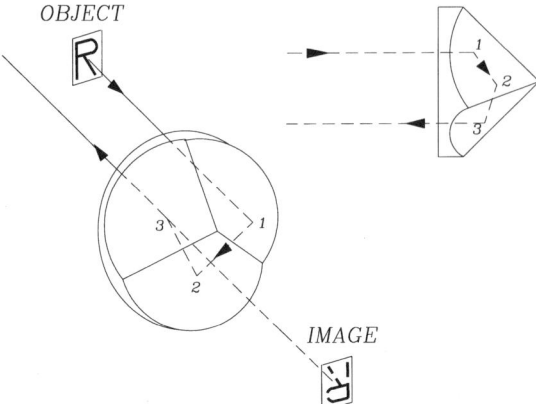

Figure 10.10 Cube corner prism.

specialized ones like the reflectors placed on the surface of the moon in the year 1969.

10.5 NONDEFLECTING TRANSFORMING PRISMS

These prisms preserve the traveling direction of the light beam, changing only the image orientation. Some of the nondeflecting transforming prisms will now be described.

10.5.1 Inverting and Reverting Prisms

In order to produce an image inversion or reversion, these prisms must have an odd number of reflections. We will consider only prisms that do not deflect the light beam. The simplest of these prisms has a single reflection, as shown in Fig. 10.11. This is a single rectangular prism, used in a configuration called a *dove prism* (for comparison with a dove tail).

The operation can be easily understood from the tunnel diagram. Although we have two refractions, there is no chromatic aberration since entrance and exiting faces act as in a plane-parallel plate. These prisms cannot be used in strongly convergent or divergent beams of light because of the spherical aberration.

An *equilateral triangle prism* can be used as an inverting or reverting prism if used as depicted in Fig. 10.12. On this configuration, we have two refractions and three reflections. Like the dove prism, this prism cannot be used in strongly convergent or divergent beams of light.

Prisms

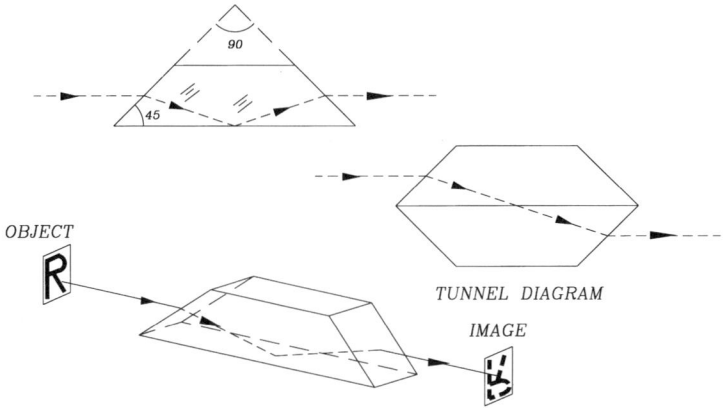

Figure 10.11 Dove prism.

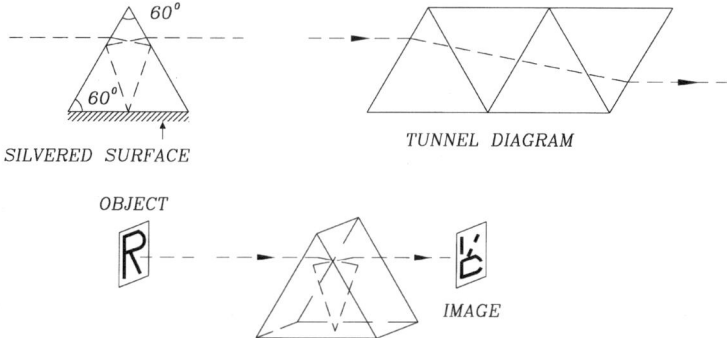

Figure 10.12 Inverting–reversing triangular prism.

Figures 10.13, 10.14, and 10.15 show three reverting prisms with three internal reflections. The first one does not shift the optical axis laterally, while in the last two the optical axis is displaced. These prisms can be used in converging or diverging beams of light. The first two prisms can be made either with two glass pieces or a single piece.

The *Pechan prism*, shown in Fig. 10.16, can be used in converging or diverging pencils of light, besides being a more compact prism than the previous ones.

10.5.2 Rotating Prisms

A *half-turn rotating prism* is a prism that produces a readable image, rotated 180°. The real image produced by a convergent lens is usually rotated 180°

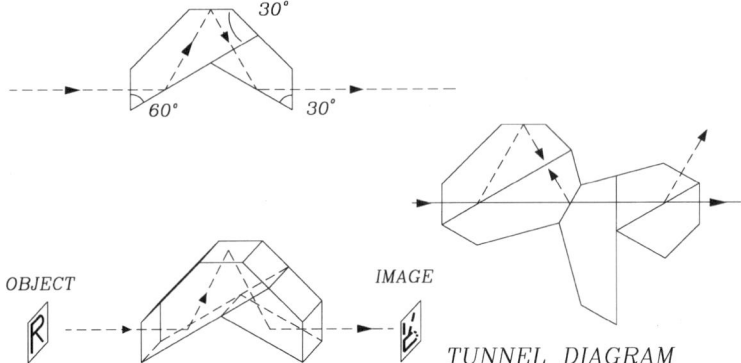

Figure 10.13   Reverting–inversing prism.

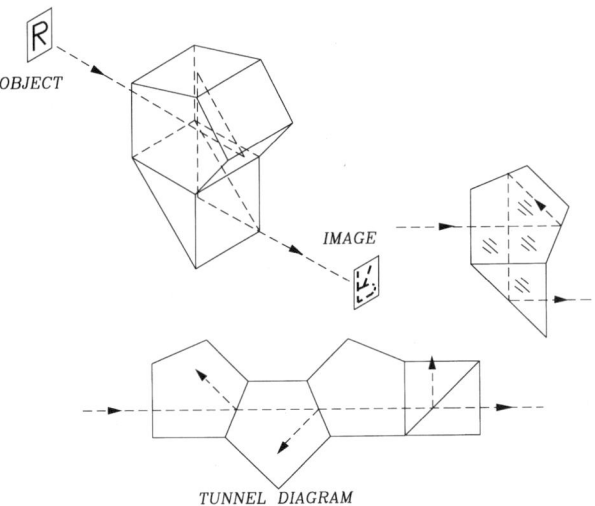

Figure 10.14   Reverting–inversing prism.

as compared with the object; hence, a rotating prism can bring back the image to the original orientation of the object. These sort of prisms are useful for monocular terrestrial telescopes and binoculars.

All of the reversing prisms previously described can be converted into rotating prisms by substituting one of the reflecting surfaces by a couple of surfaces with the shape of a roof. With this substitution the prism in Fig. 10.13 is transformed into the so-called *Abbe prism*, the one in Fig. 10.15 is transformed into the *Leman prism*, and the one in Fig. 10.16 is transformed into the *Schmidt–Pechan prism*, shown in Fig. 10.17. This last prism is used in

Prisms

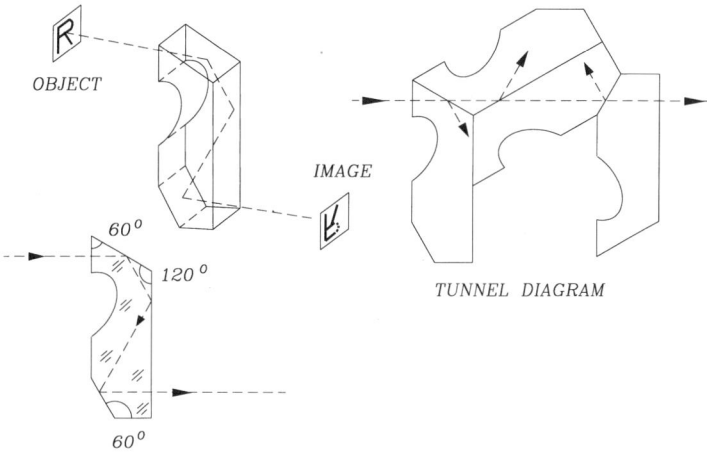

Figure 10.15 Reverting–inversing prism.

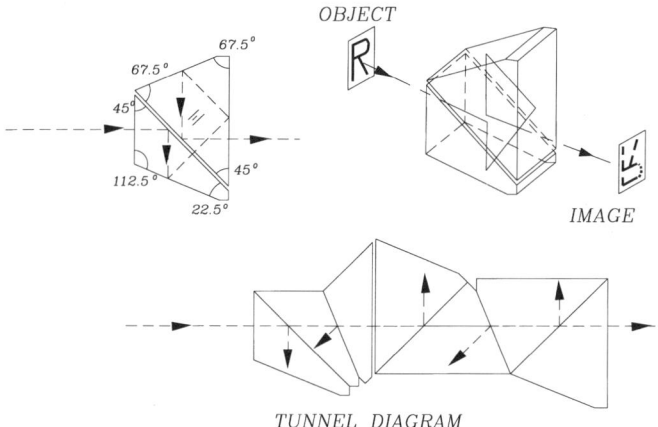

Figure 10.16 Pechan prism.

small hand telescopes. An advantage for this prism is that the optical axis is not laterally displaced.

A double prism commonly used in binoculars is the *Porro prism*, shown in Fig. 10.18.

10.6 BEAM-SPLITTING PRISMS

These prisms divide the beam of light into two beams, with the same diameter as the original one, but the intensity is reduced for both beams that now

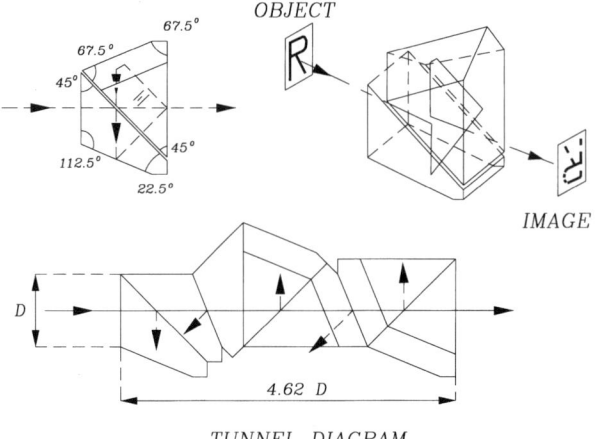

Figure 10.17 Schmidt–Pechan prism.

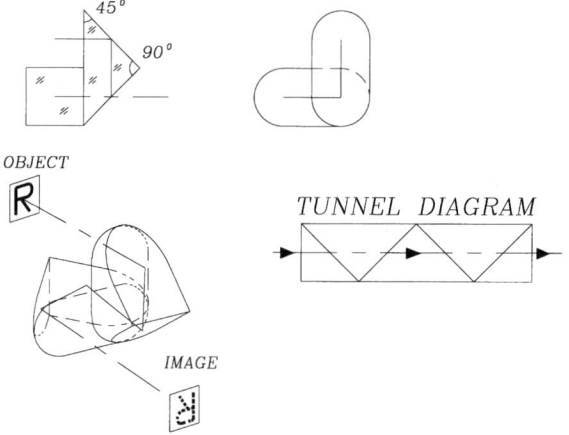

Figure 10.18 Porro prism.

travel in different directions. Beam-splitting prisms are used in amplitude division interferometers, and binocular microscopes and telescopes, where a single image must be observed simultaneously with both eyes. Basically, this prism is formed by a couple of rectangular prisms glued together to form a cube. One of the prisms has its hypotenuse face deposited with a thin reflecting film, chosen in such a way that, after cementing both prisms together, both the reflected and transmitted beam have the same intensity. Both prisms are cemented in order to avoid total internal reflection. This prism and a variant of the basic prism are shown in Fig. 10.19.

Prisms 253

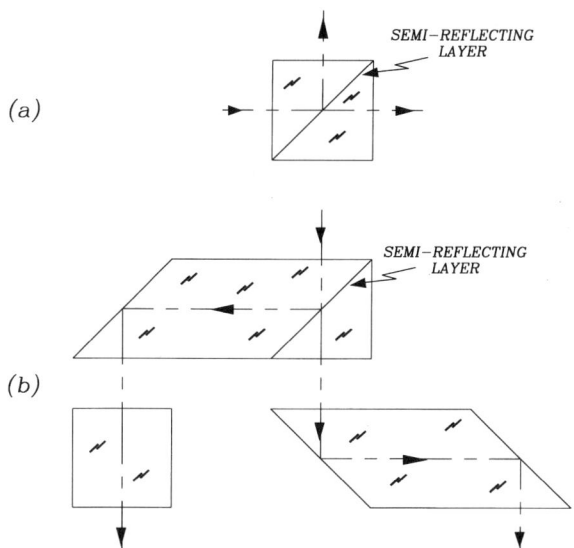

Figure 10.19 Binocular beam-splitting system. (a) A single prism and (b) a binocular prism.

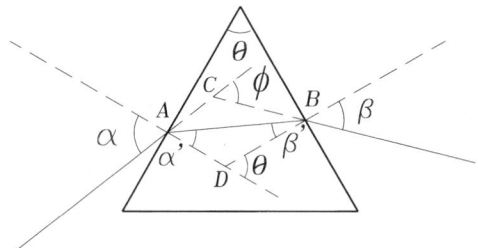

Figure 10.20 Triangular dispersing prism.

10.7 CHROMATIC DISPERSING PRISMS

As shown in Fig. 1.3, the refracting index is a function of the light wavelength and, hence, of the light color. This property is used in chromatic dispersing prisms to decompose the light into its elementary chromatic components, obtaining a rainbow, called a spectrum.

Equilateral prism. The simplest chromatic dispersing prism is the equilateral triangle prism illustrated in Fig. 10.20. This prism is usually made with flint glass, because of its large index variation with the wavelength.

As shown in Fig. 10.20, ϕ is the deviation angle for a light ray and θ is the prism angle. We can see from this same diagram that

$$\phi = (\alpha - \alpha') + (\beta - \beta') \tag{10.15}$$

also,

$$\theta = \alpha' + \beta' \tag{10.16}$$

from this we obtain

$$\phi = \alpha + \beta - \theta \tag{10.17}$$

From Snell's law, we also know that

$$\frac{\sin \alpha}{\sin \alpha'} = n \tag{10.18}$$

and

$$\frac{\sin \beta}{\sin \beta'} = n \tag{10.19}$$

From this we conclude that the deviation angle is a function of the incidence angle α, the apex angle θ, and the refractive index n. The angle ϕ as a function of the angle α for a prism with an angle $\theta = 60°$ and $n = 1.615$ is shown in Fig. 10.21.

The deviation angle ϕ has a minimum magnitude for some value of α equal to α_m. Assuming, as we can easily conclude from Fig. 10.21, that there exists a single minimum value for ϕ, we can use the reversibility principle to see that this minimum occurs when $\alpha = \beta = \alpha_m$. It may be shown that

$$\sin \alpha_m = n \sin \theta/2 \tag{10.20}$$

Assuming that for yellow light $\alpha = \alpha_m$ in a prism with $\theta = 60°$ made from flint glass, the angle ϕ changes with the wavelength λ as shown in Fig. 10.22.

Let us now suppose that the angle θ is small. It can be shown that the angle ϕ is independent from α and given by

$$\phi = (n - 1)\theta \tag{10.21}$$

Prisms

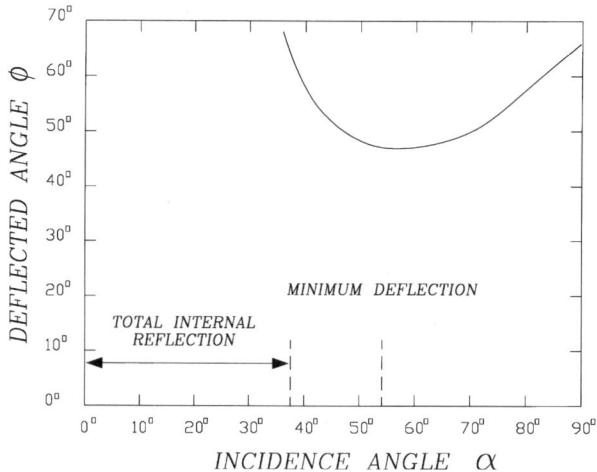

Figure 10.21 Angle of deflection versus angle of incidence in a dispersing prism.

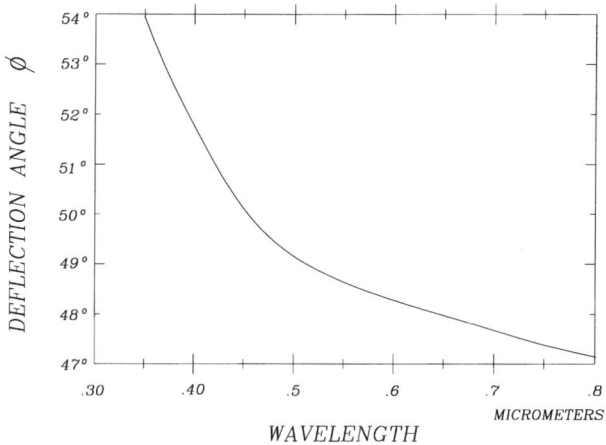

Figure 10.22 Deflection angle versus wavelength in a dispersing prism.

Constant deviation prism. Taking as an example the prism shown in Fig. 10.23. As we can see, the beam width for every color will be different and with an elliptical transverse section. The minor semiaxis for the ellipse for the refracted beam will be equal to the incident beam only when the angle α is equal to the angle β.

For precise photometric spectra measurements, it is necessary that the refracted beam width be equal to the incident beam for every wavelength.

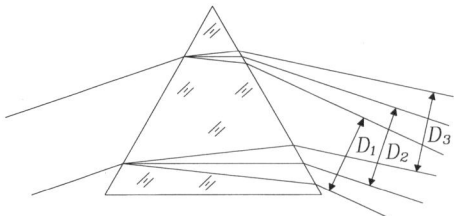

Figure 10.23 Variation in the beam width for different wavelengths in a triangular prism.

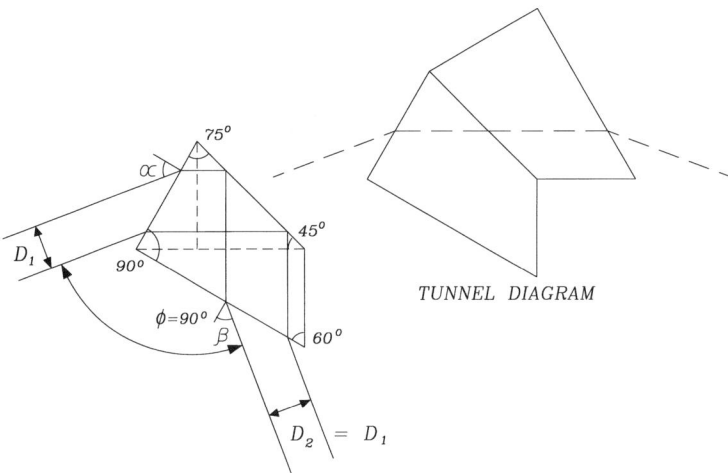

Figure 10.24 Constant deviation prism.

This condition is only met when the prism is rotated so that $\alpha = \beta$ (minimum deviation). Usually, these measurements are uncomfortable since both the prism and the observer have to be rotated.

A dispersing prism that meets the previous condition with a single rotation of the prism for every measurement and does not require the observer to move is the *constant deviation* prism, shown in Fig. 10.24. This prism is built in a single piece of glass, but we can imagine it as the superposition of three rectangular prisms, glued together as shown in the figure. The deflecting angle ϕ is constant, equal to 90°. The prism is rotated to detect each wavelength. The reflecting angle must be 45° and, hence, angles α and β must be equal.

Nondeflecting chromatic dispersing prism. Sometimes it is convenient to disperse the light chromatically without deflecting the main direction of the light beam. This can be achieved by a system of three prisms as shown in

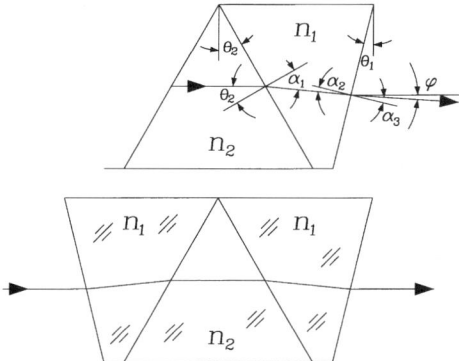

Figure 10.25 Nondeflecting chromatic dispersing prism.

Fig. 10.25. The central prism has a low Abbe number (flint glass) and the two other prisms have a high Abbe number (crown glass). The total deflection φ of the light beam can be shown to be given by

$$\varphi = \theta_1 - \arcsin\left\{n_1 \sin\left[\theta_1 + \theta_2 - \arcsin\left(\frac{n_2}{n_1}\sin\theta_2\right)\right]\right\} \tag{10.22}$$

where the angles are defined in Fig. 10.25.

REFERENCES

Berkowitz, D. A., "Design of Plane Mirror Systems," *J. Opt. Soc. Am.*, **55**, 1464–1467 (1965).

Hopkins, R. E., "Mirror and Prism Systems," in *Military Standardization Handbook: Optical Design, MIL-HDBK 141*, U.S. Defense Supply Agency, Washington, DC, 1962.

Hopkins, R. E., "Mirror and Prism Systems," in *Applied Optics and Optical Engineering*, R. Kingslake, ed., Vol. III, Chap. 7, Academic Press, San Diego, CA, 1965.

Pegis, R. J. and Rao, M. M., "Analysis and Design of Plane Mirror Systems," *Appl. Opt.*, **2**, 1271–1274 (1963).

Walles, S. and Hopkins, R. E., "The Orientation of the Image Formed by a Series of Plane Mirrors," *Appl. Opt.*, **3**, 1447–1452 (1964).

Walther, A., "Comment on the Paper: 'Analysis and Design of Plane Mirror Systems' by Pegis and Rao," *Appl. Opt.*, **3**, 543 (1964).

11
Simple Optical Systems and Photographic Lenses

11.1 OPTICAL SYSTEMS DIVERSITY

An optical system is basically formed by lenses and mirrors, but it has an extremely large number of possible configurations and requirements, as pointed out by Hilbert and Rodgers (1987). The optical system may have many different requirements, depending on its particular application, for example:

1. Speed or f-number FN
2. Field angular diameter
3. Resolution on and off-axis
4. Aperture size (entrance pupil diameter)
5. Physical size of the system
6. Construction difficulties
7. Cost, etc.

From a strictly optical point of view, the first four items are the most important. A single magnifier obviously does not have the same requirements as a microscope objective. The speed of a microscope objective is very high and that of the single magnifier is very low. The required resolution of a microscope objective is quite high but for a single lens is low.

The map in Fig. 11.1, representing the f-number FN versus the angular field size for some of the most common optical systems has been described by Hilbert and Rodgers (1987). All other characteristics, like the on and off-axis resolutions are ignored in this map, but it gives some idea of the great diversity of optical systems.

An interesting and important characteristic of imaging optical systems is the total number of image elements it produces, which depends on the f-number and on the aperture diameter. Assuming a perfect optical system, the smaller the f-number, the smaller the image element (diffraction image)

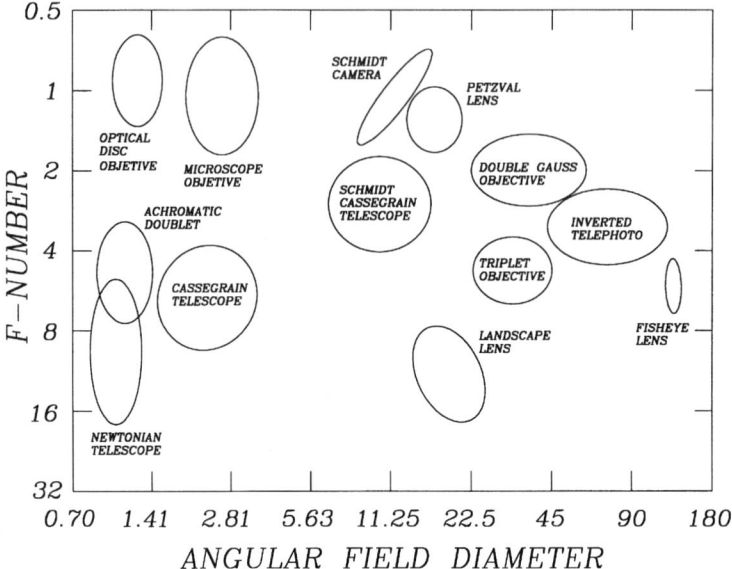

Figure 11.1 Diagram to illustrate the large diversity of optical systems.

is, as described when studying diffraction. On the other hand, given an image element size, a large field contains more image elements than a smaller field. It is easy to see, as pointed out by Hopkins (1988), that the total number of elements is equal to the square of the Lagrange invariant multiplied by $4/\lambda^2$.

The number of possible lens and mirror combinations is almost infinite. There are, however, some basic configurations that will be described in the next chapters, beginning by the simplest ones. Most optical systems may be considered as derivatives of some basic system.

11.2 SINGLE LENS

A single lens is the simplest optical instrument and has many applications. One important use is as a simple microscope or magnifier. Another application is as a simple photographic lens. These lenses will now be described.

11.2.1 Magnifiers

The apparent angular diameter of an object as measured from the pupil of the observer's eye determines the size of the image on the retina of the eye.

Optical Systems and Photographic Lenses

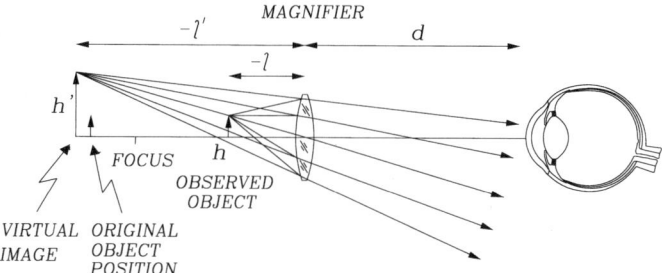

Figure 11.2 Image formation in a single magnifier.

To increase this size the distance from the object to the eye may be reduced. There is a limit, however, on this distance. As the object becomes closer, the eye has to focus by adjusting the shape of the eye lens. The younger the observer is, the shorter this distance may be. For an adult young person the average distance is about 250 mm. This distance receives the name of minimum distinct vision distance.

If we want to observe from an even closer distance, a convergent lens must be placed between the object and the observing eye. Let us consider a convergent lens as in Fig. 11.2. An object with height h is placed at a distance $-l$ in front of the lens. This distance is less than the focal length in order to obtain a virtual image (l' is also negative). Thus, we may use Eq. (2.13) to find this distance. Assuming that without the lens the observing distance is 250 mm, the angular diameter would be equal to $\alpha = h/250$, with h being in millimeters. The angular diameter for the virtual image is $\beta = h'/(-l' + d)$, where d is the distance from the eye to the lens. Then, the apparent angular magnification, or *magnifying power*, of this single microscope is

$$M = \frac{\beta}{\alpha} = \frac{250}{-l' + d}\left(-\frac{l' + 1}{f}\right) \tag{11.1}$$

where all length units are in millimeters. If we place the virtual image at infinity ($l' = \infty$), the magnifying power becomes

$$M = \frac{250}{f} \tag{11.2}$$

independently of the separation between the lens and the eye. If we try to increase the magnifying power by getting the virtual image closer to the minimum observing distance of 250 mm we obtain

$$M = \left(\frac{250}{f} + 1\right)\left(\frac{250}{250 + d}\right) \tag{11.3}$$

Since the focal length f and the distance d are less than 250 mm, in order to obtain the maximum possible magnifying power, we may see that the improvement with respect to Eq. (11.2) is negligible.

The aperture in a single magnifier is that of the observing eye, which in general is very small. Therefore, the spherical aberration, the axial chromatic aberration, and coma are negligible. Lateral chromatic aberration is unavoidable in a single lens. Thus, we are left with astigmatism, field curvature, and distortion to be corrected with the stop (the observing eye) position. As an example, let us consider the case of a plano convex lens with the following data:

Effective focal length:	100 mm
Lens diameter:	50 mm
Radius of curvature convex face:	51.67 mm
Thickness:	10 mm
Image height:	10 mm
Glass:	BK-7

If we trace an enlarged meridional plot through this lens as described in Section 9.4, using the two possible lens orientations, we obtain the graphs in Fig. 11.3. From these results we may conclude:

1. Since the two plots pass through the origin, as expected, the distortion is zero for both lens orientations when the pupil of the

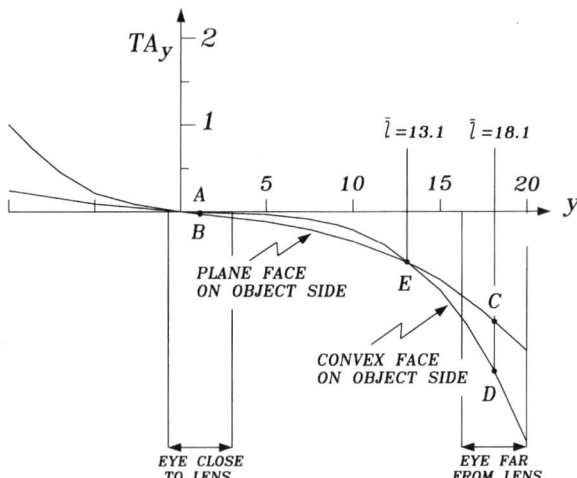

Figure 11.3 Meridional plot with an enlarged stop, in a single plano convex magnifier.

eye is in contact with the lens, and it increases with the distance of the pupil from the lens.

2. Since the eye cannot be in perfect contact with the lens, the best orientation is that with the convex face on the object side, as at point **A**. This configuration has a very small amount of coma because an inflection point is here, a flat tangential field, since the slope of the plot is zero, and a low distortion, since the distance from the point **A** to the horizontal axis is small.
3. The lens orientation at point **B**, with the plane face on the object side, has only slightly higher aberrations than those at point **A**.
4. When the observing eye is far from the lens, with the plane face on the object side, as represented by point **C** in the plot, all aberrations are higher than at points **A** and **B**.
5. When the observing eye is far from the lens, but with the convex face on the object side, as at point **D** in the plot, all aberrations are still higher than at point **C**. In conclusion, if the eye cannot be placed close to the lens, the best orientation is with the plane face on the object side.
6. There is a certain distance from the observing eye to the lens, represented by point **E**, for which both possible lens orientations produce the same amount of distortion. The only difference is that the orientation with the plane face on the object side has a slightly flatter tangential field, because of the smaller slope in the graph.

The lens configurations represented by points **A–D** in the enlarged meridional plot are illustrated in Fig. 11.4.

There are several possible designs of magnifiers that produce better images than the single lens, as shown in Fig. 11.5. These designs may be analyzed in the same manner as the single lens. As an example, Fig. 11.6 and Table 11.1 show the design of a Hastings magnifier. The image resolution is good. The most significant remaining aberration is field curvature.

11.2.2 Biocular Magnifiers

Biocular magnifiers are designed to observe the image with both eyes. They are generally placed close to the head of the observer and must have a sufficiently large diameter to permit simultaneous observation with both eyes. The diameter of the lens on the side of the observer should be at least 75 mm. The most common application of these magnifiers is to look at the image of a small cathode ray tube (CRT) or any other electronic display. This subject has been covered by several authors (Hopkins, 1946; Coulman and Petrie, 1949

264   Chapter 11

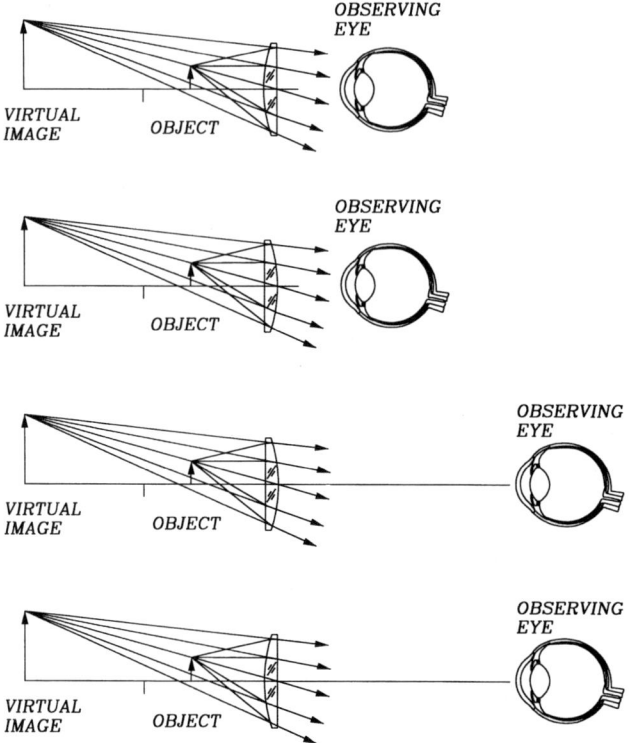

Figure 11.4 Four ways to observe with a single plano convex magnifier.

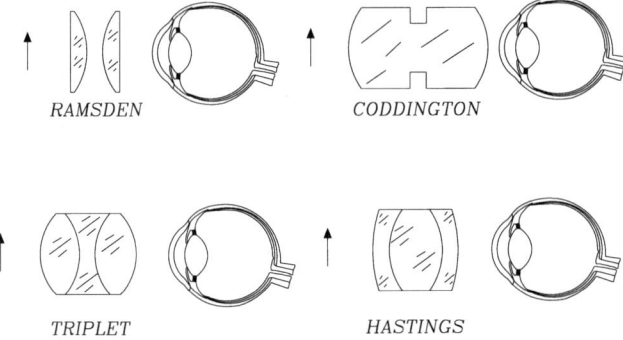

Figure 11.5 Some magnifiers.

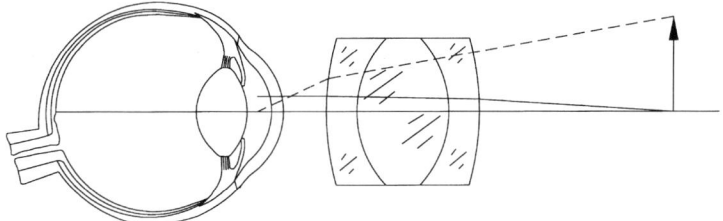

Figure 11.6 A 7 × Hastings magnifier.

Table 11.1 A 10 × Hastings Magnifier

Radius of curvature (mm)	Diameter (mm)	Separation or thickness (mm)	Material
Stop	5.0	10.00	Air
44.31	24.0	5.00	F2
17.65	24.0	15.00	SK16
−17.65	24.0	5.00	F2
−44.31	24.0	31.35	Air

Aperture (mm): 5.0
Effective focal length (mm): 40.0
Back focal length (mm): 31.35
Image height (angular semifield) (mm): 18.65 (25°)

and Rosin, 1965). A good review article on this subject with some references has been written by Rogers (1985).

The fields observed by each of the two eyes is not identical, as illustrated in Fig. 11.7. One eye sees the object from A to A' while the other eye sees it from B to B', but there is a common overlapping field. The perspective for the two eyes is different. So, if the object is not flat the virtual images provide a stereoscopic view.

The optical design is carried with a reversed orientation so that the longest conjugate is on the object side. The stop is laterally shifted with respect to the optical axis of the system, but an easier approach is to consider a large stop that covers both eye pupils. Many different designs of binocular magnifiers have been published and patented.

11.2.3 Single Imaging Lens

A single lens may be used to form real images on a screen or photographic film. The focal length is fixed, since the magnification is predetermined.

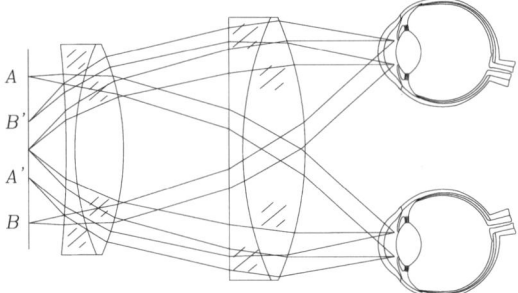

Figure 11.7 Biocular magnifier.

Then, we have only two degrees of freedom, the lens bending and the stop position. We know from third-order theory the following facts:

1. The spherical aberration cannot be completely corrected using only spherical surfaces, but may be minimized. This aberration is not important if the aperture is small, since its transverse value increases with the cube of the aperture.
2. The coma aberration for any object position, lens bending, and stop position may be calculated as described in Chap. 6. The position for the stop producing zero coma for thin lenses and the object at infinity is

$$\bar{l} = f\left(1 - \frac{SphL}{f - f_M}\right) \tag{11.4}$$

   but we must remember that this expression is valid only for lenses with spherical aberration.
3. The Petzval curvature increases linearly with the power of the lens and is independent of the lens bending. Thus, it is unavoidable in a single lens. The field may be flattened, however, if some astigmatism is introduced on purpose. A desirable condition is a flat tangential field. This has to be done by a proper selection of the stop position. The value of the longitudinal Petzval curvature is given by

$$Ptz = -\frac{h_k'^2}{2nf} \tag{11.5}$$

4. The astigmatism, when the stop is in contact with the lens, is independent of the lens bending and is directly proportional to

the power of the lens. This astigmatism is

$$AstL_s = \frac{h_k'^2}{2f} \tag{11.6}$$

If the stop is not in the lens plane, its position and the bending of the lens become useful parameters. Since the Petzval curvature is negative, the field may be flattened if the proper amount of negative astigmatism is introduced.

5. The distortion is zero if and only if the stop coincides with the lens.
6. The axial chromatic aberration is fixed, given the focal length and has a value for the longitudinal component:

$$AchrL = \frac{f(N_F - N_C)}{(N-1)} = \frac{f}{V} \tag{11.7}$$

or for the transverse aberration component:

$$AchrT = \frac{y}{V} \tag{11.8}$$

7. The magnification chromatic aberration is zero if and only if the stop is at the same position as the single lens.

Thus, we may play with only two variables, the lens bending and the stop position to obtain the desired results, according to the application of the lens. To have a feeling for the possible solutions, let us examine the meridional plots for a single lens with several lens bendings, keeping constant the focal length, as shown in Fig. 11.8. This lens has the following data:

Effective focal length:	100 mm
Lens diameter:	20 mm
Curvatures of front face:	as indicated in Fig. 11.7, in 1/mm
Image height:	36.4 mm (20°)
Thickness:	2 mm
Glass:	BK-7

The lens bending in Fig. 11.8(e) corresponds to the solution for zero coma (with the stop at the lens) and minimum spherical aberration. However, the large slope of the plot indicates tangential field curvature. Since the plot is a straight line, this lens has constant values of the tangential

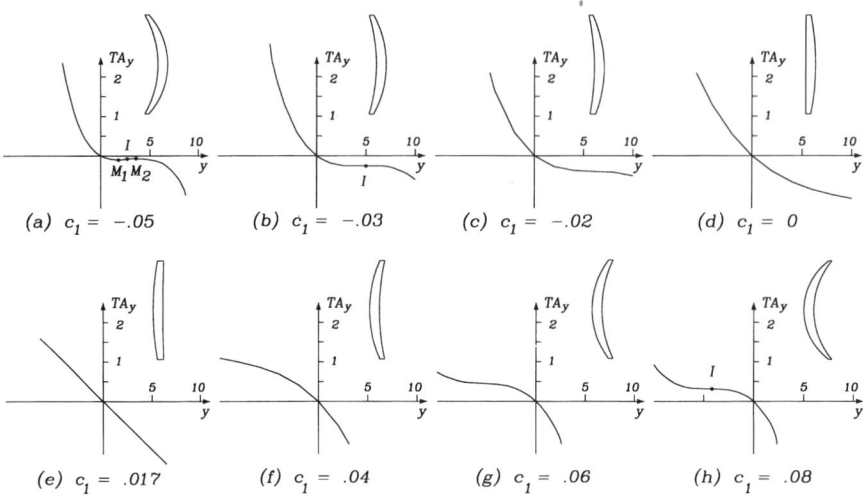

Figure 11.8 Meridional plot with an enlarged stop in a landscape lens with several different lens bendings.

field curvature and coma for all stop positions. All curves pass through the origin, indicating zero distortion when the stop is at the lens plane. All curves have the same slope at the origin, because the tangential surface has the same curvature for all bendings. This is to be expected, since the tangential curvature is given by the Petzval curvature and the astigmatism when the stop is at the lens plane, and are both independent of the lens bending.

11.2.4 Landscape Lenses

A single photographic lens is frequently called a landscape lens. This is a lens in which the stop has been shifted at the expense of some distortion. The stop has a diameter less than that of the lens, as indicated in the figures. This small stop drastically reduces the effect of the spherical and axial chromatic aberrations. A focal ratio of about 15 or larger is typical for these lenses, widely used in old photographic cameras.

Observing Fig. 11.8 we may see that any points in these plots without any curvature must be free of coma. These points are the inflection points, indicated with an **I** in Figs. 11.8(a), (b), and (h). Similarly, any points with zero slope must have a flat tangential focal surface. These points are minima, maxima, or horizontal inflection points, as indicated in Figs. 11.8(a), by M_1 and M_2, and in Figs. 11.8(b) and (h) by **I**. It should be noticed in Fig. 11.8(a) that the minimum M_1 has a larger distortion

$(TA_y$ larger) than the maximum $\mathbf{M}_2$. In these graphical results we may observe the following interesting facts

1. Only a certain amount of bending as in Figs. 11.8(b) and Fig. 11.8(h) produces an inflection point (no coma) with zero slope (flat tangential field).
2. There are two solutions, one with the stop in the front of the lens as shown in Fig. 11.8(b), and one with the stop in the back of the lens as in Fig. 11.8(h).
3. The distortion has an opposite sign for the two solutions, and larger in magnitude for the stop in the back. For these solutions, small stop shifts do not produce any change in the distortion.
4. The lens is more curved with the stop in the back of the lens, and thus its spherical aberration is larger.
5. There is some axial and magnification chromatic aberration, but not very large.

Once the height of the principal ray has been chosen from the meridional plots, the position of the stop is calculated with Eqs. (9.1) for a lens with the stop at the back or with Eq. (9.2) for a lens with the stop at the front. The stop diameter is chosen so that the spherical aberration is not noticeable. Finally, it is interesting to point out that, when the tangential surface is flat, the sagitta of the Petzval surface, or Petzval curvature Ptz, is equal to the tangential astigmatism. Since this value of Ptz is independent of the bending, the final value of the astigmatism, after making the tangential surface flat, is the same for both bending solutions. Figure 11.9 shows two designs of landscape lenses, based on the meridional plots in Fig. 11.8. The design data for a front stop landscape are presented in Table 11.2 and those for the rear stop landscape lens in Table 11.3.

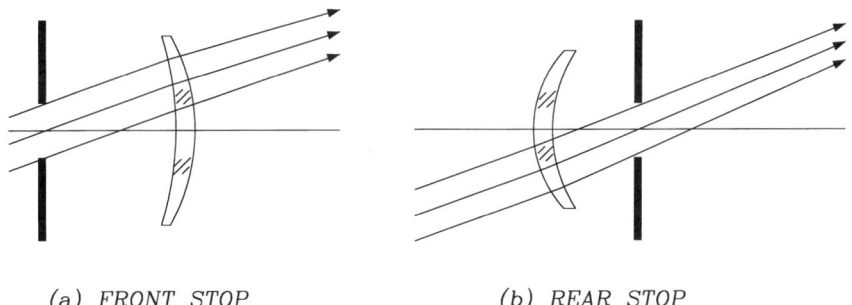

(a) FRONT STOP (b) REAR STOP

Figure 11.9 Two possible configurations for a landscape lens: (a) front stop; (b) rear stop.

Table 11.2 Front Stop Landscape Lens

Radius of curvature (mm)	Diameter (mm)	Separation or thickness (mm)	Material
Stop	5.0	13.74	Air
−33.333	20.0	2.0	BK7
−20.678	20.0	—	Air

Aperture (focal ratio) (mm): 5.0 ($F/20$)
Effective focal length (mm): 100.0
Back focal length (mm): 102.044
Object distance (mm): infinite
Image height (angular semifield) (mm): 36.4 (20°)

Table 11.3 Rear Stop Landscape Lens

Radius of curvature (mm)	Diameter (mm)	Separation or thickness (mm)	Material
12.500	17.0	2.0	BK-7
15.589	17.0	8.81	Air
Stop	5.0	—	Air

Aperture (focal ratio) (mm): 5.0 ($F/20$)
Effective focal length (mm): 100.0
Back focal length (mm): 94.548
Object distance (mm): infinite
Image height (angular semifield) (mm): 36.4 (20°)

Even though the front stop solution is optically better, the back stop solution is frequently preferred, because the lens may be easily cleaned and it is aesthetically better.

11.3 SPHERICAL AND PARABOLOIDAL MIRRORS

The first-order parameters in spherical mirror (concave or convex and spherical or paraboloidal) with the object at a finite distance l and the stop at a finite distance in front of the mirror will now be written. In Fig. 11.10 these parameters are represented for a concave surface (radius of curvature negative), but the results are valid for concave as well as for convex mirrors. After some algebraic steps using Eqs. (1.38) and (1.46) we may see that the values of i, i', u, and u' for the meridional ray and the refractive indices n and n' can be expressed by

Optical Systems and Photographic Lenses

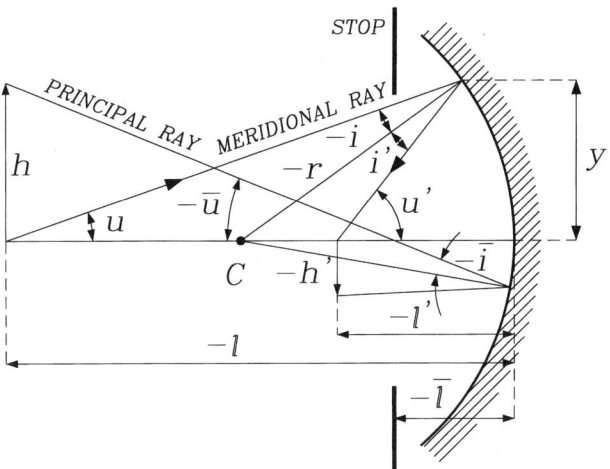

Figure 11.10 First-order parameters in a concave mirror with the object at a finite distance from the mirror.

$$i = -i' = \left(\frac{1}{r} - \frac{1}{l}\right) \quad y = \left(-\frac{1}{2f} - \frac{1}{l}\right)\frac{D}{2} \quad u = -\frac{y}{l} = -\frac{D}{2l}$$

$$u' = -\frac{(2/r - 1/l)}{(1/r - 1/l)} \quad i = -\left(\frac{2}{r} - \frac{1}{l}\right) \quad y = -\left(-\frac{1}{f} - \frac{1}{l}\right)\frac{D}{2}$$

$$n' = -n = -1 \quad (11.9)$$

When the object is at an infinite distance (see Fig. 11.11) from the mirror these expressions reduce to

$$i = -i' = \frac{y}{r} = -\frac{D}{4f}$$

$$u = 0$$

$$u' = -2i = -\frac{2y}{r} = \frac{D}{2f}$$

$$n' = -n = -1 \quad (11.10)$$

For the principal ray we may find that

$$\bar{i} = -\left(\frac{\bar{l} - r}{r}\right)\bar{u} = \left(\frac{\bar{l} - r}{r}\right)\left(\frac{2/r - 1/l}{1 - \bar{l}/l}\right)h' \quad (11.11)$$

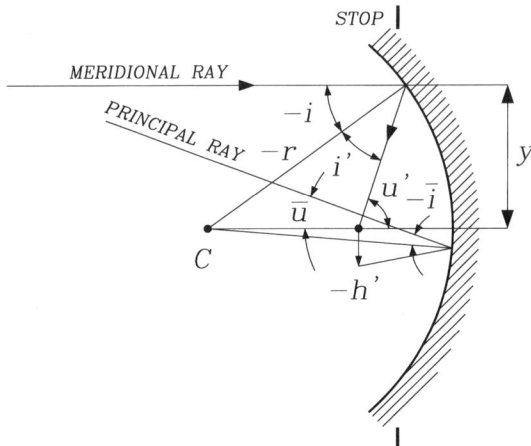

Figure 11.11 First-order parameters in a concave mirror with the object at an infinite distance from the mirror.

thus, obtaining

$$\frac{\bar{i}}{i} = -\frac{(2/r - 1/l)\left(1 - \bar{l}/r\right)}{(1/r - 1/l)\left(1 - \bar{l}/l\right)} \frac{h'}{y} \tag{11.12}$$

which, when the object is at an infinite distance from the mirror, reduces to

$$\frac{\bar{i}}{i} = 2\left(\frac{\bar{l} - r}{r}\right)\frac{h'}{y} \tag{11.13}$$

These expressions are valid for spherical as well as for paraboloidal mirrors.

11.3.1 Off-Axis Aberrations for Spherical Mirrors

Now we will find the expressions for the primary off-axis aberrations for a concave or convex spherical mirror.

Spherical Aberration

From Eq. (4.24) we can show that the primary longitudinal spherical aberration for a spherical mirror, with the object at a finite distance l in front of this mirror, is given by

$$SphL = -\frac{(1/r - 1/l)^2}{(2/r - 1/l)^2} \frac{y^2}{r} = -\left(\frac{r - l}{r - 2l}\right)^2 \frac{y^2}{r} \tag{11.14}$$

Coma

As shown in Eqs. (5.40) and (5.49) the coma and astigmatism aberrations depend on the value of the spherical aberration. In turn the value of the spherical aberration depends on the position of the object. Thus, the values of the coma and astigmatism are functions of the object position as well as the stop position. The value of the sagittal coma for an object at a distance l in front of the mirror can be shown to be given by

$$Coma_S = -\frac{(1/r - 1/l)(1 - \bar{l}/r)}{(1 - \bar{l}/l)} \frac{y^2 h'}{r} \tag{11.15}$$

Astigmatism

From Eqs. (5.49) and (11.12), the longitudinal sagittal astigmatism when the object is at a distance l in front of the mirror is given by

$$AstL_S = -\frac{(1 - \bar{l}/r)^2}{(1 - \bar{l}/l)^2} \frac{h^2}{r} \tag{11.16}$$

Petzval Curvature

From Eq. 5.23, the value of the Petzval curvature is

$$Ptz = \frac{h'^2}{r} = \frac{h'^2}{2f} \tag{11.17}$$

11.3.2 Concave Spherical Mirror

Let us analyze each of the monochromatic primary aberrations on a concave spherical mirror.

Spherical aberration

From Eq. 11.14 we can see that a spherical mirror is free of spherical aberration when the object and the image are both at the center of curvature or at the vertex of the mirror ($l=r$). If the object is at infinity ($l=\infty$ and $l'=-f=r/2$) the longitudinal spherical aberration becomes

$$SphL = \frac{y^2}{4r} = \frac{D^2}{32f} = \frac{z}{2} \tag{11.18}$$

and the transverse aberration is

$$SphT = \frac{y^3}{2r^2} = \frac{D^3}{64f^2} \tag{11.19}$$

where D is the diameter of the mirror, z is the sagitta of the surface, and r is the radius of curvature. Integrating this expression and using Eq. (1.76), with the radius of curvature of the wavefront equal to the focal length of the spherical mirror ($r/2$) we may find that the wavefront aberration is given by

$$W(y) = \frac{y^4}{4r^3} \tag{11.20}$$

It is interesting to see that this result is twice the sagitta difference $Z = B_1 S^4$ between a sphere and a paraboloid, given by Eqs. (A2.8) and (A2.9). This is to be expected, since the paraboloid is free of spherical aberration. In conclusion, the wavefront spherical aberration of a spherical mirror is twice the separation between the paraboloid and the sphere.

Coma

Restricting our analysis to the particular case of an object at an infinite distance in front of the concave mirror, from Eq. (11.15) the sagittal coma is given by

$$Coma_S = -\frac{y^2 h'(\bar{l} - r)}{r^3} \tag{11.21}$$

when the stop is at the mirror ($\bar{l} = 0$), this value of the sagittal coma is

$$Coma_S = \frac{D^2 h'}{16f^2} \tag{11.22}$$

If the exit pupil is at the center of curvature ($\bar{l} = r$), the value of the sagittal is $Coma_S = 0$.

Astigmatism

From Eq. (11.16), for the case of an object at an infinite distance, we obtain

$$AstL_S = \frac{h'^2}{r}\left(\frac{\bar{l} - r}{r}\right)^2 \tag{11.23}$$

As explained before, this result is valid only for an infinite distance from the object to the mirror. By using Eq. (11.11), the primary longitudinal sagittal astigmatism may be written as

$$AstL_S = -\left(\frac{\bar{l}-r}{r}\right)^2 Ptz \qquad (11.24)$$

The value of the primary longitudinal tangential $AstL_T$ is equal to three times this value.

Petzval Curvature

From Eq. (11.14) we see that the Petzval surface is concentric with the mirror. If the object is not at an infinite distance from the mirror the image is displaced to the corresponding conjugate distance, but its Petzval curvature remains constant.

The sagitta of the sagittal focal surface is equal to the sum of the Petzval curvature plus the longitudinal sagittal astigmatism. The sagitta of the tangential focal surface is equal to the sum of the Petzval curvature plus the longitudinal tangential astigmatism. The sagitta for the surface of best definition is given by

$$Best = \left[1 - 2\left(\frac{\bar{l}-r}{r}\right)^2\right] Ptz \qquad (11.25)$$

When the stop is at the center of curvature the astigmatism disappears, but the field has Petzval curvature, as shown in Fig. 11.12(a). If the stop is at

$$\frac{\bar{l}}{r} = \pm\frac{1}{\sqrt{3}} + 1 = 0.42; \; 1.58 \qquad (11.26)$$

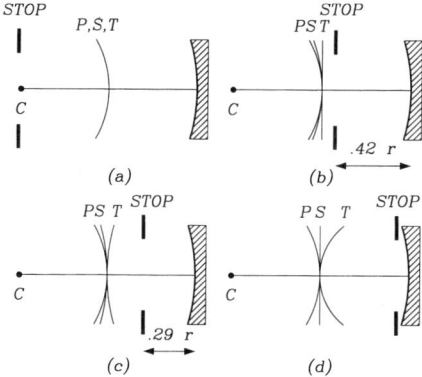

Figure 11.12 Astigmatic surfaces for a spherical mirror with four different stop positions.

the tangential surface is flat, as shown in Fig. 11.12(b). If the stop is placed at

$$\frac{\bar{l}}{r} = \pm\frac{1}{\sqrt{2}} + 1 = 0.29; 1.707 \qquad (11.27)$$

the surface of best definition, located between the sagittal and the tangential surfaces, is a plane, as shown in Fig. 11.12(c). When the stop is at the mirror the longitudinal tangential astigmatism is of opposite sign, as shown in Fig. 11.12(d). In this case the sagittal surface is flat.

When the object and the image are at the center of curvature of the concave spherical mirror, the spherical aberration, coma, and astigmatism are zero. Only the Petzval curvature exists.

11.3.3 Concave Paraboloidal Mirror

Again, let us examine each of the primary aberrations in a concave paraboloidal mirror.

Spherical Aberration

In a paraboloidal mirror there is no spherical aberration when the object is at infinity. However, if the object is at the center of curvature spherical aberration appears. We see in Eq. (A2.16) that the exact expression for the longitudinal aberration of the normals to the mirror is given by

$$SphL_{normals} = f tan^2 \varphi \qquad (11.28)$$

as illustrated in Fig. 11.13. The spherical aberration of the paraboloid when the object is at the center of curvature is approximately twice the aberration of the normals. Thus, we may write

$$SphL = 2f\frac{y^2}{r^2} = \frac{y^2}{r} = -\frac{D^2}{8f} \qquad (11.29)$$

If we compare this result with the spherical aberration for the spherical mirror with the object at infinity, we see that their absolute values are different by a factor of four and opposite in sign. This is easy to understand if we notice that their wavefront aberrations must have opposite signs and the same absolute values. In the sphere the observing plane is at a distance $-f = r/2$ while in the paraboloid it is at a distance r.

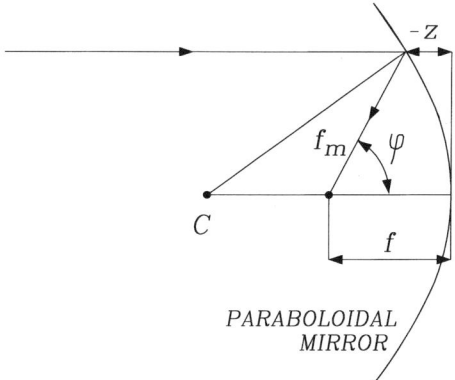

Figure 11.13 Paraboloidal mirror.

Then, the absolute values of their transverse aberrations must be different by a factor of two and the absolute values of their longitudinal aberrations by a factor of four.

Coma

Considering now the object at infinity, the paraboloid would be free of coma if the stop is at the mirror, only if the principal surface is centered at the focus, but this in not the case. The principal surface is the paraboloidal surface itself. Thus, the value of OSC (defined in Sec. 5.3.1) is given by

$$OSC = \frac{f_M}{f} - 1 \tag{11.30}$$

where f_M and f are the marginal and paraxial focal lengths, as measured along the reflected rays, as shown in Fig. 11.13. For a paraboloid, we may show that

$$f_M = f \mp z \tag{11.31}$$

where z is the sagitta, given by

$$z = -\frac{y^2}{4f} = -\frac{D^2}{16f} \tag{11.32}$$

Then, the value of the sagittal coma is equal to

$$Coma_S = OSC \cdot h' = -\frac{zh'}{f} \tag{11.33}$$

or

$$Coma_S = \frac{D^2 h'}{16f^2} \tag{11.34}$$

This value is identical to the value obtained for a spherical mirror. It may also be obtained by adding the spherical and aspherical contributions from Eqs. 5.41 and 5.85. However, it may be seen that the aspherical contribution is zero when the stop is at the mirror, which explains why the coma is the same for spherical and paraboloidal mirrors with the stop at the mirror.

Astigmatism

In a paraboloid the spherical aberration with the object at infinity is zero. If we separate the spherical and aspherical contributions of these aberrations we may write

$$SphL_{sphere} + SphL_{asphere} = 0 \tag{11.35}$$

and similarly for the longitudinal sagittal astigmatism

$$AstL_{S\ total} = AstL_{S\ sphere} + AstL_{S\ asphere} \tag{11.36}$$

which, by using Eqs. 5.50 and 5.90 is

$$AstL_{S\ total} = SphL_{sphere} \left[\left(\frac{\bar{i}}{i}\right)^2 \left(\frac{\bar{y}}{y}\right)^2 \right] \tag{11.37}$$

or

$$AstL_{S\ total} = AstL_{S\ sphere} \left[1 - \left(\frac{i\bar{y}}{\bar{i}y}\right)^2 \right] \tag{11.38}$$

Optical Systems and Photographic Lenses

The astigmatism when the stop is at the mirror is equal to the astigmatism of a spherical mirror. Then, after some algebraic manipulation, we obtain

$$AstL_{S\,total} = AstL_{S\,sphere}\left[\frac{(\bar{l}-r)^2 - \bar{l}^2}{(\bar{l}-r)^2}\right] \quad (11.39)$$

then, using Eq. (11.24), we obtain

$$AstL_{S\,total} = \left[\frac{(\bar{l}-r)^2 - \bar{l}^2}{r^2}\right]Ptz \quad (11.40)$$

Then, using Eq. (5.67) for the sagitta of the surface of best definition:

$$Best = \left[1 - 2\left(\frac{r-2\bar{l}}{r}\right)^2\right]Ptz \quad (11.41)$$

we see that the surface of best definition is flat when $\bar{l}/r = 0.25$.

11.3.4 Convex Spherical Mirror

Let us now study a convex spherical mirror. As shown in the diagram in Fig. 2.9 for diverging lenses, a convex mirror cannot produce real images with real objects. A real image can be produced only with a virtual object as in the case on the Cassegrain telescope to be studied in Chap. 15. A virtual object can produce a virtual image. We will describe here only configurations with a real object and thus producing a virtual image.

If the entrance pupil is at the center of curvature of the mirror the image will have a strong curved spherical focal surface, which is concentric with the mirror, as shown in Fig. 11.14. The only aberrations in the system are spherical aberration and Petzval curvature. Coma and astigmatism are zero. The problem with this system is that the entrance pupil is behind the convex mirror and thus this configuration is possible only if the mirror is part of a more complicated arrangement.

A more frequent configuration is when a virtual image is observed with the eye and the stop is the pupil of the observing eye. If the object is flat and infinitely extended the virtual image would be strongly curved as shown in Fig. 11.15. This convex lens acts as an extremely wide angle system covering a field of view of one-half a sphere.

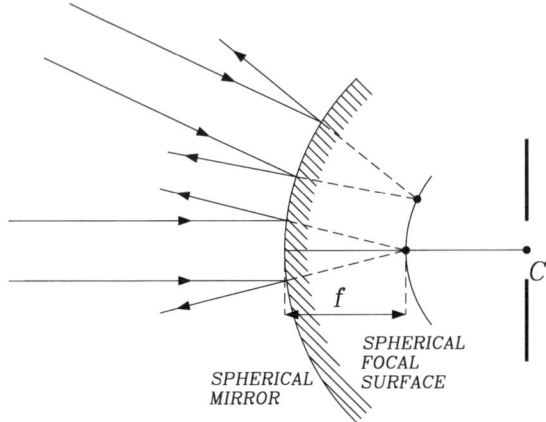

Figure 11.14 Convex spherical mirror with the stop at the center of curvature and a plane object.

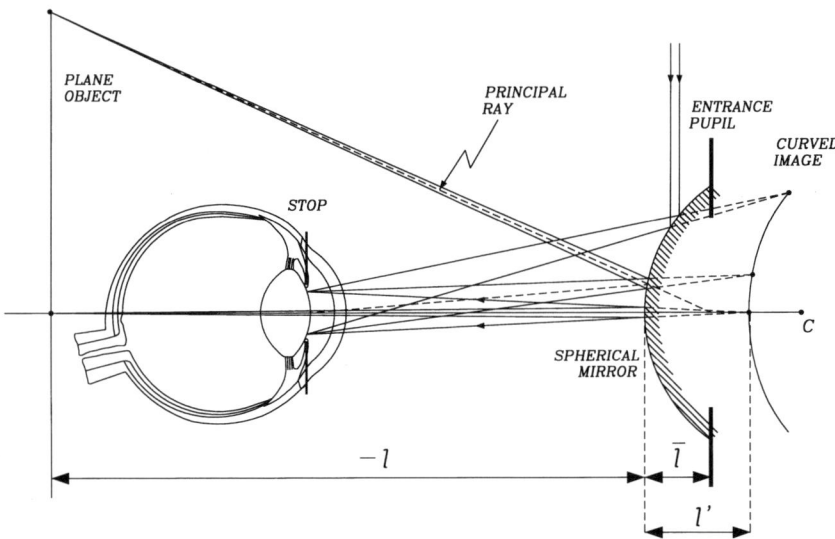

Figure 11.15 Convex spherical mirror with the stop at the center of curvature and a plane image.

When a flat virtual object is reflected on the mirror the real image has a strong curvature as shown in Fig. 11.16. For small fields the Petzval, sagittal, and tangential fields are spheres as predicted by the primary aberration theory, (see dotted lines in Fig. 11.16). For extremely large fields the shape of the image surfaces are ovoids highly resembling ellipsoids.

Optical Systems and Photographic Lenses

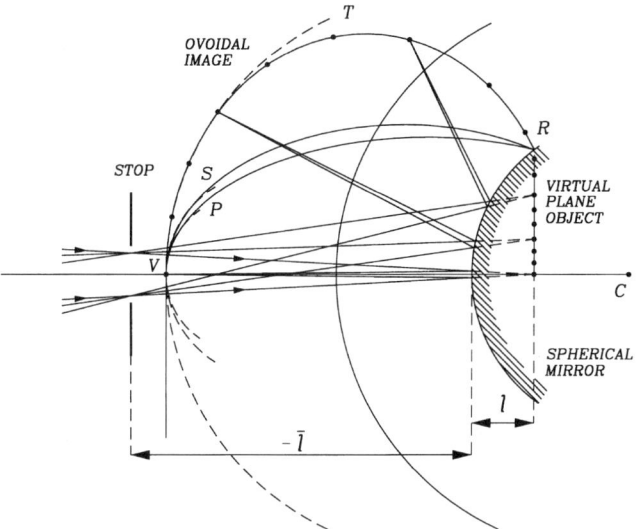

Figure 11.16 Convex spherical surface with a flat virtual object and the stop in front of the mirror.

There is a different ovoid for sagittal and tangential rays, as shown by Mejía-Barbosa and Malacara-Hernández (2001). These surfaces can be calculated with the help of the Coddington equations or with exact ray tracing. At the vertex **V** these ovoids have the same curvatures as the primary Petzval, sagittal, and tangential surfaces. These ovoids intersect at a common circle around the optical axis, indicated by the point **R** at the circular intersection of the flat object with the spherical mirror.

If the light paths are reversed the flat virtual object becomes a flat virtual image and the ovoidal images become ovoidal objects. This arrangement has been proposed by Mejía-Barbosa and Malacara-Hernández (2001) for a corneal topographer.

11.4 PERISCOPIC LENS

If two meniscus lenses are placed together with their concave surfaces facing toward each other, in a symmetrical configuration as in Fig. 11.17, we have a system invented many years ago with the trade name of periscopic lens. If the system is completely symmetric, including the object and the image distances, the coma, distortion, and magnification chromatic aberrations are automatically cancelled out. Thus, we do not have to worry about the coma correction with the lens bending. Thus, in the curve in Fig 11.8(a), we may choose a point with slope zero (flat tangential field) and we do not need to worry about the coma (plot curvature). If we examine this plot we see that there are two points M_1 and M_2 that satisfy our conditions. The point M_1 has a larger distortion, but since it is going to be canceled anyway because of the symmetry of the system, we choose this point because it produces a more compact system. The bending of the lens is stronger than for the landscape lens. The separation between the lenses is calculated from Eq. (11.2).

The complete symmetry cannot be preserved if the object is at infinity. However, we may see by ray tracing analysis that the state of correction remains surprisingly good even when the object is at infinity. The design of the periscopic lens with the object at infinity is presented in Table 11.4.

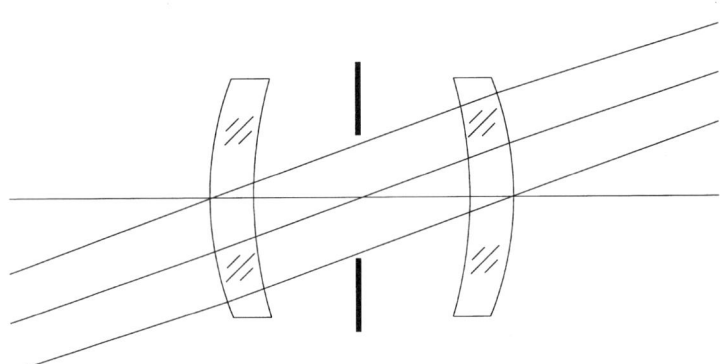

Figure 11.17 Periscopic lens.

Table 11.4 Periscopic Lens

Radius of curvature (mm)	Diameter (mm)	Separation or thickness (mm)	Material
20.000	12.0	2.0	BK-7
14.911	12.0	9.89	Air
Stop	5.0	9.89	Air
−14.911	12.0	2.0	BK-7
−20.000	12.0	—	Air

Aperture (focal ratio) (mm): 5.0 ($F/11.7$)
Effective focal length (mm): 58.46
Back focal length (mm): 44.96
Object distance (mm): infinite
Image height (angular semifield) (mm): 21.28 (20°)

Optical Systems and Photographic Lenses 283

11.5 ACHROMATIC LANDSCAPE LENSES

It is natural to think about the possibility of achromatizing the landscape lens to eliminate both chromatic aberrations. We have seen [(Eq. 6.55)] that, if the axial chromatic aberration is corrected, the stop shift will not introduce any magnification chromatic aberration, so both aberrations will be corrected. To achromatize we use two different glasses, crown and flint, as illustrated in Fig. 11.18. This achromatization may be done in many ways, e.g., by using the D–d method or by ray tracing.

The next step, as in the landscape lens, would be to bend the lens until the coma is made zero and the tangential field is made flat. However, the tangential field cannot be flattened. The reason is that the concave front surface and the crown–flint interface contribute a large positive astigmatism, making the tangential field backward curved. This type of achromatic landscape lens is called the Chevalier lens.

Unfortunately, with normal glasses the bending of the lens increases the astigmatism contribution of the glass interface while reducing the contributions of the other surfaces. Thus, this achromatization has to pay the price of increasing the astigmatism or the field curvature. In other words, the meridional plot in Fig. 11.18 cannot be made to have an inflection point I with zero slope. Then, we may correct coma by selecting the point I for the height of the principal ray, or alternatively, we may flatten the tangential surface by selecting the point M. Figure 11.19 shows an achromatic landscape lens without coma and its design data are presented in Table 11.5.

An obvious solution to the problem of the large positive astigmatism in this lens is to eliminate the contribution of the glass interface by making the positive lens with a glass with the same refractive index or even higher than that of the negative lens, but with different Abbe numbers. This

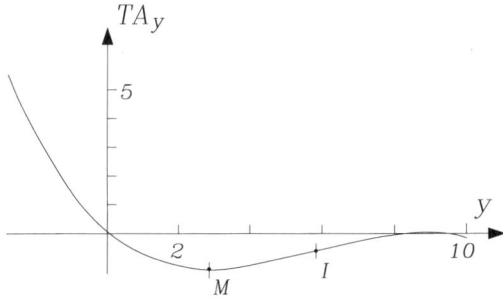

Figure 11.18 Meridional plot with an enlarged stop for one-half of the periscopic lens.

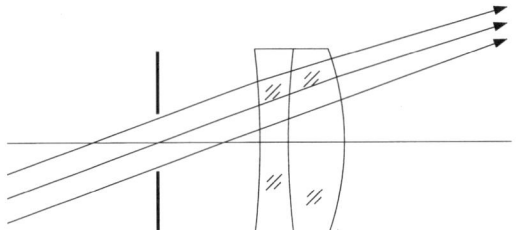

Figure 11.19 Achromatic landscape lens.

Table 11.5 Achromatic Landscape Lens

Radius of curvature (mm)	Diameter (mm)	Separation or thickness (mm)	Material
Stop	12.0	10.71	Air
−89.622	12.0	3.0	F2
89.622	12.0	6.0	BK-7
−29.933	12.0	—	Air

Aperture (focal ratio) (mm): 5.0 ($F/20.0$)
Effective focal length (mm): 100.50
Back focal length (mm): 105.00
Object distance (mm): infinite
Image height (angular semifield) (mm): 36.40 (20°)

approach requires glasses, like the barium crown glass. This is the *new achromatic landscape lens*. This glass combination reduces the Petzval curvature, increasing the field. However, the spherical aberration is worse in the new achromat than in the Chevalier lens.

11.6 ACHROMATIC DOUBLE LENS

The achromatic landscape lens could also be made with the meniscus-shaped crown positive lens on the side of the stop. This approach, proposed in 1857 by the company Thomas Grubb in Great Britain, makes the crown–flint interface very curved. The advantage is that the spherical aberration is very well corrected, but again, it is impossible to correct coma and field curvature at the same time.

A natural thing to do then is to place two of these lenses in a symmetrical configuration, to correct coma with the symmetric configuration, as shown in Fig. 11.20. This lens was given the name of Rapid

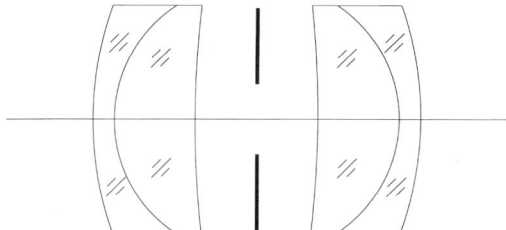

Figure 11.20 Rapid Rectilinear lens.

Rectilinear. As Kingslake (1978) points out, the Rapid Rectilinear was one of the most popular lenses ever made.

11.7 SOME CATOPTRIC AND CATADIOPTRIC SYSTEMS

A catoptric system is one formed only by mirrors. A catadioptric system is formed by both, lenses and mirrors. In this section we will study some of these systems.

11.7.1 Mangin Mirror

This mirror, illustrated in Fig. 11.21, was invented in 1876 in France by Mangin, as an alternative for the paraboloidal mirror used in searchlights. It is made with a meniscus negative lens coated with a reflective film on the convex surface. The radius of curvature of the concave front surface is the variable used to correct the spherical aberration. The coma is less than half that of a paraboloidal mirror. This system has two important advantages. One is that the surfaces are spherical not paraboloidal, making easier their construction. The second advantage is that the reflecting coating is on the back surface, avoiding air exposure and oxidation of the metal.

A Mangin mirror with a focal length F, made with crown glass BK-7, can be designed with the following formulas:

$$r_1 = 0.1540T + 1.0079F \qquad (11.42)$$

and

$$r_2 = 0.8690T + 1.4977F \qquad (11.43)$$

where T is the thickness, r_1 is the radius of curvature of the front surface, and r_2 is the radius of curvature of the back surface.

A Mangin mirror design is in Table 11.6. The system may be achromatized, if desired.

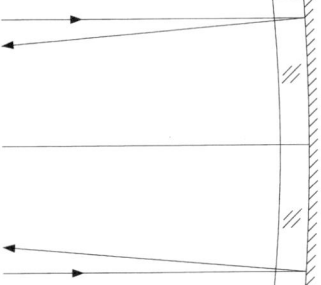

Figure 11.21 Mangin mirror.

Table 11.6 Mangin Mirror

Radius of curvature (mm)	Diameter (mm)	Separation or thickness (mm)	Material
−505.80	100.0	10.0	BK-7
−758.00	100.0	10.0	BK-7
−505.80	100.0	—	Air

Aperture (focal ratio) (mm): 100.0 ($F/5.0$)
Effective focal length (mm): 500.23
Back focal length (mm): 105.00
Object distance (mm): infinite
Image height (angular semifield) (mm): 0.0 (0°)

11.7.2 Dyson System

Unit magnification systems are very useful for copying small structures, or drawings, e.g. in photolithography in the electronics industry. In general, these systems are symmetric, automatically eliminating coma, distortion, and magnification chromatic aberration. One of these systems, illustrated in Fig. 11.22, was designed by Dyson.

The system is concentric. A marginal meridional ray on-axis leaving from **C** would not be refracted. Thus, spherical aberration and axial chromatic aberration are absent. The radius of curvature r_L of the lens is

$$r_L = \left(\frac{n-1}{n}\right) r_M \qquad (11.44)$$

where r_M is the radius of curvature of the mirror, in order to make the Petzval sum zero. The primary astigmatism is also zero, since the spherical aberration contribution of both surfaces is zero. However, the high-order

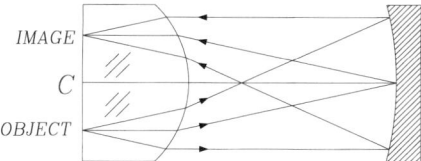

Figure 11.22 Dyson system.

Table 11.7 Dyson System

Radius of curvature (mm)	Diameter (mm)	Separation or thickness (mm)	Material
	500.00	340.72	BK-7
−340.72	500.00	659.28	Air
−1000.00	500.00	−659.28	Air
−340.72	500.00	−340.72	BK-7

Image height in mm (off-axis separation): 150.0

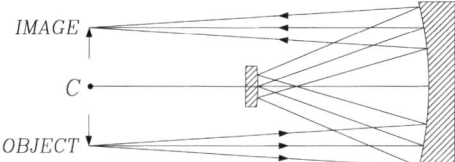

Figure 11.23 Offner system.

astigmatism appears not very far from the optical axis. Thus, all primary aberrations are corrected in this system.

It may be noticed that since the principal ray is parallel to the optical axis in the object as well as in the image medium, the system is both frontal and back telecentric. Table 11.7 shows the design of a Dyson system.

11.7.3 Offner System

The Offner system is another 1:1 magnification system, formed only by mirrors, as shown in Fig. 11.23. The system is concentric and with zero Petzval sum as in the case of the Dyson system. This system may be also corrected for all primary aberrations, but since higher order astigmatism is large in this configuration, actual Offner systems depart from this configuration. Primary and high-order astigmatism are balanced at a field zone to form a well-corrected ring where the sagittal and the tangential surfaces intersect.

11.8 FRESNEL LENSES AND GABOR PLATES

A Fresnel lens is formed by rings with different slopes and widths as shown in Fig. 11.24. We may think of such a lens as a plano convex thick lens whose thickness has been reduced by breaking down the curved face in concentric rings. The width of each ring increases with the square of its radius, as the sagitta of the thick lens. Then, the lens has constant overall thickness. Fresnel lenses have been made in many sizes, even on the micro scale (micro Fresnel lenses) (Nishihara and Suhara, 1987).

The spherical aberration of these lenses may be controlled by bending as in ordinary lenses. This is a redistribution of the power among the two surfaces, by departing from the plano convex configuration. This method, however, is seldom used, because of construction reasons it is more convenient to have the grooved surface on only one side. Another method for reducing the spherical aberration is by introducing an aspheric surface. This is done by controlling the slope of each grove in the proper way. This method is more common than the former one.

The primary aberrations of Fresnel lenses have been studied in detail by Delano (1974, 1976, 1978, 1979, 1983). Some interesting results are obtained. For example, that there exists a new kind of coma term called linear coma that does not appear in normal lenses. The coma image is shown in Fig. 11.25.

One very important difference exists between ordinary lenses and Fresnel lenses. Assuming that no aberrations are present, in ordinary lenses a spherical wavefront is produced, as shown in Fig. 11.26(a). In Fresnel lenses, however, a randomly ring-stepped wavefront is produced in general, as shown in Fig. 11.26(b). The reason is that the refracted ray direction is controlled by means of the grove slope, but the thickness is not controlled

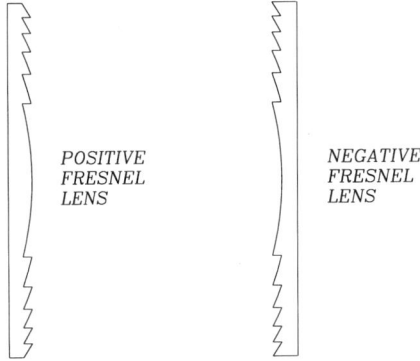

Figure 11.24 Positive and negative Fresnel lenses.

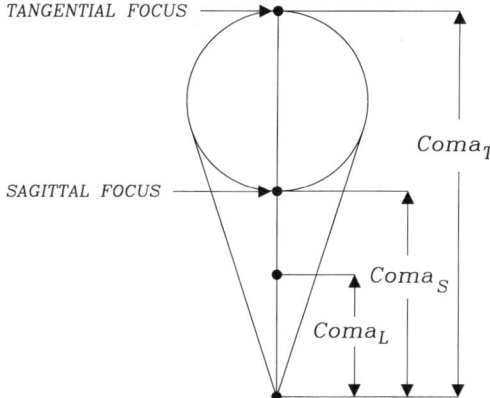

Figure 11.25 Linear coma appearing in Fresnel lenses.

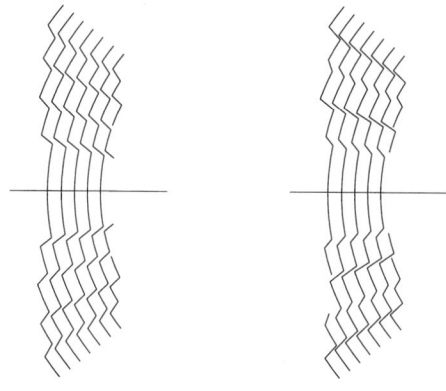

(a) COHERENT WAVEFRONT   (b) INCOHERENT WAVEFRONT

Figure 11.26 Wavefronts in coherent and incoherent Fresnel lenses: (a) coherent wavefront; (b) incoherent wavefront.

with the required precision. For a perfect spherical refracted wavefront to be produced, the thickness step on each ring has to be an integer multiple of the wavelength. The effect of this wavefront stepping is that the theoretical resolution is lower than that of a normal lens with the same aperture.

A small Fresnel lens that produces a continuous spherical or flat wavefront may be manufactured. The construction methods have to be completely different from those of ordinary Fresnel lenses. They even receive a different name. They are called Gabor plates, kinoforms, or, in general, diffractive optical elements. These elements may be produced with a very high diffraction efficiency.

REFERENCES

Coulman, C. E. and Petrie, G. R., "Some Notes on the Designing of Aspherical Magnifiers for Binocular Vision," *J. Opt. Soc. Am.*, **39**, 612–613 (1949).
Delano, E., "Primary Aberrations of Fresnel Lenses," *J. Opt. Soc. Am.*, **64**, 459–468 (1974).
Delano, E., "Primary Aberrations of Meniscus Fresnel Lenses," *J. Opt. Soc. Am.*, **66**, 1317–1320 (1976).
Delano, E., "Primary Aberration Contributions for Curved Fresnel Surfaces," *J. Opt. Soc. Am.*, **68**, 1306–1309 (1978).
Delano, E., "Stigmats Using Two Fresnel Surfaces," *Appl. Opt.*, **18**, 4187–4190 (1979).
Delano, E., "Stop and Conjugate Shift for Systems of Curved Fresnel Surfaces," *J. Opt. Soc. Am.*, **73**, 1828–1931 (1983).
Hilbert, R. S. and Rodgers, J. M., "Optical Design Issues in Electro-Optical Systems Integration," *Proc. SPIE*, **762**, 1–18 (1987).
Hopkins, R. E., "Aspheric Corrector Plates for Magnifiers," *J. Opt. Soc. Am.*, **36**, 604–610 (1946).
Hopkins, R. E., "Geometrical Optics," in *Methods of Experimental Physics, Geometrical and Instrumental Optics*, Vol. 25, D. Malacara, ed., Academic Press, San Diego, CA, 1988.
Kingslake, R., *Lens Design Fundamentals*, Academic Press, San Diego, CA, 1978.
Mejía-Barbosa, Y. and Malacara-Hernández, D., "Object Surface for Applying a Modified Hertmann Test to Measure Corneal Topography," *Appl. Opt.*, **40**, 5778–5786 (2001).
Nishihara, H. and Suhara, T., "Micro Fresnel Lenses," in *Progress in Optics*, Vol. XXIV, E. Wolf, ed., North Holland, Amsterdam, 1987.
Rogers, P. J., "Biocular Magnifiers — A Review," *Proc. SPIE*, **554**, 362–370 (1985).
Rosin, S., "Eyepieces and Magnifiers," in *Applied Optics and Optical Engineering*, Vol. III, R. Kingslake, ed., Academic Press, San Diego, CA, 1965.

12
Complex Photographic Lenses

12.1 INTRODUCTION

A large number of interesting photographic lenses have been designed since the end of the 19th century. We will describe here only a few of them, but the reader may consult the interesting books by Kingslake (1946, 1963, 1989). Additional information may also be found in the articles by Betensky (1980) and Cook (1965). Aklin (1948) made some considerations about the glass selection for photographic lenses in general.

In order to produce a good image over a large field of view, most photographic lenses are *anastigmats*. A lens is said to be anastigmat when it has a flat field free of astigmatism. In order to design an anastigmat lens it is necessary to have a very low Petzval sum, which is achieved only if the sum of the powers of the individual components (thin lenses or surfaces) is zero. This condition can be met by a large separation of the positive and negative elements. Lenses for aerial photography require a high resolution over a relatively high field (Kingslake, 1942, 1947).

The speed of a photographic objective may be changed by means of a diaphragm, variable in fixed steps to any desired value, according to any of two systems (Kingslake, 1945), as shown in the Table 12.1. From one step to the next the area of the aperture changes by a factor of two.

The image plane of a typical 35 mm objective with a 50 mm effective focal length has a diagonal equal to this focal length of the lens, producing an angular field semidiameter of about 26.5° at the corner of the rectangular field. Objectives with larger focal lengths are called telephotos and those with shorter focal lengths are called wide-angle lenses (Gardner and Washer, 1948; Thorndike, 1950).

Table 12.1 *f*-Numbers *FN* for Photographic Objectives

English system	Continental system
1.0	1.1
1.4	1.6
2.0	2.3
2.8	3.2
4.0	4.5
5.6	6.3
8.0	9.0
11.3	12.5
16.0	19.0
22.6	25.0
32.0	36.0
45.0	50.0

12.2 ASYMMETRICAL SYSTEMS

Asymmetrical anastigmats have also been designed with many different configurations. Here, some of the more important and interesting asymmetrical lenses will be described.

12.2.1 Petzval Lens

The Petzval lens is one of the oldest photographic lenses systematically, not empirically, designed. This lens consists of two achromatic doublets with the stop between them. The original Petzval lens designed in 1839, had a speed of about $f/5$. In the classic configuration the meridional ray is bent in each lens about the same amount, so that the refractive work is divided into approximately equal parts. A system with an effective focal length F has a front doublet with a focal length $f_a = 2F$, a rear doublet with a focal length $f_b = F$, and a separation between them equal to $d = F$. The back focal length is $F_B = F/2$. In the original design the stop was placed between the two components.

The Petzval sum is very large, since both components have positive power. Hence, some astigmatism must be introduced in order to produce a flat tangential field. Since the stop is in the middle, the front doublet has a tangential field curved backwards. Then, it may be proved that to flatten the field, the real doublet must have a lower power. To be able to correct the spherical aberration and the coma, and at the same time to have a flat tangential field, the crown and flint glasses must have a large Abbe number V difference. The spherical aberration and coma are corrected by a separate bending of the two elements of the rear doublet.

Over the years, many important improvements to the original design had been made. An example is the lens designed by Dallmeyer in 1860, by turning around the rear doublet, with its positive component in front. The image near the optical axis improved, but the astigmatism worsened.

More recent modifications of the Petzval design had moved the stop to the front doublet. These lenses will be described in Chap. 17, since they are mainly used for movie projectors. A Petzval lens is illustrated in Fig. 12.1, with its design data listed in Table 12.2.

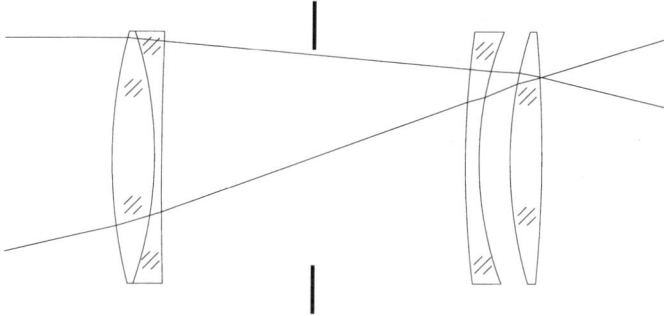

Figure 12.1 Petzval lens.

Table 12.2 Petzval Lens

Radius of curvature (mm)	Diameter (mm)	Separation or thickness (mm)	Material
55.90	36.0	7.00	K5
−56.23	36.0	2.00	F2
598.10	36.0	16.50	Air
Stop	36.0	16.50	Air
198.35	32.0	2.00	F2
38.45	32.0	3.50	Air
45.80	32.0	5.00	BK-7
−83.12	32.0	70.56	Air

Aperture (mm) (focal ratio): 10.0 ($F/10.0$)
Effective focal length (mm): 100.00
Back focal length (mm): 70.57
Object distance (mm): infinite
Image height (mm) (angular semifield): 26.8 (15°)

12.2.2 Telephoto Lens

The telephoto lens has two basic elements: a positive lens in the front and a negative lens in the back. The effective focal length of the system is larger than the total length, from the front lens to the focal plane, of the system. This kind of system is used whenever there is a need for a compact system, as compared with its focal length. A telephoto lens is the lens equivalent of a Cassegrain telescope. The telephoto ratio is defined as the ratio of the total length of the system to the effective focal length as follows:

$$k = \frac{t + F_B}{F} \quad (12.1)$$

where t is the lens separation, F is the effective focal length, and F_B is the back focal length. Then, the focal length for the front element may be written

$$f_a = \frac{Fd}{F(1-k)+d} \quad (12.2)$$

and the focal length for the second element is

$$f_b = \frac{(f_a - d)(kF - d)}{(f_a - kF)} \quad (12.3)$$

Typical values for the telephoto ratio are around 0.8.

A common problem with telephoto designs is the presence of distortion, but it may be reduced as described by Kazamaki and Kondo (1956). A telephoto lens, designed by Kingslake, redesigned by Hopkins, and reported by Smith and Genesee Optics Software (1992) is described in Table 12.3 and shown in Fig. 12.2.

An inverted telephoto lens is normally used to obtain wide-angle fields. When the field is very large, this lens is sometimes wrongly called a fisheye lens. The strong distortion of this system may be compensated by introducing a positive lens in front of the negative element.

These lenses have the property that their back focal length is longer than their effective focal length, which is useful in certain applications.

12.2.3 Cooke Triplet

H. Dennis Taylor, working for the Cooke and Sons company in York, England, in 1893, invented this famous design. This system has just enough

Table 12.3 Telephoto Lens

Radius of curvature (mm)	Diameter (mm)	Separation or thickness (mm)	Material
24.607	18.4	5.080	BK-7
−36.347	18.4	1.600	F2
212.138	18.0	12.300	Air
Stop	13.4	21.699	Air
−14.123	18.8	1.520	BK-7
−38.904	18.8	4.800	F2
−25.814	18.8	37.934	Air

Aperture (mm) (focal ratio): 18.4 ($F/5.6$)
Effective focal length (mm): 101.6
Back focal length (mm): 37.93
Object distance (mm): infinite
Image height (mm) (angular semifield): 7.44 (4.19°)

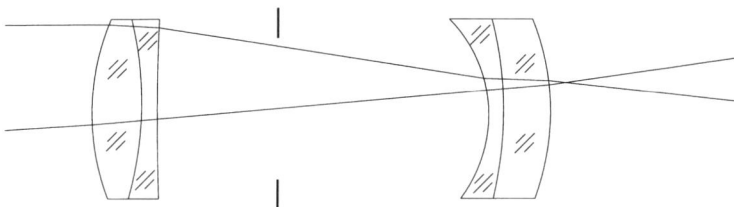

Figure 12.2 Telephoto lens.

degrees of freedom to correct all seven primary aberrations and to obtain the desired focal length. The eight degrees of freedom are the three focal lengths of the components, three lens bendings, and two air separations. To arrive at his design Taylor reasoned the following:

1. The sum of the powers of the elements has to be zero in order to have a zero Petzval sum.
2. To have low distortion and to correct the magnification chromatic aberration, the system has to be nearly symmetric. Then, a negative lens between the two positive lenses or a positive lens between two negative lenses are possible solutions. He realized later that the first solution leads to a better aberration correction.
3. To correct the axial chromatic aberration the central negative lens should be made with a flint glass and the two positive lenses with crown glass.

To find a thin-lens solution for the system we take the stop at the negative lens and then we use five variables, the three lens powers, and the two separations to solve for five conditions. These five conditions must be independent of the lens bendings. The next step is to use the three bendings to correct the remaining three conditions that must depend on the bendings. The effective focal length of the system, the axial chromatic aberration, the magnification chromatic aberration, and the Petzval curvature do not depend on the bendings. Four conditions that do depend on the bendings are the spherical aberration, the coma, the astigmatism, and the distortion.

Thus, the focal lengths (three) and the separations (two) may be fixed with the four parameters that do not depend on the lens bendings, plus another one, which may be the ratio of the lens separations or the ratio of the powers of the two positive lenses.

In the next step the lens bendings are found in order to correct the spherical aberration, the coma, and the astigmatism. Then, the distortion is calculated and if it is unacceptable, the design process is repeated, selecting a new ratio between the separations or between the powers of the two positive lenses.

The approximations used for thin lenses may give some errors when calculating the primary aberrations for very thick lenses. This error might be compensated by aiming in the thin-lens calculations to some small (nonzero) values of the primary aberrations. The best aim value is obtained by trial and error, until the real aberration becomes zero.

The design techniques for Cooke triplets have been described in detail by Conrady (1960), Cruickshank (1958, 1960), Smith (1950), and Stephens (1948). From Eq. (3.5) we may find that the total power for the lens system is

$$P = P_1 + \frac{y_2}{y_1} P_2 + \frac{y_3}{y_1} P_3 \tag{12.4}$$

If the astigmatism is zero, the Petzval surface must be flat. However, in the initial design it is frequently better to assume a small residual value different from zero, shown by Shatma and Rama Gopal (1982). However, from Eq. (5.29), if the Petzval surface is assumed to be flat we may write

$$\frac{P_1}{n_1} + \frac{P_2}{n_2} + \frac{P_3}{n_3} = 0 \tag{12.5}$$

and from Eq. (6.24), the transverse axial chromatic aberration may be written as

$$AchrT = \frac{1}{P y_1} \left(\frac{y_1^2 P_1}{V_1} + \frac{y_2^2 P_2}{V_2} + \frac{y_3^2 P_3}{V_3} \right) \tag{12.6}$$

and from Eq. (6.72), the magnification chromatic aberration is

$$Mchr = \frac{1}{P y_1} \left(\frac{y_1 \bar{y}_1 P_1}{V_1} + \frac{y_2 \bar{y}_2 P_2}{V_2} + \frac{y_3 \bar{y}_3 P_3}{V_3} \right) \qquad (12.7)$$

It is convenient to define the power ϕ of the system formed by the first two elements as follows:

$$\phi = P_1 + \frac{y_2}{y_1} P_2 \qquad (12.8)$$

We assume now that the glasses had already been selected from experience. The problem of the selection of the glasses in the triplet has been studied by Lessing (1958, 1959a,b). Then, from Eq. (12.5), and taking the glass for the first and third lenses to be the same, we find that if the Petzval surface is flat

$$P_1 + \frac{n_1}{n_2} P_2 + P_3 = 0 \qquad (12.9)$$

if the transverse axial chromatic aberration is equal to zero, from Eq. (12.6) we have

$$P_1 + \left(\frac{V_1}{V_2}\right) \left(\frac{y_2}{y_1}\right)^2 P_2 + \left(\frac{y_3}{y_1}\right)^2 P_3 = 0 \qquad (12.10)$$

and from Eq. (12.7), if the magnification chromatic aberration is also equal to zero:

$$P_1 + \left(\frac{V_1}{V_2}\right) \left(\frac{y_2}{y_1}\right) \left(\frac{\bar{y}_2}{\bar{y}_1}\right) P_2 + \left(\frac{y_3}{y_1}\right) \left(\frac{\bar{y}_3}{\bar{y}_1}\right) P_3 = 0 \qquad (12.11)$$

By observing Fig. 12.3, and assuming that the stop is in contact with the middle lens ($\bar{y}_2 = 0$), we may see that

$$\bar{y}_3 = -\frac{d_2}{d_1} \bar{y}_1 \qquad (12.12)$$

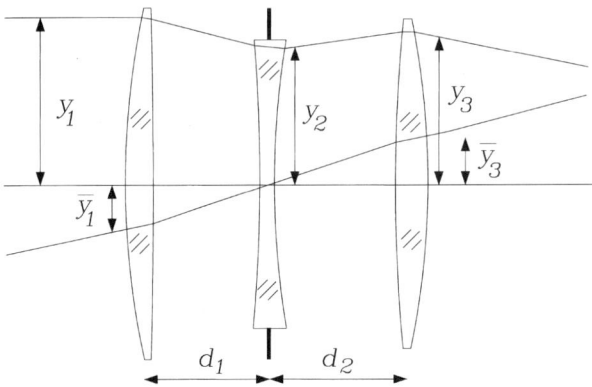

Figure 12.3 Some parameters in a Cooke triplet.

Thus, with this expression and again $\bar{y}_2 = 0$, we may write Eq. (12.11) as

$$P_1 - \left(\frac{y_3}{y_1}\right)\left(\frac{d_2}{d_1}\right) P_3 = 0 \tag{12.13}$$

From Fig. 12.3, we may obtain

$$\frac{y_2}{y_1} = 1 - d_1 P_1 \tag{12.14}$$

and similarly, assuming that ϕ is the power of the system formed by the first two elements, we have

$$\frac{y_3}{y_1} = \frac{y_2}{y_1} - d_2 \phi \tag{12.15}$$

After substituting (y_2/y_1) and (y_3/y_1) from Eqs. 12.14 and 12.15 into Eqs. (12.4), (12.8), (12.9), (12.10), and (12.13) we obtain five equations with five unknowns, namely P_1, P_2, P_3, d_1, and d_2. The power ϕ is taken as a constant, by assigning to it a tentative value. To solve the system of five equations is not simple.

Several methods have been proposed to solve this system of equations. Here, the solution described by Cruickshank (1958, 1960) will be used. The solutions for the lens powers and the separations will be written in terms of the ratio y_2/y_1. Then, a second-degree equation will define the value of this ratio. To begin, we may find from Eqs. (12.4) and (12.8) that

$$P - \phi = \left(\frac{y_3}{y_1}\right) P_3 \tag{12.16}$$

then, from Eq. (12.14), the value of the first lens separation d_1 is

$$d_1 = \frac{1 - (y_2/y_1)}{P_1} \tag{12.17}$$

and from Eqs. (12.13), (12.14), and (12.16) the second lens separation d_2 is

$$d_2 = \frac{1 - (y_2/y_1)}{P - \phi} \tag{12.18}$$

To find now the lens powers, from Eqs. (12.15) and (12.18), we write

$$\left(\frac{y_2}{y_1}\right) P - (P - \phi)\left(\frac{y_3}{y_1}\right) = \phi \tag{12.19}$$

On the other hand, from Eqs. (12.8), (12.10), and (12.16):

$$\phi = P_2\left(\frac{y_2}{y_1}\right)\left[1 - \left(\frac{V_1}{V_2}\right)\left(\frac{y_2}{y_1}\right)\right] - (P - \phi)\left(\frac{y_3}{y_1}\right) \tag{12.20}$$

Then, the power for the second lens can be found from Eqs. (12.19) and (12.20) as

$$P_2 = \frac{P}{1 - (V_1/V_2)(y_2/y_1)} \tag{12.21}$$

and the power for the third lens from Eqs. (12.16) and (12.19) as

$$P_3 = \frac{(P - \phi)^2}{(y_2/y_1)P - \phi} \tag{12.22}$$

Finally, from Eqs. (12.8) and (12.21), the power for the first lens is

$$P_1 = \phi - \left(\frac{y_2}{y_1}\right) P_2 \tag{12.23}$$

The final step now is to derive an expression that permits us to calculate the ratio y_2/y_1. Substituting in Eq. (12.9) the values of the lens powers from Eqs. (12.21), (12.22), and (12.23), we may find a second degree equation:

$$B_2 \left(\frac{y_2}{y_1}\right)^2 + B_1 \left(\frac{y_2}{y_1}\right) + B_0 = 0 \tag{12.24}$$

where

$$B_2 = P + \phi\left(\frac{y_2}{y_1}\right) \tag{12.25}$$

$$B_1 = P\left[\left(\frac{y_2}{y_1}\right) - \left(\frac{V_1}{V_2}\right)\right] - 2\phi\left[1 + \left(\frac{y_2}{y_1}\right)\right] \tag{12.26}$$

and

$$B_0 = \phi\left[\left(\frac{V_1}{V_2}\right) + 2\right] - P \tag{12.27}$$

After the solutions to this second-degree equation are obtained, the three lens powers and the two separations are easily found by selecting the solution with a negative lens between two positive lenses.

The next step is to correct the spherical aberration, coma, and astigmatism by bending of the lenses. Then, the distortion is calculated. If the distortion is not zero, a new value is assigned to the power ϕ and the whole process is repeated in an iterative process, until the distortion becomes zero. Hopkins (1962) and Wallin (1964) made complete third- and fifth-order analyses of the Cooke triplet.

It is important to know that a triplet with a large or medium aperture with a perfect correction for all six primary aberrations is quite deficient. A good design compensates high-order aberrations with the presence of some primary aberrations.

Figure 12.4 shows a Cooke triplet, with the data presented in Table 12.4.

12.2.4 Tessar Lens

The Tessar lens is a descendant of the symmetrical double meniscus anastigmats described in Section 12.1. We may consider that the front component is an air-spaced doublet and that the rear component is a new achromat. However, it may also be regarded as a modification of the Cooke triplet. With this point of view, the triplet has been modified by substituting the last element for a doublet. The negative component of this doublet has a lower index and greater dispersion than the positive component. This substitution reduces the spherical zonal aberration (spherical aberration at 0.7 of the maximum aperture) and reduces the astigmatism. Typical

Complex Photographic Lenses

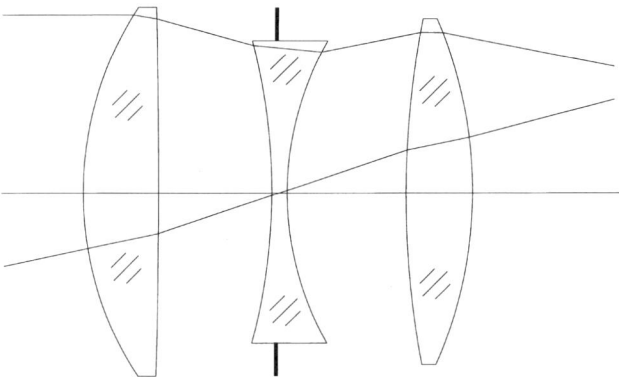

Figure 12.4 Cooke triplet lens.

Table 12.4 Cooke Triplet Lens

Radius of curvature (mm)	Diameter (mm)	Separation or thickness (mm)	Material
36.10	48.0	10.50	SK4
Flat	48.0	12.00	Air
−59.00 (stop)	36.0	2.00	SF2
−32.79	36.0	12.50	Air
95.40	48.0	11.00	SK4
−47.62	48.0	71.75	Air

Aperture (mm) (focal ratio): 40.0 (F/2.5)
Effective focal length (mm): 100.1
Back focal length (mm): 71.75
Object distance (mm): infinite
Image height (mm) (angular semifield): 26.79 (15°)

glasses for this lens are dense barium crown for the first and last lenses, a flint for the second, and a light flint for the third. The improvement over the triplet is not much, but it is noticeable.

Figure 12.5 shows a Tessar lens, described by Smith and Genesee Optics (1992), with the design data presented in the Table 12.5.

12.3 SYMMETRICAL ANASTIGMAT SYSTEMS

Many anastigmatic lenses have a symmetrical configuration about the stop, to minimize coma and distortion, even though the object and image are

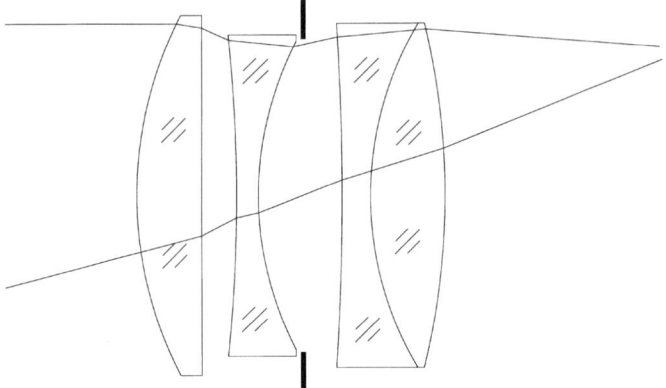

Figure 12.5 Tessar lens.

Table 12.5 Tessar Lens

Radius of curvature (mm)	Diameter (mm)	Separation or thickness (mm)	Material
24.200	22.8	4.10	BaF10
555.000	22.8	2.20	Air
−96.000	21.2	3.10	F2
21.000	21.2	3.40	Air
Stop	19.4	2.60	Air
−166.000	20.2	0.70	LF3
21.000	20.0	4.40	BaF10
−45.200	20.2	88.90	Air

Aperture (mm) (focal ratio): 22.8 ($F/4.5$)
Effective focal length (mm): 102.6
Back focal length (mm): 88.9
Object distance (mm): infinite
Image height (mm) (angular semifield): 54.36 (27.92°)

not symmetrically placed, as in the Dagor lens and the double Gauss lenses to be described here. Another interesting symmetrical lens is the Ross lens, frequently used in astrophotography (Cornejo et al., 1970).

12.3.1 Dagor Lens

One of the earliest anastigmats is the Dagor lens. This lens is symmetric or almost symmetric about the stop. As in the landscape lens, the stop and lens bendings are selected so that the astigmatism is small. If desired, the lens may be designed so that one-half of the system, if used alone, performs reasonably well. The advantage is that then we have two different focal lengths, one for the whole system and another when one-half of the system is used. If the two halves have different focal lengths, the system is not completely symmetric about the stop. This is a *hemisymmetrical* lens. If each half of the hemisymmetrical lens is allowed to be used alone, or in combination with the other half, we have three different focal lengths. A lens like this with two components that may be used alone or as a complete system is said to be *convertible*.

The two glass-to-glass interfaces in the Dagor lens contribute very little to the Petzval sum, since their power is small. Thus, the two external surfaces determine both the focal length and the Petzval sum.

The three glasses are different. If we split the central negative element of one-half into two parts, the outer part may be thought of as a new achromat and the internal part as an old achromat. A study of the possible methods to control the residual aberrations in the design of symmetrical four-element anastigmatic lenses has been made by Smith (1958). Figure 12.6 shows a Dagor lens described by Smith and Genesee Optics Software (1992), with the data presented in Table 12.6.

12.3.2 Double Gauss Lens

First, Karl Friedrich Gauss and then Alvan G. Clark suggested that a symmetrical system with two meniscus-shaped components may produce a very good lens design. A symmetrical lens with two negative lenses between

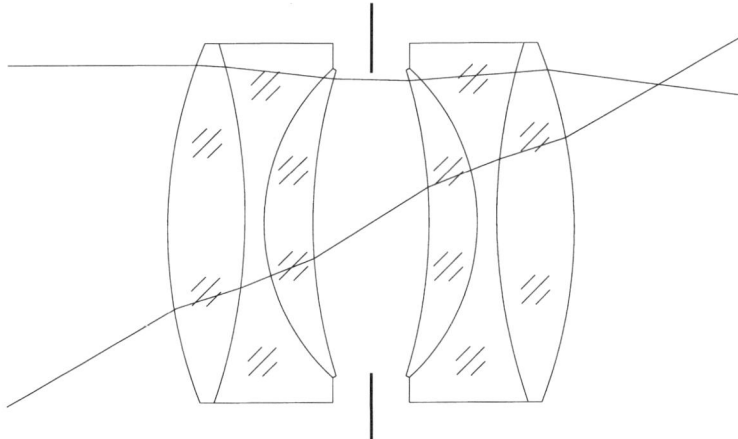

Figure 12.6   Dagor lens.

Table 12.6 Dagor Lens

Radius of curvature (mm)	Diameter (mm)	Separation or thickness (mm)	Material
19.100	14.8	3.056	SK6
−22.635	14.8	0.764	BaLF3
8.272	12.0	1.910	K4
20.453	12.0	2.292	Air
Stop	5.6	2.292	Air

Aperture (mm) (focal ratio): 12.91 ($F/8.0$)
Effective focal length (mm): 103.3
Back focal length (mm): 96.27
Object distance (mm): infinite
Image height (mm) (angular semifield): 50.99 (26.6°)

two positive lenses has five degrees of freedom, namely, two lens powers, two bendings, and the separation between the lenses. These variables are used to set the desired focal length, and to correct the Petzval sum, the spherical aberration, the axial chromatic aberration, and the astigmatism. When the two halves are mounted together, the antisymmetric wavefront aberrations are corrected.

The double Gauss lens can be considered as a modification of the triplet, where the central negative lens is split into two lenses, each of them a doublet. As in the rear doublet for the Tessar lens, the negative component of this doublet should have a lower index of refraction and a higher dispersion.

To determine the power of the lenses and their separation we need to impose three conditions. From Eq. (3.5) we may find that the total power for the lens system is

$$P = P_1 + \frac{y_2}{y_1} P_2 \tag{12.28}$$

from Eq. (5.29), the curvature of the Petzval surface is

$$\frac{1}{r_{Ptz}} = -\frac{P_1}{n_1} - \frac{P_2}{n_2} \tag{12.29}$$

and from Eq. (6.24), the transverse axial chromatic aberration may be written as

$$AchrT = \frac{1}{Py_1} \left(\frac{y_1^2 P_1}{V_1} + \frac{y_2^2 P_2}{V_2} \right) \tag{12.30}$$

The ray heights y_1 and y_2, from Eq. (12.28), are related by

$$y_2 = y_1 \frac{(P - P_1)}{P_2} \tag{12.31}$$

Since y_1 is known, these three equations may be solved to obtain the two lens powers. Then, the lens separation may be obtained from Eq. (3.45), as

$$d = \frac{(P_1 + P_2 - P)}{P_2 P_1} \tag{12.32}$$

The next step is to find the two lens bendings to correct the spherical aberration and the astigmatism. We may find many different solutions for this system, depending on the glasses being chosen.

The system just described may acquire this shape if the proper glasses are selected. These glasses have to be widely separated on the V–n diagram, e.g., dense flint and dense barium crown. In the most common case of an object at infinity, the design may depart from the exact symmetrical configuration.

A general study of double Gauss lenses and derivatives has been carried out by Kidger and Wynne (1967). The double Gauss aplanatic system, with data presented in Table 12.7 and illustrated in Fig. 12.7, was designed by Smith and Genesee Optics (1992).

Table 12.7 Double Gauss Lens

Radius of curvature (mm)	Diameter (mm)	Separation or thickness (mm)	Material
85.500	76.0	11.60	LaF2
408.330	76.0	1.50	Air
40.350	66.0	17.00	SK55
156.050	66.0	3.50	FN11
25.050	44.0	13.70	Air
Stop	42.6	8.30	Air
−36.800	44.0	3.50	SF8
55.000	52.0	23.00	LaF2
−51.500	52.0	1.00	Air
123.500	51.0	17.00	LaF2
−204.960	51.0	55.07	Air

Aperture (mm) (focal ratio): 76.0 ($F/1.35$)
Effective focal length (mm): 100.4
Back focal length (mm): 55.07
Object distance (mm): infinite
Image height (mm) (angular semifield): 23.09 (13°)

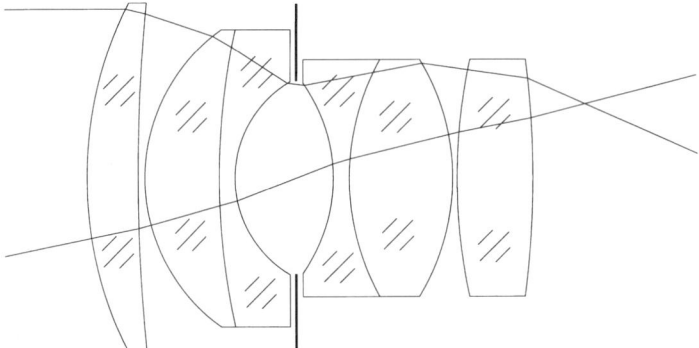

Figure 12.7 Double Gauss lens.

12.4 VARIFOCAL AND ZOOM LENSES

A varifocal lens is one whose effective focal length is variable, but the focus may have to be readjusted after each change. These lenses are useful, e.g., for slide projectors. A zoom lens, on the other hand, is a lens whose effective focal length may also be changed, but the lens remains in sharp focus while changing the focal length. These lenses are mainly used for movie or television cameras. In the past, however, the term varifocal was the only one used to name both types of lenses.

Varifocal and zoom lenses have been widely studied, e.g., by Back and Lowen (1954, 1958), Bergstein (1958), Bergstein and Motz (1957, 1962a,b,c), Clark (1973), Cook (1958), Kingslake (1960), Pegis and Peck (1962), and Wooters and Silvertooth (1965). A complete chapter on this subject by Yamaji (1967) deserves special mention.

When changing the magnification in a zoom lens, the system has to be compensated so that the focal plane does not change its position with respect to the lens holder. This focus shift may be canceled (compensated) by optical or mechanical means.

The mechanical compensation in zoom lenses is performed by simultaneously moving two lenses in the system, one to change the focal length and the other to maintain the image at a fixed plane. In general, these movements are nonlinearly related to each other, so, a complicated set of gears or cams is required.

The optical compensation is obtained when at least three points along the zoom range have the same focal plane (but different magnification). Then, the defocusing between these three points is small. This kind of compensation requires three image-forming steps, thus a minimum of three elements, as described by Back and Lowen (1954). A better system is obtained if instead of three lens elements, four are used to obtain four points without focus shift.

There are several types of zoom lenses, but most of them are designed along the same basic principles. A zoom lens system, as illustrated in Fig. 12.8, may be considered as formed by a normal photographic lens with a fixed focal length, and an afocal system with variable magnifying power (angular magnification) in front of it. The simplest and basic afocal system is formed by two positive lenses with focal length f_1, and a negative focal length with focal length f_2 between them, as shown in Fig. 12.9.

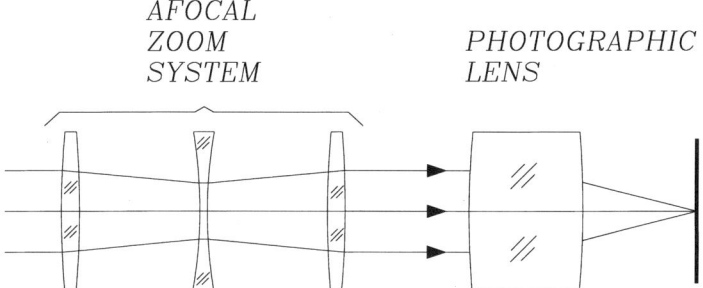

Figure 12.8 Afocal zoom lens in front of a fixed focal length lens.

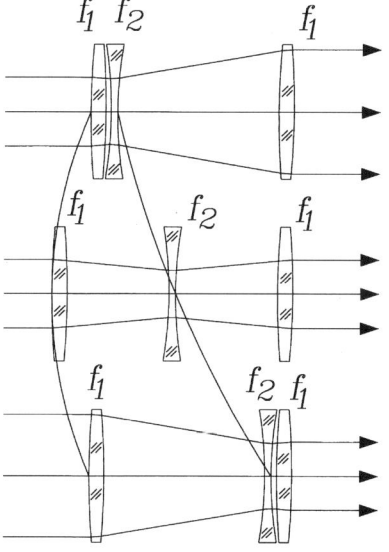

Figure 12.9 Lens movements in an afocal zoom lens.

The absolute value of the power of the negative lens is larger than twice the power of the positive lens, so that when they are in contact with each other, the power F of the combination is negative and has an absolute value higher than the power of the positive lens alone. The power of this doublet, from Eq. (3.46), with a lens separation equal to zero, is given by

$$F = \frac{f_1 f_2}{f_1 + f_2} \tag{12.33}$$

The negative lens is displaced along the optical axis, in the space between the two positive lenses. When the negative lens is in contact with the front positive lens, the system is a Galilean telescope with a magnifying power M_0, which, as will be shown in Chap. 14, is given by

$$M_0 = -\frac{F}{f_1} \tag{12.34}$$

The magnifying power is positive and less than one, since F is negative and its magnitude is less than f_1. Thus, an erect and minified image is produced. When the lens is moved to the other extreme, and placed in contact with the back lens, the angular magnification is $1/M_0$, which is positive and larger than one.

Since Galilean telescopes are afocal, the separation between the two positive lenses must be

$$d = F + f_1 \tag{12.35}$$

From these two expressions, we find that the focal length f_1 of the positive lens is

$$f_1 = \frac{d}{1 - M_0} \tag{12.36}$$

and the focal length of the negative lens is

$$f_2 = -\frac{M_0 d}{1 - M_0^2} \tag{12.37}$$

Thus, if we move the negative lens in the space between the two positive lenses, the system changes its magnifying power between the values M_0 and $1/M_0$. However, the image is clearly focused only at the ends, as in

Fig. 12.10. At intermediate positions, we have to readjust the focus by means of a small shift of the front lens. This is the so-called mechanical compensation. Let us now consider the front and middle lenses as in Fig. 12.11. When the negative lens is displaced a distance ζ from the initial position at the front, the frontal lens has to be displaced a small distance η, to keep the image in focus. These two movements have to be done maintaining the virtual image P at a fixed position in space. Then, the new separation S between the front and the middle lens is

$$\zeta = \eta + S \tag{12.38}$$

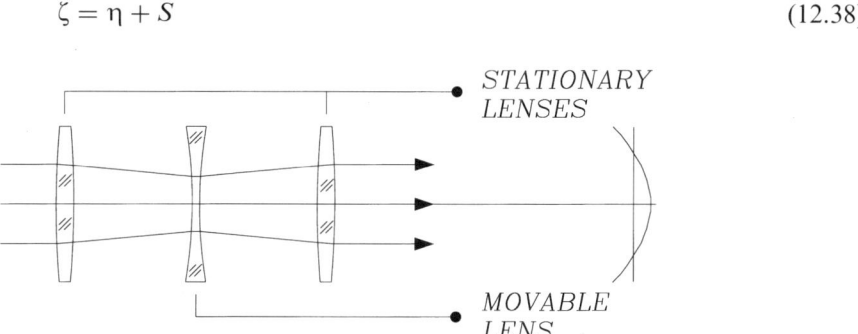

Figure 12.10 Focus shift in an afocal three-lens zoom system.

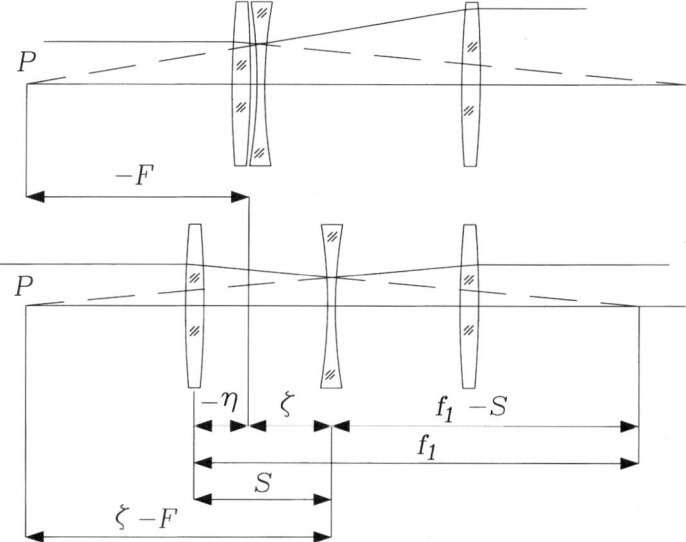

Figure 12.11 Some parameters and variables in a three-lens afocal zoom system.

Then, using Eq. (3.13) for the middle lens, we may write

$$\frac{1}{f_2} = \frac{1}{F-\zeta} - \frac{1}{f_1 - S} \tag{12.39}$$

and substituting here the value of S from Eq. (12.38), we find that, after some algebraic work,

$$\eta = \zeta\left[1 - \frac{(f_1 - f_2)^2}{f_1^2 + (f_1 + f_2)\zeta}\right] \tag{12.40}$$

which, written in terms of the minimum magnifying power M_0, becomes

$$\eta = \zeta\left[1 - \frac{d}{M_0^2 d + (1 - M_0^2)\zeta}\right] \tag{12.41}$$

This expression gives the relation between the lens displacements ζ and η, but does not allow us to calculate the values of these displacements for a given zoom magnification. The magnifying power M for any position of the middle negative lens, by using Eqs. (12.34) and (3.46) may be found to be

$$M = \frac{M_0 d}{d + (\eta - \zeta)(1 - M_0^2)} \tag{12.42}$$

The lens displacements for a magnifying power equal to one are

$$\zeta_0 = \frac{M_0 d}{1 + M_0} \tag{12.43}$$

and

$$\eta_0 = -\frac{1 - M_0}{1 + M_0} d \tag{12.44}$$

which is also the maximum displacement for η.

The system may be improved by substituting the negative lens for two negative lenses with a positive lens between them, as shown in Fig. 12.12. With the two negative lenses displaced to one extreme, the system must be afocal as is the system in Fig. 12.10. So, the image is in good focus at the two extremes. The power of the central positive lens is a degree of freedom that may be used to impose the condition that the image is also in focus when the magnifying power is one. Since the system is symmetric, the focus shift curve must also be symmetric, as shown in Fig. 12.12. Thus, if the focal plane is shifted to the plane indicated by a dashed line, we have four positions at which the image is focused. If the maximum focus shift is acceptable, the system does not need any refocusing cam and we say that the system has optical compensation.

In general, if the afocal system is symmetrical like the ones in Figs. 12.10 and 12.12, and we move n alternate symmetrically placed lenses together, we have $n+1$ different magnifications with the same focus position. A general theory of zoom systems that predicts the number of points with the same focus has been given by Bergstein (1958) and by Back and Lowen (1958).

The zoom systems we have described are all afocal and symmetric, with an imaging system fixed with respect to the focal plane, as in Fig. 12.8. The zoom system may be also asymmetric if we include in the system the last imaging lens and we move it with respect to the focal plane. An example of an asymmetric system is shown in Fig. 12.13. Yamaji (1967) has described in detail the evolution of the different types of zoom systems.

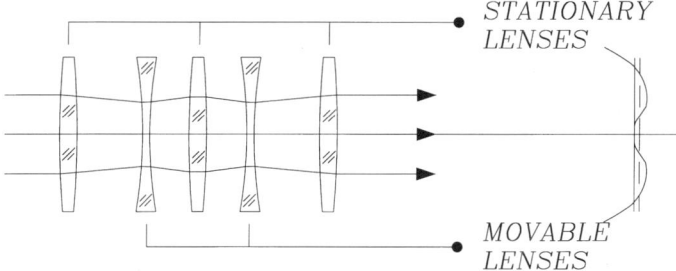

Figure 12.12 Focus shift in a five-lens afocal zoom lens system.

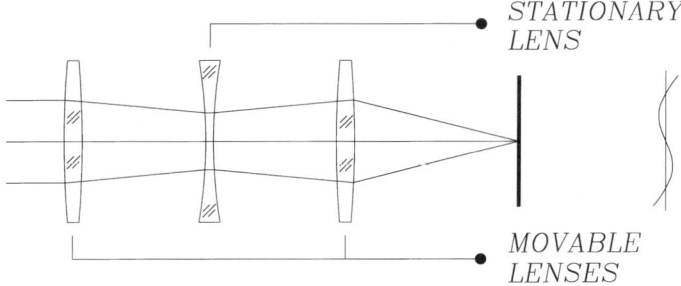

Figure 12.13 Nonafocal three-lens zoom system with two movable lenses.

The aberrations in a zoom system are corrected for at least three different focus positions. This means that we have to correct three times the number of aberrations that we have in a fixed-focus system. Thus, we need more degrees of freedom, which are obtained with the larger number of elements.

REFERENCES

Aklin, G. H., "The Effect of High Index Glasses on the Field Characteristics of Photographic Objectives," *J. Opt. Soc. Am.*, **38**, 841–844 (1948).

Back, F. G. and Lowen, H., "The Basic Theory of Varifocal Lenses with Linear Movement and Optical Compensation," *J. Opt. Soc. Am.*, **44**, 684–691 (1954).

Back, F. G. and Lowen, H., "Generalized Theory of Optically Compensated Varifocal Systems," *J. Opt. Soc. Am.*, **48**, 154–171 (1958).

Bergstein, L., "General Theory of Optically Compensated Varifocal Systems," *J. Opt. Soc. Am.*, **48**, 154–171 (1958).

Bergstein, L. and Motz, L., "Third Order Aberration Theory for Varifocal Systems," *J. Opt. Soc. Am.*, **47**, 579–593 (1957).

Bergstein, L. and Motz, L., "Two Component Optically Compensated Varifocal System," *J. Opt. Soc. Am.*, **52**, 353–362 (1962a).

Bergstein, L. and Motz, L., "Three Component Optically Compensated Varifocal System," *J. Opt. Soc. Am.*, **52**, 362–375 (1962b).

Bergstein, L. and Motz, L., "Four Component Optically Compensated Varifocal System," *J. Opt. Soc. Am.*, **52**, 376–388 (1962c).

Betensky, E. I., "Photographic Lenses," in *Applied Optics and Optical Engineering*, R. R. Shannon and J. C. Wyant, eds., Vol. VIII, Chap.1, Academic Press, San Diego, CA, 1980.

Clark, A. D., *Zoom Lenses, Monographs on Applied Optics No. 7*, Adam Hilger, London, 1973.

Conrady, A. E., *Applied Optics and Optical Design*, Part Two, Dover Publications, New York, 1960.

Cook, G. H., "Television Zoom Lenses," *J. Soc. Mot. Pic. Tel. Eng.*, **68**, 25–28 (1958).

Cook, G. H., "Photographic Objectives," in *Applied Optics and Optical Engineering*, R. Kingslake, ed., Vol. III, Chap. 3, Academic Press, San Diego, CA, 1965.

Cornejo, A., Castro, J., and Malacara, D., "Note on the Design of Two Ross Type Photographic Objectives," *Boletín de los Observatorios de Tonantzintla y Tacubaya*, **5**, 241–245 (1970).

Cruickshank, F. D., "The Design of Photographic Objectives of the Triplet Family. I: The Design of the Triplet Type 111 Objective," *Aust. J. Phys.*, **11**, 41–54 (1958).

Cruickshank, F. D., "The Design of Photographic Objectives of the Triplet Family. II: Initial Design of Compound Triplet Systems," *Aust. J. Phys.*, **13**, 27–42 (1960).

Gardner, L. and Washer, D. S., "Lenses of Extremely Wide Angle for Airplane Mapping," *J. Opt. Soc. Am.*, **38**, 421–431 (1948).
Hopkins, R. E., "Third Order and Fifth Order Analysis of the Triplet," *J. Opt. Soc. Am.*, **52**, 389–394 (1962).
Kazamaki, T. and Kondo, F., "New Series of Distortionless Telephoto Lenses," *J. Opt. Soc. Am.*, **46**, 22–31 (1956).
Kidger, M. J. and Wynne, C. G., "The Design of Double Gauss Systems Using Digital Computers," *Appl. Opt.*, **6**, 553–563 (1967).
Kingslake, R., "Lenses for Aerial Photography," *J. Opt. Soc. Am.*, **32**, 129–134 (1942).
Kingslake, R., "The Effective Aperture of a Photographic Objective," *J. Opt. Soc. Am.*, **35**, 518–520 (1945).
Kingslake, R., "A Classification of Photographic Lens Types," *J. Opt. Soc. Am.*, **35**, 251–255 (1946).
Kingslake, R., "Recent Developments in Lenses for Aerial Photography," *J. Opt. Soc. Am.*, **37**, 1–9 (1947).
Kingslake, R., "The Development of the Zoom Lens," *J. Soc. Mot. Pic. Tel. Eng.*, **69**, 534–544 (1960).
Kingslake, R., *Lenses in Photography*, Barnes, New York, 1963.
Kingslake, R., *A History of the Photographic Lens*, Academic Press, New York, 1989.
Lessing, N. V. D. W., "Selection of Optical Glasses in Taylor Triplets (Special Method)," *J. Opt. Soc. Am.*, **48**, 558–562 (1958).
Lessing, N. V. D. W., "Selection of Optical Glasses in Taylor Triplets (General Method)," *J. Opt. Soc. Am.*, **49**, 31–34 (1959a).
Lessing, N. V. D. W., "Selection of Optical Glasses in Taylor Triplets with Residual Chromatic Aberration," *J. Opt. Soc. Am.*, **49**, 872–877 (1959b).
Pegis, R. J. and Peck, W. G., "First-Order Design Theory for Linearly Compensated Zoom Systems," *J. Opt. Soc. Am.*, **52**, 905–911 (1962).
Shatma, K. D. and Rama Gopal, S. V., "Significance of Petzval Curvature in Triplet Design," *Appl. Opt.*, **21**, 4439–4442 (1982).
Smith, F. W., "Comment on Design of Triplet Anastigmat Lenses of the Taylor Type," *J. Opt. Soc. Am.*, **40**, 406–407 (1950).
Smith, W. J., "Control of Residual Aberrations in the Design of Anastigmat Objectives," *J. Opt. Soc. Am.*, **48**, 98–105 (1958).
Smith, W. J. and Genesee Optics Software, Inc., *Modern Lens Design. A Resource Manual*, McGraw-Hill, New York, 1992.
Stephens, R. E., "Design of Triplet Anastigmat Lenses of the Taylor Type," *J. Opt. Soc. Am.*, **38**, 1032–1039 (1948).
Thorndike, E. M., "A Wide Angle, Underwater Camera Lens," *J. Opt. Soc. Am.*, **40**, 823–824 (1950).
Wallin, W., "Design Study of Air Spaced Triplets," *Appl. Opt.*, **3**, 421–426 (1964).
Wooters, G. and Silvertooth, E. W., "Optically Compensated Zoom Lens," *J. Opt. Soc. Am.*, **55**, 347–351(1965).
Yamaji, K., "Design of Zoom Lenses," in *Progress in Optics*, Vol. VI, E. Wolf, ed., North Holland, Amsterdam, 1967.

13
The Human Eye and Ophthalmic Lenses

13.1 THE HUMAN EYE

The first serious studies were carried out by Helmholtz, as described in his book *Optik*. Eye studies continued in the 19th century with the pioneering work by Gullstrand. A diagram of the human eye is presented in Fig. 13.1 and its main optical constants are listed in Table 13.1. The most important optical components of the eye are:

The cornea—This is the front transparent tissue in the eye. Its normal ideal shape is nearly spherical, with a dioptric power of about 43 diopters. Any deviation from its ideal shape produces refractive errors. If it takes a toroidal shape, with different curvatures along two mutually perpendicular diameters, corneal astigmatism appears. The astigmatism is said to be with the rule if the curvature in the vertical diameter is larger than in the horizontal diameter and against the rule otherwise. A small protuberance and thinning at the center makes the cornea to have an almost conic shape, in a defect called keratoconus. These errors are measured with an ophthalmeter or a corneal topographer.

The pupil—This is the circular opening in front of the eye and it is surrounded by the iris. The pupil increases or decreases its diameter to control the amount of light entering the eye. The maximum diameter of the pupil, with low illumination levels (clear night) is around 8 mm and the minimum diameter with high illumination levels (sunny day) is near 1.5 mm. Its average diameter is about 3–4 mm (well-illuminated office).

The aqueous humor—This is the liquid between the back of the cornea and the eye lens.

The eye lens—This is a flexible lens, also called the crystalline lens, whose optical power can be modified by means of the ciliary muscles. It increases the power (accommodation) to focus near objects and relaxes its shape to focus far-away objects. The nucleus of this lens has a refractive

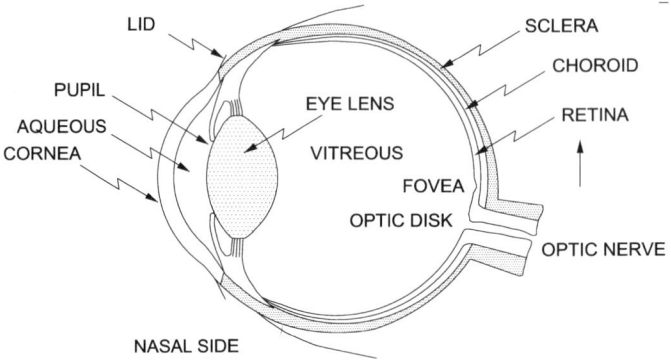

Figure 13.1 Schematics of the human eye.

Table 13.1 Average Optical Constants of the Human Eye

Total length	24.75 mm
Pupil diameter	1.5 – 8.0 mm
Effective focal length	22.89 mm
Total power (unaccommodated)	58.6 diopters
Lens power (unaccommodated)	19 diopters
Corneal power	43 diopters
Corneal radius of curvature	7.98 mm
Aqueous humor refractive index	1.336
Lens refractive index	
center	1.406
edge	1.386
Vitreous refractive index	1.337

index higher than that of its external parts. The relaxed lens has a dioptric power of about 15 diopters and can be increased (accommodation amplitude) by about 15 diopters in children or about 0.5 diopter in old people. The eye lens can lose its transparency for many reasons, producing what is known as a cataract. To correct this condition, the eye lens has to be removed. Frequently, a plastic lens is inserted to replace the eye lens.

The vitreous humor—This is the liquid filling most of the eye globe, in the space between the eye lens and the retina. Sometimes, mainly in medium or highly myopic eyes, small particles float in this medium, producing some small images that appear to float in space.

The retina—This is the light-sensitive surface of the eye, on which the images are formed. The eye retina is formed by several layers. The innermost one, in contact with the vitreous humor, is formed by cells and fiber nerves, while the last layer in the back has the light-sensitive elements, which are the rods and cones. There is a zone where the optic nerve enters the eye globe, producing a blind zone, with an angular diameter of about 5° to 7°, at 15° from the optical axis on the nasal side. The fovea is a small zone near the optical axis, where the retina becomes thinner and blood vessels are not present. The fovea contains only cones in a dense random array. Outside the fovea the main light-sensitive elements are the rods, which are responsible for the scotopic vision. They are much more sensitive to brightness than the cones, but they are not color sensitive and have very low spatial resolution. Using adaptive optics techniques, David Williams and collaborators from the University of Rochester have been able to obtain images of the cones, about 5 µm diameter from a living body (Liang and Williams, 1997; Williams, 1999).

13.1.1 Eye Aberrations

As any other optical instrument with image-forming lenses, the eye has optical aberrations that limit its optical performance (Gubisch, 1967; Liang and Williams, 1997). The off-axis aberrations are not of concern for the eye, since the image is always observed on-axis by fast scanning by continuously moving the eye. However, the on-axis aberrations are important.

Because of the axial chromatic aberration the eye focuses the different colors on different focal planes along the optical axis. Köhler (1962) has devised a simple and interesting experiment to show the presence of axial chromatic aberration in the eye. To perform this experiment, with each of your hands hold a card and place them in front of each of your eyes. Each card must be aligned with a straight edge in front of the eye covering half the eye pupil. The left edge of the right-hand card must be in front of the pupil of the right eye and the right edge of the left-hand card must be in front of the pupil of the left eye. Now look at a highly colored image, e.g., on a computer screen. It can be noticed that different color zones appear to have slightly different depths. This effect, called pseudostereopsis, can be also observed without the cards by some people. The effect arises because the pupils of the eyes are not always centered with the optical axis.

Axial chromatic aberration has been studied and measured by several researchers, e.g., Wald and Grifflin (1947) and Bedford and Wyszecki (1957). The focus shift at the blue end of the spectrum due to this aberration is as high as 2 diopters.

The spherical aberration of the eye appears because paraxial rays and marginal rays passing through the eye are focused at different planes along the optical axis. Many different experiments have been devised to measure the spherical aberration of the eye (Koomen et al., 1949; Ivanof, 1956; Williams, 1999).

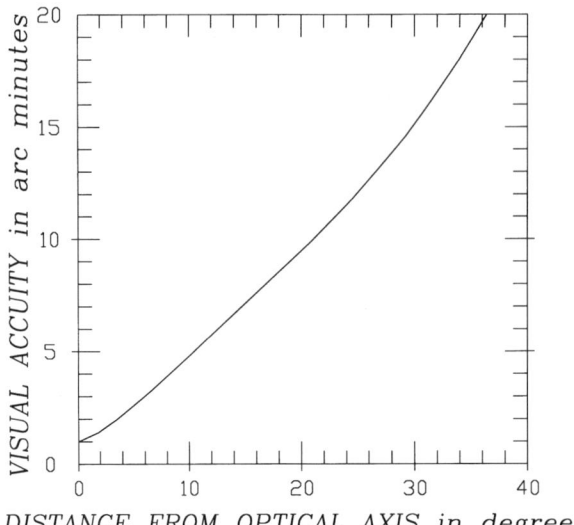

Figure 13.2 Angular resolving power of the human eye.

The contrast sensitivity of the human eye has also been the subject of many investigations. For example, Williams (1985) has studied the visibility and contrast of sinusoidal interference fringes near the resolution limit.

In a small region around the optical axis the resolving power of the eye is about 1 arcmin. However, the image quality outside the optical axis degrades quite fast with increasing angles, as plotted in Fig. 13.2 (Walker, 2000). For this reason the eye has to scan an observed image by moving the eyes quite fast in their skull cavity. When an object of interest is detected the eye is rotated quite fast to center the object on the optical axis.

13.2 OPHTHALMIC LENSES

An ophthalmic lens is a thin meniscus lens placed in front of the eye to correct its refractive defects, as shown in Fig. 13.3. This lens is mounted in a frame in such a way that the distance from the vertex of the concave surface to the cornea of the eye is 14 mm. The purpose of the lens is to form a virtual image of the observed object at the proper distance for the eye to observe it. Thus, the important parameter to describe the lens is the back focal length. The inverse of the back focal length in meters is called the vertex power,

The Eye and Ophthalmic Lenses

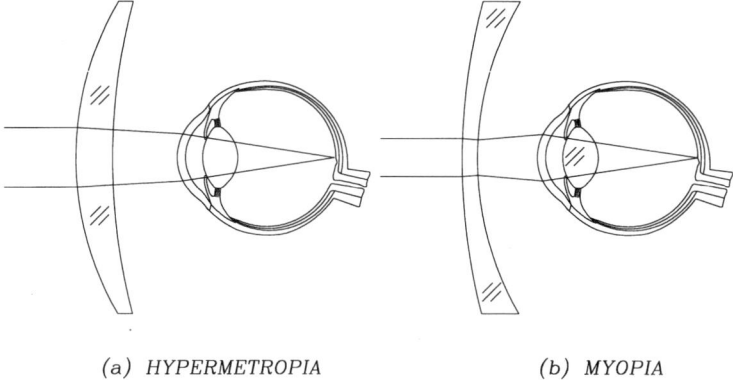

(a) HYPERMETROPIA (b) MYOPIA

Figure 13.3 (a) Hypermetropia and (b) myopia.

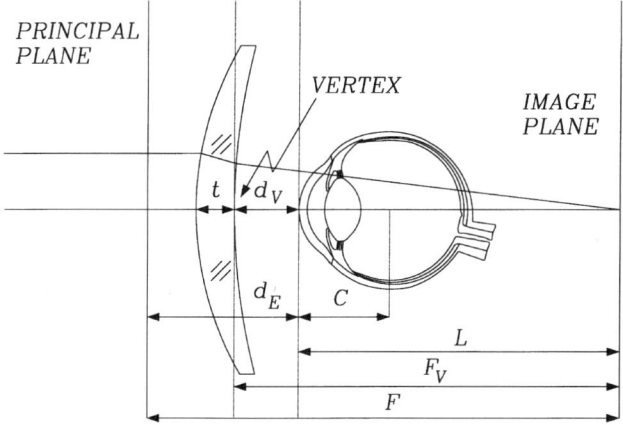

Figure 13.4 Some parameters used in the design of ophthalmic lenses.

expressed in diopters. According to the American Optometric Association Standard, the tolerance in the specified power is ±0.06 diopter.

As shown in Fig. 13.4, an eye with a refraction defect may focus an image of a very far object (without eye accommodation, i.e., with the eye lens focused for a distant object) only if the distance from the cornea to the image is L. If the distance L is positive (behind the eye), the eye is hypermetropic and, if this distance is negative (in front of the eye), the eye is myopic. To form the image at the proper place, a lens is used with a back focal length F_V such that $F_V = L + d_V$, where d_V is the distance from the vertex of the back surface of the lens to the cornea.

If the radii of curvature of the surfaces of the lens are measured in meters, the power of the surface in diopters is

$$P = \frac{(n-1)}{r} \tag{13.1}$$

It is measured with an instrument called a dioptometer (Coleman et al., 1951). The power of the frontal (convex) surface is called the base power. The refractive index for the most common ophthalmic glass is 1.523. For practical convenience, the nominal dioptral power of a grinding or polishing tool for ophthalmic lenses, with radius of curvature r, is defined as $P_n = 0.530/r$. Thus, the real power P of the surface polished with a nominal power P_n is

$$P = \frac{(n-1)}{0.530} P_n \tag{13.2}$$

From Eq. (3.37) we may show that the vertex power of the ophthalmic lens (defined as the inverse of the back focal length) is

$$P_V = \frac{P_1}{1 - P_1(t/1000n)} + P_2 \tag{13.3}$$

where the thickness t is in millimeters. This expression may be approximated by

$$P_V = P_1 + P_2 + \frac{P_1^2 t}{1000n} \tag{13.4}$$

in order to make easier all hand calculations. With a slightly greater error we may also write

$$P_V = P_1 + P_2 \tag{13.5}$$

To get an idea of the error in these formulas, let us consider as an example a glass lens with $P_1 = 9$ D, $P_2 = -4$ D, and $t = 4$ mm. Then, we obtain

$P_V = 5.2179$ with exact formula (13.3)
$P_V = 5.2127$ with approximate formula (13.4)
$P_V = 5.0000$ with approximate formula (13.5)

The effective power P_e is the inverse of the effective focal length in meters. Equation (3.39) may be written as

$$P_e = P_1 + \left(1 - \frac{P_1 t}{1000 n}\right) P_2 \qquad (13.6)$$

The effective power of the lens in the last example is $P_e = 5.094$. Thus, the relation between the effective power and the vertex power is

$$P_e = \left(1 - \frac{P_1 t}{1000 n}\right) P_V \qquad (13.7)$$

13.2.1 Ophthalmic Lens Magnifying Power

When an eye is larger than normal but the refractive components do not change their optical properties, the image of an object at infinity is defocused. This is what happens in myopia. In hypermetropia the mechanism is exactly the opposite. An ophthalmic lens corrects this defect, shifting the real image formed by the optics of the eye to the new position. If the combination of the optics of the eye with the ophthalmic lens preserves the original effective focal length, the size of the image is also preserved. In other words, the size of the images in a normal and in a corrected eye with the same effective focal length are equal. It may be easily shown that if the ophthalmic lens is in the front focus of the eye, the effective focal length of the combination remains the same. This is only approximately true, because the average eye has a cornea-to-front principal plane distance of 16.0 mm whereas a normal spectacle has a 14.5 mm distance.

When the eye is corrected with a spectacle lens, the image can change its size in a noticeable manner. This change in magnifying power is given by

$$\Delta M = \left(\frac{1}{1 - (d P_e / 1000)} - 1\right) \times 100\% \qquad (13.8)$$

where d is the distance from the principal plane of the lens to the cornea, in millimeters. This equation may be written in terms of the power P_1 of the base and the vertex power P_V, as follows:

$$M = \left(\frac{1}{(1 - (d P_V)/1000)(1 - (P_1 t)/1000 n)} - 1\right) \times 100\% \qquad (13.9)$$

Frequently, the first term in the denominator is said to be due to the power of the lens and the second term to its shape. The effect of the first term is greater than that of the second. As an approximate rule, there is a magnification of about 1.4% for each diopter in the lens.

An important property of ophthalmic lenses is that the functional power of the lens depends on its distance to the eye. If the lens-to-cornea distance is increased, the effective power decreases. Let the distance from the lens to the cornea be d_1, with its effective power P_1, and also the distance from the lens to the cornea be d_2, with its effective power P_2. Then, we have

$$\frac{P_2 - P_1}{P_1 P_2} = \frac{d_1 - d_2}{1000} \tag{13.10}$$

For example, if a 5 D lens is moved 10 mm, the effective power changes by an amount 0.25 D.

13.3 OPHTHALMIC LENS DESIGN

An optical layout used for ophthalmic lens design is shown in Fig. 13.5. The eye has a nearly spherical shape and moves in its cavity to observe objects in different directions. Thus, the stop is at the plane of the eye's pupil, which rotates about the center of rotation of the eye. The actual stop can be represented by an apparent stop located near the center of the eye. It is assumed that all observed objects in different directions are at the same distance from the eye. So, the object surface is spherical, with the center of

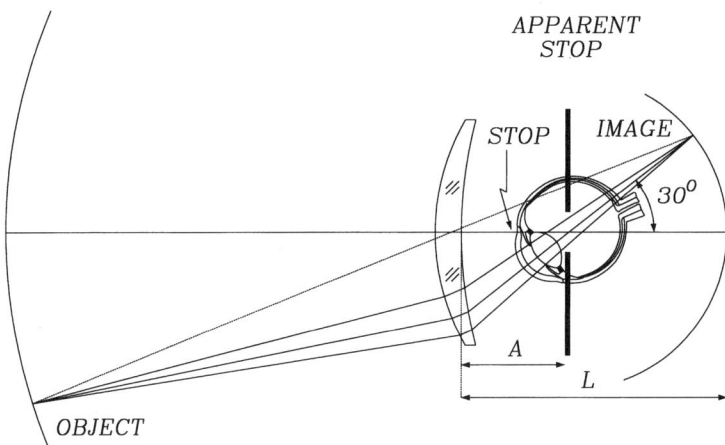

Figure 13.5 Optical schematics of an eye with its ophthalmic lens.

curvature approximately at the stop. The image formed by the ophthalmic lens should also be at a constant distance for any object in the object surface. Thus, the image surface is also spherical, with the center of curvature at the stop.

The distance C from the cornea of the eye to the center of the eye globe is called the sighting center distance, and it has been found (Fry and Hill, 1962) to be a linear function of the refractive error of the eye, which may be expressed by

$$C = -\frac{P_V}{6} + 14.5 \, \text{mm} \tag{13.11}$$

The distance from the vertex of the lens to the cornea of the eye is not the same for all observers, but has small variations, with an average of about 14 mm.

The thickness t is assumed to be constant for negative lenses and increasing linearly with the power for positive lenses, as shown in Fig. 13.6. For positive lenses the edge thickness is taken as approximately constant, equal to 1 mm, as given by

$$t = -\frac{D^2 P_V}{8000(n-1)} + 2.0 \, \text{mm} \tag{13.12}$$

Since the vertex power of the lens and the stop position are fixed, the only degree of freedom we have for the correction of aberrations is the lens bending. The spherical aberration and the axial chromatic aberration are

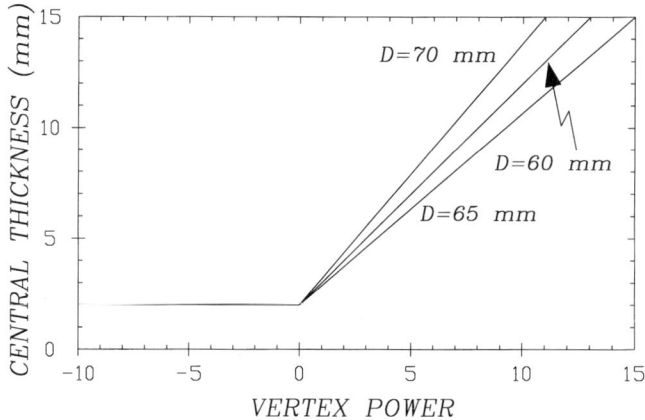

Figure 13.6 Central thickness in an ophthalmic lens.

not a problem, because the eye pupil's diameter is very small compared with the focal length of the lens. The coma is not very important compared with the astigmatism, because of the large field and the small diameter of the eye's pupil. The remaining aberrations to be corrected are astigmatism, field curvature, distortion, and magnification chromatic aberration. Distortion and magnification chromatic aberrations cannot be corrected by just lens bending. Thus, we are left with the astigmatism and the field curvature, also sometimes called peripheral power error.

Ophthalmic lenses and their design techniques have been described by several authors, e.g., Blaker (1983), Emsley (1956), Lueck (1965), Malacara and Malacara (1985a), and Walker (2000). To design an ophthalmic lens is relatively easy if we plot the curvatures of the Petzval, sagittal, and tangential surfaces as a function of the power of the front lens surface, as shown in Fig. 13.7. These curves were obtained for a thin lens ($t = 2\,\mathrm{mm}$) with a vertex power of three diopters and small field (5°), with a lens evaluation program that directly computes these curvatures. However, these may also be obtained from the slope at the origin of a meridional plot, by first calculating Δf_T with Eq. (7.40) and then calculating the tangential astigmatism with Eq. (7.19). If we assume a relatively small field, so that only the primary aberrations are present, the sagittal astigmatism is one-third of the tangential astigmatism. In Fig. 13.7 the zero for the vertical scale

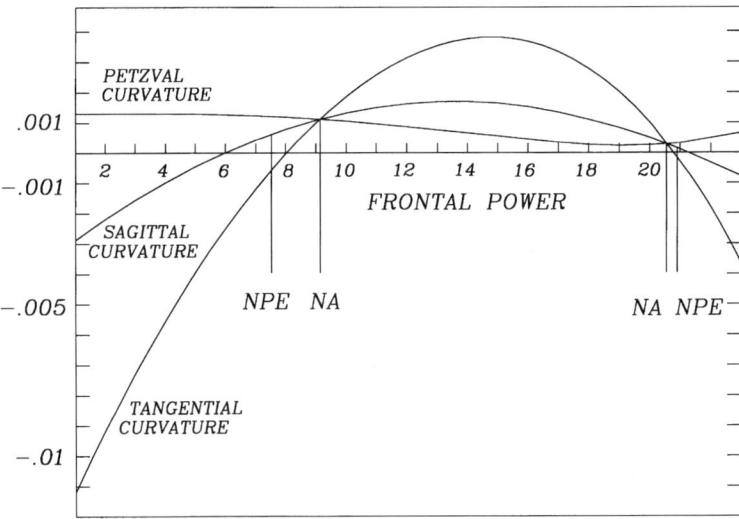

Figure 13.7 Change in the tangential and the sagittal curvatures versus the frontal power in an ophthalmic lens.

The Eye and Ophthalmic Lenses 325

is the curvature of the focal surface with center of curvature at the stop. We may see small variations in the Petzval curvature because the vertex power is constant, but the effective focal length and hence the Petzval sum is not exactly constant.

The points where the three curves in Fig. 13.7 meet are the solutions for no astigmatism (NA) and the points where the sagittal and tangential curves are symmetrical with respect to the horizontal axis are the solutions for no power error (NPE).

13.3.1 Tscherning Ellipses

We have seen that by bending we may correct either the astigmatism or the field curvature, but not both simultaneously. The second defect produces a defocusing of the objects observed through the edge of the lens. The observer may refocus the image by accommodation of the eye, but this introduces some eye strain, which may frequently be tolerated, especially by young persons. If the frontal lens power or base is used as a parameter for the bending, we may plot the total vertex power of the lens as a function of the base power that gives a lens without astigmatism and similarly for the peripheral power error. Thus, we obtain two ellipses as shown in Fig. 13.8,

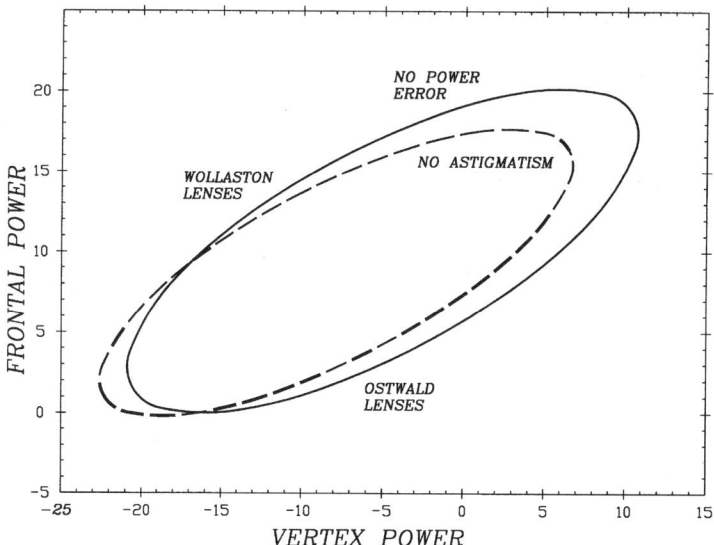

Figure 13.8 Tscherning ellipses for ophthalmic lenses free of astigmatism and power error.

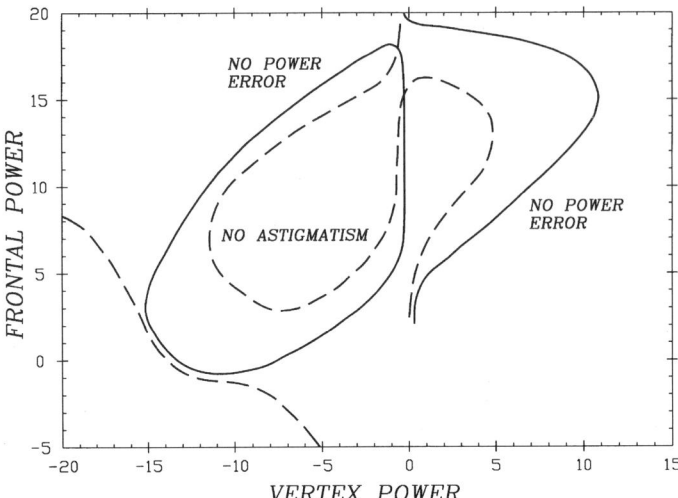

Figure 13.9 Tscherning ellipses deformed by the introduction of a finite lens thickness.

called Tscherning ellipses. We see that for each ellipse there are two solutions, one with a low frontal power (Ostwald lenses) and the other with a higher frontal power (Wollaston lenses). The Tscherning ellipses were obtained using third-order theory, with a constant very small lens thickness, a very small field, and a constant distance from the vertex of the lens to the stop, equal to 29 mm.

When the restrictions of constant thickness, constant distance from the vertex of the lens to the stop, and small field are removed, the Tscherning ellipses deform as shown in Fig. 13.9.

13.3.2 Aspheric Ophthalmic Lenses

We may see from the Tscherning ellipses that there are no solutions without astigmatism or without peripheral power error for high lens vertex powers. In this case an aspheric surface may be used in the front surface. With aspheric surfaces the Tscherning ellipses change their shape, extending the solution range to higher powers, as shown in Fig. 13.10 for the case of astigmatism and in Fig. 13.11 for the case of zero power error. Aspherical surfaces for ophthalmic lenses have been studied by Smith and Atchison (1983), by Sun et al. (2000 and 2002) using third-order theory, and by Malacara and Malacara (1985b) using exact ray tracing.

The Eye and Ophthalmic Lenses

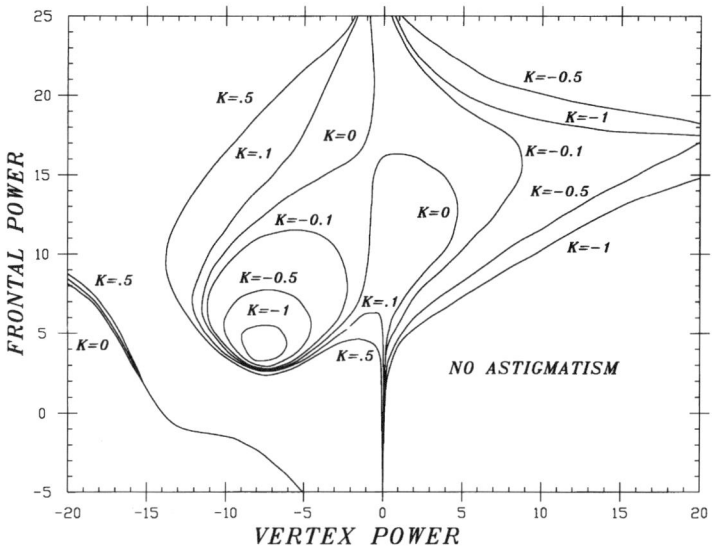

Figure 13.10 Tscherning ellipses for no astigmatism, deformed by the introduction of an aspheric surface.

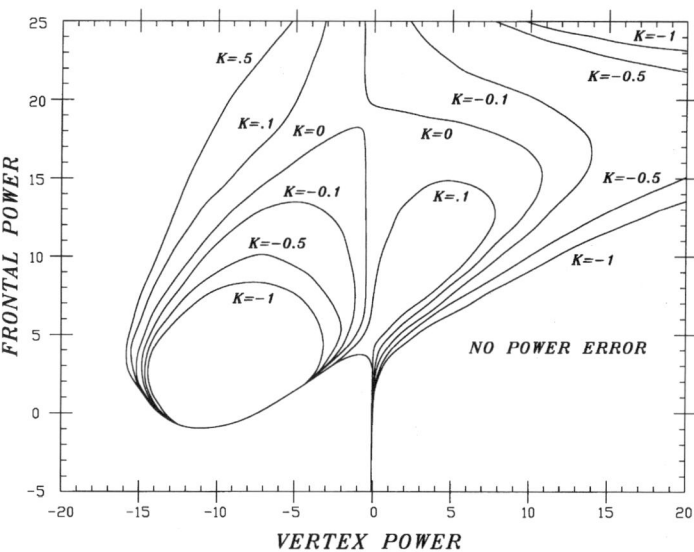

Figure 13.11 Tscherning ellipses for no power error, deformed by the introduction of an aspheric surface.

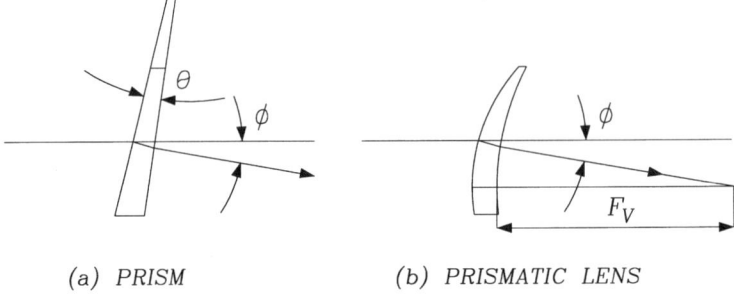

(a) PRISM　　　　(b) PRISMATIC LENS

Figure 13.12 (a) Prism and (b) prismatic lens.

13.4 PRISMATIC LENSES

The two centers of curvature of the surfaces of the lens define the optical axis. Only when the two centers of curvature coincide because the surfaces are concentric, the optical axis is not defined. If the optical axis passes through the center of a round lens, the edge has a constant thickness all around.

When the optical axis does not pass through the center of the lens, the lens is said to be prismatic because the two lens faces form an angle between them at the center of the lens. Then, a ray of light passing through the center of the lens is deviated by an angle ϕ. If a prism deviates a ray of light by an angle ϕ, as shown in Fig. 13.12, the prismatic power P_P in diopters is given by

$$P_P = 100 \tan \phi \qquad (13.13)$$

thus, a prism has P_P diopters if a ray of light passing through the center of the lens is deviated P_P centimeters at a distance of 1 m.

If the angle between the two faces of the lens is θ, the angular deviation of the light ray is

$$\tan \phi = \frac{\sin \theta}{n - \cos \theta} \qquad (13.14)$$

or approximately, for thin prisms:

$$\phi = \frac{\theta}{n - 1} \qquad (13.15)$$

The Eye and Ophthalmic Lenses

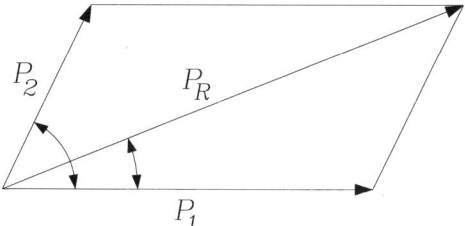

Figure 13.13 Vector addition of prisms.

If two thin prisms with prismatic powers P_1 and P_2 are superimposed, forming at their bases an angle α between them, the resulting combination has a prismatic power P_R given by

$$P_R^2 = P_1^2 + P_2^2 + 2P_1 P_2 \cos \alpha \tag{13.16}$$

and its orientation is

$$\sin \beta = \frac{P_2}{P_R} \sin \alpha \tag{13.17}$$

this result may also be obtained graphically, as shown in Fig. 13.13.

A lens with vertex power P_V and prismatic power P_P is a lens whose optical axis is deviated from the center of the lens by an amount Δy given by

$$P_P = \frac{P_V \Delta y}{10} \tag{13.18}$$

where the decentration Δy is in millimeters.

13.5 SPHEROCYLINDRICAL LENSES

A spherocylindrical lens has a toroidal or spherocylindrical surface. The lens does not then have rotational symmetry and an axial astigmatism is introduced, to compensate that of the eye. Optically, these lenses may be considered as the superposition of an spherical lens (with rotational symmetry) and a cylindrical lens (power in only one plane). As shown in Fig. 13.14, a spherocylindrical lens is defined by (1) its spherical power, (2) its

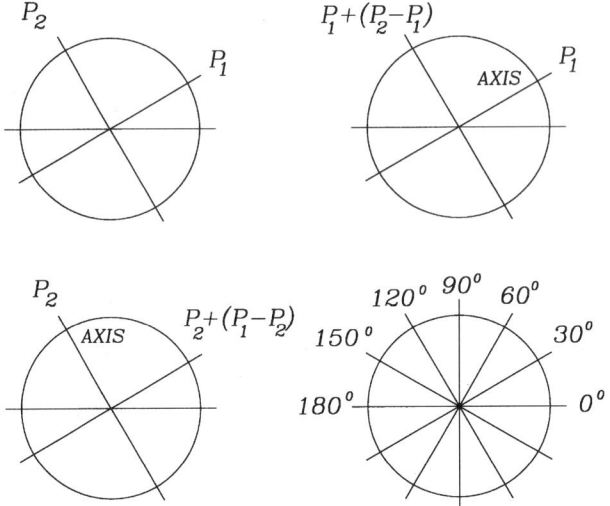

Figure 13.14 Powers and axis orientation in spherocylindrical lenses.

cylindrical power, and (3) its cylinder orientation. If a lens has a power P_1 in one diameter, at an angle ϕ with the horizontal, and a power P_2 in the perpendicular diameter, at an angle $\phi + 90°$, we may specify the lens as

>Spherical power $= P_1$
>Cylindrical power $= P_2 - P_1$
>Axis orientation $= \phi$

or as

>Spherical power $= P_2$
>Cylindrical power $= P_1 - P_2$
>Axis orientation $= \phi + 90°$

The two specifications are identical. To pass from one form to the other is said to be to transpose the cylinder. A cylinder transposition is done in three steps, as follows:

1. A new spherical power value is obtained by adding the spherical and cylindrical power values.
2. A new cylindrical power value is obtained by changing the sign of the old value.
3. The new axis orientation is obtained by rotating the old axis at an angle equal to 90°.

The Eye and Ophthalmic Lenses

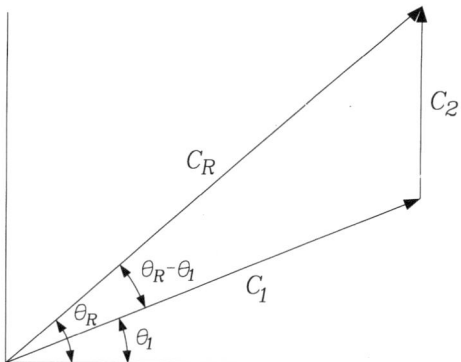

Figure 13.15 Vector addition of spherocylindrical lenses.

The power P_θ of a spherocylindrical lens along a diameter at an angle θ may be found with the expression:

$$P_\theta = P_C \sin^2(\theta - \phi) + P_S \tag{13.19}$$

where P_C is the cylindrical power, P_S is the spherical power, and ϕ is the cylinder orientation. If two spherocylindrical lenses are superimposed, the combination has a cylindrical power P_{CR} given by

$$P_{CR}^2 = P_{C1}^2 + P_{C2}^2 + 2P_{C1}P_{C2}\cos 2(\theta_2 - \theta_1) \tag{13.20}$$

an axis orientation θ_R:

$$\tan 2(\theta_R - \theta_1) = \frac{P_{C2} \sin 2(\theta_2 - \theta_1)}{P_{C1} + P_{C2} \cos 2(\theta_2 - \theta_1)} \tag{13.21}$$

and a spherical power P_{SR} given by

$$P_{SR} = P_{S1} + P_{S2} + \frac{P_{C1} + P_{C2} - P_{CR}}{2} \tag{13.22}$$

Graphically, these expressions may be represented as in Fig. 13.15.

REFERENCES

Bedford, R. E. and Wyszecki, G., "Axial Chromatic Aberration of the Human Eye," *J. Opt. Soc. Am.*, **47**, 564–565 (1957).

Blaker, J. W., "Ophthalmic Optics," in *Applied Optics and Optical Engineering*, R. R. Shannon and J. C. Wyant, eds., Vol. IX, Chap. 7, Academic Press, San Diego, CA, 1983.

Coleman, H. S., Coleman, M. F., and Fridge, D. S., "Theory and Use of the Dioptometer," *J. Opt. Soc. Am.*, **41**, 94–97 (1951).

Emsley, H. H., *Aberrations of Thin Lenses*, Constable and Co., London, 1956.

Fry, G. A. and Hill, W. W., "The Center of Rotation of the Eye," *Am. J. Optom. Arch. Am. Acad. Optom.*, **39**, 581–595 (1962).

Gubisch, R. W., "Optical Performance of the Human Eye," *J. Opt. Soc. Am.*, **57**, 407–415 (1967).

Ivanof, I., "About the Spherical Aberration of the Eye," *J. Opt. Soc. Am.*, **46**, 901–903 (1956).

Köhler, I., "Experiments with Goggles," *Scientific American*, **206** (May), 63–72 (1962).

Koomen, M. J., Tousey, R., and Scolnik, R., "The Spherical Aberration of the Eye," *J. Opt. Soc. Am.*, **39**, 370–376 (1949).

Liang, J. and Williams, D. R., "Aberrations and Retinal Image Quality of the Normal Human Eye," *J. Opt. Soc. Am. A*, **14**, 2873–2883 (1997).

Lueck, I., "Spectacle Lenses," in *Applied Optics and Optical Engineering*, R. Kingslake, ed., Vol. III, Chap. 6, Academic Press, San Diego, CA, 1965.

Malacara, Z. and Malacara, D., "Tscherning Ellipses and Ray Tracing in Ophthalmic Lenses," *Am. J. Opt. Phys. Opt.*, **62**, 447–455 (1985a).

Malacara, D. and Malacara, Z., "Tscherning Ellipses and Ray Tracing in Aspheric Ophthalmic Lenses," *Am. J. Opt. Phys. Opt.*, **62**, 456–462 (1985b).

Smith, G. and Atchison, D. A., "Effect of Conicoid Asphericity on the Tscherning Ellipses of Ophthalmic Spectacle Lenses," *J. Opt. Soc. Am.*, **73**, 441–445 (1983).

Sun, W.-S., Tien, C.-L., Sun, C.-C., Chang, M.-W., and Chang, H., "Ophthalmic Lens Design with the Optimization of the Aspherical Coefficients," *Opt. Eng.*, **39**, 978–988 (2000).

Sun, W.-S., Chang, H., Sun, C.-C., Chang, M.-W., Lin, C.-H., and Tien, C.-L., "Design of High-Power Aspherical Ophthalmic Lenses with a Reduced Error Budget," *Opt. Eng.*, **41**, 460–470 (2002).

Wald, G. and Grifflin, D. R., "The Change in Refractive Power of the Human Eye in Dim and Bright Light," *J. Opt. Soc. Am.*, **37**, 321–336 (1947).

Walker, B. H., *Optical Design for Visual Systems*, SPIE Press, Bellingham, WA, 2000.

Williams, D. R., "Visibility of Interference Fringes Near the Resolution Limit," *J. Opt. Soc. Am. A*, **2**, 1087–1093 (1985).

Williams, D. R., "Wavefront Sensing and Compensation for the Human Eye," in *Adaptive Optics Engineering Handbook*, R. K. Tyson, ed., Marcel Dekker, New York, 1999.

14
Astronomical Telescopes

14.1 RESOLUTION AND LIGHT GATHERING POWER

The subject of astronomical telescopes has been treated by many authors, e.g., the books by Dimitroff and Baker (1945), Linfoot (1955a,b), Maxwell (1972), and Schroeder (1987, 1993) are excellent references. In this chapter, astronomical telescopes will be studied in some detail.

To begin, let us consider the resolution of a telescope, which is limited by several factors, like diffraction and atmospheric turbulence. The light gathering power is another important characteristic in a telescope. In the first sections of this chapter these important concepts will be reviewed.

14.1.1 Diffraction Effects and Atmospheric Turbulence

The atmosphere has large inhomogeneities in the index of refraction due to variations in the pressure, to air currents, and to variations in the temperature. These inhomogeneities are continuously changing and produce wavefront distortions in the light coming from the stars, as shown in Fig. 14.1. The effects of atmospheric turbulence in the stellar images are mainly of two kinds, *scintillation* and *seeing*. Scintillation is a random change in the light intensity and seeing is a random change in the direction of the light arriving at the telescope.

Scintillation is observed only with small apertures, mainly with the naked eye, producing what is commonly known as *twinkling*. The larger the telescope, the smaller the effect of scintillation.

The effect of seeing depends on the aperture. As may be understood by examination of Fig. 14.1, for small apertures the wavefront distortion is not observed, but only a continuous change in the direction of propagation of the wavefront. Thus, in telescopes with a small aperture, less than about 150 mm, the image moves very fast about a mean position, with excursions of the order of 1 arcsec. If the aperture is larger than about 1 m, the image movements are not seen. Only a large blurred image is observed.

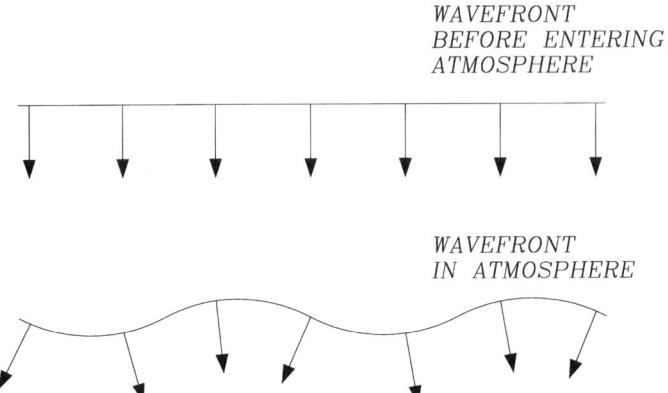

Figure 14.1 Undistorted and distorted wavefronts traveling in the atmosphere.

The diameter of the seeing enlarged image depends on the astronomical site. In a medium size city it may be as large as 2 or 3 arcsec or even more. In the best astronomical sites in a high mountain in the best nights it may be of the order on one-tenth of an arcsec. It is the easy to notice that under average conditions the diffraction image size is smaller than the seeing enlarged image, if the telescope aperture is larger than about 150–200 mm. In other words, the angular resolving power of a telescope is limited by the atmosphere, not by diffraction. Hence, a telescope does not need to be diffraction limited unless it is located outside the atmosphere, in orbit, like the Hubble telescope. Another exception is when some techniques are used to compensate the wavefront deformations as in the new adaptive optics techniques.

14.1.2 Visual Limit Magnitude of a Telescope

The magnitude of a star is an indication of its brightness. The greater the magnitude, the fainter the star. The magnitude of a star is an arbitrary scale invented by the Greeks. According to them, the brightest stars in the sky had magnitude 1 and the faintest had magnitude 6. The same basic definition is now used, but with a more formal and mathematical meaning. Now we know that, according to the psychophysical law of Fetchner, the optical sensation in the eye is directly proportional to the logarithm of the luminous excitation. Based on this effect, John Herschel in 1830 defined that the first magnitude star is 100 times brighter than the sixth magnitude star. Thus, the brightness of a star one magnitude higher is $(100)^{1/5} = 2.512$ times larger.

Astronomical Telescopes

If two stars with magnitudes m and n have brightness B_n and B_m, respectively, we may write

$$\frac{B_n}{B_m} = 2.512^{m-n} \tag{14.1}$$

whence, taking logarithms on both sides:

$$\log B_n - \log B_m = (m-n)\log 2.512 = 0.4(m-n) \tag{14.2}$$

Thus, the magnitude difference between two stars is directly proportional to the difference in the logarithm of their brightnesses. The first magnitude star was then defined arbitrarily, as the magnitude of one star close to the brightest stars.

When a star is observed through a visual telescope, the apparent brightness is increased because the amount of light forming the image in the retina is larger and the size of the image is not increased. This amount of light is increased as the ratio of the square of the diameters of the telescope D_t and the naked eye pupil's, D_0, as follows:

$$\log\left(\frac{B_t}{B_0}\right) = -2\log\left(\frac{D_t}{D_0}\right) = 0.4(m_0 - m_t) \tag{14.3}$$

where m_0 is the largest magnitude that may be observed with the eye and m_t is the largest magnitude that may be observed through the telescope. Hence, we may write

$$m_t - m_0 = -5\log\left(\frac{D_1}{D_0}\right) \tag{14.4}$$

The diameter of the pupil of the eye is different for different observers, and it changes with the amount of light entering the eye, even for a single observer. However, we may assume an average value of D_0 equal to 6 mm. Thus, the limiting visual magnitude when observing through a telescope is

$$m_v = 7.10 + 5\log D \tag{14.5}$$

14.1.3 Photographic and CCD Limit Magnitude of a Telescope

The limiting magnitude in astronomical photography is larger than in visual observation and depends not only on the aperture of the telescope, but also on the exposure time. The sensitivity of the photographic emulsion is also a factor, but not very important, since the range of common sensitivities is not so large. The reason is that the resolving power and the sensitivity have an inverse relation to each other. If we assume an average emulsion, an empirical relation for the limiting photographic magnitude is

$$m_f = 4 + 5\log D + 2.15\log t \tag{14.6}$$

where D is the diameter of the telescope aperture and t is the exposure time. It is easy to see that the limiting visual and photographic magnitudes are approximately equal for an exposure time of 28 min.

The sky background is not absolutely dark, so the ideal maximum exposure time would be that such that the darkening of the background in the photographic plate is not larger than a certain limit. The sky brightness in the image of the telescope may be shown to be inversely proportional to the square of the focal ratio or f-number FN.

The angular diameter of all stars is the same for all telescopes in a given astronomical site, due to the atmospheric turbulence. Thus, the size of the image is directly proportional to the effective focal length. On the other hand, the amount of energy forming the image is directly proportional to the square of the aperture. Then, the energy per unit area in the image is also inversely proportional to the square of the focal ratio FN. However, the grain in the photographic plate is chosen to be at least as large as the image of the star. The blackening of an emulsion grain depends not on the distribution of the light over the grain but on the total amount of energy falling on it. So, the image may be considered to be a point, if all the light falls on a single grain. This leads us to the conclusion that the ratio of the effective brightness of the star image over the brightness of the background depends only on the f-number FN.

An important conclusion is that the f-number determines the maximum exposure time, producing the maximum allowed background blackening. This is approximately given by the following empirical relation:

$$\log t = 0.6 + 2.325 \log FN \tag{14.7}$$

where the exposure time is in minutes, but obviously this maximum exposure time cannot be larger than about 300 min (5 h). This means that for 5-h exposures the optimum f-number is equal to 6.4. It is then easy to see that the limiting photographic magnitude is

$$m_f = 7.29 + 5\log F \tag{14.8}$$

for an exposure time of 5 h. This means that the focal length has to be as large as possible, but this also means a large diameter, since the focal ratio is fixed.

A more general method for treating the problem of limiting magnitudes has been given by several authors (Schroeder, 1987). The evident conclusion is that the aperture of the telescope should be as large as possible and that the observing site should be as transparent and seeing free as possible.

To finish this section, we must know that modern astronomical telescopes do not use photographic emulsions any more, but a much better photoelectric device called a *CCD* (*coupled charge device*). This is an electro-optical detector, exploiting semiconductor manufacturing technology, that permits the telescope to reach the maximum theoretically possible sensitivity (Mallama, 1993).

14.2 CATADIOPTRIC CAMERAS

In astronomical photography it is necessary to have a large aperture. Thus, most astronomical cameras are either reflective (catoptric) or catadioptric. A catadioptric camera is one formed by mirrors as well as by lenses. The most popular of these are the Schmidt and the Maksutov cameras.

14.2.1 Schmidt Camera

The Schmidt camera, invented by Bernard Schmidt, is extremely popular for wide-field, high-speed astronomical cameras and has been fully described by many authors, e.g., Baker (1940b), Lucy (1940, 1941), Synge (1943), Benford (1944), Linfoot (1955a), Linfoot and Wolf (1949), Wormser (1950), Linfoot (1955a), Bowen (1960), Cornejo et al. (1970), Buchroeder (1972), and many others. We have seen that a spherical mirror with the stop at the center of curvature is free of coma, astigmatism, and distortion. It is easy to see that this is due to the symmetry of the system about the center of curvature. This symmetry also explains that the field is curved and also concentric with the center of curvature of the mirror. The only problems with large apertures are the spherical aberration and the spherochromatism. If a parabolic shape is given to the mirror the spherical aberration is corrected, but the spherical symmetry of the system about the center of curvature is lost. Schmidt corrected then the spherical aberration by introducing a thin aspheric correcting plate at the stop, as shown in Fig. 14.2. The off-axis aberrations introduced by this correcting plate are negligible, due to its small power and location at the stop.

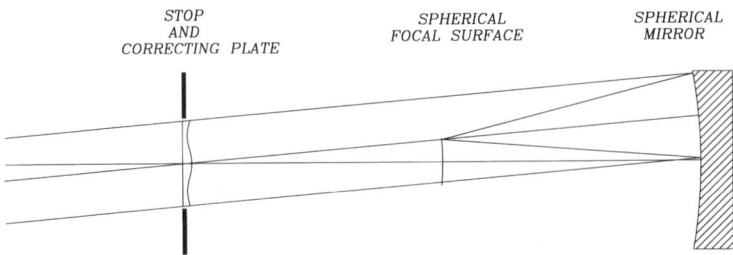

Figure 14.2 Schmidt camera.

There are many ways of calculating the shape of this correcting plate, but one simple approximate method is by considering that a flat wavefront is deformed to obtain the shape:

$$W(S) = (n - 1)Z(S) \tag{14.9}$$

when passing through the plate with shape $Z(S)$ on its face. On the other hand, the desired wavefront deformation is opposite to that of the spherical wavefront deformation of a spherical mirror, as given in Eq. (11.20). Thus,

$$Z(S) = \frac{S^4}{4(n-1)r_M^3} \tag{14.10}$$

where r_M is the radius of curvature of the spherical mirror. This is a surface shape as illustrated in Fig. 14.3(a). This is not the optimum shape, however, because there is a noticeable spherical aberration for wavelengths differing from the reference. This aberration may be minimized by introducing a small curvature on the glass plate (a defocusing term) to minimize the maximum slope on the surface of the correcting plate, as in Fig. 14.3(b). Then, the new sagitta is

$$Z(S) = \frac{S^4}{4(n-1)r_M^3} + \frac{S^2}{2r_p} \tag{14.11}$$

where r_p is the vertex curvature of the glass plate and r_M is the radius of curvature of the spherical mirror. Then, r_p is calculated by setting

$$\left(\frac{dZ(S)}{dS}\right)_{0.707 S_{\max}} = 0 \tag{14.12}$$

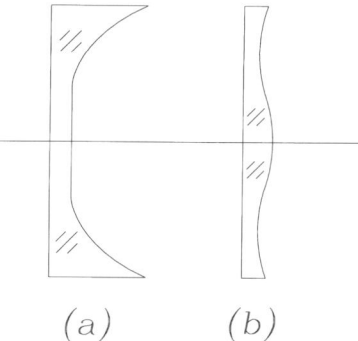

Figure 14.3 Schmidt corrector plates.

where S_{max} is the value of S at the edge of the plate (semidiameter) obtaining for the radius of curvature of the plate a value:

$$\frac{1}{r_p} = \frac{S_{max}^2}{2(n-1)r_M^3} \tag{14.13}$$

and the shape of the plate becomes

$$Z(S) = \frac{S^2}{4(n-1)r_M^3}(S^2 - S_{max}^2) \tag{14.14}$$

as shown in Fig. 14.3(b). This result may be improved by exact ray tracing.

Since the focal plane is curved, with its radius equal to the system focal length, we have to adopt either of two possible solutions: (1) the photographic plate is curved on the plate holder by pushing it on its center from the back, or (2) by flattening the field with a lens flattener.

14.2.2 Bouwers Camera

During the Second World War, Baker (1940a), Bouwers (1946), and Maksutov (1944) independently developed cameras with a similar approach to that of the Schmidt camera. All of them have the stop at the center of curvature, but there are two main differences: (1) the correcting element is not at the stop and (2) the correcting element has only spherical surfaces, with a meniscus shape, concentric with the stop, as shown in Fig. 14.4.

The spherical aberration of the correcting plate may be found from Eq. (4.32), which is valid for thick lenses. However, although the correcting

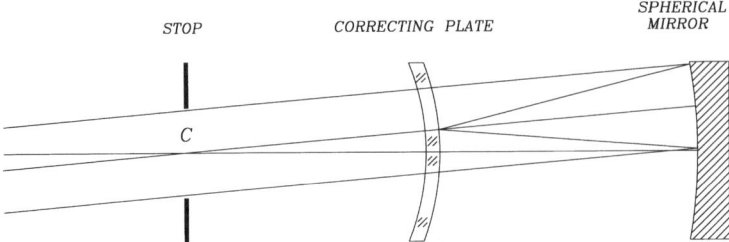

Figure 14.4 Concentric camera.

lens has strongly curved surfaces, its thickness is very small compared with the effective focal length. Thus, we may use a thin lens approximation. If we take the object distance as infinity ($v_1 = 0$) we obtain from Eq. (4.33):

$$SphT = F\kappa S^3(G_1 \kappa^2 - G_2 \kappa c_1 + G_4 c_1^2) \tag{14.15}$$

where F is the effective focal length of the correcting lens (not of the whole system), given by Eq. (3.35). Since the faces are strongly curved and the focal length is large, we may safely assume that $\kappa \gg c_1$. Thus, if we neglect the first two terms we obtain

$$SphT = \frac{(n+2) S^3}{2n r_1^2} \tag{14.16}$$

Then, using Eq. (1.77), the wavefront aberration $W(S)$ is given by

$$W = \frac{1}{F} \int_0^S SphT \, dS = \frac{(n+2) S^4}{8Fn r_1^2} \tag{14.17}$$

As in the Schmidt camera, the desired wavefront deformation must be opposite to the spherical wavefront deformation of a spherical mirror, as given in Eq. (11.17). Thus, equating this result with the wavefront aberration for the spherical mirror in Eq. (11.20) we obtain the result that the correcting lens must satisfy the condition:

$$F r_1^2 = \frac{(n+2) r_M^3}{2n} \tag{14.18}$$

The lens thickness t has not been defined, and may thus be used as an extra degree of freedom.

Astronomical Telescopes

Bouwers selected a value of the thickness t and a value of r_1 such that the whole system is concentric about the center of curvature **C**, as shown in Fig. 14.4. Then all off-axis aberrations become zero. From Eq. (3.35), the effective focal length of a concentric lens ($r_2 = r_1 - t$) is

$$F = \frac{n r_1 (r_1 - t)}{(n-1)t} \qquad (14.19)$$

The only remaining degree of freedom is the radius of curvature r_1, because the correcting meniscus may be placed at any distance from the stop, with the only restriction that the system is concentric. The closer the meniscus is to the center of curvature, the more curved and thinner the plate is. It is important to see that the zonal spherical aberration has different degrees of correction at different values of y. To achieve the maximum well-corrected field of view and axial color correction, a weak lens element (or zero power doublet) at the stop is needed; $f/1.0$, $30°$ total field of view designs is common.

14.2.3 Maksutov Camera

A slightly different system to that of Bouwers was designed by Maksutov (1944). The system is based on the same concentricity principle. Maksutov, however, deviates a little from concentricity in order to correct the achromatic aberration. The magnification chromatic aberration is automatically corrected with the concentricity of the surfaces. However, a small amount of axial chromatic aberration is present, but it may be corrected with the principle described in Section 6.2.4, by selecting the radii of curvature such that

$$\frac{r_2}{r_1} = \frac{[1 - (n-1)t/n r_1]^2}{[1 - (n-1)^2 t/n^2 r_1]} \qquad (14.20)$$

where t is the thickness of the corrector. When the axial color is corrected in this way the field of excellent correction is smaller than for the concentric element approach.

14.3 NEWTON TELESCOPE

The Newton (or Newtonian) telescope, illustrated in Fig. 14.5, is just a paraboloid with a small diagonal mirror near the focus to deviate the light beam to one side. The aberrations of this telescope are those of a single paraboloid, as studied in Section 11.3.3.

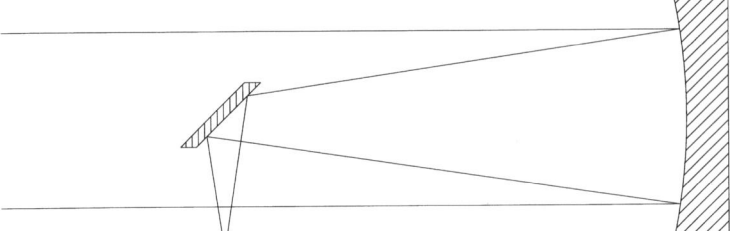

Figure 14.5 Newton telescope.

If the focal length is large compared with its diameter (large focal ratio, $FN = f/D > 10$), the spherical aberration of a spherical mirror becomes negligible. Then, the telescope can be made with a spherical instead of a parabolic mirror. The decision about when to use a spherical mirror must be taken according to the criteria in Section 4.7 if a diffraction-limited telescope is desired. This might be the case for a small diameter telescope (less than about 20 cm), otherwise the limiting factor is the atmospheric turbulence or *seeing*.

14.4 REFLECTING TWO-MIRROR TELESCOPES

A two-mirror telescope is formed by a large concave mirror, called the primary mirror, and a small concave or convex mirror, called the secondary mirror, in front of it, to reflect back the light towards the primary mirror. The image is formed behind the primary mirror, with the light passing through a hole in its center. If the secondary mirror is convex, as in Fig. 14.6, we have a Cassegrain telescope, but if the secondary mirror is concave we have a Gregory (or Gregorian) telescope, as in Fig. 14.11. We may easily see that the effective focal length *F* is positive for the Cassegrain and negative for the Gregory telescope. These telescopes have been studied and described by Bouwers (1946), Yoder et al. (1953a,b), Jones (1954), Baker (1963), DeVany (1963), Malacara (1965), Schulte (1966a,b), Bowen (1967), Wynne (1968), Meinel (1969), Wetherell and Rimmer (1972), Cornejo and Malacara (1973, 1975), Gascoine (1973), Shafer (1976), and Schroeder (1978, 1987).

14.4.1 First-Order Design of Two Mirror Telescopes

We will study in some detail these telescopes, but the first step is to calculate the curvatures and separation between the mirrors, using first-order theory. As a first step we may see that the focal lengths for the primary and

Astronomical Telescopes

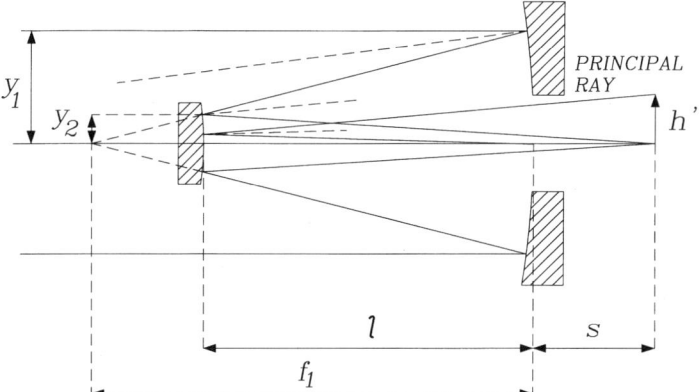

Figure 14.6 Some important parameters in a two-mirror telescope.

secondary mirrors are f_1 and f_2 and are related to the radii of curvature r_1 and r_2 by

$$r_1 = \frac{1}{c_1} = -2f_1 \qquad (14.21)$$

and

$$r_2 = \frac{1}{c_2} = 2f_2 \qquad (14.22)$$

According to our sign convention both radii of curvature are negative; hence, f_1 is positive and f_2 is negative. The image formed by the primary mirror is used by the secondary mirror as an object, to form a magnified image at the secondary focus. This lateral magnification m_s introduced by the secondary mirror is equal to

$$m_s = \frac{F}{f_1} = \frac{l+s}{f_1-l} \qquad (14.23)$$

where F is the effective focal length of the telescope and s is the distance from the vertex of the primary mirror to the secondary focus. For any two mirror telescope we may find the effective focal length with Eq. (3.43). However, it is simpler if, as shown in Fig. 14.6, we write

$$\frac{y_1}{y_2} = \frac{F}{l+s} = \frac{f_1}{f_1-l} = \pm\frac{D_1}{d_2} \qquad (14.24)$$

where l is the separation between the mirrors, $y_1 = D_1/2$ is the height of the meridional ray on the primary mirror (half the diameter of the primary mirror), and $y_2 = \pm d_2/2$ is the height of the meridional ray on the secondary mirror (half the diameter of the reflected conic light beam, on the secondary mirror, with a plus sign for the Cassegrain or a minus sign for the Gregory). From the second and third terms in this expression we may obtain the separation l between the mirrors as

$$l = \frac{F-s}{1+F/f_1} = \frac{m_s f_1 - s}{1+m_s} \tag{14.25}$$

and substituting back into Eq. (14.24):

$$\frac{y_1}{y_2} = \frac{f_1 + F}{f_1 + s} \tag{14.26}$$

On the other hand, applying Eq. (2.13) to the secondary mirror, we have

$$\frac{1}{f_2} = \frac{1}{l+s} - \frac{1}{f_1 - l} \tag{14.27}$$

and, from Eq. (14.26), the ray height y_2 is

$$y_2 = y_1 \left(\frac{f_1 + s}{f_1 + F} \right) = \frac{D_1}{2} \left(\frac{f_1 + s}{f_1 + F} \right) \tag{14.28}$$

If the height of the image on the secondary focal plane is h', the size of the image at the prime focus is equal to $h'/m_s = h' f_1/F$. Then, the height of the principal ray on the secondary mirror is

$$\bar{y}_2 = h' m_s \left(\frac{l}{f_1} \right) = h' \left(\frac{l}{F} \right)$$
$$= \frac{f_1(F-s)}{F(f_1 + F)} h' \tag{14.29}$$

From Eqs. (14.24) and (14.27), the focal length of the secondary mirror is

$$\frac{1}{f_2} = \frac{y_1}{y_2} \left(\frac{1}{F} - \frac{1}{f_1} \right) \tag{14.30}$$

and using Eq. (14.26), the focal length f_2 may be written as

$$f_2 = \left(\frac{F}{f_1}\right)\left(\frac{f_1+s}{1-(F/f_1)^2}\right) = f_1\frac{m_s(1+s/f_1)}{1-m_s^2} \tag{14.31}$$

The diameter of the secondary mirror is

$$D_2 = 2(\bar{y}_2 + y_2) = \frac{(f_1-l)D_1}{f_1} + 2\left(\frac{l}{F}\right)h' \tag{14.32}$$

To complete the first-order analysis of two-mirror telescopes, it is convenient to find the expressions for some parameters that will be needed when studying the aberrations of some particular systems. One of these quantities is the ratio between the principal ray height and the meridional ray height at the secondary mirror. From Eqs. (14.28) and (14.29), this ratio is

$$\frac{\bar{y}_2}{y_2} = \left(\frac{f_1(F-s)}{y_1 F(f_1+s)}\right)h' \tag{14.33}$$

The angle of incidence of the meridional ray at the primary mirror has a value:

$$i_1 = -\frac{y_1}{2f_1} \tag{14.34}$$

which is negative, since the ray arrives to the mirror above the normal. The angle of incidence at the primary mirror for the principal ray is

$$\bar{i}_1 = \frac{h'}{F} \tag{14.35}$$

which is positive, since the ray arrives to the mirror below the normal. The angle of incidence of the meridional ray at the secondary mirror, from Fig. 14.6, may be shown to be

$$i_2 = y_2\left(\frac{1}{(l+s)} - \frac{1}{2f_2}\right) \tag{14.36}$$

which is positive, since the ray arrives from the right side to the mirror, above the normal. Also, from examination of the same figure, the angle of incidence at the primary mirror for the principal ray is

$$\bar{i}_2 = -\bar{y}_2\left(\frac{1}{l} - \frac{1}{2f_2}\right) \tag{14.37}$$

which is negative, since the ray arrives at the mirror from the right side, below the normal. Then, after some algebraic work, we may find that these two angles may be expressed as

$$i_2 = -\frac{(f_1 + F)}{2f_1 F} y_1 \tag{14.38}$$

and

$$\bar{i}_2 = -\frac{(f_1 + F)(F + s)}{2F^2(f_1 + s)} h' \tag{14.39}$$

It is interesting to know that a system of two spherical mirrors can be designed to be anastigmat, i.e., with spherical aberration, coma, and astigmatism corrected. This system, formed by two concentric spherical reflecting surfaces as described by Erdös (1959), is used for an object at a finite distance. So, this configuration is not useful for telescopes, but for imaging a small object located at a relatively small distance.

14.4.2 Cassegrain Telescope

The Cassegrain telescope will now be analyzed by studying the five Seidel aberrations.

Spherical Aberration

In a Cassegrain telescope the spherical aberration is corrected separately on each of the mirrors and hence on the complete system. The primary mirror is a paraboloid and since its eccentricity is equal to 1, its conic constant is

$$K_1 = -1 \tag{14.40}$$

and the secondary mirror is a hyperboloid. The eccentricity e of this paraboloid, from analytic geometry, may be found to be

$$e = \frac{f_1 + s}{2l - f_1 + s} \tag{14.41}$$

thus, the conic constant is

$$K_2 = -\left(\frac{f_1+s}{2l-f_1+s}\right)^2 = -\left(\frac{m_s+1}{m_s-1}\right)^2 \quad (14.42)$$

By aspherizing the mirror with the conic constants, the spherical aberration is corrected by introducing an aspherical contribution that cancels the spherical contribution. It is useful to compute the spherical contribution, i.e., the aberration for spherical mirrors, using Eq. (4.29), because they will be used several times later when computing other aberrations. For a primary spherical primary mirror this aberration contribution, using the value of the angle i_1 from Eq. (14.34) we find that

$$SphTC_1 = -\left(i_1 + \frac{y_1}{f_1}\right)Fi_1^2$$
$$= -\frac{y_1^3 F}{8f_1^3} \quad (14.43)$$

and similarly for the spherical secondary mirror, using Eqs. (14.26) and (14.38):

$$SphTC_2 = \frac{y_2}{y_1}(Fi_2 - y_1)i_2^2$$
$$= \frac{(f_1+s)(F+f_1)(F-f_1)}{8f_1^3 F^2}y_1^3 \quad (14.44)$$

Coma

The coma of a Cassegrain telescope may be found by adding the spherical and aspherical contributions to the coma aberration for both mirrors, as follows:

$$Coma_S = Coma_S\, C_1 + Coma_S\, C_2 + Coma_S\, C_{asph1} + Coma_S\, C_{asph2} \quad (14.45)$$

and using now Eqs. (5.41) and (5.85) with the conditions that $\bar{y}_1$ (stop in contact with primary mirror) and that $SphTC + SphTC_{asph} = 0$ (each mirror is individually corrected for spherical aberration), we find that

$$Coma_S = SphTC_1\left(\frac{\bar{i}_1}{i_1}\right) + SphTC_2\left[\left(\frac{\bar{i}_2}{i_2}\right) - \left(\frac{\bar{y}_2}{y_2}\right)\right] \quad (14.46)$$

Now we consider the spherical aberration contributions in Eqs. (14.43) and (14.44) and substitute into this expression the other required values. After some algebraic work, we may prove that the sagittal coma of a Cassegrain telescope is given by

$$Coma_S = \left(\frac{y_1}{2F}\right)^2 h' \qquad (14.47)$$

which is equal to the sagittal coma of a paraboloid with the same effective focal length. Hence, the principal surface of a Cassegrain telescope is a paraboloid with a focal length equal to the effective focal length of this telescope. Since the principal plane is a hyperboloid, we may obtain, using Eq. (11.30):

$$Coma_S = OSCh' = \left(\frac{F_m}{F} - 1\right)h' \qquad (14.48)$$

where F_m is the marginal effective focal length for the paraboloid.

It is also possible to prove in a direct manner that the principal surface is a paraboloid with focal length F. Then, it is an immediate conclusion that the coma of the Cassegrain telescope is equal to the coma of the equivalent paraboloid.

Astigmatism

As in the case of the sagittal coma, the transverse sagittal astigmatism for the Cassegrain telescope is

$$AstL_S = AstL_S\, C_1 + AstL_S\, C_2 + AstL_S\, C_{asph1} + AstL_S\, C_{asph2} \qquad (14.49)$$

and using Eqs. (6.49) and (6.90) with the conditions that $\bar{y}_1$ and that $SphTC + SphTC_{asph} = 0$, as in the calculation of the sagittal coma, we find that

$$AstL_S = -SphTC_1 \left(\frac{\bar{i}_1}{i_1}\right)^2 \left(\frac{F}{y_1}\right) - SphTC_2 \left[\left(\frac{\bar{i}_2}{i_2}\right)^2 - \left(\frac{\bar{y}_2}{y_2}\right)^2\right]\left(\frac{F}{y_1}\right)$$

$$(14.50)$$

Then, after some algebraic manipulation, the astigmatism of a Cassegrain telescope may be proved to be given by

$$AstL_S = \frac{1}{2F}\left[\frac{m_s^2 + S/f_1}{m_s(1 + S/f_1)}\right] h'^2 \qquad (14.51)$$

If S/f_1 is very small as compared with m_s, as Meinel (1960) points out, the astigmatism of the Cassegrain telescope is equal to that of the equivalent paraboloid, multiplied by the magnification of the secondary mirror. However, the limiting factor in the size of the image is the coma and not the astigmatism.

Petzval Curvature

The longitudinal Petzval aberration Ptz may be shown to be

$$Ptz = \left(\frac{1}{f_1} + \frac{1}{f_2}\right)\frac{h'^2}{2}$$
$$= -\left[\frac{m_s(m_s - s/f_1) - (m_s + 1)}{m_s(f_1 + s)}\right]\frac{h'^2}{2} \tag{14.52}$$

This Petzval curvature and the astigmatism produce a curved focal surface of best definition, which is convex, as seen from the observer's side. The surface of best definition has a curvature given by

$$c_{best} = -\frac{(m_s^2 - 2)(m_s - s/f_1) + m_s(m_s + 1)}{m_s^2(f_1 + s)} \tag{14.53}$$

We see that this curvature increases with the magnification of the secondary mirror, but the field size also decreases, compensating this curvature increase.

Distortion

The distortion is, in general, extremely small, of the order of a few hundredths of an arcsecond. This magnitude is less than the atmospheric seeing size and hence it is not a problem.

Figure 14.7 shows a Cassegrain telescope with the characteristics listed in Table 14.1 (Cornejo and Malacara, 1973). The spot diagrams for this telescope are shown in Fig. 14.8; the image height is 11 cm. We may observe that the image has a large amount of coma, but also some astigmatism. The left-hand side diagram corresponds to the flat focal surface and the right-hand side diagram to the focal surface of best definition.

14.4.3 Ritchey–Chrètien Telescope

A Ritchey–Chrètien telescope is aplanatic. That is, it is corrected for spherical aberration and coma. The price is that both mirrors have to be

SECONDARY MIRROR　　　　　　PRIMARY MIRROR

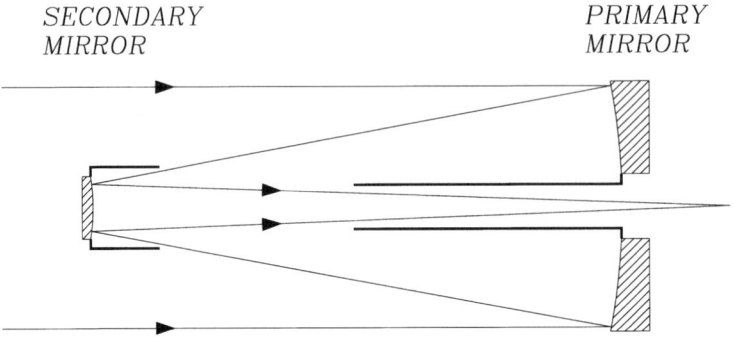

Figure 14.7 Cassegrain telescope.

Table 14.1 Cassegrain Telescope

Radius of curvature (cm)	Conic constant	Diameter (cm)	Separation or thickness (cm)
−1134.000	−1.0000	211.5	−446.327
−311.416	−2.4984	53.0	536.327

Aperture (cm) (focal ratio): 211.5 (12)
Effective focal length (cm): 2520.0
Primary focal length (cm): 567.0
Secondary magnification m_s: 4.44
Distance s from primary vertex to secondary focus (cm): 90.0
Object distance: infinite
Curvature of best focal surface in 1/cm: −0.00772
Image height (cm) (semifield): 11.4 (15′)

hyperboloids and thus the spherical aberration is not corrected on the primary focus. Most modern astronomical telescopes are of this type.

We may calculate the conic constants for this telescope by taking the Cassegrain telescope as a starting point and modifying the conic constants to correct the sagittal coma, maintaining the correction for the spherical aberration. The transverse spherical aberration for the complete telescope may be written as

$$SphT = SphTC_1 + SphTC_2 + SphTC_{asph1} + SphTC_{asph2}$$
$$= SphT_{cassegrain} + \Delta SphT$$
$$= \Delta SphT \qquad (14.54)$$

Astronomical Telescopes

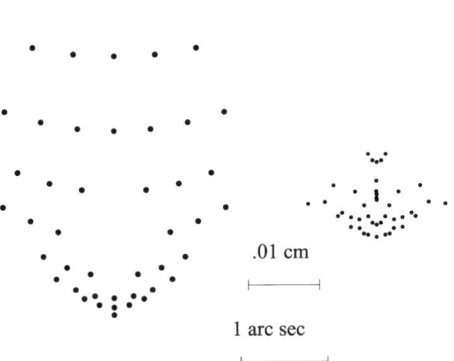

Figure 14.8 Spot diagrams for Cassegrain telescope.

where the subscript 1 is for the primary mirror and the subscript 2 is for the secondary mirror. The last term is the spherical aberration introduced by the change in the conic constants for the mirrors. This term has to be zero in order to keep the spherical aberration corrected. Then, from Eq. (4.49), but writing y instead of S, we obtain

$$\Delta SphT = -\left(\frac{y_1^4 c_1^3}{u_2'}\right)\Delta K_1 + \left(\frac{y_2^4 c_2^3}{u_2'}\right)\Delta K_2 = 0 \tag{14.55}$$

Now, using Eqs. (14.21)–(14.25) and then Eq. (14.31), we may find the following condition for the correction of spherical aberration in a two-mirror telescope:

$$\begin{aligned}\Delta K_2 &= -\frac{f_1 f_2^3}{(f_1-l)^4}\Delta K_1 \\ &= \frac{m_s^3(m_s+1)^4}{(m_s^2-1)^3(1+s/f_1)}\Delta K_1\end{aligned} \tag{14.56}$$

where ΔK_1 and ΔK_2 are the changes in the conic constants with respect to those in the Cassegrain telescope. This is a completely general condition for the correction of spherical aberration in any two-mirror telescope, not only for the Ritchey–Chrètien.

The sagittal coma of the two mirror telescope may be written as

$$\begin{aligned}Coma_S &= Coma_S C_1 + Coma_S C_2 + Coma_S C_{asph1} + Coma_S C_{asph2} \\ &= Coma_{S\,cassegrain} + \Delta Coma_S\end{aligned} \tag{14.57}$$

where $\Delta Coma_S$ is the change in the sagittal coma due to the changes in the conic constants, obtained from Eq. (6.85) and (14.55) as

$$\Delta Coma_S = -\left(\frac{y_1^4 c_1^3}{u_2'}\right)\left(\frac{\bar{y}_1}{y_1}\right)\Delta K_1 + \left(\frac{y_2^4 c_2^3}{u_2'}\right)\left(\frac{\bar{y}_2}{y_2}\right)\Delta K_2 \qquad (14.58)$$

Since the stop is in contact with the primary mirror, the first term is zero and adding the sagittal coma obtained for the Cassegrain telescope in Eq. (14.47), we find for the two-mirror telescope:

$$Coma_S = \left(\frac{y_1}{2F}\right)^2 h' + \left(\frac{y_2^4 c_2^3}{u_2'}\right)\left(\frac{\bar{y}_2}{y_2}\right)\Delta K_2 \qquad (14.59)$$

Using Eqs. (3.2), (14.22), and (14.24):

$$Coma_S = \left(\frac{y_1}{2F}\right)^2 \left[h' - \frac{(l+s)^3}{2f_2^3}\bar{y}_2\,\Delta K_2\right] \qquad (14.60)$$

hence, using Eq. (14.29):

$$Coma_S = \left(\frac{y_1}{2F}\right)^2 \left[1 - \frac{l(l+s)^3}{2f_2^3 F}\Delta K_2\right]h' \qquad (14.61)$$

Using Eq. (14.23) and (14.31) we finally obtain

$$Coma_S = \left(\frac{y_1}{2F}\right)^2 \left[1 + \frac{(m_s - s/f_1)(m_s - 1)^3}{2m_s(m_s+1)}\Delta K_2\right]h' \qquad (14.62)$$

For the particular case of a Ritchey–Chrètien telescope $Coma_S = 0$, then, from this relation we may now find, after some algebraic steps, that

$$\Delta K_2 = \frac{2F(f_1+f_2-l)^3}{l(f_1-l)^3} = -\frac{2m_s(m_s+1)}{(m_s-s/f_1)(m_s-1)^3} \qquad (14.63)$$

Hence, the conic constant for the secondary mirror of the Ritchey–Chrètien telescope is

$$\begin{aligned}K_2 &= -\left(\frac{f_1+s}{2l-f_1+s}\right)^2 + \frac{2F(f_1+f_2-l)^3}{l(f_1-l)^3}\\ &= -\left(\frac{m_s+1}{m_s-1}\right)^2 - \frac{2m_s(m_s+1)}{(m_s-s/f_1)(m_s-1)^3}\end{aligned} \qquad (14.64)$$

and substituting the conic constant increment from Eq. (14.41) into Eq. (14.37) we may find that

$$K_1 = \frac{2(f_1-l)f_1^2}{F^2 l} - 1 - \frac{2(f_1+s)}{m_s^2(m_s f_1 - s)} - 1 \tag{14.65}$$

The astigmatism for this telescope, as pointed out by Schroeder (1987), is equal to

$$AstL_S = \frac{1}{2F}\left[\frac{(2m_s+1)m_s + s/f_1}{2m_s(1+s/f_1)}\right] \tag{14.66}$$

and the curvature for the surface of best definition is

$$c_{best} = -\frac{(m_s+1)}{m_s^2(f_1+s)}\left[m_s^2 - (m_s-1)\left(\frac{s}{f_1}\right)\right] \tag{14.67}$$

Figure 14.9 shows the spot diagrams for a Ritchey–Chrètien telescope with the same dimensions as the Cassegrain telescope, but with conic constants:

$K_1 = -1.02737$
$K_2 = -2.77476$

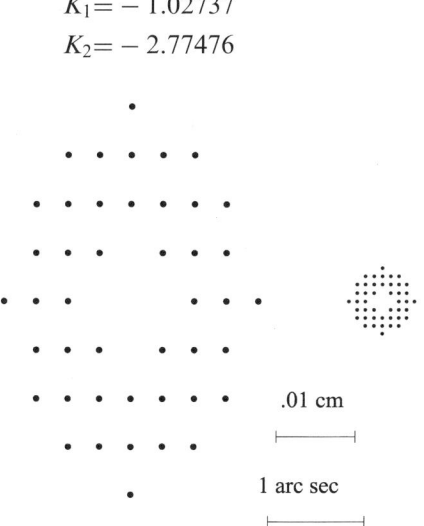

Figure 14.9 Spot diagrams for Ritchey–Chrètien telescope.

and the curvature of the surface of best definition equal to -0.00806. The image height for these spot diagrams is 11 cm. The left-hand side diagram is taken at the focal plane and the one on the right-hand side at the surface of best definition, between the sagittal and the tangential focal surfaces.

14.4.4 Dall–Kirham Telescope

The construction of the hyperbolic secondary mirror in the Cassegrain and the Ritchey–Chrètien telescopes is not easy. It presents many problems, especially for amateur telescope makers. The Dall–Kirham telescope solves this difficulty by using a spherical convex secondary instead of a hyperboloid. In order to correct the spherical aberration the primary mirror becomes an ellipsoid. This is another advantage, since it is easier to test an ellipsoidal mirror than a hyperboloidal mirror. Another important advantage is that the spherical shape of the secondary makes the telescope virtually insensitive to misalignments.

Then, the secondary mirror has a conic constant $K_2 = 0$ and hence ΔK_2 is equal to minus the value of K_2 for the Cassegrain configuration, as follows:

$$\Delta K_2 = \left(\frac{m_s+1}{m_s-1}\right)^2 \tag{14.68}$$

Substituting this value into Eq. (14.55), we obtain

$$K_1 = -1 + \frac{(m_s^2-1)(1+s/f_1)}{m_s^2} \tag{14.69}$$

After some algebraic work we may find that the sagittal coma for this telescope is

$$Coma_S = \left(\frac{y_1}{2F}\right)^2 \left[1 + \frac{(m_s - s/f_1)(m_s^2-1)}{2m_s}\right] h' \tag{14.70}$$

which, by assuming that $m \gg s/f_1$, becomes approximately:

$$Coma_S = \left(\frac{y_1}{2F}\right)^2 \left(\frac{m_s^2+1}{2}\right) h' \tag{14.71}$$

There are some disadvantages to this telescope. The most important is that the coma and the astigmatism are very large compared with the other telescopes. For example, the coma is $(m_s^2 + 1)/2$ times larger than in the Cassegrain. Thus, the diameter of the useful field of view is reduced by the same factor. Another difference is that the secondary mirror produces some high-order spherical aberration that the conic constant of the primary mirror cannot compensate. The spot diagrams at the focal plane for this telescope, with the same data as the Cassegrain telescope, are shown in Fig. 14.10, but with the conic constants:

$K_1 = -0.7525$

$K_2 = 0.0000$

14.4.5 Gregory Telescope

The Gregory (or Gregorian) telescope, shown in Fig. 14.11, uses a concave ellipsoidal secondary mirror instead of the hyperboloid. The advantage is that the elliptical concave mirror is much easier to construct and test than the convex hyperboloid. The main disadvantage is that, given a primary mirror focal length, the Gregory telescope is longer than the Cassegrain and Ritchey–Chrétien telescopes.

As with the Cassegrain, the Gregory telescope may be made aplanatic by properly selecting the conic constants for the two mirrors. It is interesting to know that in this case the primary mirror becomes ellipsoidal instead of hyperboloidal as in the Ritchey–Chrétien telescope.

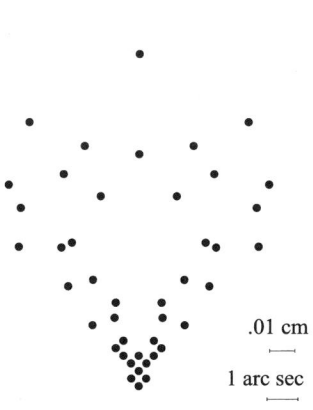

Figure 14.10 Spot diagrams for a Dall–Kirham telescope.

14.4.6 Coudé and Nasmythe Focus Configurations

In large astronomical telescopes the secondary focus obviously moves around with the telescope, as it points to different stars. It might be convenient in very large telescopes to have the secondary focus at a fixed position in space. If the distance S is made large, the light may be directed by means of mirrors along the polar axis, as shown in Fig. 14.12. The only inconvenience might be in some cases that the effective focal length becomes very large. Another disadvantage is that the image rotates as the telescope moves following the star.

In a similar manner, if the telescope has an altitude–azimuth mounting, the light beam may be directed along the altitude axis, with the same advantages and disadvantages as the Coudé focus.

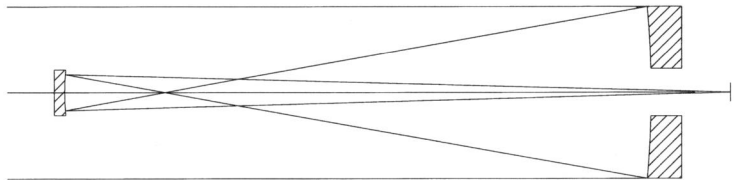

Figure 14.11   Gregory telescope.

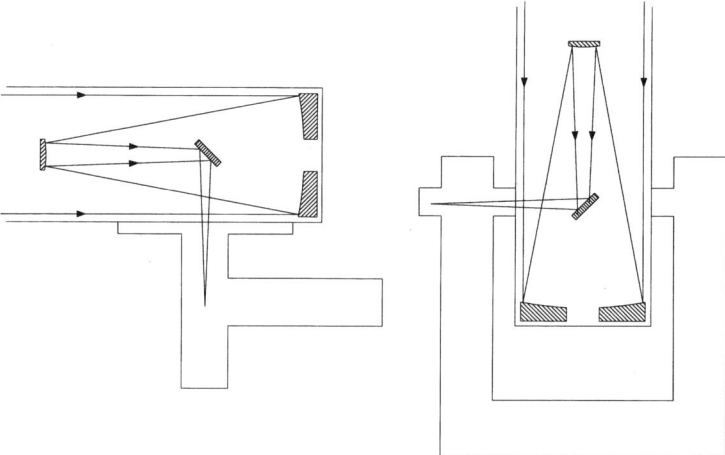

Figure 14.12   Coudé and Nasmythe focus configurations.

14.4.7 Cassegrain Light Shields

In a two-mirror telescope an adequate shielding is required in order to avoid direct sky or environment light to fall on the focal plane. The problem of designing a good set of light shields for two-mirror telescopes has been treated by several authors, e.g., Young (1967), Cornejo and Malacara (1968), Prescott (1968), Davies (1987), LaVaughn Hales (1992), and Song et al. (2002). A large field with the shielding introduces two problems, i.e., vignetting and central obscuration. The central obscuration is due to the large diameter of the secondary shield. The edge vignetting is due to both shields. A compromise has to be found to minimize the light losses.

Let us assume that the angular field diameter is 2α. As shown in Fig. 14.13, the light ray **AB** from a star enters the telescope with a small angle α with respect to the optical axis. Upon reflection on the primary mirror, this ray follows the path **BC** and **CD**, arriving to the focal plane at a point with a height h'. The image height on the primary focal plane is h_p. A light ray **ED** from an extraneous light source also arrives to the focal point **D**, passing through the rims **M** and **N** of the light shields. These points **M** and **N** define the dimensions of the shields. The algebraic steps to find the light shields are not given here, but the interested reader can consult the references.

As an example, the light shields for the Cassegrain telescope designed in this chapter have the dimensions shown in Table 14.2.

14.5 FIELD CORRECTORS

Field correctors for telescopes have been designed to improve the quality of the image by correcting some aberrations near the focal surface. These correctors have been studied by many authors, e.g., Ross (1935), Rosin (1961, 1964, 1966), Wynne (1965), and Schulte (1966b).

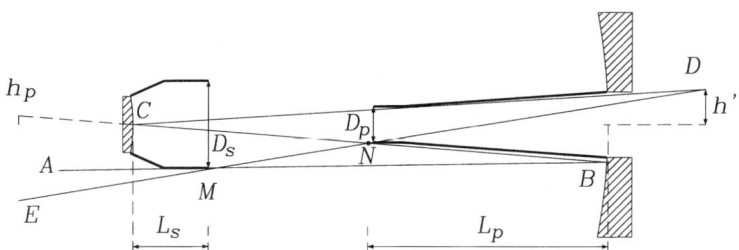

Figure 14.13 Light shields for a Cassegrain or Ritchey–Chrètien telescope.

Table 14.2 Cassegrain Light Shields

	Primary shield	Secondary shield
Diameter	37.6	68.9
Length	224.7	56.2

Central obscuration (area): 10.77%

14.5.1 Single Field Flattener

A field lens or a field flattener is a single negative or positive lens placed almost at the image plane. The contribution of this lens is zero or very small to all aberrations, with the only exception of the Petzval curvature. As we have seen, the Petzval curvature for a system of thin lenses is

$$\frac{1}{r_{Ptz}} = \sum_{j=1}^{k} \frac{1}{n_j f_j} \qquad (14.72)$$

In general, we do not want the Petzval surface to be flat, due to the presence of astigmatism. The usual requirement is to have a flat surface of best definition. If r_{end} and r_{sys} are the desired and initial values of the radius of curvature of the Petzval surface, we may write

$$\frac{1}{r_{end}} = \frac{1}{r_{sys}} + \frac{1}{nf} \qquad (14.73)$$

where n and f are the refractive index and the focal length of the field flattener. A field flattener for the Ritchey–Chrètien telescope designed in this chapter is shown in Fig. 14.14 and has the construction parameters presented in the Table 14.3.

Figure 14.15 shows the spot diagrams for the Ritchey–Chrètien telescope using this field flattener. We may clearly see the presence of the small magnification chromatic aberration introduced.

14.5.2 Ross Corrector

The coma of a parabolic mirror or a Cassegrain telescope may be removed by means of a pair of lenses placed near the focus, as suggested by Ross (1935). This lens is an air-spaced doublet of nearly zero power, as shown in Fig. 14.16. The coma and field curvature are greatly reduced with this

Astronomical Telescopes

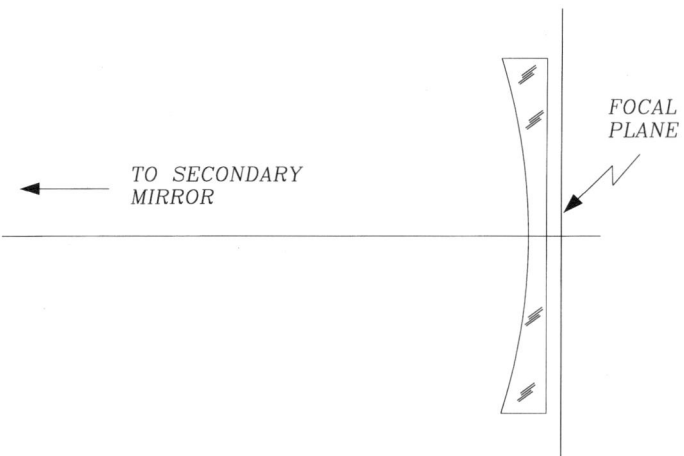

Figure 14.14 Single-field flattener.

Table 14.3 Field Flattener for Ritchey–Chrètien Telescope

Radius of curvature (cm)	Diameter (cm)	Separation or thickness (cm)	Material
−42.02	25.0	1.27	Fused silica
Flat	25.0	1.00	Air

Distance from secondary mirror of telescope: 534.48
Image height (cm) (semifield): 11.0

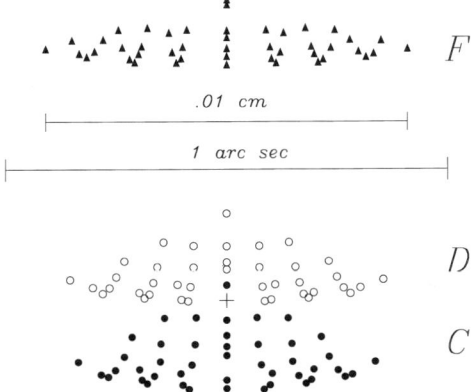

Figure 14.15 Spot diagrams with a single-field flattener.

system, without introducing much spherical aberration. Since the power is nearly zero, a good chromatic correction may be obtained using the same glass for both lenses. The spherical aberration may be reduced by moving the system close to the focal plane. However, as the lenses get closer to the focus, the curvatures become larger, increasing the high-order aberrations. A solution is to use a system of three lenses as described by Wynne (1965), achieving a good correction of the spherical aberration.

14.5.3 Wynne Corrector

The image at the secondary focus in a Ritchey–Chrètien telescope is free of spherical aberration and coma, but has astigmatism and a relatively large amount of field curvature. Wynne (1965) corrected these aberrations without introducing any other aberrations, by means of a system of two lenses with almost zero power. As shown in Fig. 14.17, a convergent lens is followed by a divergent lens with the shape of a meniscus. Table 14.4 shows

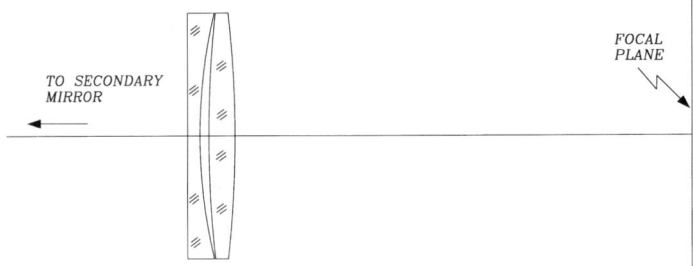

Figure 14.16 Field Ross corrector.

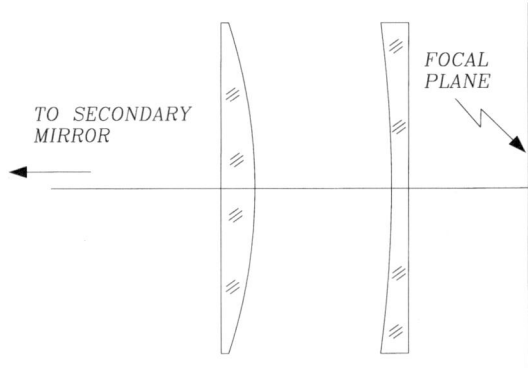

Figure 14.17 Ritchey–Chrètien field Wynne corrector.

Astronomical Telescopes

Table 14.4 Wynne Corrector for Ritchey–Chrètien Telescope

Radius of curvature (cm)	Diameter (cm)	Separation or thickness (cm)	Material
Flat	25.0	2.54	Fused silica
−94.89	25.0	10.38	Air
−40.45	25.0	1.27	Fused silica
−1479.79	25.0	7.93	Air

Distance from secondary mirror of telescope: 514.38
Image height (cm) (semifield): 11.0

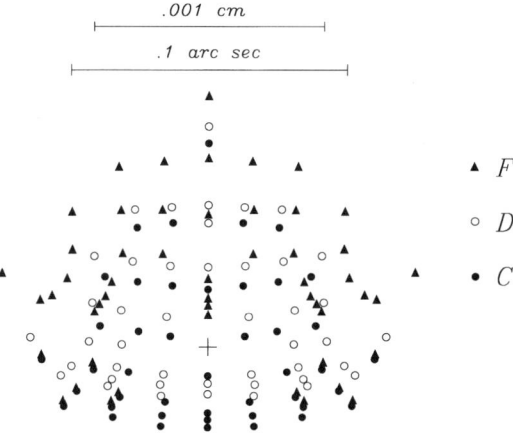

Figure 14.18 Spot diagrams for a Wynne corrector.

a corrector of this type for the Ritchey–Chrètien telescope described before. Figure 14.18 shows the spot diagrams.

14.5.4 Aspheric Correctors

As explained before, in the Ritchey–Chrètien telescope the only significant aberrations are the astigmatism and the field curvature. Schulte (1966b) and Gascoigne (1973) showed that an aspheric plate, similar to a Schmidt plate, near the secondary focus, removes the astigmatism and most of the field curvature without introducing any coma or spherical aberration. This system is simpler than the Wynne corrector, but more difficult to construct.

14.6 CATADIOPTRIC TELESCOPES

A catadioptric telescope is formed by mirrors as well as lenses (Churilovskii and Goldis1964; Villa, 1968; Maxwell, 1972). The telescopes studied in this section are of the Cassegrain configuration, with two mirrors, but a Schmidt correcting plate has been added to eliminate the spherical aberration. Several variations of the basic configuration are possible (Linfoot, 1955b) as will be seen.

14.6.1 Anastigmatic Concentric Schmidt–Cassegrain Telescope

A concentric Schmidt–Cassegrain telescope is formed by two concentric spherical mirrors (DeVany, 1965), with a Schmidt correcting plate and the stop placed at the common center of curvature of the mirrors (Fig. 14.19). The concentric configuration avoids all off-axis aberrations with the exception of the field curvature. The correcting plate, as in the Schmidt camera, eliminates the spherical aberration. In conclusion, the only remaining aberrations in this telescope are the Petzval curvature and spherochromatism.

The condition for concentricity may be written as

$$f_2 + f_1 = \frac{l}{2} \tag{14.74}$$

but substituting the value of f_2 from this expression into Eq. (14.23) we obtain

$$\frac{l}{2} - f_1 = \frac{m_s(f_1 + s)}{(1 - m_s^2)} \tag{14.75}$$

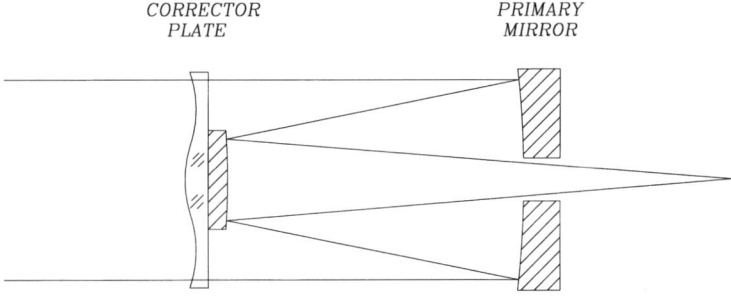

Figure 14.19 Schmidt–Cassegrain telescope.

Then, substituting here the value of l from Eq. (14.25) we may obtain

$$(m_s - \frac{s}{f_1} - 2)(m_s + 1) = 0 \qquad (14.76)$$

The solution for $m_s = -1$ has no physical interest for us. So, the concentric telescope must have a secondary magnification equal to

$$m_s = \frac{F}{f_1} = 2 + \frac{s}{f_1} \qquad (14.77)$$

Then, in this telescope the ratio between the effective focal length and the primary mirror focal length is fixed to a value slightly greater than 2, depending on the value of s.

14.6.2 Flat-Field Anastigmatic Schmidt–Cassegrain Telescopes

The Schmidt–Cassegrain telescope may deviate from the concentric configuration. Then, the off-axis aberrations appear, but they may be corrected by aspherizing the mirrors with a nonzero value for the conic constants. However, the spherical aberration is not completely corrected with the conic constants and the correcting plate has still to be used. The condition for a flat Petzval surface is

$$f_1 = -f_2 \qquad (14.78)$$

after some straightforward algebra, using Eq. (14.31), we may find

$$-f_s = \frac{m_s(f_1 + s)}{(1 - m_s^2)} \qquad (14.79)$$

with a solution:

$$m_s = \frac{(1 + s/f_1) + \sqrt{5 + 2(s/f_1) + (s/f_1)^2}}{2} \qquad (14.80)$$

which may be approximated by

$$m_s = \left(\frac{F}{f_1}\right) = 1.118 + 0.2236\left(\frac{s}{f_1}\right) + 0.1006\left(\frac{s}{f_1}\right)^2 \qquad (14.81)$$

This value of the magnification of the secondary mirror insures a flat Petzval surface, but we now have to correct the spherical aberration, coma, and astigmatism. The Schmidt correcting plate does not introduce astigmatism nor coma due to its low power. Thus, the spherical aberration correction is left to the end, by means of this plate. So, we have to correct two aberrations, namely, coma and astigmatism, by means of three degrees of freedom: two conic constants and the stop position. We will consider two possible configurations, studied by Baker (1940a).

Stop at Primary Focal Plane

In this configuration, shown in Fig. 14.20(a), the stop is fixed at the location of the focus for the primary mirror. Then, coma and astigmatism are corrected by means of the conic constants. The two mirrors become strongly elliptical. At the end, the correcting plate is calculated to correct the spherical aberration.

Spherical Secondary

In this second configuration, shown in Fig. 14.20(b), the secondary mirror is made spherical. Then, the conic constant for the primary mirror and the stop position are used to correct coma and astigmatism. The primary mirror becomes elliptical, but very close to a sphere. As in the first solution, at the end the spherical aberration is corrected by means of the correcting plate.

It is important to notice that in an anastigmatic system the spherical aberration, coma, and astigmatism are zero. Then, this correction is independent of the stop position, but not of the corrector plate position. However, it is desirable that the corrector plate is always located at the stop.

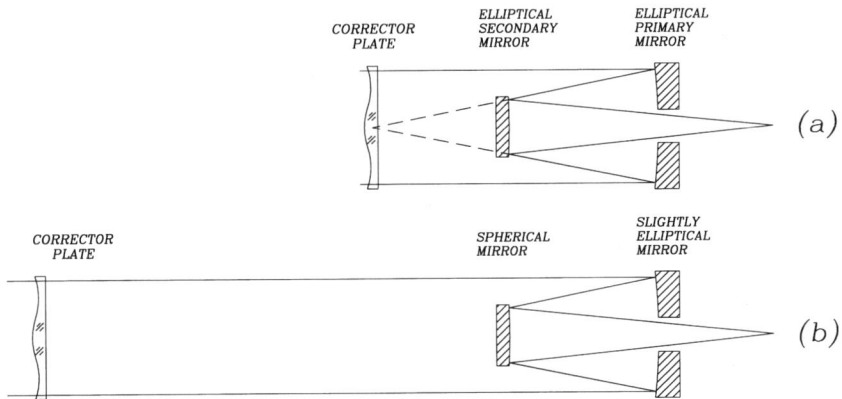

Figure 14.20 Two flat-field anastigmatic Schmidt–Cassegrain telescope.

14.6.3 Aplanatic Schmidt–Cassegrain Telescope with Spherical Mirrors

If the system is to be anastigmatic, with both mirrors being spherical, the only solution is the concentric system, but the Petzval surface would be curved, as described before. A Schmidt–Cassegrain system is strictly aplanatic, by definition, only if the spherical aberration and coma are simultaneously corrected. With this definition, it may be proved that an aplanatic Schmidt–Cassegrain is possible only if it is also anastigmatic. However, Linfoot (1955b) defines an aplanatic Schmidt–Cassegrain system as one that has a small coma, a small astigmatism, and the surface of best definition is flat.

This system requires a small Petzval curvature, so that the flattening of the field does not require a large astigmatism. Thus, the focal lengths for the two mirrors must have almost the same magnitude, but not be exactly equal. A ratio f_2/f_1 of about -0.95 is reasonable.

14.6.4 Maksutov–Cassegrain Telescope

The Maksutov–Cassegrain telescope (Waland, 1961; Malacara, 1975) is similar to the Schmidt–Cassegrain telescope, with the difference that the corrector element is a meniscus lens as in the Maksutov camera, as in Fig. 14.21. The secondary mirror may be a separate element or a small reflecting area at the center of the convex face of the correcting plate.

14.7 MULTIPLE MIRROR TELESCOPES

Instead of using a single large, thick, and heavy mirror, multiple-mirror telescopes are formed by an array of smaller mirrors, as shown in Fig. 14.22. The light-collecting capacity of the telescope is equal to that of a single

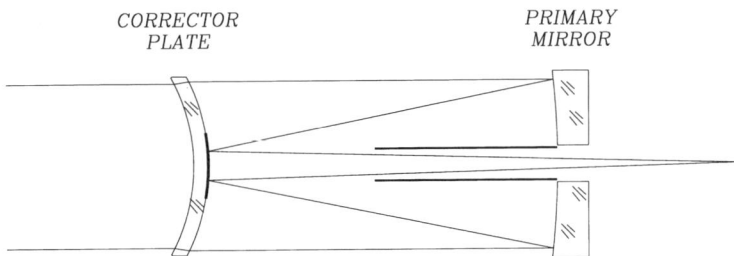

Figure 14.21 Maksutov–Cassegrain telescope.

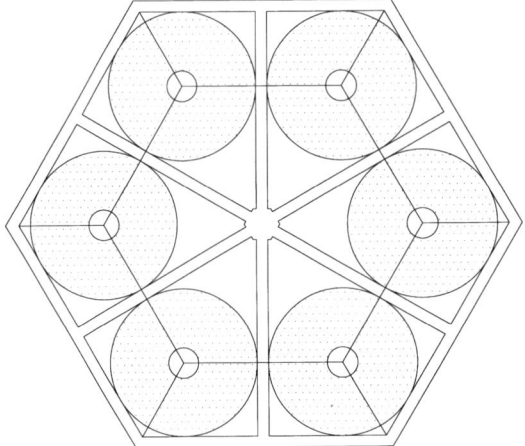

Figure 14.22 Mirror array in a multiple-mirror telescope.

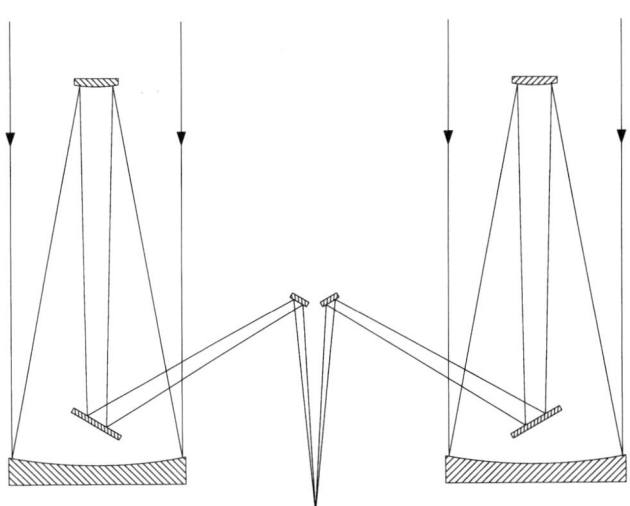

Figure 14.23 Optical layout in a multiple-mirror configuration.

mirror, multiplied by the number of mirrors. The light from these mirrors is brought to a common focus by means of small flat mirrors, as shown in Fig. 14.23.

Each mirror produces its own diffraction image. When these images are superimposed, they add their amplitudes. These amplitudes, however, may have wavefronts that are in phase or out of phase. If the phase of the wavefronts from the mirrors is not accurately controlled, the resulting image approximates the sum of the intensities. Then, the final image is about the same size as the image of each individual mirror. This is an incoherent multiple-mirror telescope. The resolving power is not greater than that of a single mirror, but the light gathering power is that of the sum of the areas of the mirrors.

In some cases, however, it might be desirable to have a coherent telescope by superimposing all the light beams with the same phase. This is possible by mechanically phasing the mirrors within a small fraction of the wavelength. Then, the combined image is smaller than the individual images, increasing the resolving power of the telescope, approximating that of a large telescope with a diameter equal to the diameter of the whole array. Most modern multiple-mirror telescopes are of this type. Meinel et al. (1983) have published a detailed and complete study of the diffraction images produced by many types of coherent arrays of apertures. The reader is encouraged to examine these pictures in detail.

14.8 ACTIVE AND ADAPTIVE OPTICS

The wavefront forming the image of a point light source (star) may be deformed due to imperfections in the optical components or to atmospheric turbulence. These wavefront deformations may be eliminated by introducing the opposite deformations in a procedure called active or adaptive optics. This subject is relatively new, but has already been treated in detail by some authors, e.g., Tyson and Ulrich (1993), where many additional references may be found.

In order to measure the wavefront shape, so that the appropriate compensation may be introduced, a reference wavefront from a point source near the observed object is required. Thus, an isolated bright star has to be located in the vicinity of the observed objects. This is not always possible, but this star may be artificially produced by sending a strong laser beam in the direction pointed to by the telescope. The laser beam produces a fluorescent spot in the upper atmosphere, acting as the required reference star. Several laser pulses are sent at equal time intervals in order to measure the wavefront frequently enough.

The active optics device to compensate for the atmospheric turbulence is schematically shown in Fig. 14.24. The light from the reference star, after passing through the telescope, enters a specially designed optical system to collimate the light. This optical system also forms the image of the exit pupil of the telescope on the compensating mirrors.

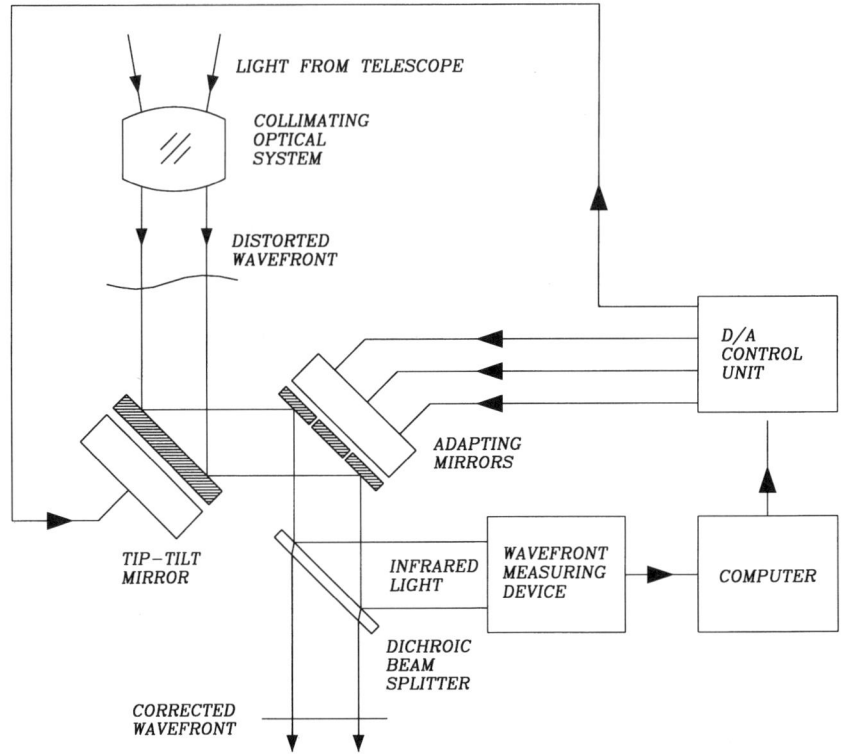

Figure 14.24 Adaptive optics basic arrangement.

The wavefront deformations are compensated on the system formed by the tip–tilt mirror and the adapting mirrors. These mirrors move to compensate the deformations, but they cannot easily compensate a wavefront tilt. This is done by means of the tip–tilt mirror.

The dichroic mirror reflects the infrared reference wavefront from the incoming light beam in order to send it to the wavefront-measuring device. The information from this device is sent to a computer. Then, the computer controls the adapting mirrors and the tip–tilt mirror by means of a digital/analog control unit.

Obviously, adapting mirrors compensate not only the atmospheric turbulence, but also any optics imperfections. If the system has a large time constant and it is not fast enough to follow the atmospheric disturbances, only the optics imperfections are compensated. Then, the system is said to be *adaptive*, not *active*.

REFERENCES

Baker, J. G., "A Family of Flat-Field Cameras, Equivalent in Performance to the Schmidt Camera," *Proc. Am. Philos. Soc.*, **82**, 339–349 (1940a).
Baker, J. G., "The Solid-Glass Schmidt Camera and a New Type Nebular Spectrograph," *Proc. Am. Philos. Soc.*, **82**, 323–338 (1940b).
Baker, J. G., "Planetary Telescopes," *Appl. Opt.*, **2**, 111–129 (1963).
Benford, J. R., "Design Method for a Schmidt Camera with a Finite Source," *J. Opt. Soc. Am.*, **34**, 595–596 (1944).
Bouwers, A., *Achivements in Optics*, Chap. 1, Elsevier, Amsterdam, 1946.
Bowen, I. S., "Schmidt Cameras," in *Stars and Stellar Systems, I. Telescopes*, G. P. Kuiper and B. Middlehurst, eds., The University of Chicago Press, Chicago, 1960.
Bowen, I. S., *Annual Review of Astronomy and Astrophysics*, Vol. 5, p. 45, Annual Reviews, Palo Alto, CA, 1967.
Buchroeder, R., "Proposed 131 cm. f/3.5 Achromatic Schmidt Telescopes," *Appl. Opt.*, **11**, 2968–2971 (1972).
Cornejo, A. and Malacara, D., "Direct Design Solution for Cassegrain Shields,". *Boletín de los Observatorios de Tonantzintla y Tacubaya*, **4**, 246–252 (1968).
Cornejo, A. and Malacara, D., "Design of a Ritchey-Chrétien Telescope for the INAOE," *Boletín del Instituto de Tonantzintla*, **1**, 35–44 (1973).
Cornejo, A. and Malacara, D., "Required Accuracy in the Radius of Curvature of a Primary Astronomical Telescope Mirror," *Boletín del Instituto de Tonantzintla*, **1**, 293–294 (1975).
Cornejo, A., Malacara, D., and Cobos, F., "A Schmidt Cassegrain Camera for Use with an Image Intensifier Tube," *Boletín de los Observatorios de Tonantzintla y Tacubaya*, **5**, 319–323 (1970).
Churilovskii, V. N. and Goldis, K. I., "An Apochromatic Catadioptric System Equivalent to a Parabolic Mirror," *Appl. Opt.*, **3**, 843–846 (1964).
Davies, P. K., "Baffle Design for Telescopes with Tiltable Secondary Mirrors," *Proc. SPIE*, **766**, 163–168 (1987).
DeVany, A. S., "Optical Design for Two Telescopes," *Appl. Opt.*, **2**, 201–204 (1963).
DeVany, A. S., "Schmidt–Cassegrain Telescope System with a Flat Field," *Appl. Opt.*, **4**, 1353 (1965).
Dimitroff, G. Z. and Baker, J. J., *Telescopes and Accessories*, Blakiston, Philadelphia, PA, 1945.
Erdös, P., "Mirror Anastigmat with Two Concentric Spherical Surfaces," *J. Opt. Soc. Am.*, **49**, 877–886 (1959).
Gascoine, S. C. B., "Recent Advances in Astronomical Optics," *Appl. Opt.*, **12**, 1419–1429 (1973).
Jones, R. C., "Coma of a Modified Gregorian and Cassegrainian Mirror System," *J. Opt. Soc. Am.*, **44**, 623–630 (1954).
LaVaughn Hall, W., "Optimum Cassegrain Baffle Systems," *Appl. Opt.*, **31**, 5341–5344 (1992).

Linfoot, E. H., "The Schmidt Camera," in *Recent Advances in Optics*, Chap. III, Oxford University Press, London, 1955a.

Linfoot, E. H., "Plate Diagram and its Analysis," in *Recent Advances in Optics*, Chap. IV, Oxford University Press, London, 1955b.

Linfoot, E. H. and Wolf, E., "On the Corrector Plates of Schmidt Cameras," *J. Opt. Soc. Am.*, **39**, 752–756 (1949).

Lucy, F. A., "Exact and Approximate Computation of Schmidt Cameras: I. Classical Arrangement," *J. Opt. Soc. Am.*, **30**, 251 (1940).

Lucy, F. A., "Exact and Approximate Computation of Schmidt Cameras: II. Some Modified Arrangements," *J. Opt. Soc. Am.*, **30**, 358 (1941).

Maksutov, D. D., "New Catadioptic Meniscus Systems," *J. Opt. Soc. Am.*, **34**, 270–284 (1944).

Malacara, D., "Design of Telescopes of the Cassegrain and Ritchey–Chrétien Types," *Boletín de los Observatorios de Tonantzintla y Tacubaya*, **4**, 64–72 (1965).

Malacara, D., "Design of a Cassegrain–Maksutov Telescope," *Boletín del Instituto de Tonantzintla*, **1**, 221–225 (1975).

Mallama, A., "The Limiting Magnitude of a CCD Camera," *Sky and Telescope*, 84 (Feb. 1993).

Maxwell, J., *Catadioptric Imaging Systems*, American Elsevier, New York, 1972.

Meinel, A. B., "Design of Reflecting Telescopes," in *Stars and Stellar Systems*, *I. Telescopes*, G. P. Kuiper and B. Middlehurst, eds., The University of Chicago Press, Chicago, 1960.

Meinel, A. B., "Astronomical Telescopes," in *Applied Optics and Optical Engineering*, R. Kingslake, ed., Vol. V, Chap. 6, Academic Press, San Diego, CA, 1969.

Meinel, A. B., Meinel, M. P., and Woolf, N. J., "Multiple Aperture Telescope Diffraction Images," in *Applied Optics and Optical Engineering*, R. R. Shannon and J. C. Wyant, eds., Vol. IX, Chap. 5, Academic Press, San Diego, CA, 1983.

Prescott, R., "Cassegrainian Baffle Design," *Appl. Opt.*, **7**, 479–481 (1968).

Robb, P., "Three Mirror Telescopes," *Appl. Opt.*, **17**, 2677–2685 (1978).

Rosin, S., "Optical Systems for Large Telescopes," *J. Opt. Soc. Am.*, **51**, 331–335 (1961).

Rosin, S., "Corrected Cassegrain System," *Appl. Opt.*, **3**, 151–152 (1964).

Rosin, S., "Ritchey–Chrétien Corrector System," *Appl. Opt.*, **5**, 675–676 (1966).

Ross, F. E., "Lens Systems for Correcting Coma of Mirrors," *Astrophys. J.*, **81**, 156 (1935).

Schroeder, D., "All Reflecting Baker–Schmidt Flat-Field Telescopes," *Appl. Opt.*, **17**, 141–144 (1978).

Schroeder, D., *Astronomical Optics*, Academic Press, San Diego, CA, 1987.

Schroeder, D., *Selected Papers on Astronomical Optics*, SPIE Milestone Series, Vol. MS 73, SPIE Optical Engineering Press, Bellingham, WA, 1993.

Schulte, D., "Anastigmatic Cassegrain Type Telescopes," *Appl. Opt.*, **5**, 309–311 (1966a).

Schulte, D., "Prime Focus Correctors Involving Aspherics," *Appl. Opt.*, **5**, 212–217 (1966b).
Shafer, D. R., "New Types of Anastigmatic Two-Mirror Telescopes," *J. Opt. Soc. Am.*, **66**, 1114 (1976).
Song, N., Yin, Z., and Hu, F., "Baffles Design for an Axial Two-Mirror Telescope," *Opt. Eng.*, **41**, 2353–2356 (2002).
Synge, J. L., "The Theory of the Schmidt Telescope," *J. Opt. Soc. Am.*, **33**, 129–136 (1943).
Tyson, R. K. and Ulrich, P. B., "Adaptive Optics," in *The Infrared & Electro-Optical Systems Handbook*, Vol. 8: *Emerging Systems and Technologies*, Chap. 2, S. R. Robinson, ed., Infrared Information Analysis Center, Ann Arbor, MI; SPIE, Optical Engineering Press, Bellingham, WA, 1993.
Villa, J., "Catadioptric Lenses," *Spectra*, **1**, 57 (March–April), 49 (May–June) (1968).
Waland, R. L., "Flat Field Maksutov–Cassegrain Optical Systems," *J. Opt. Soc. Am.*, **51**, 359–366 (1961).
Wetherell, W. B. and Rimmer, M., "General Analysis of Aplanatic Cassegrain, Gregorian and Schwarzschild Telescopes," *Appl. Opt.*, **11**, 2817–2832 (1972).
Wormser, E. M., "On the Design of Wide Angle Schmidt Optical Systems," *J. Opt. Soc. Am.*, **40**, 412–415 (1950).
Wynne, C. G., "Field Correctors for Large Telescopes," *Appl. Opt.*, **4**, 1185–1192 (1965).
Wynne, C. G., "Ritchey–Chrétien Telescopes and Extended Field Systems," *Astrophys. J.*, **152**, 675–694 (1968).
Yoder, P. R., Patrick, F. B., and Gee, A. E., "Analysis of Cassegrain-Type Telescopic Systems," *J. Opt. Soc. Am.*, **43**, 1200–1204 (1953a).
Yoder, P. R., Patrick, F. B., and Gee, A. E., "Permitted Tolerance on Percent Correction of Paraboloidal Mirrors," *J. Opt. Soc. Am.*, **43**, 702–703 (1953b).
Young, A. T., "Design of Cassegrain Light Shields," *Appl. Opt.*, **6**, 1063–1067 (1967).

15

Visual Systems, Visual Telescopes, and Afocal Systems

15.1 VISUAL OPTICAL SYSTEMS

15.1.1 Exit Pupil Location in Visual Optical Systems

The final image in a visual instrument is formed at the retina of a human eye. If the eye is emmetropic, i.e., if it does not have any refractive errors, the virtual image provided by the instrument has be located at an infinite distance. The coupling of the visual instrument to the eye is done by locating the eye close to the exit pupil of the instrument. The distance from the last optical surface of the system to the exit pupil is called the *eye relief*.

The eye relief should be at least 10 mm to provide enough space for eyelashes, 15 mm for a more comfortable viewing or even 20 mm for people wearing eyeglasses. For rifle sights the eye relief should be even larger, at least 60 mm to give space for the rifle recoil. A larger eye relief requires a larger eyepiece, making the instrument more expensive.

Another consideration that should be made regarding the exit pupil in visual instruments is its optimum location with respect to the observing eye. It is commonly stated that the exit pupil of the instrument should be at the same plane as the pupil of the observing eye. However, in some instruments a better location is at the center of the eye globe.

To study this problem let us consider Fig. 15.1 where the exit pupil of the visual system and the observing eye are shown. In Figs. 15.1(a) and (b) the exit pupil of the system and the pupil of the eye are at the same plane. If the object of interest is small and it is located at the center of the field, the whole field is observed, but only the object at the center is clearly defined. When an object is at the periphery it is observed as in Fig. 15.1(b), rotating the eye globe about its center. Now, the object at the edge of the field is clearly seen but much dimmer, unless the head is slightly moved laterally to center again the exit pupil of the system with the pupil of the eye. Thus, this position for the observing eye with respect to the exit pupil is

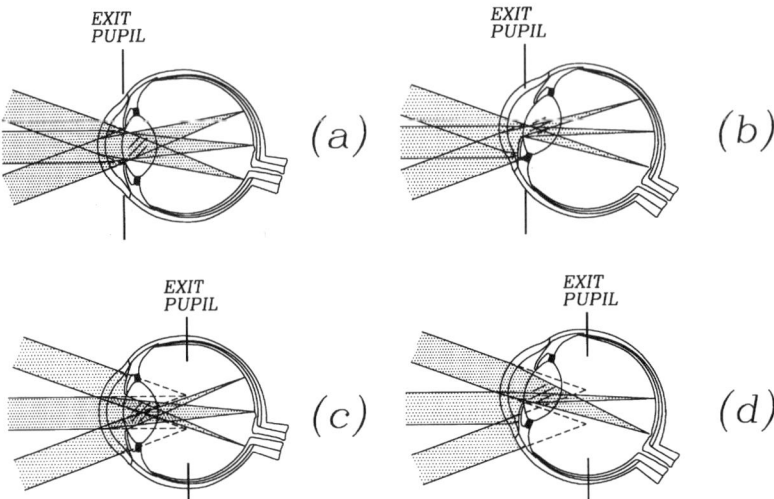

Figure 15.1 Different observing eye positions with respect to the exit pupil of the visual system.

correct if the objects of interest are small and in the vicinity of the optical axis. A typical example is a visual astronomical telescope.

Let us now consider the case when the objects of interest are uniformly distributed over the whole field and the exit pupil of the system is at the center of the eye globe, as shown in Figs. 15.1(c) and (d). Now the eye has continuously to move in its eye socket to observe the object of interest at a given time. All the observed objects at any position within the field will have the same luminous efficiency. However, the objects located at the opposite side of the optical axis of the object of interest will be much dimmer or even invisible. They will appear bright and clear if the eye is rotated to observe in that direction. This position for the exit pupil of the system is appropriate for ophthalmic lenses or systems with a wide field where the objects of interest are over the whole field. If the exit pupil of the optical system is much smaller than the pupil of the eye, the tolerance in the position of the aye along the optical axis is much greater.

All preceding considerations were made assuming that the exit pupil position of the visual optical instrument is defined independently of the pupil of the observer. This is not true in some optical systems where the stop of the system is the pupil of the observer's eye, as in a magnifier, an ophthalmic lens or a Galilean telescope. In these cases, clearly the exit pupil should be considered at the center of rotation of the eye globe.

15.1.2 Optical Models of the Human Eye

It is sometimes desirable in the computer analysis of the optical design of visual optical systems to incorporate a model of the human eye. Walker (2000) has described a model, illustrated in Fig. 15.2 with the data presented in Table 15.1. The pupil diameter is variable from about 2 mm up to 6.0 mm, depending on the light illumination, but in this model a fixed average value of 4.0 mm is assumed. The refractive index of the eye lens is not uniform. In this model the back surface of the eye lens is taken with a hyperboloidal shape with a conic constant equal to -4.5 in order to simulate the nonhomogeneous refractive index of the eye lens.

When numerically tracing rays in a computer to analyze the off-axis performance of a visual system to which this model of the eye has been attached, the eye model has to be rotated about the center of the exit pupil of the system being analyzed.

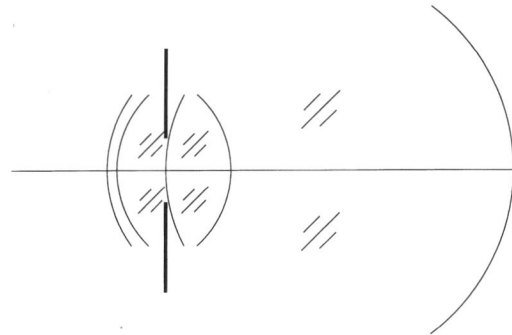

Figure 15.2 Optical model of a human eye.

Table 15.1 Optical Model of the Human Eye

Radius of curvature (mm)	Conic constant	Separation or thickness (mm)	Refractive index ($V_d = 55.0$)
7.8	0.0	0.6	1.377
6.4	0.0	3.0	1.336
10.1 (5.95)	0.0	4.0	1.411
-6.1 (-4.50)	-4.5	17.2	1.337
-12.5	0.0	—	—

Pupil diameter (average): 4.0.
Numbers in parentheses are for the accommodated eye.

An alternative to avoid rotating the eye model for each off-axis point is to use as an eye model a system that produces good quality optical images on-axis as well as off-axis. The main requirements for this eye model is that the image size has an angular diameter much less than a minute of arc and second. An example is a concave spherical mirror with a stop at its center of curvature and a spherical focal surface concentric with the mirror, as shown in Fig. 11.12(a). The spherical aberration has to be small enough so that the angular diameter of the image is smaller than 1 arcmin. A mirror with a radius of curvature equal to 30.184 mm ($f = 15.092$ mm) and 4 mm diameter produces an image with a transverse spherical aberration equal to 1 arcmin. With a radius of curvature equal to 687.4 mm the spherical aberration is 0.044 arcmin and a transverse aberration of 0.1 mm will correspond to an angular aberration equal to 1 arcmin.

15.2 BASIC TELESCOPIC SYSTEM

A visual telescope consists of two lenses as shown in Fig. 15.3. The lens closer to the object is called the *objective* and the lens closer to the eye is the *eye lens* or *eyepiece*. The objective forms a real image of the object on the focal plane of this objective. Then, the eye lens acts as a magnifier, forming a virtual image, to be observed by the eye. If the object as observed with the naked eye has an angular height α, observed through the telescope it has an angular height β. Then, the *angular magnification* or *magnifying power* of the telescope is defined by

$$M = -\frac{\tan \beta}{\tan \alpha} \tag{15.1}$$

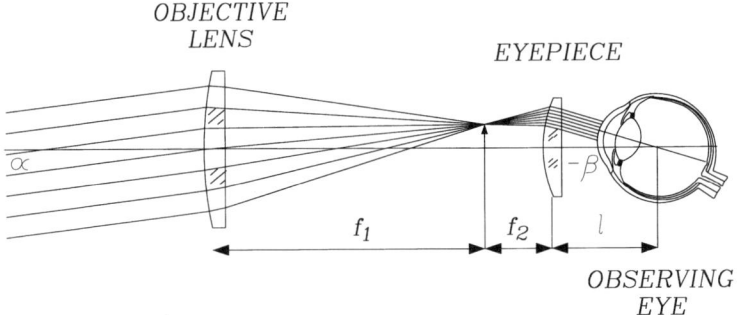

Figure 15.3 Basic telescope arrangement.

where the angles α and β are positive if their slope in Fig. 15.3 is positive. This expression is to be expected from the definition of angular magnification given in Chap. 1. Let us assume that the pupil of the telescope is in contact with the objective. If f_1 is the focal length of the objective, f_2 is the focal length of the eye lens, and l is the distance from the eyepiece to the exit pupil, we may write

$$\frac{1}{f_2} = \frac{1}{l} - \frac{1}{f_1 + f_2} \tag{15.2}$$

assuming that the observed object is at infinity, so that the separation between the objective and the eye lens is equal to the sum of the focal lengths of both lenses. Thus,

$$l = \frac{f_2}{f_1}(f_1 + f_2) \tag{15.3}$$

Now, from Fig. 15.3 we may see that the angular magnification is given by

$$M = -\frac{\tan \beta}{\tan \alpha} = -\frac{f_1 + f_2}{l} \tag{15.4}$$

Hence, substituting in this expression the value of l we obtain

$$M = -\frac{f_1}{f_2} \tag{15.5}$$

If the diameter of the objective (entrance pupil) is D_1 and the diameter of the exit pupil is D_2, we may find that the magnitude of the magnifying power is also given by (see Fig. 15.4)

$$|M| = \frac{D_1}{D_2} \tag{15.6}$$

where, by definition, the signs of these diameters D_1 and D_2 must be the same as those of the angles α and β, respectively. The distance from the eyepiece to the exit pupil of the telescope is a very important parameter called *eye relief*. As in ophthalmic lenses (Section 13.2) the observer rotates the eye globe in its skull socket to scan the whole image. Thus, ideally the exit pupil of the telescope must coincide not with the pupil of the observer's eye but with the center of rotation of the eye globe. The eye relief must be large enough to allow space for the eyelashes. In telescopic gun sights the eye relief must be considerable larger to avoid hitting the eye during the recoil of the gun. A convergent lens, named a field lens, may be placed at the image plane of the objective, to reduce the eye relief, as shown in Fig. 15.5. If we assume that the focal length of the objective is much larger than the focal length of the eye lens, the exit pupil would be located approximately at the focus of the eye lens. The distance from this focus to the eye lens or eye relief is equal to the back focal length of the field lens–eye lens combination. It is interesting to notice that if the field lens is exactly at the image plane, the back focal length is reduced but the effective focal length is not changed by this lens. [See Eqs. (3.38) and (3.42).]

The combination of the eye lens and the field lens is known as an *eyepiece* or *ocular*. They will be studied in Section 15.6.

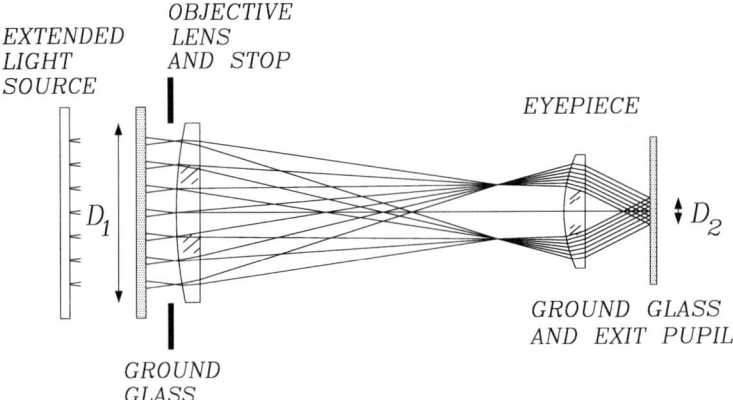

Figure 15.4 Measurement of the size of the exit pupil of a telescope.

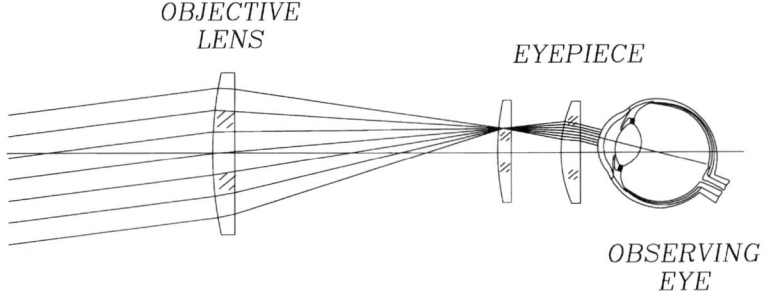

Figure 15.5 Telescope with an eyepiece having a field lens.

15.3 AFOCAL SYSTEMS

An afocal system is one that has an effective focal length equal to infinity. A good review of the properties of these systems has been published by Wetherell (1987). A telescope focused on an object placed at an infinite distance, and with the virtual image also at an infinite distance, has a separation between the objective and the eye lens equal to the sum of the focal lengths of both lenses. Thus, it is a special kind of afocal system. With afocal systems we may form an image (real or virtual) of a real object. Let us consider the afocal system shown in Fig. 15.6. The exit pupil is a real image of the entrance pupil. The angles α and β satisfy Eq. (15.4) for the magnifying power of the system. If we have an object H at a distance X from the entrance pupil with diameter D_1, its real image is at H', at a distance X' from the exit pupil with diameter D_2. Thus, we may see that the magnifying power is given by

$$M = \frac{\tan \beta}{\tan \alpha} = \frac{D_2}{D_1} \frac{X}{X'} = \frac{1}{M} \frac{X}{X'} \qquad (15.7)$$

thus

$$M^2 = \frac{X}{X'} \qquad (15.8)$$

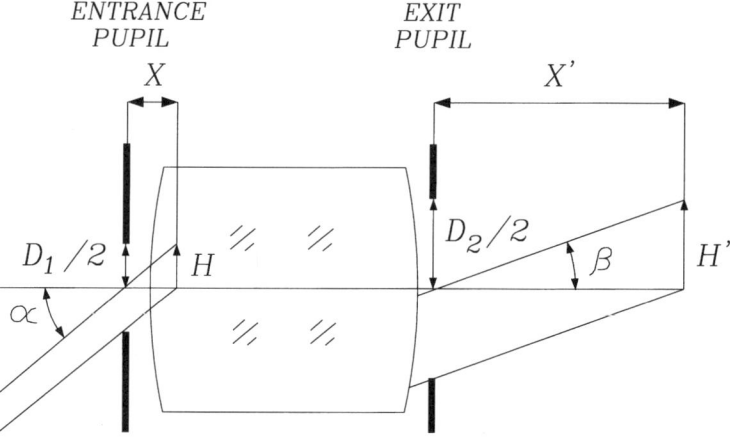

Figure 15.6 Image formation in an afocal system.

The distance X is positive when the object is to the right of the entrance pupil and the distance X' is positive when the object is to the right of the exit pupil. It is interesting to notice that X and X' will always have the same sign. Let us assume that the separation between the entrance pupil and the exit pupil is L_P. Then, we may see that the distance between the object and its image is L_I, given by

$$L_I = L_P + X' - X = L_P - X\left(1 - \frac{1}{M^2}\right) \tag{15.9}$$

Since in Fig. 15.6 we have $H = D_1/2$ and $H' = D_2/2$, we may see that the lateral magnification is equal to the inverse of the magnifying power of the system. Afocal systems are very interesting, and as Wetherell (1987) points out, they have three important properties:

1. If the stop is at the intermediate focus, as shown in Fig. 15.7, the system is both front and back telecentric. Then, the lateral magnification is constant, even if the image is defocused.
2. If the magnifying power M is equal to ± 1, the distance L_I is equal to the distance L_P, since X is equal to X'. Then, if the afocal system is shifted along the optical axis in the fixed space between the object and the image, the lateral magnification is always unitary and the image position remains constant.
3. If the magnifying power M is not equal to ± 1, the distance L_I is not equal to the distance L_P, unless the object is at the entrance pupil. Then, the lateral magnification is constant and, if the object and the image are fixed, the image focusing may be adjusted by moving the afocal system, without modifying the lateral magnification.

These properties have very important practical applications, especially in microlithography. Let us now examine two other interesting properties of afocal systems.

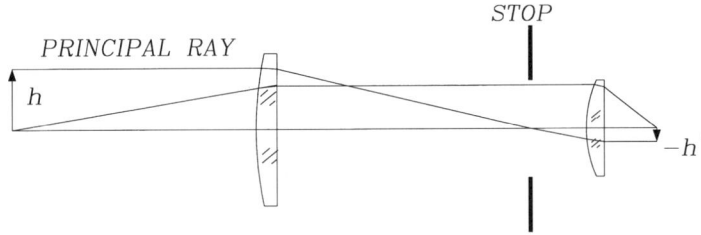

Figure 15.7 Image formation in an afocal system with the stop at the common focus of objective and eyepiece.

A diverging light beam becomes even more diverging after passing through a terrestrial telescope. The angle of divergence is increased by a factor equal to the magnification of the telescope. If we are observing a near object (at a distance X in front of the entrance pupil) through a telescope, the observing eye has to focus on the image formed by the afocal system (at a distance X' in front of the exit pupil). Thus, if M is greater than one, the apparent distance X' is less than the actual distance X. On the other hand, if M is less than one, the apparent distance X' is greater than the actual distance X. In conclusion, the depth of field is reduced when looking through a telescope. Obviously, when observing with an inverted terrestrial telescope the depth of field is increased. For this reason a myopic person wearing noncontact ophthalmic lenses has a greater apparent depth of field than an emetropic person of the same age (Malacara and Malacara, 1991), reducing the effect of presbyopia.

Another interesting effect when looking through a telescope is that the objects do not only look larger and closer, but also thinner. The apparent compression along the line of sight is directly proportional to the square of the magnifying power. Hence, if the magnifying power is greater than one, the objects look too compressed in depth.

15.3.1 Two-Mirror Afocal Systems

Two-mirror afocal systems can be constructed with a similar arrangement to that used for two-mirror telescopes, using a concave and a convex mirror, as illustrated in Fig. 15.8. The Mersene system, formed by two paraboloids,

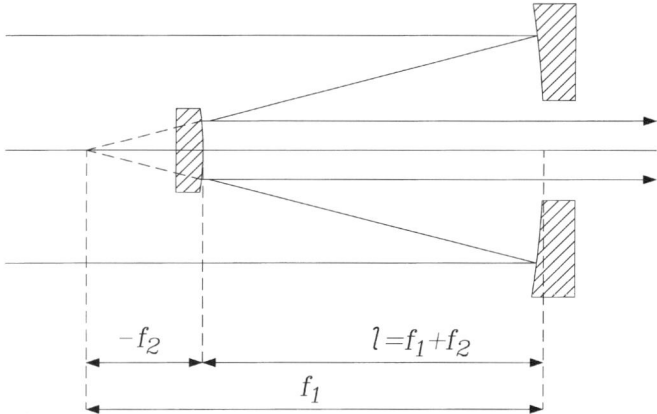

Figure 15.8 Two-mirror afocal system.

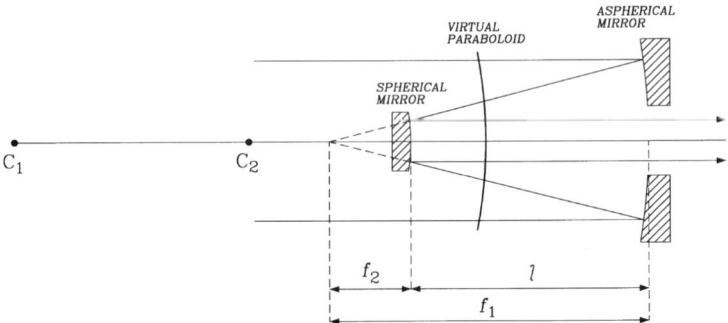

Figure 15.9 Virtual paraboloid in two-mirror afocal system with a spherical secondary mirror and an aspherical primary mirror.

can be considered as a Cassegrain telescope with an infinite effective focal length. From Eqs. (14.47) and (14.51) we see that in the Cassegrain telescope the coma and the astigmatism are inversely proportional to the effective focal length. Thus, this afocal system is free of spherical aberration, coma, and astigmatism. The distortion is small and only the Petzval curvature remains.

Puryayev (1993) has considered a modification of this afocal system, substituting one of the two paraboloidal mirrors by a spherical mirror. Then, the shape of the other mirror is modified to correct the spherical aberration of the system. Strictly speaking, the shape of this modified mirror is not a hyperboloid, although it is close. Puryayev has shown that the figure of this mirror has the same evolute (same caustic for the spherical aberration of the normals) as a paraboloidal surface virtually located at some point along the optical axis.

Considering the case of a spherical secondary mirror, the virtual paraboloid with the same evolute as the concave aspherical mirror has a focal length equal to the separation between the mirrors and it is concentric with the primary mirror, as shown in Fig. 15.9. Let us imagine the surface of the aspheric mirror to be a wavefront. When this aspheric wavefront propagates in space its shape is continuously changing along its trajectory, but at some point in space it acquires the exact paraboloidal shape. This is the virtual paraboloid.

15.4 REFRACTING OBJECTIVES

The typical configuration of a refracting telescope objective as an achromatic doublet has been described in Section 6.2, but it can be more complicated (Fulcher, 1947). The design of a doublet presents many options, depending on the aberrations that have to be corrected as described by several authors (Hopkins and Lauroesh, 1955; Hopkins, 1959, 1962a,d; Korones and Hopkins, 1959). We will study in this section these options and their design procedures.

15.4.1 NonAplanatic Doublet

Given the glass types and the effective focal length, the degrees of freedom are the three curvatures, in order to have the desired focal length and to correct only two aberrations, namely, the spherical aberration and the axial chromatic aberration. The ratio of the power of the two lenses is used to correct the axial chromatic aberration, and the spherical aberration is corrected by properly bending the lens to the right shape. The thicknesses are chosen so that the lens is neither too thin nor too thick.

A very general graphical method that may be of great help in the design of optical systems has been described by Kingslake (1978) and ascribed by him to H. F. Bennett. This method may be used when an automatic lens improvement program is not available, or when a good understanding of the influence on the design of each of the available variables is desired. It is interesting to use this method even when using ray-tracing programs, in order to get a good feeling of the influence of the variables on the aberrations. It may be used whenever we have two available parameters to modify and two functions (aberrations) to correct.

As an example, let us consider an achromatic doublet in which the starting point is calculated with the first-order formulas in Chap. 6 with the shape of the positive lens being equiconvex, with the following characteristics:

Diameter	20.00 mm
Effective focal length	100.00 mm
First radius of curvature	43.68 mm ($c_1 = 0.022893$)
Second radius of curvature	−43.68 mm ($c_2 = -0.022893$)
Third radius of curvature	−1291.00 mm
Thickness of first lens	4.00 mm
Thickness of second lens	2.00 mm
Glasses	BK7 and F2

As shown in Fig. 15.10 the two functions to correct are the primary spherical aberration $SphT$ and axial chromatic aberration $AchrT$. The variables in this case are the two curvatures c_1 and c_2 of the positive lens. The third (last) curvature c_3 is determined by the effective focal length of the doublet. If our initial trial solution is at **A**, we change the front curvature of the positive lens by an amount Δc_1, adjust the third curvature to preserve

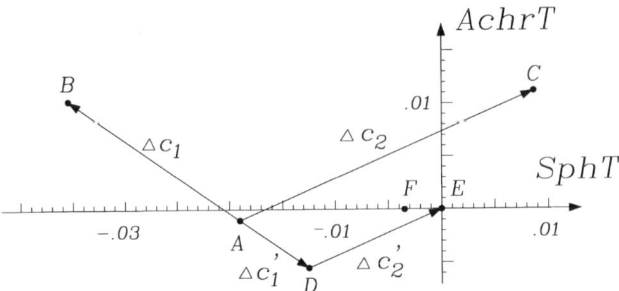

Figure 15.10 Graphical solution for an achromatic doublet corrected for spherical aberration.

the focal length, evaluate the aberrations, and find that a new lens is at **B**. Then, we go back to the initial point and change c_2 by an amount Δc_2, evaluate again, and find a new lens solution at **C**.

The final step is to assume a linear variation of the aberrations with the parameters being modified. Graphically, as illustrated in Fig. 15.10, we find that the desired solution at **E** may be obtained by changing the radii of curvature lenses by amounts $\Delta c'_1$ and $\Delta c'_3$, given by

$$\Delta c'_1 = \frac{AD}{AB} \Delta c_1 \tag{15.10}$$

and

$$\Delta c'_3 = \frac{DE}{AC} \Delta c_3 \tag{15.11}$$

We may notice several interesting features in this diagram in Fig. 15.10:

1. Due to nonlinearities, the predicted solution **E** is not where the actual solution **F** is; however, it is very close for all practical purposes.
2. Both parameters c_1 and c_2 have about the same influence on both aberrations. If the index of refraction n_D is the same for both glasses, the glass interface c_2 would not have any influence on the spherical aberration. Then, the line **AC** would be vertical. The slope of this line is thus smaller for a large difference in the refractive indices.
3. The lines **AB** and **AC** are not parallel to each other, permitting the possibility of a solution.

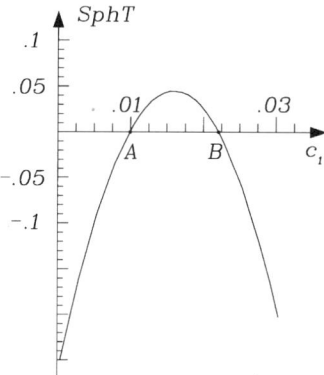

Figure 15.11 Variation of the transverse spherical aberration versus the front curvature of an achromatic doublet.

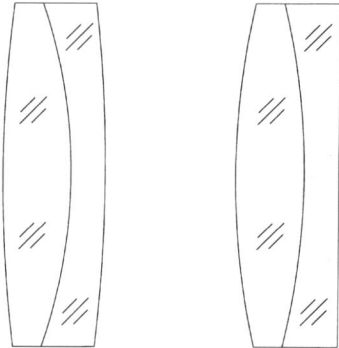

(a) LEFT SOLUTION   (b) RIGHT SOLUTION

Figure 15.12 Two solutions for an achromatic doublet free of spherical aberration: (a) left solution; (b) right solution.

We may find solutions for zero axial chromatic aberration, without requiring correction of the spherical aberration. This produces the graph in Fig. 15.11. We may see that there are two solutions, **A** and **B**, for zero spherical aberration. The point **B** corresponds to the solution just found. These two solutions are illustrated in Fig. 15.12.

The axial plots for the solution **B** are in Fig. 15.13. We may notice the following:

  1. Fifth-order spherical aberration and defocus is introduced, to compensate the primary spherical aberration, making the total

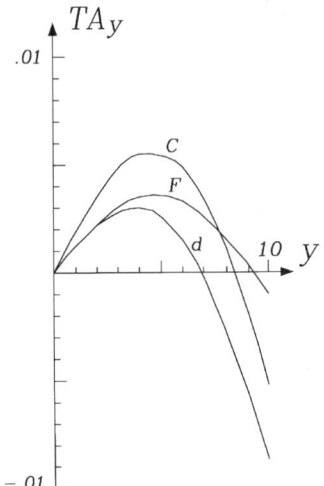

Figure 15.13 Axial plots for lens solution B in Fig. 15.11.

      spherical aberration close to zero at the edge of the pupil. The maximum transverse aberration is about 0.006 mm (6 μm), which is close to the diameter of the Airy disk.
2. The F and C curves cross each other near the edge of the pupil, indicating a good correction for the primary axial chromatic aberration. This curve for D colors does not cross the curves for C and F colors near the edge of the aperture due to the presence of secondary color.
3. The curves F and C have different amounts of spherical aberration, due to the spherochromatism.

    Figure 15.14 shows the meridional and sagittal plots at 5° off-axis ($h' = 8.75$ mm). The curves for the three colors are almost identical, indicating the absence of magnification chromatic aberration. As expected, there is a large coma aberration, indicated by the symmetric component of the meridional ray-trace plot. The antisymmetric component of this curve is a linear function with a slope due to the curved tangential field. The different slopes for the meridional and sagittal plots indicate the presence of some astigmatism. This lens design is presented in Table 15.2. If used as a collimator, this lens produces a flat wavefront with a peak-to-valley (P–V) error equal to 0.22 wavelength in yellow light.

    We have seen, as in Chap. 5, that, if the glasses for two thin lenses in contact have a different index of refraction, there are two solutions for zero

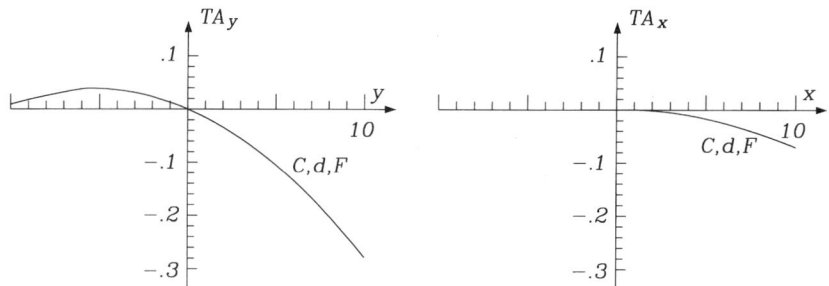

Figure 15.14 Meridional and sagittal plots at 5° off-axis for lens solution B in Fig. 15.11.

Table 15.2 $F/5$ Nonaplanatic Achromatic Doublet

Radius of curvature (mm)	Diameter (mm)	Separation or thickness (mm)	Material
44.14 (stop)	22.0	4.0	BK7
−45.56	22.0	2.0	F2
−1216.80	—	95.74	Air

Aperture (mm) (focal ratio): 20.0 ($F/5.0$).
Effective focal length (mm): 100.0.
Back focal length of doublet (mm): 273.47.
Back focal length of system (mm): 95.74.
Object distance (mm): infinite.
Image height (angular semifield): 8.75 (5°).

spherical aberration. Thus, by selection of the glasses, or other methods later described, we may place the zero spherical aberration solution at any desired position to achieve coma correction. A system corrected for spherical aberration and coma is said to be aplanatic.

15.4.2 Broken Contact Doublet

In this type of objective, described by many authors, e.g., Rosin (1952), the two elements of the doublet are separately bent to the optimum shape to obtain simultaneous full correction of the primary spherical aberration and coma. Thus, given a focal length and the requirement for zero axial chromatic aberration, the focal lengths of the two components are fixed. The two bendings are then the two degrees of freedom needed to correct the spherical aberration and coma. To modify our last design we first separate the two lenses by introducing an air space between them equal to 0.15 mm. This starting design is then evaluated. The primary sagittal coma aberration is calculated with the usual coefficients. The spherical aberration, however, is calculated using exact ray tracing for an axial ray passing through the edge of the pupil. The transverse aberration TA_y is then the value of the transverse axial spherical aberration, including all high-order terms.

The next step is to bend the positive lens. This is done by changing the curvature c_1 by an amount $\Delta c_1 = -0.002$ and then adjusting the curvature c_2 to preserve the same effective focal length. This new configuration is evaluated like the starting design. The third step is to go back to the original design and to bend the negative lens by the same amount and method as the first lens. Again, this configuration is evaluated. In this manner, the graph in Fig. 15.15 is obtained, where A is the original design, and B and C are the configurations with the two lens bendings.

With the method previously described, a solution is found at a point near **E**, but not exactly there. Due to nonlinearities this point is not at the origin as desired, but a small bending of the negative lens brings the design to a very good solution. Figure 15.16 and Table 15.3 show a doublet of the broken contact type.

Figure 15.17 shows the axial plots for this lens, where we may observe that:

1. The spherical aberration is quite similar to that of the cemented doublet. Again, high-order spherical aberration and defocusing is present to make the total aberration close to zero near the edge of the aperture.

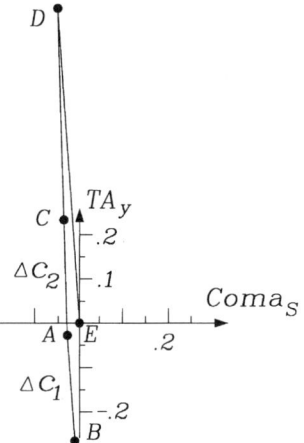

Figure 15.15 Graphical solution of a broken contact doublet.

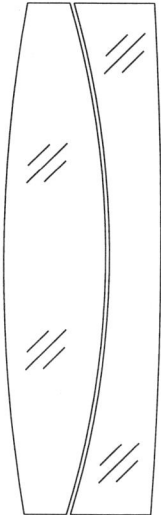

Figure 15.16 Broken contact doublet.

Table 15.3 $F/5$ Broken Contact Aplanatic Achromatic Doublet

Radius of curvature (mm)	Diameter (mm)	Separation or thickness (mm)	Material
58.39 (stop)	22.0	4.0	BK7
−36.29	22.0	0.15	Air
−36.67	22.0	2.0	F2
−161.92	—	96.55	Air

Aperture (mm) (focal ratio): 20.0 ($F/5.0$).
Effective focal length (mm): 100.0.
Back focal length (mm): 96.55.
Object distance (mm) : infinite.
Image height (angular semifield):8.75 (5°).

2. The primary chromatic aberration is also quite similar to that of the cemented doublet.

The meridional and sagittal plots in Fig. 15.18 for an off-axis point object with a height of 8.75 mm (5°) show a straight line, indicating the absence of coma. The large slope is due to the curved tangential field. Notice that the slope at the origin in this plot and in Fig. 15.14 is the same.

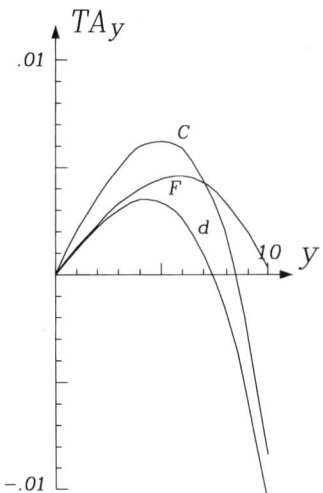

Figure 15.17 Axial plots for a broken contact doublet.

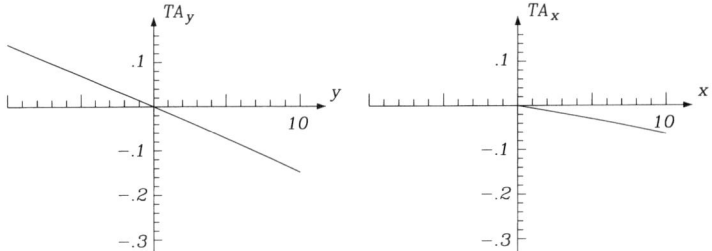

Figure 15.18 Meridional plot for a broken contact doublet.

The different slopes of the meridional and sagittal plots indicate the presence of some astigmatism. Obviously, a lens system like this can not be cemented and has to be very carefully mounted to avoid misalignments.

15.4.3 Parallel Air-Space Doublet

In the parallel air-space doublet the two inner radii of curvature are equal, to reduce manufacturing costs. Thus, the two adjusting parameters are the bending of the whole lens and the spacing between the two lenses.

The only problem with the design is that according to the principles explained in Chap. 4, the high-order spherical aberration may become too large, producing a large zonal aberration. Thus, the primary and the high-order spherical aberrations have to be compensated with each other. In general, the broken-type design is superior to the parallel air-space system.

15.4.4 Cemented Aplanatic Doublet

As explained at the beginning of the Chap. 4, the solution for zero spherical aberration may be placed at the bending value such that the coma is also zero. In this lens system we have three degrees of freedom, namely, the bending of the whole lens and the powers of the two components. We may use these variables to obtain the desired focal length, to correct the spherical aberration and to correct the coma, given the refractive indices of the lenses. The Abbe numbers may then be chosen to correct the chromatic aberration. For example, the same design in Fig. 15.12 can be improved, reducing the coma by substituting the glass F2 by SF5 or SF9. With these two last glasses the coma has opposite signs.

15.4.5 Apochromatic Lenses

The secondary color may be reduced by means of any of the following methods:

    1. Choosing the right glasses, so that the partial dispersions are equal or at least close to each other for the two glasses. Unfortunately, this is not a good solution since this requires the use of special glasses.

    2. Using three different glasses, forming a triplet as shown in Chap. 6. As mentioned there, the area of the triangle formed by the points representing the three glasses in a P–V diagram must be different from zero in order to have a solution. On the other hand, in order to have lens components with low power this area must be as large as possible. A good selection of glasses, as pointed out by Kingslake (1978), is a crown for the first element, a short flint or lanthanum crown for the central element, and a very dense flint for the last element.

15.4.6 Laser Light Collimators

It is frequently necessary to produce a well-collimated beam of laser light. A normal telescope objective with a large *f-number* may be used, but an important requirement is that the focal length is short. An *f-number* as low as possible is convenient. This imposes the need for an extremely good spherical aberration correction, with a low zonal aberration. As described by Hopkins (1962d) and Korones (1959), the zonal aberration may be

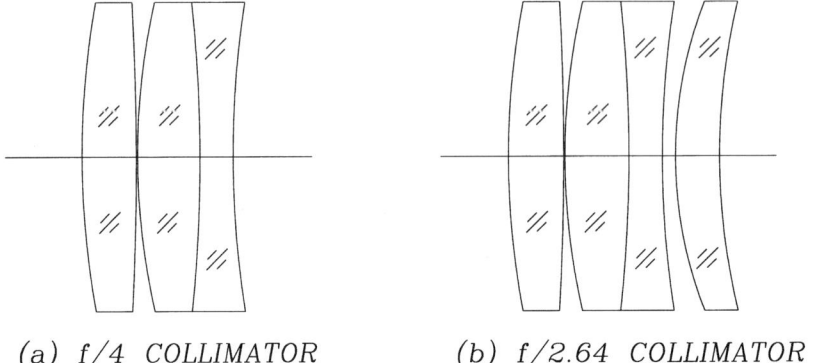

(a) f/4 COLLIMATOR   (b) f/2.64 COLLIMATOR

Figure 15.19 Two He–Ne laser collimators. $f/4$ collimator; (b) $f/2.64$ collimator.

Table 15.4 $F/4$ Laser Light Collimator

Radius of curvature (mm)	Diameter (mm)	Separation or thickness (mm)	Material
136.4	50.0	8.886	BK7
−465.6	50.0	0.193	Air
111.4	50.0	10.274	BK7
−236.3	50.0	5.401	SF15
166.8	50.0	—	Air

Aperture (mm) (focal ratio): 50.0 ($F/4$).
Effective focal length (mm): 200.0.
Back focal length (mm): 177.2.
Object distance (mm): infinite.

reduced by any of four methods:

1. Choosing the proper glasses
2. Using an air space
3. Introducing an aspherical surface
4. Splitting the positive lens into two

The requirement for good chromatic correction is not necessary in a laser collimator, but the spherical aberration should be small for a range of wavelengths. A refocusing may be necessary when changing the color. Another important requirement is coma correction, so that small misalignments are tolerated. Malacara (1965) has designed two such collimators using the last approach, shown in Fig. 15.19, with the data in Table 15.4.

Table 15.5 $F/2.64$ Laser Light Collimator

Radius of curvature (mm)	Diameter (mm)	Separation or thickness (mm)	Material
136.4	50.0	8.886	BK7
−465.6	50.0	0.193	Air
111.4	50.0	10.274	BK7
−236.3	50.0	5.401	SF15
166.8	50.0	2.100	—
69.6	50.0	7.000	—
108.5	50.0	—	—

Aperture (mm) (focal ratio): 50.0 ($F/2.64$).
Effective focal length (mm): 132.0.
Back focal length (mm): 108.5.
Object distance (mm): infinite.

In the second collimator, with data in Table 15.5, a meniscus lens was added as in the *duplex front* of an immersion microscope objective, in order to reduce the focal length.

The maximum wavefront deviation from flatness at 632.8 μm is $\lambda/40$.

15.5 VISUAL AND TERRESTRIAL TELESCOPES

Unlike professional astronomical telescopes, amateur astronomical telescopes have an eyepiece to observe the image visually. Professional astronomers almost never make direct visual observations.

We have seen that the ratio of the entrance pupil to the exit pupil is equal to the angular magnification of the system. Let us assume that the exit pupil of the telescope has a diameter equal to the diameter of the pupil of the eye, which is about 6 mm at night. With this condition the angular magnification of the telescope is equal to $D_1/6$, where D_1 is the diameter of the entrance pupil (the objective diameter for a Keplerian telescope). However, this is not the case in most telescopes since the angular magnification can be larger or smaller as follows:

1. If the angular magnification is larger than this value the size of the exit pupil becomes smaller than the pupil of the eye. Then, the brightness of a star is the same for any magnification since no light is lost. We have seen in Chap. 14 that the limit magnitude is a function only of the diameter of the entrance pupil, assuming that no light is lost. However, the larger the

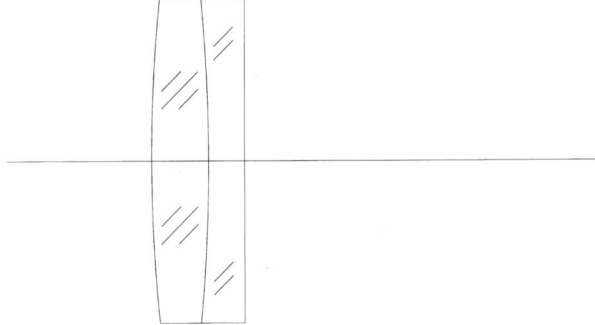

Figure 15.20 An $F/10$ telescope objective.

magnification is, the smaller the field becomes, reducing the number of stars within the field of view.

2. If the angular magnification is smaller than this value, the size of the exit pupil becomes larger than the pupil of the eye and not all the light entering the objective reaches the retina of the observer. Then, the effective entrance pupil diameter is reduced. This condition increases the field but decreases the resolution of the telescope and the star brightness. A conclusion is that the maximum number of observed stars is obtained when the magnification has the optimum value equal to $D_1/6$. A telescope with this magnification is sometimes called the richest field telescope.

Terrestrial telescopes must also present to the observer an erect image. This is accomplished by means of a prism or lens erector. The simplest and more common type of lens erector uses an inverting eyepiece as will be described later in this chapter. Other erecting systems with better image quality use erecting prism systems, such as those described in Chap. 10.

A typical visual refractive astronomical telescope has a larger focal ratio that the doublets designed in the previous sections, close to $F/10$. An example of an $F/10$ telescope objective is illustrated in Fig. 15.20 and Table 15.6.

15.5.1 Galilean Telescopes

Although the first uses of Galilean telescopes were for direct astronomical observations beginning with Galileo himself, now they are almost exclusively used for terrestrial observations.

Galilean telescopes are now mainly used as theater binoculars and for improving the visual capacity of low-vision persons. The optics literature

Visual Systems, Visual Telescopes, and Afocal Systems

Table 15.6 $F/10$ Telescope Objective

Radius of curvature (mm)	Diameter (mm)	Separation or thickness (mm)	Material
433.70 (stop)	102.0	16.0	BK7
−459.90	102.0	10.0	F2
20000.00	102.0	981.1	Air

Aperture (mm) (focal ratio): 100.0 ($F/10.0$).
Effective focal length (mm): 1000.0.
Back focal length (mm): 981.1.
Object distance (mm): infinite.
Image height (angular semifield):17.45 (1°).

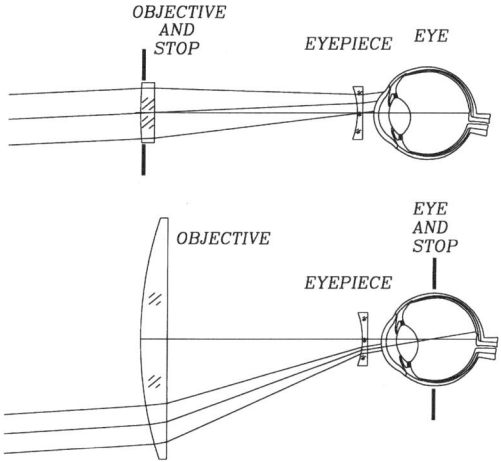

Figure 15.21 Galilean telescopes with stop at the objective and at the observer's eye.

about this topic is scarce (Bertele, 1983; Rusinov et al., 1983; Menchaca and Malacara, 1988). Now, we will briefly describe these systems.

In a Galilean telescope, shown in Fig. 15.21, the pupil may be at the observer's eye or at the objective, according to the magnification of the telescope and the objective diameter. If the telescope has a small diameter objective and a large magnification, as in the case of the original telescope made by Galileo, the stop is at the objective's plane, as shown in Fig. 15.21(a). In this case the exit pupil is far from the observer's eye; hence, the size of the field of view is very small and limited by the pupil of the observer. In this type of telescope the observer has the sensation of observing through a long and narrow tube.

If the ratio d_{in}/M is much greater than the diameter of the pupil of the eye or, in order words, if the telescope has a large objective and a small magnification, the stop would be at the observer's eye, as shown in Fig. 15.21(b). In this case the entrance pupil is not at the objective and its size has no relation to the diameter of this objective, but it is equal to the stop diameter times the telescope magnification power. The size of the objective lens determines the field diameter. Thus, to have a reasonable field of view, the objective should be large. As explained at the beginning of this chapter, the exit pupil, which is defined by the position of the observer's eye, should be considered at the center of rotation of the eye globe and not at the pupil of the eye.

On-axis aberrations, namely, spherical aberration and axial chromatic aberration, in low-power telescopes, are not a serious problem, because the diameter of the entrance pupil is small.

The correction of the magnification chromatic aberration is extremely important because even a small amount is noticeable, due to the large objective size, especially for large magnifying powers. This correction may be achieved at the objective by using two different glasses. However, if the axial chromatic aberration is also to be corrected, the eyepiece has also to be a doublet.

The coma should be corrected as well as possible, but just below the resolving power of the eye. The field curvature should be corrected only if a focusing of a point on the edge of the field cannot be done with a small amount of eye accommodation. In other words, a small concavity (from the observer's point of view) of the field is acceptable, but never a convexity of the field.

The distortion cannot be completely eliminated and will always be present and noticeable. In afocal systems like this the distortion is defined by the ratio of the slopes of the principal rays after exiting the system and before entering. If this ratio is a constant there is no distortion.

When using a ray-tracing program to design a telescope like this, it is advised to design it with a reversed orientation, with the light entering the stop. The reason is that with its normal orientation the system may have strong chromatic aberration, producing several entrance pupils, one for each color.

A $5.0 \times$ Galilean telescope designed by Menchaca and Malacara (1988) is illustrated in Fig. 15.22 and the data are presented in Table 15.7.

A human eye can accommodate by refocusing the eye lens in order to see near objects clearly. The amplitude of accommodation decreases with age, mainly after 40 years of age. The depth of field can be defined as the maximum angle of convergence that an emmetropic human eye can focus the observed objects.

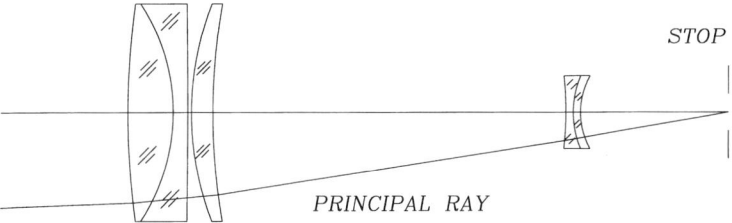

Figure 15.22 Galilean telescope.

Table 15.7 A 5× Galilean Telescope

Radius of curvature (mm)	Diameter (mm)	Separation or thickness (mm)	Material
111.4	30.0	6.0	SSK2
−27.2	30.0	2.0	F2
−958.0	30.0	0.5	Air
40.9	30.0	3.0	SSK2
85.5	30.0	48.5	Air
−50.4	10.0	2.0	BK7
12.8	10.0	1.0	F2
10.3	10.0	20.0	Air
Stop	3.0	—	—

Angular magnification power: 5.0×.
Entrance pupil diameter: 15.0.
Angular semifield: 2.0°.
Eye relief (mm): 20.0.

A diverging light beam becomes even more diverging after passing through a Galilean telescope. The angle of divergence is increased by a factor equal to the magnification of the telescope. A consequence is that when observing through the telescope the depth of field is reduced by the same amount. Obviously, when observing with an inverted Galilean telescope the depth of field is increased. For this reason a myopic person wearing noncontact ophthalmic lenses has a greater depth of field than an emmetropic person of the same age (Malacara and Malacara, 1991).

15.5.2 Design of a Terrestrial Telescope Objective

A terrestrial telescope objective has to be designed to include a block of glass with plano parallel faces between the objective and the eyepiece. The reason

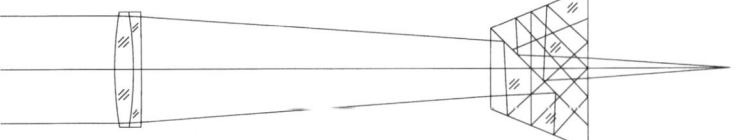

Figure 15.23 An $F/10$ terrestrial telescope objective.

Table 15.8 $F/10$ Telescope Objective

Radius of curvature (mm)	Diameter (mm)	Separation or thickness (mm)	Material
134.51 (stop)	43.0	7.0	BK7
−132.88	43.0	3.0	F2
−1230.50	43.0	130.0	Air
Flat	30.0 × 30.0	138.6	BK7
Flat	—	52.1	Air

Aperture (mm) (focal ratio): 40.0 ($F/7.0$).
Effective focal length (mm): 280.0.
Back focal length (mm): 52.1.
Object distance (mm): infinite.
Image height (angular semifield): 4.88 (1°).

is that the erecting prism system is used in converging light and thus spherical aberration and axial chromatic aberrations are introduced. Those aberrations have to be compensated in the design of the telescope objective. An example is the system shown in Fig. 15.23 and in Table 15.8 using a Schmidt–Pechan prism.

15.6 TELESCOPE EYEPIECES

An eyepiece has its entrance and exit pupils outside the system (Hopkins, 1962c). Normally, the entrance pupil is located at the same plane as the exit pupil of the telescope objective. To observe the image, the center of the eye globe of the observer is located at the exit pupil of the eyepiece. The transverse axial aberrations (spherical and chromatic) have to be smaller than those of the telescope objective. The focal length of the eyepiece is smaller than the focal length of the objective by a factor equal to the magnification of the telescope. This means that the angular axial aberration may be as large as those of the objective, times the telescope magnification. In conclusion, the axial aberrations do not represent any problem.

Off-axis aberrations, on the other hand, must be more carefully controlled. In general, the magnitude of all transverse (on- and off-axis) aberrations have to be small enough so that the eye cannot detect them. Since the resolving power of the eye is about 1 arcmin, the corresponding value of the transverse aberrations is

$$TA = \frac{\pi}{10,800} F \tag{15.12}$$

where F is the effective focal length of the eyepiece.

Since most elements in an eyepiece are positive and with a short focal length, in general, the Petzval curvature is large and negative. Then, a large positive astigmatism has to be introduced, so that the sagittal surface becomes flat. This means that the tangential surface would be curved towards the observer. Then, the eye would be relaxed for images on-axis, but it has to accommodate for off-axis images. A field curved away from the observer is very unpleasant to the observer, because off-axis images cannot be accommodated when on-axis images are focused at infinity. Thus, off-axis images will always be defocused.

To have a feeling for the type of lens capable of introducing the desired positive astigmatism, let us consider some typical configurations for plano convex lenses, as shown in Fig. 15.24. The first four lenses (a)–(d) are candidates for the field lens, since the stop is far away and the principal ray arrives almost parallel to the optical axis. These four lenses have two possible orientations, with the object in front and behind this lens. The last two lens configurations (e) and (f) are candidates for the eye lens, because the stop is located close to the lens, on the side of the collimated beam. The astigmatism surface contributions, from Eqs. (4.23) and (5.51), may be found to be

$$AstL_S C = \frac{y(n/n')(n - n')(i + u')\bar{i}^2}{2n'_k u^2_k} \tag{15.13}$$

thus, applying this expression to the lens configurations in Fig. 15.24 we find that these surface contributions are as indicated in the figure. Here, we discard all lenses with a negative astigmatism and so we are left with lenses (a), (b), and (e). For the eye lens we have only one possibility, with a small astigmatism. Thus, all the desired astigmatism must come from the field lens. Regarding the field lens, we have two possibilities. We must notice that in the two lenses (a) and (b), the astigmatism would be extremely small if the object coincided with the lens. Hence, the object should be at a certain distance from the lens and on the side of the flat face.

In conclusion, the two possible eyepiece configurations with single lenses are as shown in Fig. 15.24(a) and (b). Bending the lenses from the

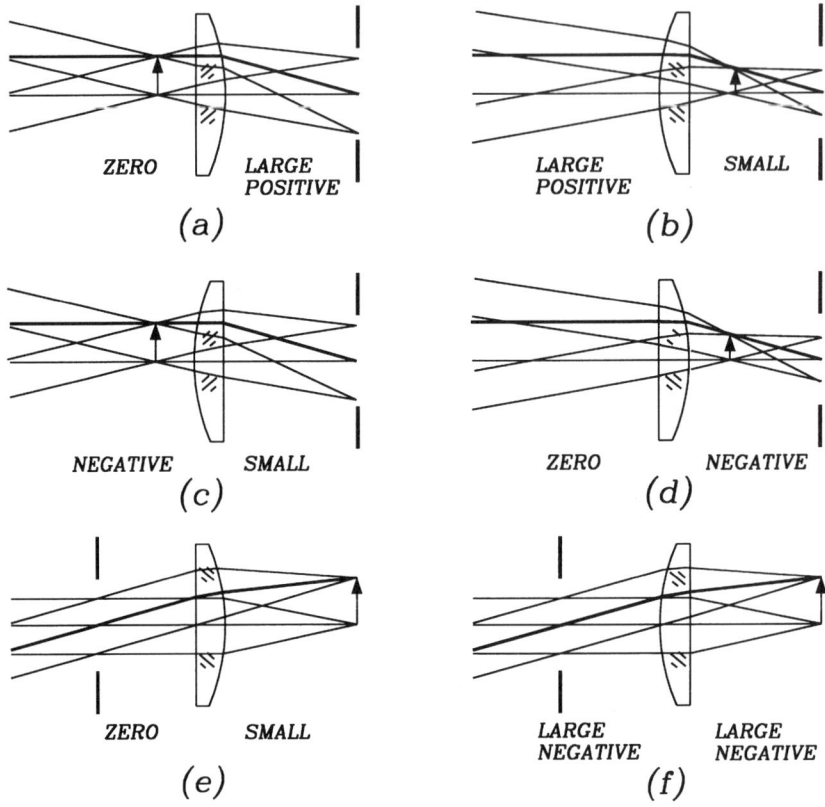

Figure 15.24 Several orientations and stop positions for a plano convex lens.

plano convex configuration does not much improve the correction. As to be expected, due to the position of the pupil, the distortion is normally quite high and of the pincushion type. These two designs receive the names of Huygens and Ramsden eyepieces.

It is important to point out that eyepieces are evaluated in lens design programs with the light entering through the long conjugate that is in the opposite direction of the actual use. However, the designs described here are shown with the correct orientation.

15.6.1 Huygens and Ramsden Eyepieces

The magnification chromatic aberration as explained in Section 6.4.2 may be corrected in a system of two lenses made with the same glass when the

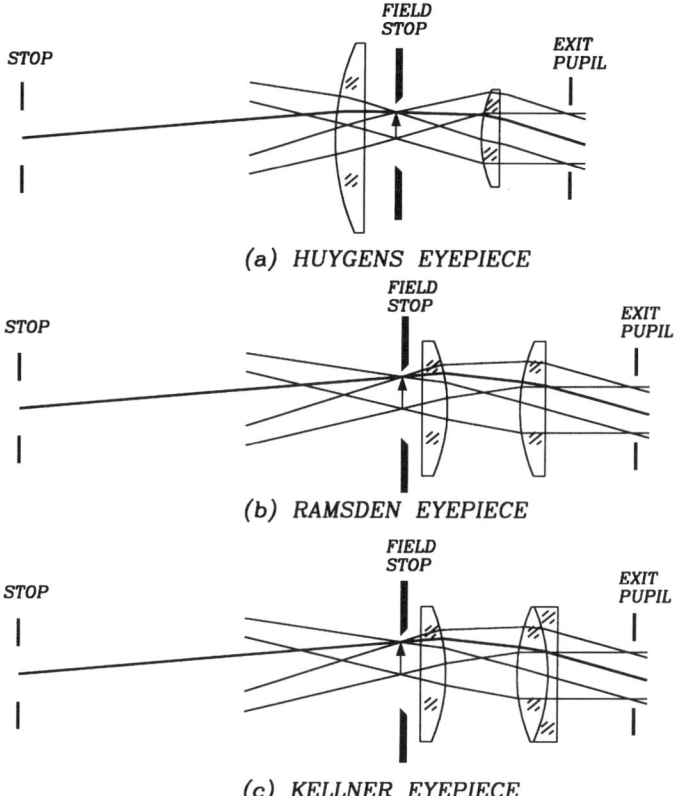

Figure 15.25 Three popular telescope eyepieces: (a) Huygens; (b) Ramsden; (c) Kellner.

stop is at infinity with a separation d between them equal to the average of their focal lengths. In the case of telescopes the stop is not at infinity, but it is far enough for all practical purposes. The two focal lengths cannot be made equal, because the separation between them is then equal to the focal length of the lenses, and the object would be located at the field lens plane. The ratio between the two focal lengths may be used as a variable to obtain the best possible off-axis image. For the case of the Huygens eyepiece, shown in Fig. 15.25(a), a typical condition is $f_b/f_a = 2$, where f_a and f_b are the focal lengths for the field lens and the eye lens, respectively.

A Huygens eyepiece has a relatively small apparent field of view of about $\pm 15°$. The magnification chromatic aberration is corrected for the whole eyepiece, but not for the eye lens alone. Thus, a reticle in the image

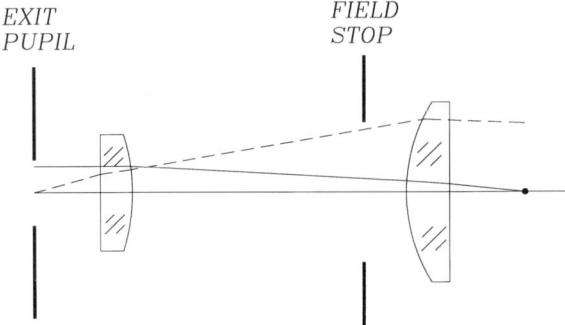

Figure 15.26 A 25 mm telescope Ramsden eyepiece.

plane would not be color free. The eye relief is relatively short, in telescopes even shorter than in microscopes. For this reason, they are rarely used in telescopes. More details on the design of Huygens eyepieces will be given in Chap. 16.

In the Ramsden eyepiece, Fig. 15.25(b), the achromatism is incompatible with the condition that the back focal length is greater than zero, so that the focus is outside of the system. This means that the Ramsden eyepiece cannot be corrected for the magnification chromatic aberration. The aberrations and field width are similar in magnitude to those in the Huygens eyepiece.

The Ramsden eyepiece may be used with a reticle, since the field stop (object) is outside the system. The eye relief is short, but greater than in the Huygens eyepiece. A 25 mm Ramsden eyepiece is shown in Fig. 15.26 and its design is presented in Table 15.9. The exit pupil (telescope exit pupil) is assumed to be at a distance of 1 m.

15.6.2 Erecting Eyepiece

The terrestrial telescope shown in Fig. 15.27 has an erecting eyepiece, which consists of a relay system to erect the image and a Huygens eyepiece. The erecting part may be considered as a projecting Ramsden eyepiece. However, there is one important difference, that the stop is between the two lenses and not outside. Since the stop is inside, the lens orientation also has to be changed, so that the flat face of the lens is on the side of the stop as in the Ramsden eyepiece. There is a magnification in this system to increase the magnification of the telescope, to compensate for the increase in length.

An erecting eyepiece can also take the configuration of a Huygens eyepiece if desired.

Table 15.9 A 25 mm Telescope Ramsden Eyepiece

Radius of curvature (mm)	Diameter (mm)	Separation or thickness (mm)	Material
Stop	5.0	6.4	Air
Flat	12.0	3.0	BK7
−18.5	12.0	22.0	Air
16.5	20.0	4.0	BK7
Flat	20.0	7.0	Air
Field stop	13.4	—	—

Angular semifield: 15.0°.
Exit pupil diameter (mm): 5.0.
Effective focal length (mm): 25.0.
Eye relief (mm): 6.4.

15.6.3 Kellner Eyepiece

As pointed out before, the Ramsden eyepiece cannot be completely corrected for the magnification chromatic aberration. The system may be achromatized, substituting the single eye lens for a doublet, as shown in Fig. 15.25(c). The system is correctly achromatized when the two principal rays for colors C and F, crossing the exit pupil, are parallel to each other.

The Kellner eyepiece has a greater field of view than that of the Huygens and Ramsden eyepieces, of about $\pm 20°$.

15.6.4 Symmetric or Plössl Eyepiece

A good eyepiece design has been described by Kingslake (1978), using two doublets, as shown in Fig. 15.28(a). The field is as large as $\pm 25°$. In this system we may not only flatten the field, but we may also correct the coma.

A similar eyepiece may be obtained by placing two identical doublets in a symmetrical configuration. This is the symmetrical or Plössl eyepiece, shown in Fig 15.28(b). Two important characteristics of this eyepiece are its long eye relief and its wide field of view of about $\pm 25°$ with a very good image. A design of a 25 mm telescope symmetric eyepiece is presented in the Table 15.10.

15.6.5 Orthoscopic Eyepiece

The orthoscopic eyepiece has a low Petzval sum, a long eye relief, and a very good color correction. The field of view is about $\pm 25°$. The best and most expensive instruments have this kind of eyepiece. A 25 mm telescope

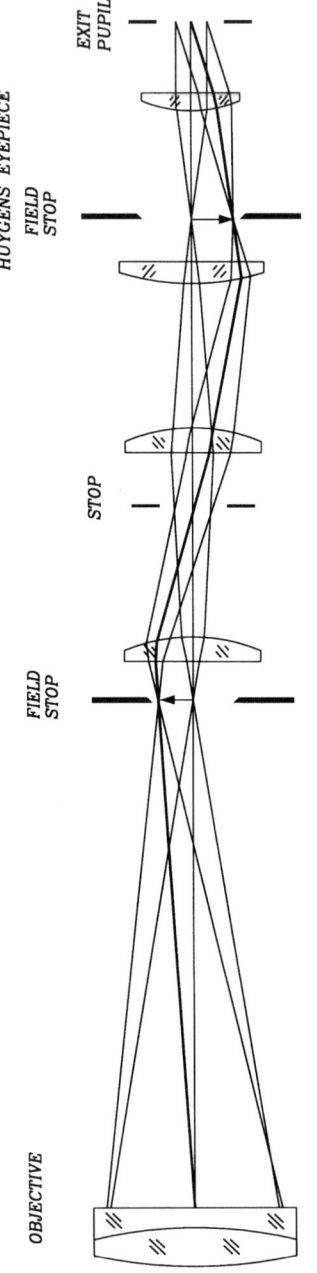

Figure 15.27 Terrestrial telescope with an inverting eyepiece.

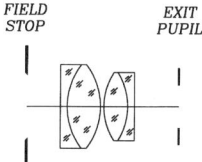

a) KINGSLAKE EYEPIECE

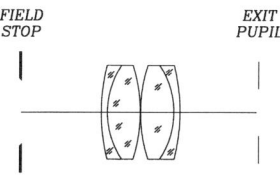

b) SYMMETRICAL EYEPIECE

Figure 15.28 (a) Kingslake and (b) symmetric telescope eyepieces.

Table 15.10 A 25 mm Telescope Symmetric Eyepiece

Radius of curvature (mm)	Diameter (mm)	Separation or thickness (mm)	Material
Stop	5.00	19.00	Air
48.86	24.00	1.5	SF2
22.19	24.00	7.9	BK7
−35.97	24.00	0.1	Air
35.97	24.00	7.9	BK7
−22.19	24.00	1.5	SF2
−48.86	24.00	18.50	Air
Field stop	18.2	—	—

Angular semifield: 20.0°.
Exit pupil diameter (mm): 5.0.
Effective focal length (mm): 25.0.
Eye relief (mm): 19.0.

orthoscopic eyepiece is shown in Fig. 15.29 and its design is presented in Table 15.11.

15.6.6 Erfle Eyepiece

The Erfle eyepiece was designed by H. Erfle in 1921, and has the largest field of all, approaching ±28°, as shown in the design in Fig. 15.30 and Table 15.12.

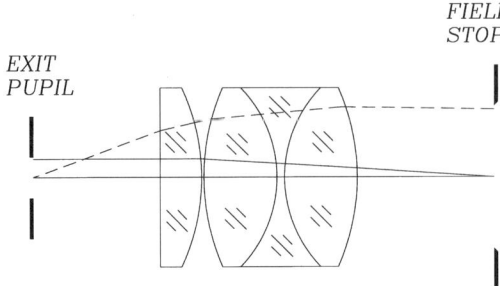

Figure 15.29 A 25 mm telescope orthoscopic eyepiece.

Table 15.11 A 25 mm Telescope Orthoscopic Eyepiece

Radius of curvature (mm)	Diameter (mm)	Separation or thickness (mm)	Material
Stop	5.0	17.80	Air
Flat	24.0	5.70	SK4
27.52	24.0	0.30	Air
29.81	24.0	9.80	BK7
−17.24	24.0	1.00	F4
17.24	24.0	9.80	BK7
−29.81	24.0	12.90	Air
Field stop	18.2	—	—

Angular semifield: 20.0°.
Exit pupil diameter (mm): 5.0.
Effective focal length (mm): 25.0.
Eye relief (mm): 17.8.

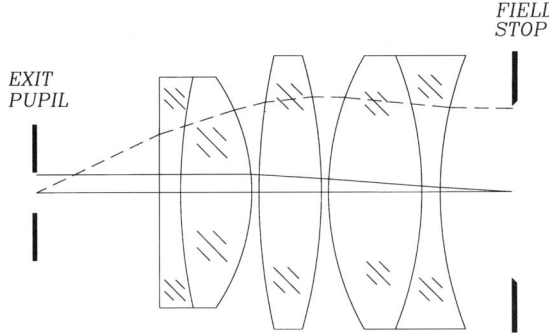

Figure 15.30 A 25 mm telescope Erfle eyepiece.

Table 15.12 A 25 mm Telescope Erfle Eyepiece

Radius of curvature (mm)	Diameter (mm)	Separation or thickness (mm)	Material
Stop	5.0	17.3	Air
Flat	32.0	3.0	F4
73.60	32.0	10.0	BK7
−28.20	32.0	1.0	Air
87.50	38.0	8.7	SK4
−68.50	38.0	1.0	Air
38.80	38.0	13.0	SK4
−51.06	38.0	2.5	SF2
56.68	38.0	10.01	Air
Field stop	11.66	—	—

Angular semifield: 25.0°.
Exit pupil diameter (mm): 5.0.
Effective focal length (mm): 25.0.
Eye relief (mm): 17.3.

15.7 RELAYS AND PERISCOPES

Relays like the lens erector described in Section 15.6.2 are frequently used, not only in terrestrial telescopes, but also in periscope systems or photocopiers. As illustrated in Fig. 15.31, periscopes are formed by relay lenses to transfer the image from one plane to the next with a magnification equal to one, and field lenses located at the image planes, to form the image of the exit pupil of the preceding system, on the entrance pupil of the next system (Hopkins, 1949, 1962b). A good aberration correction may be obtained if each relay system is formed by a pair of doublets in a symmetrical configuration. The periscope can be considered as formed by a series of symmetrical unit angular magnification systems, each of them with a relay lens at each end and a field lens at the center. The relay lenses have to be corrected for longitudinal chromatic aberration, but the field lenses do not need to be achromatic.

A practical disadvantage with the preceding configuration is that the field lens is exactly located at the image plane, and any surface imperfections and dirt on this lens can be clearly seen on top of the image. A common solution is to split the field lens into two lenses, one on each side of the stop, as illustrated in Fig. 15.32. Then, the unit relay system is formed by a relay lens and a field lens. This unit relay system has to be corrected for chromatic aberration as a whole because the field lens has some

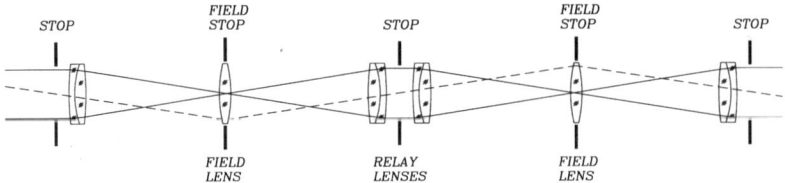

Figure 15.31 Optical relay system.

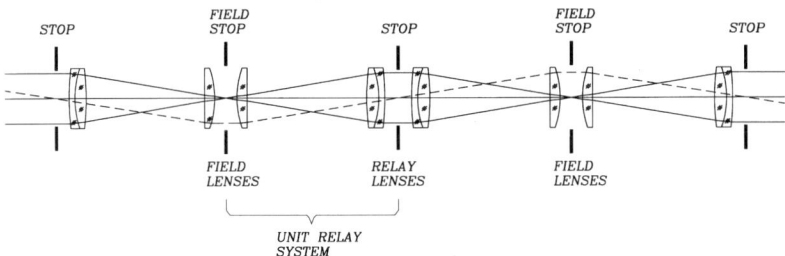

Figure 15.32 Optical relay system with split field lenses.

contribution to this aberration, which has to be compensated with the opposite aberration in the doublet.

The symmetrical unit angular magnification system is formed by two unit relay systems front to front in a symmetrical configuration, each of them with a relay lens at each end and a pair of field lenses at the center.

An important problem when designing relay systems is that all elements have a positive power and hence the Petzval sum is always positive, inward curving the final image surface. A solution is to compensate with the proper amount of astigmatism and to look for a flat tangential surface and/or to reduce the Petzval sum as much as possible by means of thick elements. These optical systems are quite important and useful since they are the basic building components for many instruments, like periscopes and endoscopes, as will be described later in this chapter.

15.7.1 Indirect Ophthalmoscope

Ophthalmoscopes are designed as periscopic afocal systems to observe the retina of the eye. They have the following characteristics:

1. The entrance pupil of the instrument has the same position and a smaller diameter than the eye pupil of the observed patient. This

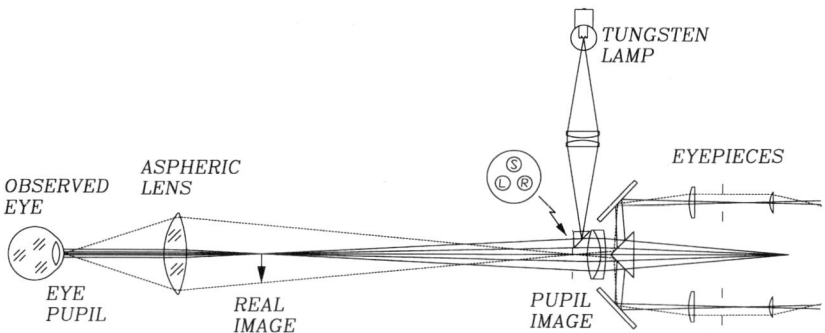

Figure 15.33 Indirect ophthalmoscope.

condition allows room for illuminating the retina of the patient through the edge of its pupil.
2. In order to have a wide observed field of the retina of the observed eye, the angular magnification should be smaller that one (typically equal to 1/3). Hence, the exit pupil of the instrument is larger that its entrance pupil.

The are several versions of the basic system, illustrated in Fig. 15.33. The first lens, located at the front of the instrument, forms the image of the retina at its back focal plane. The diameter of the entrance pupil is much smaller than the lens, so that the on-axis aberrations of this lens are very small. The height of the principal ray on the first lens is large since the field must be wide enough to observe most of the retina of the patient. This lens forms an image of the retina as well as an image of the pupil of the observed eye. To form good images, the front surface of this lens has to be aspheric.

A diaphragm is located at the image of the pupil of the observed eye, which has three small apertures. Two of these windows, on opposite sides of the optical axis, provide a stereoscopic view of the retina by sending the light from each window to a different observing eye. The third window, on top of the other two, is used to illuminate the retina of the observed eye.

An achromatic doublet, located at the image of the pupil of the observed eye just behind the diaphragm, forms an erected image of the retina of the observed eye at the focal planes of the eyepieces. The final image is observed with a pair of Huygens eyepieces.

The angular magnification M of this ophthalmoscope, which has to be smaller than one, is given by

$$M = \frac{\tan \beta}{\tan \alpha} = m \frac{f_a}{f_e} \qquad (15.14)$$

where f_a is the effective focal length of the aspheric lens, f_e is the effective focal length of the eyepiece, and m is the lateral magnification of the achromatic lens located at the image of the pupil of the observed eye.

To be able to use the full field provided by the aspheric lens, the tangent of the angular field semidiameter α_e of the eyepiece should be equal to the tangent of the angular field semidiameter β of the aspheric lens, multiplied by the angular magnification M, as follows:

$$\tan \beta = M \tan \alpha_e \tag{15.15}$$

15.7.2 Endoscopes

Endoscopes are also afocal periscopic systems designed to observe the interior of the human body. They take different names, depending on which part of the body they are used, e.g., gastroscope for the digestive system, laparoscope for the abdominal cavity, etc. These instruments are characterized for their extreme length in comparison with their diameter. The complete endoscope system must have an optical path for the transmission of the image but it must also have a parallel path, most of the times with an optical fiber, to illuminate the object under study. Basically there are three different methods to design these optical systems as will now be described.

Traditional endoscope—With a periscope formed by a series of basic unit angular magnification relay systems using relay and field lenses as described before.

Hopkins endoscope—Another popular design is one due to H. H. Hopkins (1966, 1976). This system is a modification of the basic system, where the space in the unit relay system between the relay doublet and the field lens is filled with a glass rod as shown in Fig. 15.34. These lenses and rod are cemented together into a single block. An important advantage may be obtained with this modification. If the rod length is made equal to the

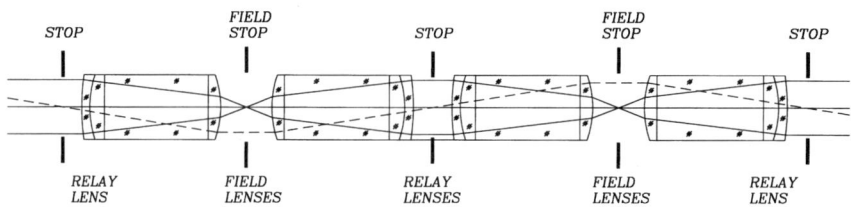

Figure 15.34 H. H. Hopkins endoscope system.

original field lens to relay lens distance, the light throughput and hence the image brightness is increased by a factor equal to the square of the refractive index of the rod. As in the standard system, each unit system has to be corrected for axial chromatic aberration. An additional advantage is that this system is easier to assemble.

Gradient index endoscopes—Endoscopes can also be constructed with a series of long radial gradient index rods. The radial index gradient in these rods is symmetric about the optical axis and is represented by

$$n(r) = N_{00} + N_{10}\, r^2 + N_{20}\, r^4 + \cdots \qquad (15.16)$$

where r is the radial distance from the optical axis. In a rod lens a light ray entering the rod follows a curved sinusoidal path as illustrated in Fig. 15.35. The wavelength L of this wavy sinusoidal is given by

$$L = 2\pi \left(-\frac{N_{00}}{2\, N_{20}} \right)^{1/2} \qquad (15.17)$$

If the rod has a length L, an object located at the front surface is imaged free of spherical aberration with unit magnification on the rear surface without any spherical aberration. The image is sharp and with good contrast. These properties make these rods ideal for endoscopic relays (Tomkinson et al., 1996).

Fiber optics endoscope—With a coherent bundle of optical fibers that transmit the image from one end of the fiber to the other. Their great advantage is flexibility. Their disadvantages are a relatively low resolution and also a low light efficiency.

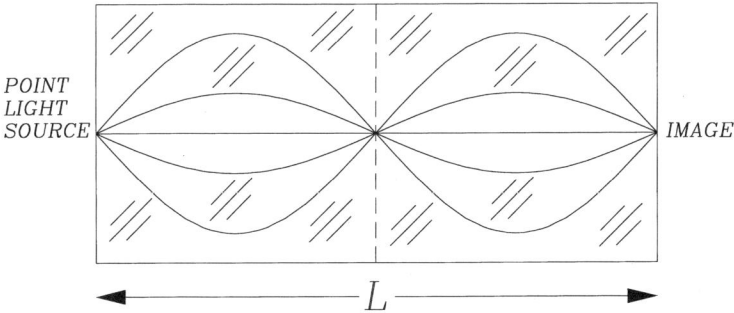

Figure 15.35 Radial gradient rod lens.

Electronic endoscope—A similar system transfers the image electronically with a microscopic television camera and electrical wires to the eyepiece where a tiny monitor is located.

All of these instruments have a light source to illuminate the observed object through an optical fiber that runs along the optical system.

REFERENCES

Bertele, L., "Galilean Type Telescope System," U.S. Patent No. 4,390,249 (1983).
Fulcher, G. S., "Telescope Objective Without Spherical Aberration for Large Apertures Consisting of Four Crown Glasses," *J. Opt. Soc. Am.*, **37**, 47 (1947).
Hopkins, H. H., U.S. Patent 3,857,902, "Optical System Having Cylindrical Rod-Like Sense", (1946).
Hopkins, H. H., "Optical Principles of the Endoscope," in *Endoscopy*, G. Berci, ed., Appleton Century Crofts, New York, 1976.
Hopkins, R. E., "Secondary Color in Optical Relay Systems," *J. Opt. Soc. Am.*, **39**, 919–921 (1949).
Hopkins, R. E., "Telescope Doublets," *J. Opt. Soc. Am.*, **49**, 200–201 (1959).
Hopkins, R. E., "Eyepieces," in *Military Standardization Handbook*: *Optical Design*, *MIL-HDBK 141*, U.S. Defense Supply Agency, Washington. DC, 1962.
Hopkins, R. E., "Telescope Objectives," in *Military Standardization Handbook*: *Optical Design*, *MIL-HDBK 141*, U.S. Defense Supply Agency, Washington, DC, 1962a.
Hopkins, R. E., "Lens Relay Systems," in *Military Standardization Handbook*: *Optical Design*, *MIL-HDBK 141*, U.S. Defense Supply Agency, Washington, DC, 1962b.
Hopkins, R. E., "Eyepieces," in *Military Standardization Handbook*: *Optical Design*, *MIL-HDBK 141*, U.S. Defense Supply Agency, Washington, DC, 1962c.
Hopkins, R. E., "Complete Telescope," in *Military Standardization Handbook*: *Optical Design*, *MIL-HDBK 141*, U.S. Defense Supply Agency, Washington, DC, 1962d.
Hopkins, R. E. and Lauroesh, J., "Automatic Design of Telescope Doublets," *J. Opt. Soc. Am.*, **45**, 992–994 (1955).
Kingslake, R., *Lens Design Fundamentals*, Academic Press, New York, 1978.
Korones, H. D. and Hopkins, R. E., "Some Effects of Glass Choice in Telescope Doublets," *J. Opt. Soc. Am.*, **49**, 869–871 (1959).
Malacara, D., "Two Lenses to Collimate Red Laser Light," *Appl. Opt.*, **4**, 1652–1654 (1965).
Malacara, D. and Malacara, Z., "An Interesting Property of Inverted Galilean Telescopes and Their Relation to Myopic Eyes," *Opt. Eng.*, **30**, 285–287 (1991).
Menchaca, C. and Malacara, D., "Design of GalileanType Telescope Systems," *Appl. Opt.*, **27**, 3715–3718 (1988).
Puryayev, D. T., "Afocal Two-Mirror System," *Opt. Eng.*, **32**, 1325–1327 (1993).
Rosin, S., "A New Thin Lens Form," *J. Opt. Soc. Am.*, **42**, 451–455 (1952).

Rusinov, M. M., Judova, G. N., Kudryashov J. V., and Aguror, P. Y., "Galilean Type Telescope System," U. S. Patent No. 4,390,249 (1983).

Tomkinson T. H., Bentley, J. L., Crawford, M. K., Harkrider, C. J., Moore D. T., and Ronke, J. L., "Rigid Endoscopic Relay Systems: A Comparative Study," *Appl. Opt.*, **35**, 6674–6683 (1996).

Walker, B. H., *Optical Design for Visual Systems*, SPIE Press, Bellingham, WA, 2000.

Wetherell, W. B., "Afocal Lenses," in Applied Optics and Optical Engineering, *Vol. X*, Academic Press, San Diego, CA, 1987.

16
Microscopes

16.1 COMPOUND MICROSCOPE

We have seen in Chap. 11 that the magnifying power of a single lens may be increased by decreasing the focal length, but there is a practical limit to this procedure. Then, the lens becomes too small and during observation, object, lens, and eye must be brought very close together. A practical limit is a magnifying power of about 100×, with a focal length of about 2 mm. Anthony Van Leeuwenhoek in the 17th century in Holland made several microscopes using a minute single lens with very strong curvatures, obtaining magnifying powers of the order of 100. Two problems with this simple design are the low numerical aperture (0.2 NA) and the small field. With these simple microscopes, he discovered the first microorganisms.

A compound microscope, as illustrated in Fig. 16.1, is a possible solution. The lens closer to the object is known as the objective and the lens on the eye side is the eyepiece. There are two equivalent methods for the interpretation of this optical system, as we will now see.

First Method

We may think of this system as one formed by two thin lenses, the objective, with focal length f_o and the eyepiece with focal length f_e. Then, if the separation between the two lenses is $l_2 + f_e$, the effective focal length is, from Eq. (3.42),

$$F = \frac{f_o f_e}{f_o + f_e - (l_2 + f_e)} = \frac{f_o f_e}{f_o - l_2} \tag{16.1}$$

Thus, from Eq. (11.2), the magnifying power of the system is

$$M = \frac{250(f_o - l_2)}{f_o f_e} \tag{16.2}$$

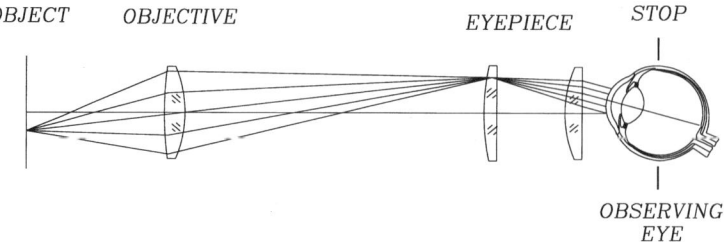

Figure 16.1 Basic microscope arrangement.

Second Method

An alternative but equivalent way of obtaining the magnifying power of the microscope is as follows. The objective in the system shown in Fig. 16.1 forms a real image of the object. The magnification m_o of the objective, from Eq. (3.9), is given by

$$m_o = 1 - \frac{l_2}{f_o} \tag{16.3}$$

If we write this expression in terms of the distance X' from the back focus of the objective to the image, an alternative expression for the magnification m_o of the objective may be found from Newton's equation (3.10) and Eq. (3.9) as follows:

$$m_o = -\frac{nX'}{f_o} \tag{16.4}$$

where n is the refractive index in the object media and f_o is the effective focal length of the objective on the object side. If the distance X' is defined as the optical tube length, and made a constant for all objectives, the objective magnification would be a function only of its focal length. This is why this distance has been standardized by most manufacturers. The two most common values are either 160 or 170 mm.

The magnifying power M_e of the eyepiece is given by Eq. (11.2). Thus, since the microscope magnifying power is the product of the magnification m_o of the objective multiplied by the magnifying power M_e of the eyepiece, using Eq. (16.3) we may find that

$$M = m_o M_e = \left(1 - \frac{l_2}{f_o}\right)\frac{250}{f_e} \tag{16.5}$$

Microscopes

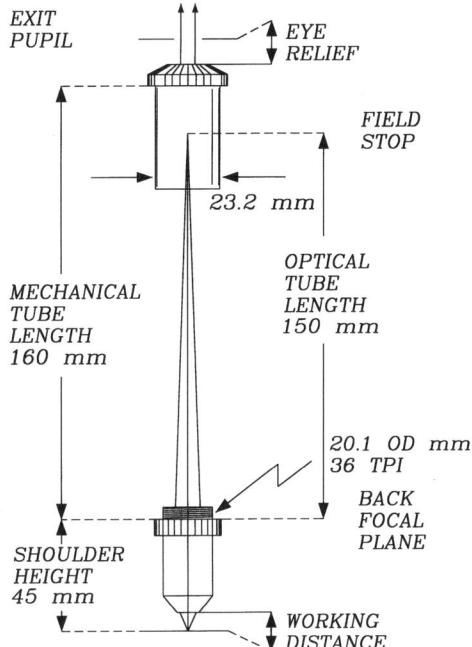

Figure 16.2 Some standard physical dimensions in a microscope for the DIN standard.

which may be shown to be equal to Eq. (16.2). However, using Eq. (16.3) and a value of $X' = 160$ mm, we have

$$M = m_o M_e = \frac{250 n X'}{f_o f_e} = \frac{40,000 n}{f_o f_e} \tag{16.6}$$

Mechanically, the microscope is assembled as in Fig. 16.2, where the following terms are defined:

Optical Tube Length. As previously explained, this is the distance from the back focal plane to the focus. This length has been standardized to a value of 160 mm by the Deutsche Industrie Normen (DIN) and to a value of 170 mm by the Japanese Industry Standard (JIS).

Shoulder Height. This is the height of the shoulder of the objective with respect to the object plane. In order to be able to interchange microscope objectives without a large refocusing, this distance has been standardized to a value of 45 mm by the DIN and to a value of 36 mm by the JIS. Since the objective lens design may have many configurations, this standard value automatically places the back focal plane at many different possible positions with respect to the shoulder.

If a series of objectives have the same shoulder height and the same shoulder-to-image distance, the objectives may be interchanged (e.g., by means of a rotating turret) without any refocusing. These objectives are said to be *parfocal*.

Mechanical Tube Length. This is the distance from the end of the thread in the objective mount (shoulder) to the end of the eyepiece. This distance has not been standardized, due to the variations in the position of the objective's back focal plane with respect to the shoulder. As a consequence, the optical tube length is not preserved when changing objectives and eyepieces. Thus, the mechanical tube length has to be adjusted for optimum performance. This mechanical tube length has to be carefully adjusted, to use the objective's proper conjugate positions, for three reasons: (1) to obtain the prescribed magnification, (2) to fine tune the minimum spherical aberration, and (3) to keep the image in focus when parfocal objectives are used and the magnification is changed.

Working Distance. This is the distance from the upper surface of the object cover glass to the vertex of the lowest optical surface of the microscope objective. The distance X from the object to the focus on the object side may be found from Eqs. (3.11) and (3.14) as

$$X = -\frac{f_o^2}{nX'} \qquad (16.7)$$

where n is the refractive index in the object medium (air or oil). Hence, the working distance S is

$$S = X + f_{oF} - n_c T = -\frac{f_o^2}{nX'} + f_{oF} - n_c T \qquad (16.8)$$

where f_{oF} is the front focal length (object side) of the microscope objective, T is the thickness of the object cover glass, and n_c is its refractive index.

16.1.1 Microscope Aperture and Resolving Power

The relative aperture of telescope objectives was defined as the focal ratio or *f*-number *FN*. In the case of microscope objectives this

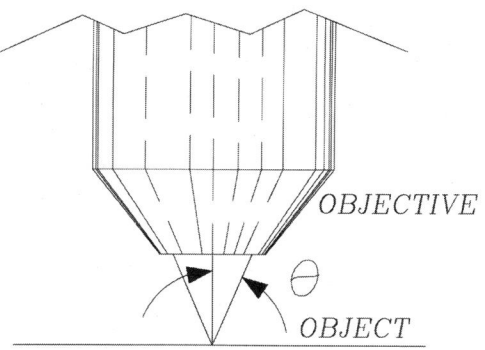

Figure 16.3 Numerical aperture of a microscope objective.

aperture is defined by the numerical aperture, illustrated in Fig. 16.3, as follows:

$$NA = n \sin \theta \qquad (16.9)$$

where n is the refractive index in the object media and θ is the angular semidiameter of the entrance pupil as observed from the object. The resolving power of a diffraction-limited microscope objective is determined by the diameter of the diffraction image of a point source. As will be described in more detail in Chap. 17, in the microscope, as well as in projectors, the object in general is illuminated with partially coherent light. The minimum separation of two-point images is a function of the degree of coherence, given by

$$d = \kappa \frac{\lambda_0}{NA} \qquad (16.10)$$

where λ_0 is the wavelength in vacuum and the constant κ has a value equal to 0.61 for fully incoherent illumination and a value 0.77 for totally coherent illumination (Born and Wolf, 1964). In the case of the microscope the illumination is partially coherent. Abbe's theory of image formation is developed, assuming a totally coherent illumination.

Observing these two images in a microscope with magnifying power M, the apparent separation of the virtual images at a distance of 250 mm from the observer's eye is

$$d = \kappa \frac{M \lambda_0}{NA} \qquad (16.11)$$

On the other hand, the resolving power of the eye is nearly one arcmin. At a distance of 25 mm this gives a separation of about 0.07 mm. If we take a value of κ equal to 0.7 we may easily conclude that the magnifying power needed to just match the resolving powers of the eye and that of the objective is

$$M = 182\, NA \qquad (16.12)$$

However, this is not a practical limit for the highest magnifying power. A more realistic limit is about five times this value, hence, we may write

$$M_{\max} = 1000\, NA \qquad (16.13)$$

any higher magnifying power will not provide any more detail and the image will look worse. This is what is called *empty magnification*.

It is interesting to notice that using Lagrange's theorem the objective magnification may be written as

$$m_o = \frac{NA}{NA_i} \qquad (16.14)$$

where NA and NA_i are numerical apertures in the object and image spaces, respectively.

Most microscope objectives have the dimensional characteristics illustrated in Fig. 16.4. A series of objectives is frequently designed so that the exit pupil has a nearly constant value of about 6.4 mm. This gives a nearly constant value of the numerical aperture in the image space, equal to $NA_i = 0.02$ (except for the high-power objectives, where this diameter may be

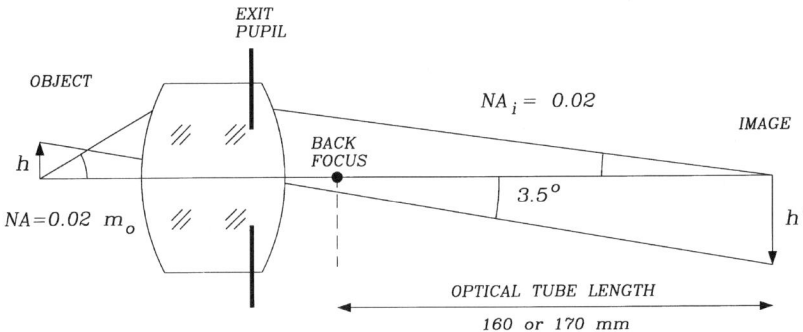

Figure 16.4 Optical schematics for a microscope objective.

Microscopes

as low as 0.0125). Then, from Eq. (16.14), we see that the numerical aperture in the object space is approximately equal to $NA = 0.02\, m_o$. The numerical aperture in the object space obviously determines the size of the exit pupil of the whole microscope. If d is the diameter of this pupil, using Eq. (11.2), we may show that

$$d = \frac{500\, NA_i}{M_e} \qquad (16.15)$$

hence, using Eq. 16.6 we find that

$$M = \frac{500\, NA}{d} \qquad (16.16)$$

Since the average diameter of the exit pupil with bright illumination is about 2 mm, making the diameter of the exit pupil of the microscope equal to the diameter of the eye, we may find an alternative expression to Eq. (16.12) for the maximum microscope magnifying power, given by

$$M = 250 NA \qquad (16.17)$$

Any magnifying power higher than this value will produce an exit pupil smaller than the eye's pupil.

16.2 MICROSCOPE OBJECTIVES

Reviews of microscope optics, in particular objectives and eyepieces has been given by Bennett (1943, 1962, 1963), Cruickshank (1946), Foster and Thiel (1948), Foster (1950), Benford (1965), Benford and Rosenberger (1967, 1978), Laikin (1990), Broome (1992), and Smith and Genesee Optics Software (1992).

The optical tube length, as we pointed out before, is standardized to a value of 160 or 170 mm. This imposes a limit to the mechanical tube length. However, sometimes it is necessary to increase substantially the mechanical tube length to be able to bend the tube for a more comfortable observation, or in order to introduce a prism system for binocular vision. In this case a relay lens system must be used. This relay system may take the form of a zoom lens, like the ones in Chap. 12 as described by Benford (1964). A microscope with a zoom relay lens and a binocular prism system is illustrated in Fig. 16.5.

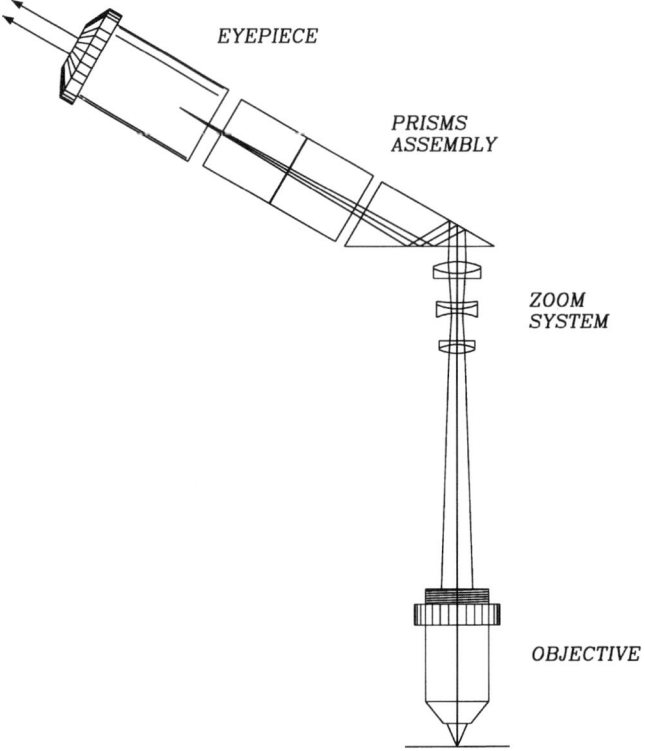

Figure 16.5 Microscope with a zoom assembly and deflecting binocular prisms.

In order to obtain diffraction-limited resolution it is very important to obtain a high level of correction of the chromatic aberration in microscope objectives. They can be classified in several different categories according to the degree of chromatic correction and field curvature, as described by Broome (1992).

Achromats are microscope objectives with a limited spectral correction and moderate field of view. These are the most popular objectives, with many design variations.

Semiapochromats are designed to achieve nearly apochromatic correction by reducing the secondary spectrum, where calcium fluoride or *fluorite* was introduced into microscope objectives by Ernst Abbe. Fluorite has a low refractive index equal to 1.43 and an unusual dispersion, permitting a large reduction in the chromatic aberration. Figure 16.6 shows the longitudinal chromatic aberration for a normal achromatic objective and for one made with fluorite.

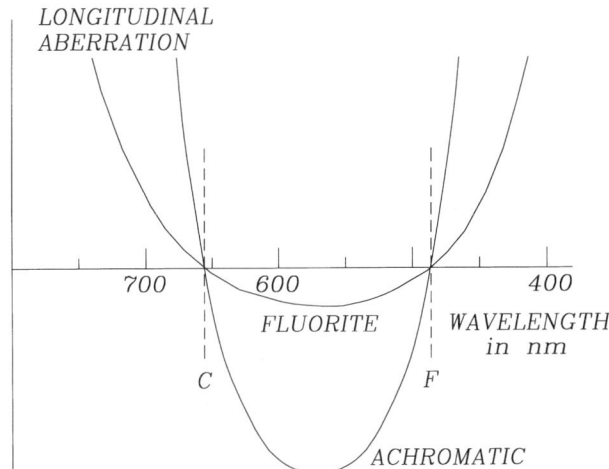

Figure 16.6 Longitudinal achromatic aberration in conventional achromatic and fluorite objectives.

Apochromats or true apochromatic microscope objectives have also been designed using calcium fluoride positive elements cemented to two negative lenses from the LAK glass series instead of the doublet used in the design of achromats.

Plan-achromats: the main characteristic of these objectives is their flat field achieved with a low Petzval curvature and a small astigmatism. Their chromatic correction is that of an acromat.

Plan-Apochromats: these objectives, like the plan-achromats, have a flat field with an excellent correction of Petzval curvature and astigmatism. Their chromatic correction is that of an apochromat.

In microscope objectives with large magnifications the magnification chromatic aberration is difficult to correct, although the remaining aberration is not very large. However, this aberration may be noticeable if all other aberrations are well corrected, especially in apochromatic objectives. In some high-quality microscopes called *compensated microscopes*, the residual magnification chromatic aberration of the objective is corrected in the eyepiece. The problem then is that objectives and eyepieces are not interchangeable. To solve this difficulty, a residual magnification chromatic aberration is intentionally introduced in low-power objectives, so that all objectives could be used with all eyepieces. However, as pointed out by Hopkins (1988), this compensation necessarily introduces some axial chromatic aberration and the only good solution is to correct fully the chromatic aberration in both the objective and the eyepiece.

When designing a microscope objective it should be pointed out that it is easier to design and evaluate it if the light rays enter the lens in an opposite direction to the actual one. Thus, when using ray-tracing programs it is convenient to interchange the object and image positions.

There are many types of microscope objectives, whose design depends on the desired magnification. In the next sections some of the most common types will be described. Table 16.1 lists some of the main characteristics of microscope objectives with the DIN standard.

Figure 16.7 shows some of the most common objective achromatic microscope objectives and Fig. 16.8 shows some apochromatic objectives,

Table 16.1 Some Microscope Objectives (DIN)

Type	Power	Focal length (mm)	NA	Field (mm)	Working distance (mm)
Doublet	4.0 ×	30.60	0.08	4.50	15.80
Low power	3.5 ×	30.00	0.07	5.20	25.40
Lister	10 ×	16.60	0.20	1.80	6.30
Lister	20 ×	8.78	0.40	0.90	1.50
Amici	40 ×	4.50	0.70	0.45	0.45
Oil immersion	100 ×	1.86	1.25	0.18	0.13

NA = numerical aperture.

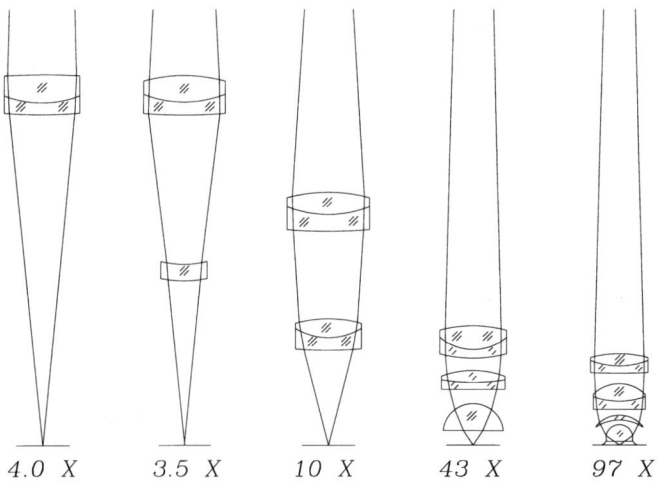

Figure 16.7 Some microscope objectives. (From Benford, 1965.)

Microscopes

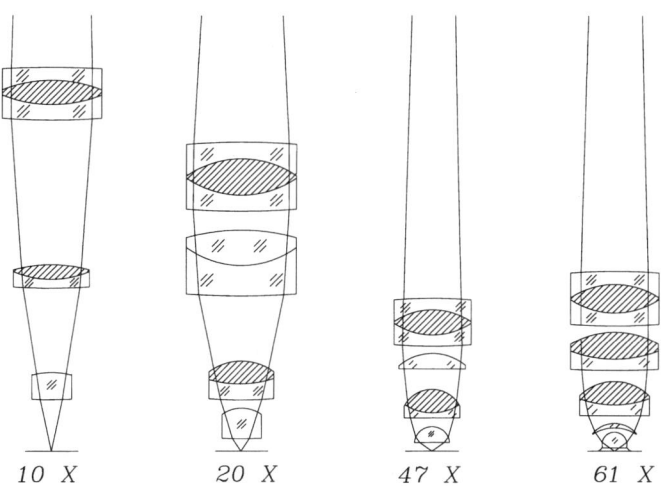

Figure 16.8 Some fluorite microscope objectives. (From Benford, 1965.)

made with fluorite. These figures are adapted from a publication by Benford (1965). The field of a microscope objective is very small compared with that of a photographic lens. This field is approximately constant for all objectives, with a diameter of about 7°.

16.2.1 Low-Power Objectives

The simplest objective is a doublet. The procedure for the design of this lens is identical to the one used in Chap. 14 for telescope objectives, with the only difference that high-index glasses are frequently used to obtain a better spherical aberration and axial chromatic aberration correction. Sometimes low-power objectives have an extra single thick meniscus element to reduce the Petzval curvature.

16.2.2 Lister Objectives

Medium power or Lister microscope objectives are formed by a pair of widely separated doublets, originally designed as separable, so that they could be used alone or as a system. In the separable system both elements are free of spherical aberration and coma, so that no astigmatism or field curvature correction is possible.

The principle used in the design of these objectives was first used by Lister in 1830 when he discovered that a plano convex lens has two pairs of object and image positions for which these conjugate points are aplanatic. One pair occurs when both the object and the image are real. In the other pair the object is real and the image is virtual. Thus, Lister placed the first doublet (the lens closer to the object) with the real object at the aplanatic location and its image at the virtual aplanatic point. The second doublet is then placed with its real aplanatic object at the virtual image of the first doublet. The final image is real.

A better correction is obtained if, as in the Petzval lens described in Chap. 12, the system is not to be separated. Then, the large separation between the two elements, which is equal to about the effective focal length of the objective, allows the designer to correct partially the field curvature with the proper amount of astigmatism in the cemented interface. It should be noticed, however, as pointed out by Hopkins (1988), that by reducing the field curvature in this manner increases the secondary color. This is the main reason for the use of fluorite in these lenses.

In order to find the lens bendings that corrects both the spherical aberration and the coma in the system, Conrady (1960) and Kingslake (1978) describe a graphical method called *the matching principle*. They trace rays through both components, individually, the first component from left to right and the second component from right to left. Then, the system is assembled and the rays from both components in the space between them should match each other. This method was especially useful when computers for tracing rays were not widely available. Four different bending combinations may be found to produce a system with spherical aberration and coma corrected. One of these solutions corresponds to the two components individually corrected for both aberrations, thus producing a separable system. Unfortunately, this is the form with the astigmatism producing the strongest field curvature. The solution with the largest curvature on the surface closest to the image gives a system with the less strong curvatures on average. This is the best of the four solutions.

In a typical design the two elements have about the same power contribution, so that the refractive work is equally shared by the two components. Normally, the separation between the two lenses is equal to the focal length of the second element (this is the lens closer to the image). The stop is assumed to be at the plane of the second element, so that the objective is telecentric in the object space, as described by Broome (1992).

16.2.3 Amici Objectives

The Amici objective, first designed in 1850, is obtained from the Lister objective by adding an almost aplanatic hyperhemispherical lens on the object side. It might be thought at first that the hyperspherical surface should be made perfectly aplanatic. However, the front flat surface introduces some spherical aberration, since it is not in contact with the object. Thus, the radius of curvature of the spherical surface is made slightly larger than the aplanatic solution in order to compensate the aberrations introduced by the first surface. This solution not only introduces a small amount of spherical aberration that compensates that introduced by the plane surface, but also a small astigmatism that tends to flatten the field. An important problem with the Lister objective is its short working distance.

As in the Lister objective, the stop is placed at the position that produces a telecentric objective in the object space.

16.2.4 Oil Immersion Objectives

An oil immersion objective is often a Lister objective with an *aplanatic front* system as shown in Fig. 16.9. The flat object is protected with a thin *cover glass* with an approximate thickness of 0.18 mm. Then, a thin layer of oil is placed between the cover glass and the front flat face in the objective. The refractive index of immersion oil is 1.515, a value that nearly matches the refractive index of the cover glass and the first lens.

If the radius of curvature of the hyperhemispherical surface is r, the distance L from the object to this surface, according to the Abbe aplanatic condition described in Chap. 4 is

$$L = \left(1 + \frac{1}{n}\right)r \tag{16.18}$$

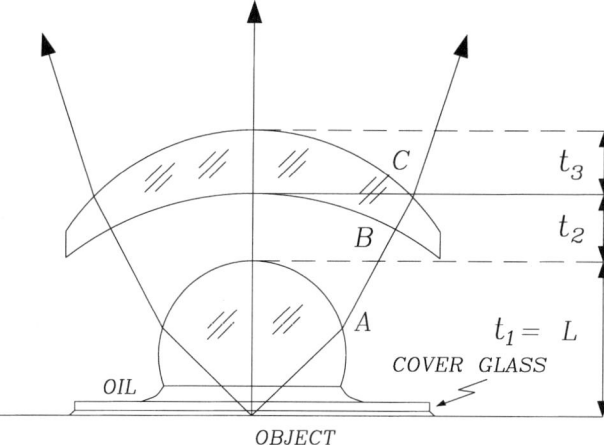

Figure 16.9 Duplex front of an oil immersion microscope objective.

and the image would then be at a distance L' from the surface, given by

$$L' = (1 + n)r \tag{16.19}$$

Then, the radius of curvature of the front face of the meniscus lens should be equal to $t_2 + L'$, so that the object for this surface is at its center of curvature. The second radius of curvature of the meniscus lens is also aplanatic, satisfying the Abbe condition in Eq. (16.18), where L' is the distance from this surface to the object position. There is some chromatic aberration and field curvature introduced by the aplanatic front, but they are compensated in the rest of the system.

The effect of each Abbe aplanatic surface is to reduce the numerical aperture of the cone of rays by a factor equal to the refractive index n of the glass. Since the aplanatic front system contains two Abbe surfaces, the numerical aperture of the cone of rays is reduced by a factor n^2.

16.2.5 Other Types of Objectives

Many other objectives of microscope have been designed. One improvement that can be made is to reduce the field curvature and the astigmatism. This is not an easy problem and very complicated designs may result, as described, e.g., by Claussen (1964) and Hopkins (1988). The basic principle used in these designs is the separation of elements with positive and negative power contributions, so that the Petzval sum could be made small. These systems resemble inverted telephoto lenses, with the negative elements closer to the image and the positive elements closer to the object.

A recent innovative change (Muchel, 1990) has been made by designing the objective with the image at infinity, so that the output beam is collimated. Then, another lens at the end of the tube is used to form the image at the eyepiece. This approach has two main advantages: (1) the mechanical tube length can have any magnitude and (2) the magnification chromatic aberration is fully corrected in the objective even for high magnifications.

16.2.6 Reflecting Objectives

Around the year 1904 Karl Schwarzschild discovered that the two-mirror system in Fig. 16.10(a) is free of spherical aberration, coma, and astigmatism. The system is formed by two concentric spherical mirrors

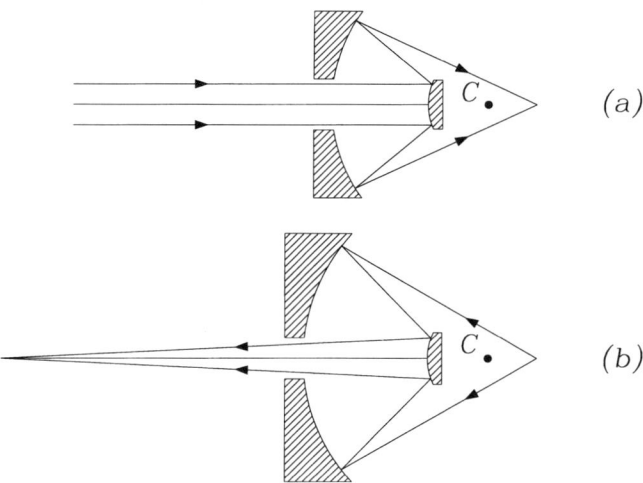

Figure 16.10 Reflecting microscope objective.

and a collimated beam of light (point light source at infinity) enters the system on the convex mirror. The radii of curvature of the two mirrors must be in the ratio:

$$\frac{r_2}{r_1} = \frac{\sqrt{5}+1}{\sqrt{5}-1} = 2.618034 \qquad (16.20)$$

as shown by Erdös (1959).

This system may be easily modified for use as a microscope objective (Grey and Lee, 1949a,b; Grey, 1950, 1952; Norris, et al., 1951; Thornburg, 1955) by moving the object closer to the system, as shown in Fig. 16.10(b). This introduces some aberrations, but we may still correct them by modifying the ratio between the two radii of curvature, preserving the concentricity. In an example given by Kingslake (1978), for a 10× objective the new ratio is equal to 3.07787.

This type of design is limited to a maximum magnification of about 30× with a numerical aperture $NA = 0.5$. For higher magnifications quartz refractive components have to be added to the system. These objectives may be improved with the use of aspherical surfaces (Miyata et al., 1952).

These microscope objectives are useful with ultraviolet or infrared light, where normal objectives cannot be used due to the absorption of such radiation by the glass.

16.2.7 Compact Disk Objectives

Compact disk objectives are very similar to microscope objectives, with the important difference that they are used with monochromatic laser light. An excellent review on this subject by Broome (1992) is recommended, who points out that there are five basic configurations for these objectives, namely: a double Gauss derivative, a Petzval derivative, a triplet derivative, a doublet, and a single lens. These lenses may be designed for collimated light, if a collimator in front of the laser is used. Otherwise, a $5\times$ to $20\times$ magnification is used. Many popular designs use aspheric lenses or gradient index glass to reduce the number of lenses.

16.3 MICROSCOPE EYEPIECES

A microscope with an eyepiece without a field lens has the disadvantage that the exit pupil of the microscope is far from the eyepiece, making the observation uncomfortable. A field lens may be introduced, as shown in Fig. 16.1, at the plane where the intermediate real image is formed. If the field lens is very thin and exactly at the real image plane, no aberrations are introduced (with the exception of Petzval curvature). Only the exit pupil is moved closer to the eyepiece. Another more important reason for introducing the field lens is the correction of some aberrations. The system formed by the eyepiece and the field lens is simply called an eyepiece. The effective focal length of the eyepiece is then obtained from Eq. (4.41). If the field lens is exactly at the plane of the observed image, we may easily see that the effective focal length is equal to the focal length of the lens closer to the eye, without any influence from the field lens. However, in general, the field lens is not at the plane of the observed image.

The exit pupil of the microscope is located close to the eye lens and the observer's eye pupil must be placed there. If this is not so, the whole field will not be observed. The distance from the eye lens to the exit pupil of the eyepiece is called the *eye relief*. A large eye relief may be important if the observer wears spectacles.

It is important to remember when designing an eyepiece that the aberrations do not need to be corrected better than a normal eye can detect them. This means that the angular aberrations do not need to be smaller than about 1 arcmin.

Table 16.2 lists the main characteristics of some microscope eyepieces, illustrated in Fig. 16.11. The design principles of these eyepieces will be described in the next sections.

Microscopes 431

Table 16.2 Some Commercial Microscope Eyepieces (DIN)

Type	Power	Focal length (mm)	Field (degrees)	Eye relief (mm)
Huygens	5×	50.00	19.00	14.00
Huygens	10×	25.00	13.00	8.50
Huygens	15×	16.70	8.00	7.00
Wide field	10×	25.00	18.00	15.50
Wide field	15×	16.70	13.00	12.60
Wide field	20×	12.50	10.00	9.80

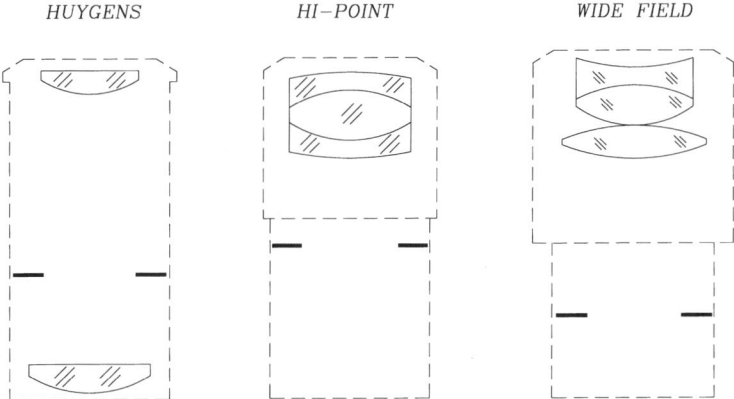

Figure 16.11 Some microscope eyepieces.

16.3.1 Huygens Eyepieces

Huygens eyepieces (Stempel, 1943) are the simplest and more common ones for telescopes as well as for microscopes. The design of Huygens eyepieces for telescopes has been described in Chap. 15. The main difference for microscope eyepieces is that the entrance pupil for the eyepiece is closer to the eyepiece than in the case of telescopes. The magnification chromatic aberration is not exactly corrected by a separation of the two lenses equal to the average of the focal lengths. In this case the separation must be selected so that the incident white principal ray is split into two parallel colored rays after exiting the eyepiece, as shown in Fig. 16.12. This means that the condition for the correction of the magnification chromatic aberration is

$$\bar{u}'_{kC} = \bar{u}'_{kF} \qquad (16.21)$$

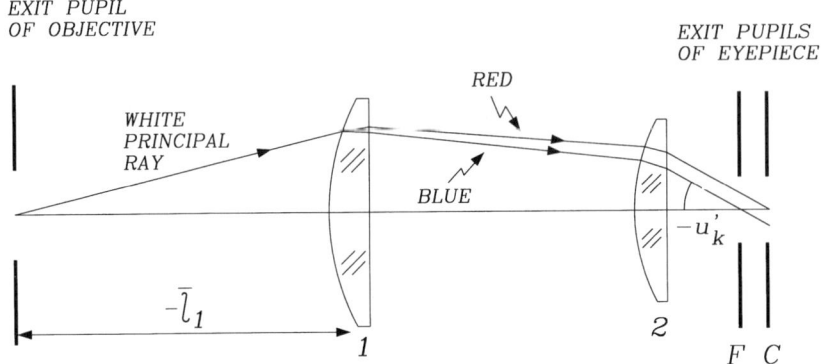

Figure 16.12 Design of Huygens eyepieces.

Another way of expressing this condition is by means of a generalization of Eq. (6.69), for the case of a near entrance pupil, as follows (Conrady, 1957):

$$d = \frac{f_1 + f_2}{2 + (f_1/\bar{l}_1)} \tag{16.22}$$

where l_1 is the distance from the stop to the first lens, as in Fig. 16.12. If the eyepiece power is large, this expression approaches Eq. 6.69. If the stop is close to the eyepiece, the distance d between the two lenses becomes larger, since l_1 is negative.

Having fixed the shape of the lenses to the plano convex form, as pointed out in Chap. 15, the only available degrees of freedom are the two powers and the lens separation to obtain the desired effective focal length, to correct the coma and the magnification chromatic aberration, and to minimize the field curvature. As a compromise, if necessary, it is better to correct the coma as well as possible, even if some residual chromatic aberration remains.

Eyepieces are typically designed with the light entering the system in reverse so that the object is at an infinite distance. Then, the curvature of the last surface can be set so that the system has the required effective focal length. Since the eye relief is another important parameter to take care of, which should be as large as possible, the conclusion is that we have the convex curvature of the first lens and the separation between the lenses in order to correct the coma and to obtain a good eye relief. If the coma aberration is corrected, solutions may exist for a range of values of the radius of curvature of the eye lens. As this radius of curvature increases, the focal length of the eye lens gets longer until it finally reaches the field lens (for a 10 × eyepiece this occurs for a radius of curvature close to a value of about −12 mm). These solutions are plotted in Fig. 16.13 for a 10 × eyepiece. Some designers prefer to sacrifice a good correction for coma to obtain a reasonably large eye relief.

To obtain a better correction we can make the eye lens with crown glass and the field lens with flint glass. It is found that a good ratio of the focal lengths of the two lenses is about 2.3 for high magnifying powers in telescopes, but may be as low as 1.4 for low microscope magnifying powers. For example, a typical microscope eyepiece has a ratio f_1/f_2 equal to 1.5 for a 5 × power and equal to 2.0 for a 10 × power.

A 10 × Huygens eyepiece for microscopes is shown in Fig. 16.14 and its design is presented in Table 16.3.

16.3.2 Wide-Field Eyepieces

Wide-field eyepieces are designed to provide a large field of view. Very good correction of most aberrations is obtained as well as a large eye relief.

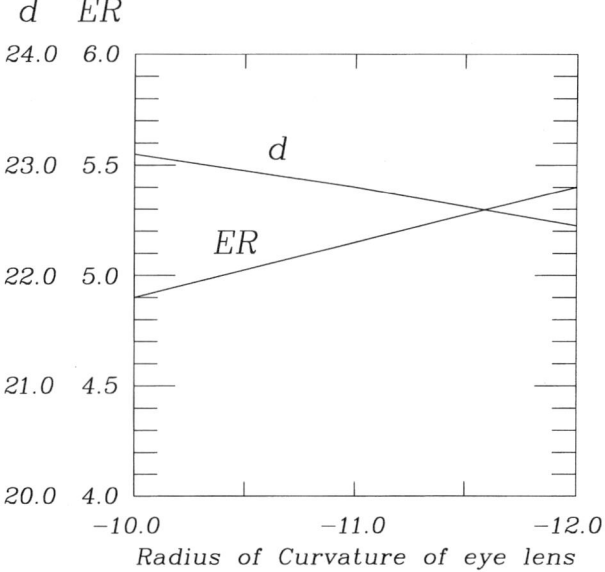

Figure 16.13 Values of the lens separation d and the eye relief ER versus the convex radius of curvature of the eye lens for a Huygens eyepiece corrected for coma.

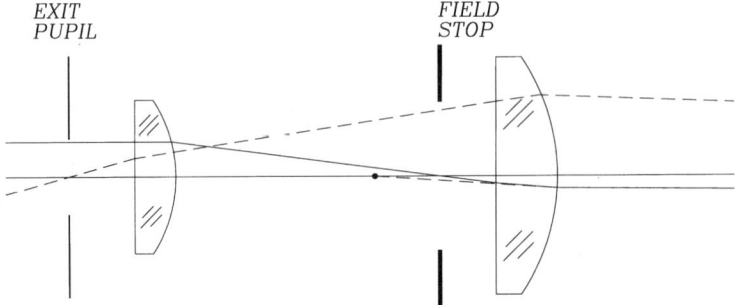

Figure 16.14 A 10× Huygens eyepiece.

Table 16.3   10 × Huygens Eyepiece

Radius of curvature (mm)	Diameter (mm)	Separation or thickness (mm)	Material
Stop	5.0	4.8	Air
Flat	11.0	3.0	BK7
−10.0	11.0	19.1	Air
Field stop	10.4	4.0	Air
Flat	17.0	4.4	BK7
−15.21	17.0	−8.59	Air

Angular field radius: 15.0°.
Exit pupil diameter (mm): 5.0.
Effective focal length (mm): 25.0.
Eye relief (mm): 4.8.

These designs are often a Kellner eyepiece, like the one shown in Fig. 16.15, whose design is presented in Table 16.4. The field width in this case is the same as in the Huygens eyepiece, but with a much better off-axis image and a larger eye relief.

The Hi-Point eyepiece has a reasonable aberration correction, with a large eye relief. Its symmetrical configuration makes it easy to construct.

16.4   MICROSCOPE ILLUMINATORS

The illuminating systems (Dempster, 1944) in a microscope are of the *Koehler* or *critical* types. The first one is also used in projectors; hence, its description is left to Chap. 17. Critical illuminators have the basic

Microscopes

Table 16.4 A 10 × Microscope Wide-Field Eyepiece

Radius of curvature (mm)	Diameter (mm)	Separation or thickness (mm)	Material
Stop	5.0	18.9	Air
−32.8	20.0	3.1	SF1
22.0	20.0	7.1	BK7
−17.1	20.0	0.2	Air
33.25	24.0	5.8	BK7
−33.25	24.0	27.45	Air
Field stop	10.4	–	–

Angular field radius: 5.0°.
Exit pupil diameter (mm): 5.0.
Effective focal length (mm): 25.0.
Eye relief (mm): 18.9.

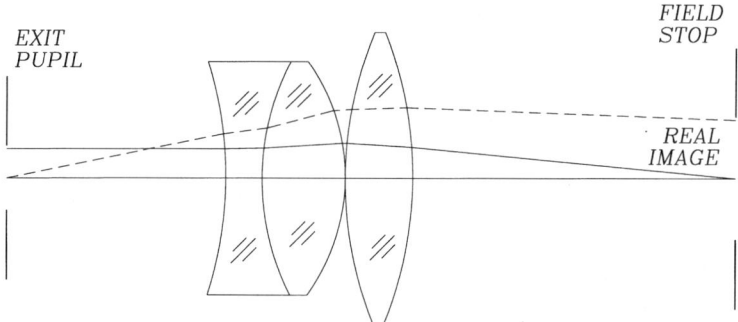

Figure 16.15 A 10× microscope wide-field eyepiece.

arrangement shown in Fig. 16.16. The tungsten filament of the light source is imaged into the *stop* by means of a condenser. Then, the *substage condenser* images the *field stop* on the object plane. The *stop*, also called an *aperture diaphragm* or *substage iris* determines the numerical aperture of the illuminating system. The numerical aperture of this illuminating system must be as large as the largest numerical aperture of the microscope objective. The focus of the substage condenser must be critically adjusted. Once this condenser is carefully focused, the stop or substage iris diameter has to be adjusted to match the numerical aperture of the condenser with that of the objective. A diameter of the aperture greater than 30 mm is needed to achieve a reasonable focal length and a large numerical aperture.

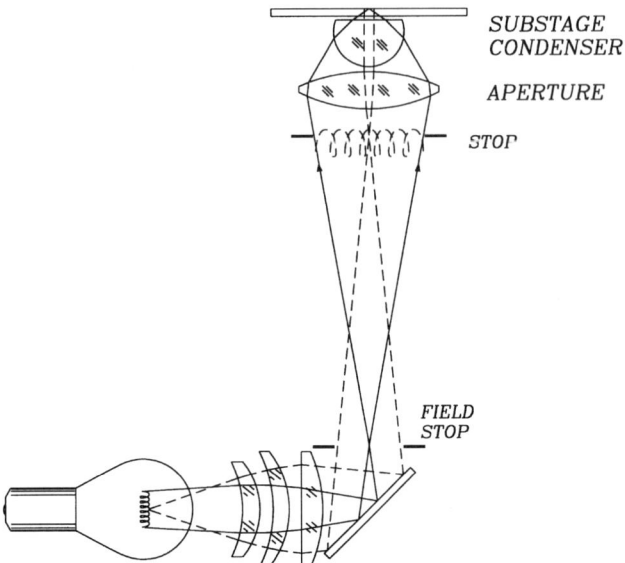

Figure 16.16 A microscope critical illuminator. The dashed lines represent the principal rays.

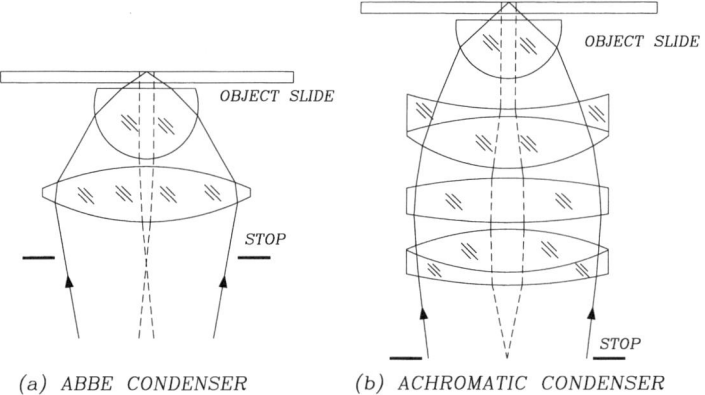

(a) ABBE CONDENSER   (b) ACHROMATIC CONDENSER

Figure 16.17 (a) Abbe and (b) achromatic condensers for microscope illuminators.

The simplest substage condenser is the Abbe condenser that consists of a nearly hemispherical lens and a convergent lens, as shown in Fig. 16.17(a). A more complicated design permits the correction of the chromatic aberration, as shown in Fig. 16.17(b).

REFERENCES

Benford, J. R., "Recent Microscope Developments at Bausch and Lomb," *Appl. Opt.*, **3**, 1044–1045 (1964).
Benford, J. R., "Microscope Objectives," in *Applied Optics and Optical Engineering*, R. Kingslake, ed., Vol. III, Chap. 4, Academic Press, San Diego, CA, 1965.
Benford, J. R. and Rosenberger, H. E., "Microscopes," in *Applied Optics and Optical Engineering*, R. Kingslake, ed., Vol. IV, Chap. 2, Academic Press, San Diego, CA, 1967.
Benford, J. R. and Rosenberger, H. E., "Microscope Objectives and Eyepieces," in *Handbook of Optics*, W. G. Driscoll and W. Vaughan, eds., Chap. 6, McGraw Hill, New York, 1978.
Bennett, A. H., "The Development of the Microscope Objective," *J. Opt. Soc. Am.*, **33**, 123–128 (1943).
Bennett, A., "Microscope Optics," in *Military Standardization Handbook: Optical Design, MIL-HDBK 141*, U.S. Defense Supply Agency, Washington, DC, 1962.
Bennett, A., "Microscope Optics," *Appl. Opt.*, **2**, 1227–1231 (1963).
Born, M. and Wolf, E., *Principles of Optics*, 2nd ed., Pergamon Press, Oxford (1964).
Broome, B. G., "Microscope Objectives and Their Evolution to Optical Disks Objectives," in *Lens Design. Critical Reviews of Optical Science and Technology*, Vol. CR41, p. 325, SPIE, Bellingham, WA, 1992.
Claussen, H. C., "Microscope Objectives with Plano-Correction," *Appl. Opt.*, **3**, 993–1003 (1964).
Conrady, A. E., *Applied Optics and Optical Design*, Part One, Dover Publications, New York, 1957.
Conrady, A. E., *Applied Optics and Optical Design*, Part Two, Dover Publications, New York, 1960.
Cruickshank, F. D., "The Trigonometrical Correction of Microscope Objectives," *J. Opt. Soc. Am.*, **36**, 296–298 (1946).
Dempster, W. T., "The Principles of Microscope Illumination and the Problem of Glare," *J. Opt. Soc. Am.*, **34**, 695–710 (1944).
Erdös, P., "Mirror Anastigmat with two Concentric Spherical Surfaces," *J. Opt. Soc. Am.*, **49**, 877 (1959).
Foster, L. V., "Microscope Optics," *J. Opt. Soc. Am.*, **40**, 275–282 (1950).
Foster, L. V. and Thiel, E. M., "An Achromatic Ultraviolet Microscope Objective," *J. Opt. Soc. Am.*, **38**, 689–692 (1948).
Grey, D. S., "A New Series of Microscope Objectives: III. Ultraviolet Objectives of Intermediate Numerical Aperture," *J. Opt. Soc. Am.*, **40**, 283–290 (1950).
Grey, D. S., "Computed Aberrations of Spherical Schwarzchild Reflecting Microscope Objectives," *J. Opt. Soc. Am.*, **41**, 183–192 (1952).
Grey, D. S. and Lee, P. H., "A New Series of Microscopes Objectives: I. Catadioptric Newtonian Systems," *J. Opt. Soc. Am.*, **39**, 719–723 (1949a).
Grey, D. S. and Lee, P. H., "A New Series of Microscopes Objectives: II. Preliminary Investigation of Catadioptric Schwarzchild Systems," *J. Opt. Soc. Am.*, **39**, 723–728 (1949b).

Hopkins, R. E., "The Components in the Basic Optical System," in *Geometrical and Instrumental Optics*, D. Malacara, ed., Academic Press, Boston, MA, 1988.

Kingslake, R., *Lens Design Fundamentals*, Academic Press, New York, 1978.

Laikin, M., *Lens Design*, Marcel Dekker, New York, 1990.

Miyata, S., Yamagawa, S., and Noma, M., "Reflecting Microscope Objectives with Nonspherical Mirrors," *J. Opt. Soc. Am.*, **42**, 431–432 (1952).

Muchel, F., "ICS—A New Principle in Optics," *Zeiss Information*, **30**, 100, 20–26 (1990).

Norris, K. P., Seeds, W. E., and Wilkins, M. H. F., "Reflecting Microscopes with Spherical Mirrors," *J. Opt. Soc. Am.*, **41**, 111–119 (1951).

Smith, W. J. and Genesee Optics Software, Inc., *Modern Lens Design. A Resource Manual*, McGraw Hill, New York, 1992.

Stempel, W. M., "An Empirical Approach to Lens Design—The Huygens Eyepiece," *J. Opt. Soc. Am.*, **33**, 278–292 (1943).

Thornburg, W., "Reflecting Objective for Microscopy," *J. Opt. Soc. Am.*, **45**, 740–743 (1955).

17
Projection Systems

17.1 SLIDE AND MOVIE PROJECTORS

A slide projector or home movie projector has an optical arrangement as illustrated in Fig. 17.1, and uses a special lamp as shown in Fig. 17.2, with a tungsten filament coiled over the area of a small square. Behind the tungsten filament, outside the lamp, but sometimes inside it, a small metallic spherical mirror is placed. The purpose of this mirror is to reflect back the light to the lamp, forming an image of the filament on the same plane. Ideally, the mirror should be aligned so that the image of the filament coil falls in the filament spaces, not over them. Unfortunately, this mirror is easily misaligned, making it completely useless.

After the lamp, a lens, called the condenser is placed in order to form on the entrance pupil of the projection lens an image of the filament. This image of the filament should fill the lens aperture without losing any light on the edges. This type of system in which the image of the light source is imaged on the entrance pupil of the projection lens is known by the name of *Köhler* illuminator. If the light source is assumed to have small dimensions as compared with the condenser aperture so that it can be considered essentially a point light source, the total luminous flux φ in lumens passing through the condenser is given by

$$\phi = LA_{\text{source}}\Omega = MA_{\text{source}} \quad (17.1)$$

where L is the luminance in lumens per square meter per steradian emitted by the light source with area A_{source}, M is the luminous emittance in lumens per square millimeter, and Ω is the projected solid angle subtended by the condenser as seen from the center of the light source.

On the other hand, if the total flux needed on the screen is equal to the required illuminance E in lumens per square meter (lux), multiplied by the area A_{screen} of the screen, the following relation has to be satisfied for

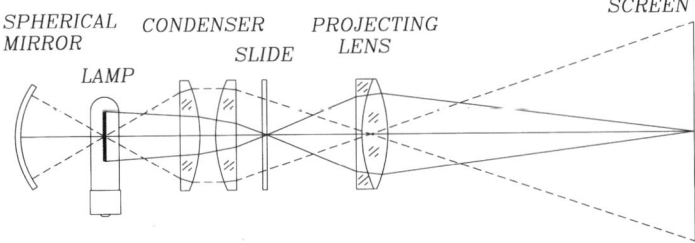

Figure 17.1 Basic optical arrangement for a projector.

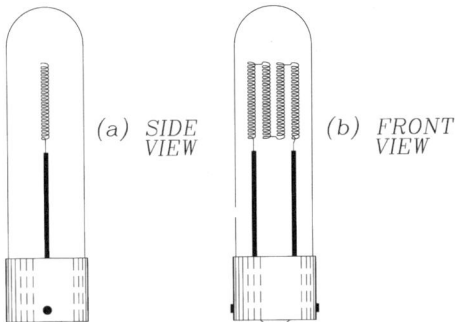

Figure 17.2 (a) Side and (b) front views of a projection lamp.

the flux passing through the condenser:

$$\phi = \phi_T \left(\frac{\Omega}{4\pi}\right) = M\,A_{\text{source}} = L\,A_{\text{source}}\,\Omega = E\,A_{\text{screen}} \qquad (17.2)$$

where ϕ_T is the total flux in lumens emitted by the lamp, and the solid angle Ω is given by

$$\Omega = \pi \sin^2 \theta \qquad (17.3)$$

If the light source is not small, which is the normal case, these formulas may still produce approximate results, as long as the light source may be considered diffuse, or in more precise words, lambertian.

17.2 COHERENCE EFFECTS IN PROJECTORS

To understand the effect of the spatial coherence of the light source let us first consider a projector with a point light source (Hopkins, 1988).

Then, the image of the light source on the entrance pupil of the projecting lens would be a point (assuming a perfect condenser). If the object being projected is a diffraction grating, from the Abbe theory of image formation we know that the fidelity of the image increases with the aperture of the projecting lens, since more spatial frequencies would be allowed to pass through the lens. Since the lens has a finite entrance pupil, the image of the grating is never perfect, but has some spurious oscillations (or fringes). If the point light source is substituted by an extended source, each point of the light source produces an image of the grating. However, each image is incoherent with each other, since the light from different points of the light source are mutually incoherent. Then, the irradiances of the images, and not the field amplitudes, are added to form the resulting composed image. This final image does not have the spurious fringes and looks smoother.

Given an entrance pupil diameter, the ideal thing is to fill the aperture of the entrance pupil with the image of the light source formed by the condenser. However, then the light source image points near the edge of the pupil would not form a good image because some of the high spatial frequencies would be cut by the rim of the pupil. In conclusion, the ideal situation is when the image of the extended light source does not completely fill the aperture. Another important reason to avoid filling the lens pupil is the scattering and diffraction of light on the edge, this drastically reduces the image contrast.

17.3 MAIN PROJECTOR COMPONENTS

17.3.1 Lamp

It has been found empirically that almost all tungsten–halogen lamps have the same luminous emittance L of about 30 lumens per steradian per square millimeter. The difference between lamps of different power is only the area of the filament. The total flux ϕ_T in lumens of a lamp with a power W in watts is approximately given by

$$\phi_T = 3W \qquad (17.4)$$

Another common type of lamp used in projectors uses a conic (parabolic or elliptical) reflector behind it, with a large collecting solid angle, that focuses the light at some distance in front of the lamp (Malacara and Morales, 1988). A typical lamp focuses the light at a distance of 5.6 inches (14.2 cm) from the rim of the reflector. These lamps have a corrugated (or formed by many small plane facets) reflector and a tungsten filament coiled along the axis of the lamp. These lamps produce an extended and diffuse spot of light in front of them.

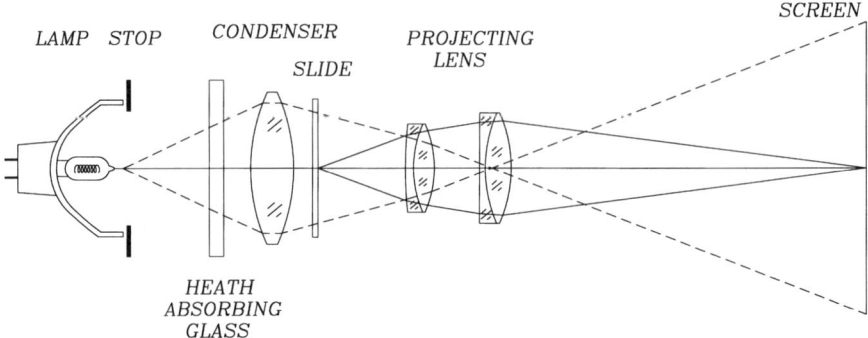

Figure 17.3 Optical arrangement of a modern slide projector.

We may then think, in the projector shown in Fig. 17.3, that we have an extended light source in the plane of the stop in front of the lamp. However, most of the light passing through the stop is traveling towards the condenser. In other words, this virtual light source is not a diffuse (lambertian) light source. Then, to collect most of the light from this source a condenser with a large collecting solid angle is not needed. Thus, the assumptions made in Eq. (17.1) are not valid. Most modern slide projectors use this kind of lamp.

17.3.2 Condenser

The spherical mirror on the back of the lamp in Fig. 17.1 duplicates the collecting solid angle, by forming a real image of the filament, back on itself. Unfortunately, it is almost never well aligned. Some lamps have this spherical mirror included as part of the glass envelope, reducing the possibility of misalignments. To have a large collecting solid angle in the condenser, its focal length must be as short as possible. The usual requirement is a lens with a focal length so short that the spherical aberration becomes extremely large and some rays may not even enter the projecting lens. Then, the slide regions being illuminated by these rays would be dark on the screen.

To obtain a short focal length and with a tolerable spherical aberration the condenser may be designed in several ways, as illustrated in Fig. 17.4. The simplest condenser is a pair of plano convex lenses as in Fig. 17.4(a), but the collecting solid angle cannot be made very large without introducing a large amount of spherical aberration. An improvement is in Fig. 17.4(b) where the refractive work is divided among three lenses. The first two lenses have aplanatic surfaces, and the third one is equiconvex. Another solution is

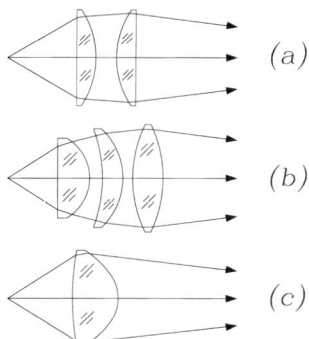

Figure 17.4 Three condensers for projectors.

to use aspheric lenses, as shown in Fig. 17.4(c). The problem with aspheric lenses is that they must be produced by pressing in large quantities to obtain them at a reasonable low price.

The light from the condenser passes through the slide, illuminating it as evenly as possible. It is desirable to use a heath absorbing glass before the slide to avoid damaging it.

17.3.3 Projecting Lens

The projection lens forms the image of the slide over the screen, and should be well corrected for aberrations. There are some important differences of projecting lenses with the photographic lenses. One difference is that their focal lengths are normally longer and the other is that their f-numbers are larger, of the order of 3.5. The point spread function must be less than 1 arcmin (the resolving power of the eye) for the observer being closest to the screen. However, commercial projecting lenses in general are not as well corrected as good photographic lenses.

The aberrations of projecting lenses should be corrected, taking into account any curvature of the slide. Film slides are always slightly curved, unless they are sandwiched between two thin glasses. When purchasing a high-quality projection lens it is necessary to specify if film or glass slides are to be used. The screen may also be cylindrically curved. Then, it is better to take into account this curvature, even if some small defocusing is produced in the upper and lower parts of the screen. If this defocusing is not tolerated, the screen must have a spherical shape instead of cylindrical.

It should be noticed that the effective entrance pupil is only as large as the image of the light source being formed by the collimator.

A frequently used design for projection lenses is the Petzval lens, which differs from with the classical Petzval design studied in Chap. 12 in that the stop is on the first lens (on the lens closer to the long conjugate) or on the second lens for the case of a projecting lens. Since the lens with the stop in contact contributes negative astigmatism to the system, the other lens must contribute with positive astigmatism. The astigmatism of this lens is a function of many factors, but one of them is the cemented surface if the power of both glasses is different. The field may be flattened by means of a negative lens close to the focal plane. An example of a Petzval projection lens with a field flattener, described by Smith and Genesee Optics Software (1992), is shown in Fig. 17.5 and its design parameters are listed in Table 17.1, where the closest Schott glass has been written in parentheses.

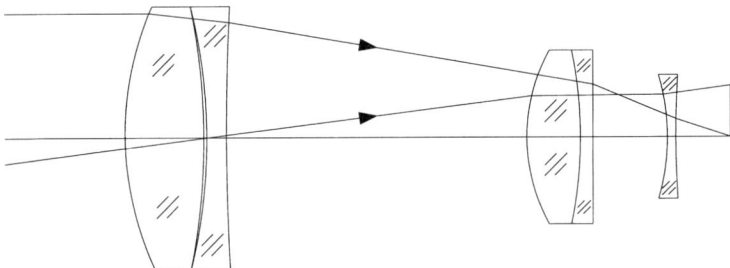

Figure 17.5 Petzval projection lens with a field flattener.

Table 17.1 A Petzval Projecting Lens

Radius of curvature (mm)	Diameter (mm)	Separation or thickness (mm)	Material
73.962	63.2	18.550	DBC1 (SK6)
−114.427	63.2	0.776	Air
−99.163	63.2	5.300	EDF3 (SF10)
680.831	63.2	59.678	Air
55.173	46.2	15.900	DBC1 (SK6)
−228.329	42.2	19.769	Air
44.891	30.8	2.650	EDF3 (SF10)
−2130.600	30.8	15.724	Air

Aperture (mm) (focal ratio): 54.0 (1.6)
Effective focal length (mm): 100.70
Back focal length (mm): 15.72
Object distance (mm): infinite
Image height (mm) (field): 16.12 (9.09°).

Asymmetrical double Gauss lenses such as those described in Chap. 12 are also frequently used as projecting lenses.

17.4 ANAMORPHIC PROJECTION

An anamorphic compression consists in producing an image with different x and y magnifications, the x magnification being smaller. To see this image with the correct proportions, it has to be projected with anamorphic expansion. This imaging process has found in the past a successful application in the motion picture industry (Benford, 1954), to produce images on a wide screen with a normal film format.

A method to produce this anamorphic imaging is by means of focal attachments, in front of the imaging or projection lens, just as in the case of zoom lenses. Figure 17.6 shows an focal Galilean anamorphic attachment. Another method to produce this anamorphic imaging is by means of a system of two prisms, as invented by Brewster in 1831, as shown in Fig. 17.7. Using two identical prisms the chromatic dispersion and the angular deviation produced by the first prism are compensated in the second prism. On the other hand, the anamorphic compression (or expansion) in one prism is $\cos \theta$ but for the two prisms the effect is doubled, producing an effect equal to $\cos^2 \theta$. Changing the prisms' orientation, a continuously variable prism anamorphoser may be constructed.

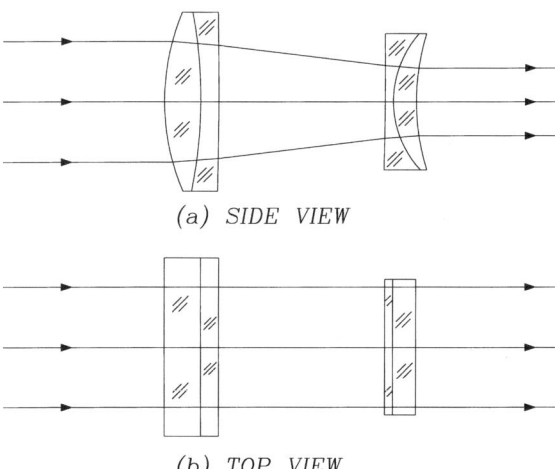

Figure 17.6 Anamorphic lens projector afocal attachment: (a) side view; (b) top view.

17.5 OVERHEAD PROJECTORS

An overhead projector is represented in Fig. 17.8. The light source is a small tungsten filament. The condenser is a pair of plastic Fresnel lenses, on top of

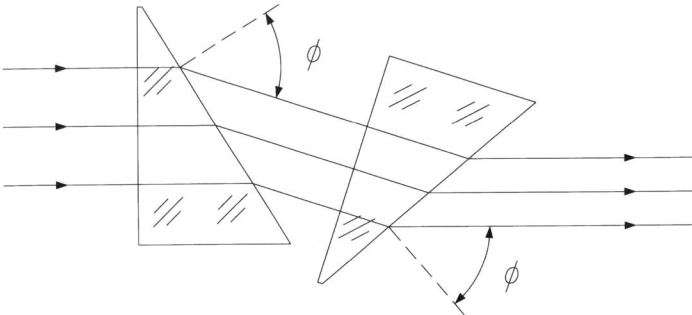

Figure 17.7 Anamorphic prism system.

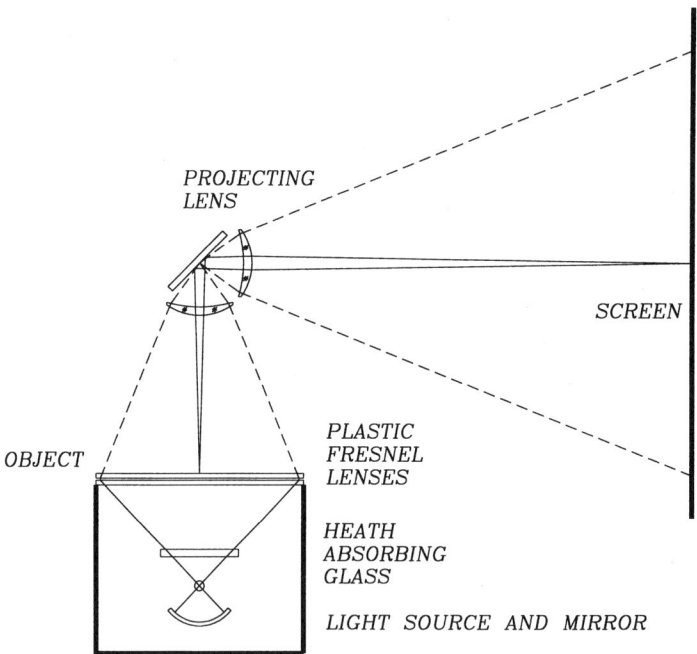

Figure 17.8 Overhead projector attachment.

which the object slide is placed. The image of the light source on the projecting lens is only a small spot of light. The size of this image determines the size of the stop, hence, the effective diameter of the entrance pupil of the lens is also small. The projecting lens is basically a symmetric landscape lens as described in Chap. 11, with most aberrations corrected, due to the symmetry and low aperture of the system. However, large aberrations may appear if the system is not properly aligned.

Frequently the lens head of the overhead projector is tilted to project the image on to a screen much higher that the overhead projector. Then, the ideally square image is strongly distorted to a trapezoidal shape with its largest side up and the smallest side down. This effect is known by the name of keystone aberration.

17.6 PROFILE PROJECTORS

In high-precision mechanical and instrumentation shops or laboratories a profile projector is used to examine small parts. The projecting lens, as illustrated in Fig. 17.9, has to be telecentric, i.e., with the entrance pupil at infinity. There are two important reasons for this telecentricity. One is that a volume object, as a sphere, may be examined and its profile measured only if the principal ray is parallel to the optical axis. The other reason is that small defocusings by displacements of the object do not produce any change in the magnification.

For the same reasons, the illuminating beam has to be collimated, by placing the light source at the focus of the collimator. This is the normal Köhler illumination used in projectors, where the light source is imaged over the entrance pupil of the projecting lens. However, in this case the entrance pupil is at infinity. More detail on projectors may be found in the book by Habell and Cox (1948).

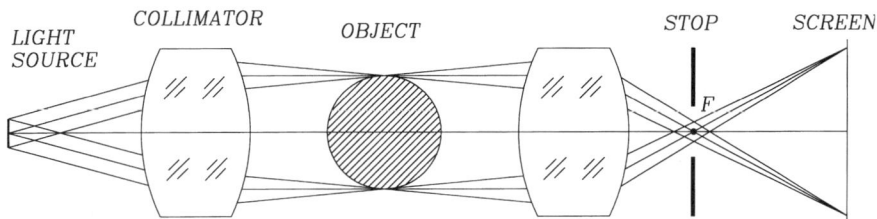

Figure 17.9 Telecentric projection system.

17.7 TELEVISION PROJECTORS

The projection of television images is made in many different ways. Here, we will describe only two methods; the first one is illustrated in Fig. 17.10 (Habell and Cox, 1948). In this system, three images tubes are normally used, one for each color. Then, the three colors are superimposed on the screen. A Schmidt system (Friedman, 1947) is used with great advantage, due to its large aperture. Plastic or glass lens objectives can also be used with success. Frequently, the lenses have to be aspheric. A few important facts have to be remembered when designing these lenses (Osawa *et al.*, 1990), for example:

1. The f-number has to be small, of the order of $f/1$ due to the need for a bright image.
2. The field of view angle has to be large, around 30° or more to shorten the projecting distance.
3. A good correction of distortion is not so important, since it can be electronically corrected on the image tubes.
4. If three independent image tubes are used a good color correction is not necessary.
5. Field curvature can be tolerated and compensated by curving the surface of the image tubes or by using a liquid field flattener.

In another method commonly used in portable television projectors, the color image is formed in a transparent liquid crystal display. Then, the image is projected as in the conventional slide projector depicted in Fig. 17.1. Since there is only one colored image to project, the objective has to be well corrected for chromatic aberrations.

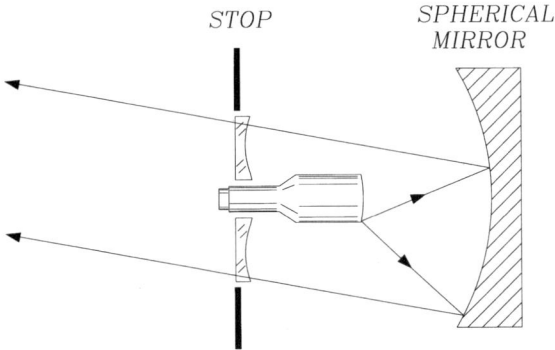

Figure 17.10 Television projection Schmidt system.

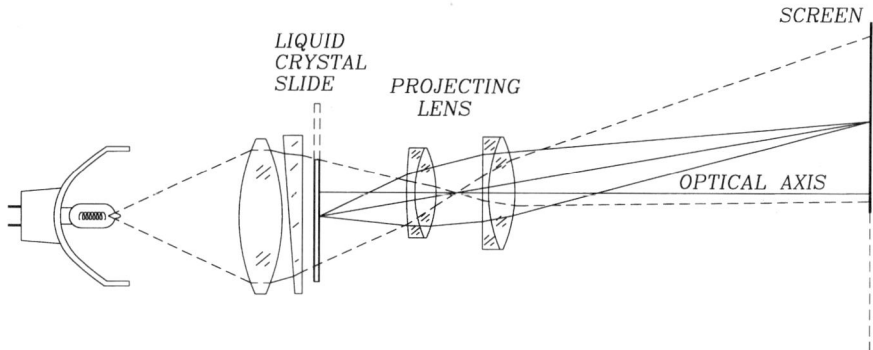

Figure 17.11 Optical layout in a portable television projector.

When a slide projector is on top of a table and the screen is not located with its center at the optical axis of the projector, the projector has to be tilted upwards by an angle to point to the center of the screen. This tilting produces keystone distortion. That is, the screen has a trapezoidal shape with its largest side on top. To eliminate this problem the optical system has to be designed as illustrated in Fig. 17.11. We can see that the distortion free field has to be greater than the actual field being used. The optical axis of the projecting lens does not pass through the center of the slide nor the center of the screen.

REFERENCES

Benford, J. R., "The Cinemascope Optical System," *J. Soc. Mot. Pic. Tel. Eng.*, **62**, 64–70 (1954).

Friedman, H. S., "Method of Computing Correction Plate for Schmidt System for Near Projection, with Special Reference to System for Television Projection," *J. Opt. Soc. Am.*, **37**, 480–484 (1947).

Habell, K. J. and Cox, A., *Engineering Optics*, Sir Isaac Pitman and Sons, London, 1948.

Hopkins, R., "The Components in the Basic Optical System," in Geometrical and Instrumental Optics (Methods of Experimental Physics, *Vol. 25*), D. Malacara, ed., Academic Press, Boston, MA, 1988.

Malacara, Z. and Morales, A., "Light Sources," in Geometrical and Instrumental Optics (Methods of Experimental Physics, *Vol. 25*), D. Malacara, ed., Academic Press, Boston, MA, 1988.

Osawa, A., Fukuda, K., and Hirata, K., "Optical Design of High Aperture Aspherical Projection Lens," *Proc. SPIE*, **1354**, 337–343 (1990).

Smith, W. J. and Genesee Optics Software, Inc., *Modern Lens Design. A Resource Manual*, McGraw Hill, New York, 1992.

18
Lens Design Optimization

18.1 BASIC PRINCIPLES

Up to the late 1940s, all optical designs had to be made by hand calculation, tracing rays with a logarithmic table. This method was extremely tedious and slow. To trace only one ray through an optical system took many hours of work and patience. Probably the earliest use of a computer to trace rays was by James G. Baker in 1944, who used the Mark I calculator at Harvard to trace rays. To trace only one skew ray through an optical surface on this machine took 120 sec. At the Institute of Optics of the University of Rochester, the first use of a computer to trace rays was made in 1953 for the IBM 650, by Robert E. Hopkins. Another leader in this field, at the Eastman Kodak Co., was Donald P. Feder. This computer work saved a lot of time in the design process, but the programs were just tools to make the whole process faster. Both methods were essentially the same. However, this work set up a solid foundation for future results. A good review of the history of automatic lens design is found in the article by Feder (1963). The methods being used are quite numerous (Brixner 1964a,b,c; Feder, 1951, 1957a, 1962; Grey, 1963a,b; Holladay, 1960; Hopkins, 1962a,b,c; Hopkins and Spencer, 1962; Meiron and Volinez, 1960; Peck, 1980; Stavroudis, 1964; Wynne, 1963). Present lens design programs are much better, but they are still far from being completely automatic. To operate them correctly a good optical design knowledge is absolutely necessary.

In lens design the *merit function* is a function of the parameters that describe the quality of the system (how close it is to the ideal solution). The greater the merit function, the worse the system is. The goal of all optimization methods is to reduce this merit function as much as possible. Some authors have proposed to call this function the *error function*, because the smaller the final value of this function, the closer the optical system is to the desired state.

Lens optimization programs need a starting design to make several iterations until a good design is found. Good sources of starting designs are the books by Smith and Genesee Optics Software (1992), Laikin (1990) and

Cox (1964). Some programs have been made so automatic that the starting point may be a set of parallel glass plates (Brixner, 1963a), but even so, an operator with experience is required.

The starting design, as described by Hilbert et al. (1990), may be set up in several different ways:

1. Similar existing designs in the technical literature. The books by Cox (1964), Laikin (1990), Smith and Genesee Optics Software (1992), and many others are good sources for these designs.
2. Scaling of an existing design, with the same f-number FN and similar main characteristics.
3. Substantially modifying a design with characteristics at least close to the ones desired.
4. First-order designing the system from scratch and then using third-order aberration theory to obtain an approximate design.

18.2 OPTIMIZATION METHODS

Many different schemes have been developed for lens design optimization, with different advantages and disadvantages. Several good reviews on the subject have been written, e.g., by Feder (1963) and by Hayford (1985).

The main problem in lens design is that the functions forming the merit function are not linear. Then, the solution very much depends on the starting point. To approach the solution, several mathematical methods have been used. We will now describe some of them.

1. One of the early methods used is called the *optimum gradient method*, the *steepest descent method*, or the *zig-zag method*. These are different names for a method first proposed by Cauchy in 1847. An initial point in the parameter space is taken and its merit function and partial derivatives with respect to each parameter are calculated at that point. Then, a new point is taken in the opposite direction to the gradient in order to reduce the merit function. This point is moved in a straight line until the merit function ceases to decrease and begins increasing. At this point the calculations are repeated and a new gradient direction is found. It may be shown that the new gradient is perpendicular to the former; hence, the name of zig-zag method. This is probably the simplest method, but after a few successful iterations the convergence to a solution becomes extremely slow. This problem is present in the method, even if the functions are exactly linear, which is not usually the case.

2. The *conjugate gradient method* is quite similar to the optimum gradient method. The difference is that the new direction after each iteration

is not the new gradient, but slightly different. So, this new direction is not perpendicular to the previous path, but at an angle less than 90° with respect to it. The consequence of this difference is that the speed of convergence is greatly improved. For example, the solution to a system of N linear equations is always found in N iterations. Mathematical details for this method are found in the article by Feder (1963).

3. *Grey's optimization method*, developed by Grey (1963a,b), is based on the construction of an aberration theory with orthogonal aberration coefficients, to simplify aberration balancing.

4. The *Glatzel adaptive method*, described in detail by Glatzel and Wilson (1968), has been independently proposed by Glatzel (1961) and Hopkins and Spencer (1962). This method resembles the path followed by a designer before the lens optimization programs were used. A merit function is not used. Instead, the individual aberrations are corrected, a few at a time, until all are corrected. In a certain manner, this method resembles the graphic method described in Chap. 11. We will describe this method in some detail in this chapter.

5. Another procedure applies the *least squares method* invented by Legendre in 1805. This method was devised to solve a system of linear equations by minimizing the sum of the square of the residuals. An extension of this method to a system with nonlinear functions was made by approximating them to linear functions by means of a Taylor expansion, neglecting high-order terms. The first ones to apply this method to lens design were Rosen and Eldert (1954). A merit function is defined as the sum of the squares of the aberrations. The aberrations are the linearized functions. Since the functions are not really linear, the process has to be repeated in successive iterations until a solution is found when the merit function is not reduced any more. A problem soon found with this method is that the functions are so far from being linear that frequently a solution is not found and the iterations just oscillate wildly about the minimum without ever approaching it.

6. The *damped least squares* method tries to force the least squares method to find a solution by damping the magnitude of the jumps on each iteration. This method was originally proposed by Levenverg in 1944, and will be described here later in detail (Meiron, 1959, 1965; Rosen and Chung, 1956). This is the most common method used in commercial lens design programs.

18.3 GLATZEL ADAPTIVE METHOD

In this method, as described in detail by Glatzel and Wilson (1968), the designer selects a small number of aberrations, not exceeding the number of independent variables and assigns a target value to them. For the other aberrations only a maximum limit is assigned. Thus, we may write for this system a set of simultaneous linear equations, for the aberrations f_n, with the variables p_j, as follows:

$$\frac{\partial f_n}{\partial p_1}(p_1 - p_{o1}) + \frac{\partial f_n}{\partial p_2}(p_2 - p_{o2}) + \cdots + \frac{\partial f_n}{\partial p_J}(p_J - p_{oJ}) = f_n - f_{on}$$

$$n = 1, \ldots, N \quad (18.1)$$

assuming that the functions are linear in the vicinity of the initial point in the variables space $\{p_{o1}, p_{o2}, \ldots, p_{oN}\}$. At this initial point the aberrations have the values f_{on}.

The optimization for the whole optical system is accomplished in the following steps:

1. The matrix of the system of equations (18.1) is evaluated.
2. A solution $(p_j - p_{oj})$ is calculated using the system matrix and the values $(f_n - f_{on})$ of the changes required in the aberrations.
3. A new system is proposed, moving the variables only a fraction k of the calculated change $(p_j - p_{oj})$, but in that direction. The increments $k(p_j - p_{oj})$ for each variable are stored in memory for later use.
4. The system is evaluated at this new point and the calculated values for the aberrations are compared with the expected values, taking into account that only a fraction k of the calculated step length was made. This permits an estimation of the nonlinearities.
5. The same system matrix obtained in step 1 is used to calculate a new solution that produces the desired changes in the aberrations.
6. Another variable change is thus calculated. The increment for the variable that had the largest change is compared with the increment obtained and stored in memory for the same variable in step 3. Then, the step length is adjusted in such a way that the largest of the two compared increments is used for the variable just considered. If the largest of all variable increments is zero, we go to step 8, otherwise we go to step 4 to repeat the process from that point.
7. If the largest proposed increment for the variables is zero, we go to step 8, otherwise we should go back to step 4.
8. A new set of aberrations with new target values is selected and then we go back to step 1.

The optimization procedure is finished when all desired aberrations are corrected within the prescribed limits. Further details may be found in the article by Glatzel and Wilson (1968).

Lens Design Optimization

18.4 CONSTRAINED DAMPED LEAST SQUARES OPTIMIZATION METHOD

This section is written closely following the treatment of Spencer (1963a,b). The system variables, like lens curvatures, thicknesses, separations, etc., are represented by J variables $p_1, p_2, p_3, \ldots, p_J$. Some functions $f_1(p_1, p_2, \ldots, p_J)$, $f_2(p_1, p_2, \ldots, p_J), \ldots, f_K(p_1, p_2, \ldots, p_J)$, some times called *operands*, are defined in terms of these variables. These functions define the characteristics of the optical system and are the effective focal length, the back focal length, the spherical aberration, the coma, the Petzval curvature, the optical transfer function, etc., or any other lens characteristics chosen by the lens designer. These functions aim to a certain value, s_k, not necessarily zero. Then, the problem reduces to finding a simultaneous solution to the set of equations:

$$f_k(p_1, \ldots, p_J) = s_k; \quad k = 1, \ldots, K \tag{18.2}$$

where the s_k are constants representing the goal values for the functions f_k.

A simultaneous solution to the set of equations does not always exist. For instance, frequently the number of functions is larger than the number of variables ($K > J$). Then, instead of looking for a solution, the goal is to minimize the function:

$$\varphi = \sum_{k=1}^{K} w_k^2 (f_k - s_k)^2 \tag{18.3}$$

where the w_k are weight factors. These factors are used to set the relative priority in the minimization of the functions. If a simultaneous solution exists, the minimization will correspond to this solution. The definition of this function, called the *merit function*, is one of the most critical steps in the lens design process (Feder, 1957b).

In general, some of these functions f_k require minimization, e.g., the primary spherical aberration, but some others require an exact solution, e.g., the final value of the effective focal length. Thus, it is convenient to separate these functions into two groups, one:

$$g_m(p_1, \ldots, p_j); \quad m = 1, \ldots, M \tag{18.4}$$

requiring minimization, and another:

$$h_n(p_1, \ldots, p_j); \quad n = 1, \ldots, N < J \tag{18.5}$$

requiring an exact solution. Then, the following merit (or error) function is minimized:

$$\varphi = \sum_{m=1}^{M} w_m^2 (g_m - s_m)^2 \qquad (18.6)$$

where s_m is the target value of g_m and the following set of simultaneous equations is solved:

$$h_n(p_1, \ldots, p_j) = t_n; \quad n = 1, \ldots, N \qquad (18.7)$$

where the t_n are the desired values for the functions h_n.

18.4.1 Linearization of the Problem

The great problem in lens design is that the functions (*operands*) are not linear with changes in the system variables. If the changes in the variables are not large, however, they may be approximated by a linear function. This is done by expanding each function in a Taylor series about the initial point $(p_{01}, p_{02}, \ldots, p_{0J})$ and taking only the constant and linear terms. The final solution is found by successive iterations of this procedure. Then, we may write the functions g_m and h_n as

$$g_m = g_{om} + \frac{\partial g_m}{\partial p_1}(p_1 - p_{o1}) + \frac{\partial g_m}{\partial p_2}(p_2 - p_{o2}) + \cdots + \frac{\partial g_m}{\partial p_J}(p_J - p_{oJ}) \qquad (18.8)$$

and

$$h_n = h_{on} + \frac{\partial h_n}{\partial p_1}(p_1 - p_{o1}) + \frac{\partial h_n}{\partial p_2}(p_2 - p_{o2}) + \cdots + \frac{\partial h_n}{\partial p_J}(p_J - p_{oJ}) \qquad (18.9)$$

We now define the following variables to simplify the notation:

$$\begin{aligned} d_m &= s_m - g_{0m} \\ e_n &= t_n - h_{0n} \\ q_j &= p_j - p_{0j} \\ a_{mj} &= \frac{\partial g_m}{\partial p_j} \\ b_{nj} &= \frac{\partial h_n}{\partial p_j} \end{aligned} \qquad (18.10)$$

Then, using the linear approximations in expressions (18.6) and (18.7), the problem becomes one of minimizing the merit (error) function:

$$\varphi = \sum_{m=1}^{M} w_m^2 \left(\sum_{j=1}^{J} a_{mj} q_j - d_m \right)^2 \tag{18.11}$$

and at the same time obtaining a solution to the set of N simultaneous equations of constraint:

$$\sum_{j=1}^{J} b_{nj} q_j = e_n; \quad n = 1, \ldots, N \tag{18.12}$$

18.4.2 Use of the Lagrange Multipliers

Let us now describe in a general manner the method of Lagrange multipliers. We have a merit function $\varphi(q_1, q_2, \ldots, q_J)$ that we want to minimize. This minimum is found with the condition:

$$d\varphi = \left(\frac{\partial \varphi}{\partial q_1}\right) dq_1 + \left(\frac{\partial \varphi}{\partial q_2}\right) dq_2 + \cdots + \left(\frac{\partial \varphi}{\partial q_J}\right) dq_J = 0 \tag{18.13}$$

If the variables $(q_1, q_2, \ldots, q_J)$ are all independent, the solution to this expression is simply obtained with

$$\frac{\partial \varphi}{\partial q_j} = 0; \quad j = 1, 2, \ldots, J \tag{18.14}$$

however, in our system there are N equations of constraint that we may represent by

$$u_n(q_1, q_2, \ldots, q_J) = e_n; \quad n = 1, \ldots, N < J \tag{18.15}$$

So, instead of J independent variables, only $J - N$ are independent and Eq. (18.13) is not true.

If before minimizing the merit function we look for a solution to the equations of constraint, we will find that there is a continuous set of points satisfying these constraints, in a space of $(J - N)$ dimensions, because there are J variables and $N < J$ equations of constraint. The region in space where the constraint equations are satisfied, assuming linearity with respect to all variables, may be considered a $(J - N)$-dimensional plane in a space of N dimensions.

Out of the possible solutions, in the space with only $(J - N)$ independent variables, we must find one solution that minimizes the merit function by satisfying Eq. (18.12). We have explained why we cannot use Eq. 18.13, but the method of undetermined Lagrange multipliers may help us.

Now, let us assume that the point $\{q_1, q_2, \ldots, q_J\}$ is one of the desired solutions and that $\{q_1 + dq_1, q_2 + dq_2, \ldots, q_J + dq_J\}$ is another solution, if the differentials satisfy the conditions:

$$du_n = \left(\frac{\partial u_n}{\partial q_1}\right)dq_1 + \left(\frac{\partial u_n}{\partial q_2}\right)dq_2 + \cdots + \left(\frac{\partial u_n}{\partial q_J}\right)dq_J = 0;$$
$$n = 1, \ldots, N \tag{18.16}$$

so that the value of u_n does not change. A geometrical interpretation for this expression is that each equation of constraint is a plane (in a small neighborhood of the point where the derivatives are taken, so that the linearity assumptions remain valid) with $J-1$ dimensions, in a space of J dimensions, with a vector normal to this plane, given by

$$\left(\frac{\partial u_n}{\partial q_1}\right), \left(\frac{\partial u_n}{\partial q_2}\right), \ldots, \left(\frac{\partial u_n}{\partial q_J}\right) \tag{18.17}$$

The gradient of the merit function must be perpendicular to the $(J-N)$-dimensional plane where all constraints are satisfied. In this plane the minimum value, or to be more precise, a stationary for the merit function, must be located. Obviously, at this minimum, the gradient in Eq. (18.17) must be perpendicular to the $(J-N)$-dimensional plane. Hence, it is possible to write this gradient as a linear combination of the vectors perpendicular to each of the constraint planes, obtaining

$$\left(\frac{\partial \varphi}{\partial q_k}\right) + \sum_{n=1}^{N} \lambda_n \left(\frac{\partial u_n}{\partial q_k}\right) = 0; \quad k = 1, \ldots, J \tag{18.18}$$

where the unknown constants λ_n are called the Lagrange multipliers.

The constraint equations (18.15) and (18.18) form together a set of $N+J$ equations with $N+J$ unknowns, $\lambda_1, \ldots, \lambda_N, q_1, \ldots, q_J$. The solution satisfies both the constraint conditions and the extremum value of the merit function.

Since the constraint conditions are given by Eqs. (18.12) and (18.15), we obtain

$$\frac{\partial u_n}{\partial q_k} = b_{nk} \tag{18.19}$$

Lens Design Optimization

and from Eq. (18.11):

$$\frac{\partial \varphi}{\partial q_k} = 2\left[\sum_{m=1}^{M}\sum_{j=1}^{J} w_m^2 a_{mk} a_{mj} q_j - \sum_{m=1}^{M} w_m^2 a_{mk} d_m\right] \quad (18.20)$$

Thus, Eq. (18.19) becomes:

$$\sum_{m=1}^{M}\sum_{j=1}^{J} w_m^2 a_{mk} a_{mj} q_j + \sum_{n=1}^{N} b_{nk} v_n = \sum_{m=1}^{M} w_m^2 a_{mk} d_m; \quad k=1,\ldots,J \quad (18.21)$$

where $v_n = \lambda_n/2$. Equations (18.12) and (18.21) form a set of $N+J$ linear equations with $N+J$ unknowns that may be solved with standard methods. As pointed out by Spencer (1963b) this method obtains a minimum of the merit function, and rules out the possibility of a maximum. However, sometimes this minimum is not uniquely defined, since the minimum may be a small area and not a point. Then, this ambiguity is easily removed by adding to the merit function the sum:

$$S = \sum_{j=1}^{J} (c_j q_j)^2 \quad (18.22)$$

where the c_j are weight factors. This sum, besides removing the ambiguity, allows control over the influence of the change of the different variables over the solution. A large value of c_j forces the system to produce a small change in that parameter. In other words, the solution is taken in the small region of the minima, at the point closest to the initial solution. Taking into account this term, Eq. (18.22) becomes

$$\sum_{m=1}^{M}\sum_{j=1}^{J} w_m^2 a_{mk} a_{mj} q_j + c_k^2 q_k + \sum_{n=1}^{N} b_{nk} v_n = \sum_{m=1}^{M} w_m^2 a_{mk} d_m; \quad k=1,\ldots,J$$

$$(18.23)$$

18.4.3 Matrix Representation

Continuing along Spencer's lines, including his notation, we will now represent the system of Eqs. (18.12) and (18.23) in matrix form by defining the following matrices, where J is the number of parameters, used as variables,

M is the number of functions (aberrations) appearing in the merit function, and N is the number of constraints:

$$A = \begin{bmatrix} a_{11} & \cdots & a_{1J} \\ \vdots & & \vdots \\ a_{M1} & \cdots & a_{MJ} \end{bmatrix} \quad (18.24)$$

$$B = \begin{bmatrix} b_{11} & \cdots & b_{1J} \\ \vdots & & \vdots \\ b_{N1} & \cdots & b_{NJ} \end{bmatrix} \quad (18.25)$$

$$C = \begin{bmatrix} c_1^2 & 0 & \cdots & 0 \\ 0 & c_2^2 & \cdots & 0 \\ \vdots & \vdots & & \vdots \\ 0 & 0 & \cdots & c_J^2 \end{bmatrix} \quad (18.26)$$

$$W = \begin{bmatrix} w_1 & 0 & \cdots & 0 \\ 0 & w_2 & \cdots & 0 \\ \vdots & \vdots & & \vdots \\ 0 & 0 & \cdots & w_M \end{bmatrix} \quad (18.27)$$

$$d = \begin{bmatrix} d_1 \\ \vdots \\ d_M \end{bmatrix} \quad (18.28)$$

$$e = \begin{bmatrix} e_1 \\ \vdots \\ e_N \end{bmatrix} \quad (18.29)$$

$$q = \begin{bmatrix} q_1 \\ \vdots \\ q_J \end{bmatrix} \quad (18.30)$$

$$v = \begin{bmatrix} v_1 \\ \vdots \\ v_N \end{bmatrix} \quad (18.31)$$

Lens Design Optimization

Since the matrix A and the vector d always appear multiplied by W, we may for convenience define

$$M = WA \qquad (18.32)$$

and

$$r = Wd \qquad (18.33)$$

Also, a matrix G and a vector q may be defined by

$$G = M^t M + C \qquad (18.34)$$

and

$$g = M^t r \qquad (18.35)$$

where the superscript t represents the transpose.
Hence, Eqs. (18.23) and (18.12) become

$$Gq + B^t v = g \qquad (18.36)$$

and

$$Bq = e \qquad (18.37)$$

respectively.

18.4.4 Numerical Calculation of Matrix Solution

Expressions (18.36) and (18.37) form together a system of $(N+J)$ equations with $(N+J)$ unknowns, namely N values of v_n and J values of q_j. Then, the inversion of a matrix of $(N+J) \times (N+J)$ elements is required. Fortunately, the system may be separated into two smaller systems, increasing the computational accuracy. This separation is achieved as follows. From Eq. (18.37) we may write

$$q = G^{-1}(g - B^t v) \qquad (18.38)$$

and substituting this result into Eq. (18.38) and solving for v:

$$v = E^{-1}(BG^{-1} - e) \qquad (18.39)$$

where

$$E = BG^{-1}B^t \qquad (18.40)$$

The calculations are now carried on by following the next steps:
1. The inverse of the matrix G with $J \times J$ elements is calculated.
2. The matrix E with $N \times N$ elements is calculated using Eq. (18.40) and then inverted.
3. The vector v is found with Eq. (18.39).
4. The solution vector q is finally calculated with Eq. (18.38).

The two matrices G and E are symmetric, requiring less numeric operations to invert than ordinary matrices.

After obtaining the solution, the new lens is taken as the initial point and the whole procedure is started again. The final result is obtained in an iterative manner. It is important, however, that the lens changes in each iteration are kept within reasonable limits, so that the linear approximations remain valid.

18.4.5 Use of the Weight Factors

There are two sets of weight factors w_m and c_j that may be used to control the nature of the solution. They are so important that their value may determine if a solution is found or not. The weights w_m define the relative importance of the various aberrations forming the merit function. Their values depend on many factors, as described in Section 18.5.

The weight factors c_j control the influence of each variable on the search for a solution. As the factor c_j is made larger, the associated variable q_j is forced to change less. Thus, variables with smaller weights will do most of the work. These weights are also called damping factors. It is convenient to separate these damping factors into the product of two factors. One is the general damping factor, with a common value for all variables, and the other is another damping factor for each variable.

It should be mentioned that sometimes two variables may not be completely linearly independent from the other. Then, the system matrix is nearly singular and designer intervention is needed to remove one of the variables, or to introduce different factors c_j to these variables to remove the singularity, before any further progress can be made. Some modern lens design programs, however, perform this function automatically.

Frequently, a *stagnation* point is encountered, where very small improvements in the merit function are achieved in each iteration and the program never converges to a solution. This situation occurs when the merit function change is buried in the numerical noise. In this case a variable may be changed, to begin the optimization at some other initial point. The variables' weights are useful in this case.

Even when all the individual damping factors are equal, it is logical to expect that different variables have different effects on the search for a solution. Thus, equal individual factors are not convenient, because then the work may be done with a large change of some variables and a small change of some others. Several solutions have been proposed to counteract this artificial weighting of the variables in a lens system (Spencer, 1963a). Buchele (1968) proposes modifying the damping factors continuously, by trial and error, as the iterations go on. Rayces and Lebich (1988) have made a careful study of the effect of different choices for the damping factor. Sometimes a change in the damping factor is not sufficient to increase the speed of convergence towards a solution when the solution is being approached extremely slowly. In this case, Robb (1979) has proposed to alter the direction of the next proposed solution.

Many practical comments about the use of weights and variables in an optimization program are found in Chap. 2 of the book by Smith and Genesee Optics Software (1992).

18.5 MERIT FUNCTION AND BOUNDARY CONDITIONS

The definition of the merit (or error) function and the boundary conditions are two very important steps in lens design. Next, we will briefly review the two problems.

18.5.1 Merit Function

The merit function may be defined in many different manners, e.g., by:

1. The geometrical spot size or mean square size of the image, as defined in Section 9.2.
2. The root mean square wavefront deviation as described in Section 9.3.
3. The modulation transfer function (MTF), optimized at some desired spatial frequencies range, as explained by Rimmer et al. (1990).
4. An appropriate linear combination of primary and high-order aberration coefficients.

As pointed out in Section 9.5.1, the image mean square size minimization optimizes the MTF for low spatial frequencies and as shown in

Section 9.5.1, the minimization of the root mean square wavefront deviation is related to the high spatial frequencies.

The decision about the type of merit function to be used depends on the application of the optical system being designed, as well as on the personal preferences and experience of the lens designer. It should be noticed that in a perfect (diffraction-limited system) all definitions of the merit function are simultaneously minimized to a zero value. In a real imperfect system, the choice of the merit function affects the final performance of the lens.

Let us assume that the geometrical spot size is selected for the evaluation. The next important decision is how important is the color correction, since each color produces a different image. To use the image size as the merit function, let us write Eq. (10.12) as follows:

$$TA_{\text{rms}}^2(\lambda,h) = \sum_{k=1}^{N_\theta} \sum_{j=1}^{N_\rho} w_j(\rho) TA_\rho^2(\rho_j,\theta_k,\lambda,h) - \overline{TA}^2(\lambda,h) \tag{18.41}$$

where the dependence on the wavelength λ of the light and the image height h has been made explicit. A complete merit function representation must take into account many different factors as will now be seen.

Color Averaging

This image size may be averaged over all the colors in many different ways. For example, for a visual instrument the chromatic response of the eye has to be taken into account. The easiest but less accurate manner of taking this average for a visual instrument is by just considering that the visual efficiency of different wavelengths is as shown in Fig. 18.1, and assigning

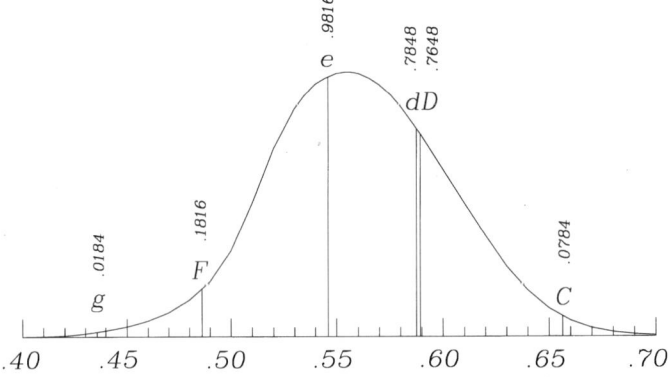

Figure 18.1 Sensitivity curve for the standard human eye.

Lens Design Optimization

a proper weight to the spot diagram taken with each color. The weights $w_i(\lambda)$ take into account the relative importance of each wavelength λ. The weights taken from this figure, must be normalized so that their sum is equal to one. Thus, the color averaged value of $TA_{\text{rms}}^2(\lambda, h)$ is

$$TA_{\text{rms}}^2(h) = \sum_{i=1}^{N_\lambda} w_i(\lambda) TA_{\text{rms}}^2(\lambda_i, h) \tag{18.42}$$

hence, substituting this relation into Eq. (18.41) we obtain

$$TA_{\text{rms}}^2(h) = \sum_{k=1}^{N_\theta} \sum_{j=1}^{N_\rho} w_j(\rho) \sum_{i=1}^{N_\lambda} w_i(\lambda) TA_\rho^2(\rho_j, \theta_k, \lambda_i, h)$$

$$- \sum_{i=1}^{N_\lambda} w_i(\lambda) \overline{TA}^2(\lambda_i, h) \tag{18.43}$$

As an example, if we obtain three spot diagrams ($N_\lambda = 3$), for colors C, d, and F, the three color weights are:

$w_C = 0.0784/(0.0784 + 0.7848 + 0.1816) = 0.0750$
$w_d = 0.7848/(0.0784 + 0.7848 + 0.1816) = 0.7511$
$w_F = 0.1816/(0.0784 + 0.7848 + 0.1816) = 0.1738$

However, this is not an accurate method, since the whole function in Fig. 18.1 is not taken into account. The error using this procedure is higher than 10%. Forbes (1988) developed a highly accurate method based on gaussian integration. The details of this method may be read in his paper. Equation (18.43) remains valid, but with the important difference that the color for the calculation of the spot diagrams must be those listed in Table 18.1 with the weights given there. These weights are normalized so that their sum is equal to one.

Table 18.1 Gaussian Integration Parameters for Color Averaging of a Visual Instrument, Using Three Colors ($N_\lambda = 3$)

i	λ_i	$w_i(\lambda)$
1	0.434658	0.006963
2	0.518983	0.054553
3	0.614795	0.038483

With this method, an accuracy of about 0.01% may be obtained. These results are valid for a visual instrument, but not for one whose detector has a more uniform chromatic response (panchromatic). If this response is assumed to be constant from 0.4 to 0.7 μm, the gaussian constants in Table 18.2 must be used.

Obviously, if the system is to be used with monochromatic light, no color averaging should be made.

Field Averaging

The image size may also be averaged over the field using gaussian or Radau integration, as described by Forbes (1988). Radau integration is chosen, so that the on-axis image is included. Then, the mean square size of the image with color and field averaging is

$$TA_{\text{rms}}^2 = \sum_{k=1}^{N_\theta} \sum_{j=1}^{N_\rho} w_j(\rho) \sum_{i=1}^{N_\lambda} w_i(\lambda) \sum_{n=1}^{N_h} w_h TA_\rho^2(\rho_j, \theta_k, \lambda_i, h)$$

$$- \sum_{i=1}^{N_\lambda} w_i(\lambda) \sum_{n=1}^{N_h} w_n(h) \overline{TA}(\lambda_i, h) \qquad (18.44)$$

where the image heights to be used with their corresponding normalized weights are as listed in Table 18.3, where the maximum image height has been normalized to one.

Table 18.2 Gaussian Integration Parameters for Color Averaging of a Panchromatic Instrument, Using Three Colors ($N_\lambda = 3$)

i	λ_i	$w_i(\lambda)$
1	0.418886	0.164853
2	0.505546	0.412843
3	0.644536	0.422307

Table 18.3 Radau Integration Parameters for Field Averaging, Using Three Image Heights ($N_h = 3$)

n	h_n	$w_n(h)$
1	0.000000	0.14000
2	0.564842	0.57388
3	0.893999	0.28612

If the lens is to be used only on the optical axis, no field averaging should be used. However, even if the whole field is going to be used, the off-axis images do not in general have the same priority for correction as the on-axis image. As an example given by Forbes (1988), let us consider the off-axis weight function in Fig. 18.2. For this case the Radau parameters for three image points, one on-axis and two off-axis, are listed in Table 18.3.

Distortion

The merit function based on the image size in Eq. (18.44) does not take into account any possible distortion in the optical system, since any shift of the image centroid from the gaussian image position is compensated. The distortion would be taken into account in the merit function if the centroid shift is not compensated, by making the last term in Eq. (18.44) equal to zero ($\overline{TA}_y = 0$).

18.5.2 Boundary Conditions

Some of the possible boundary conditions used in lens optimization programs are the following.

Axial Optical Thickness

If the light ray travels from a surface j in the optical system to optical surface $j+1$, the traveled optical path is positive for both traveling

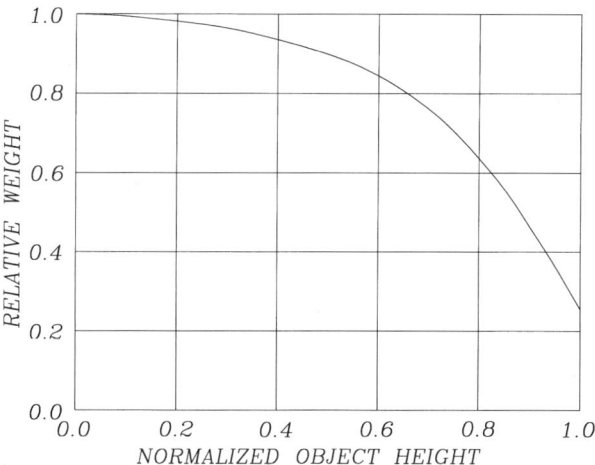

Figure 18.2 Proposed relative weight versus the normalized object height.

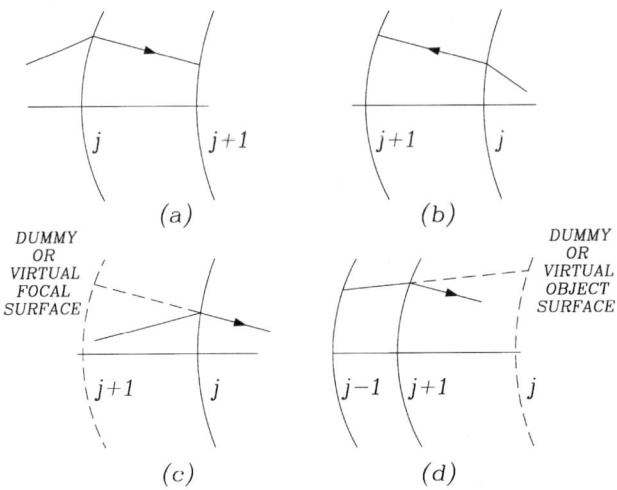

Figure 18.3 Boundary conditions in lens design.

directions, left to right as well as right to left, as shown in Figs. 18.3(a) and (b). Thus, we may write

$$n_j t_j \geq 0 \qquad (18.45)$$

This optical path nt may be negative only in the following cases:

1. The ray exits from surface j and does not travel towards surface $j+1$, but in the opposite direction. This is possible only if surface $j+1$ is a dummy surface (not an actual refractive or reflective surface) ($n_j = n_{j+1}$) or a virtual focal surface ($j+1 = k$), as in Fig. 18.3(c).
2. The ray travels towards surface $j+1$, but it is also approaching surface j, not getting away from it. This is possible only if surface j is a dummy surface ($n_{j-1} = n_j$) or it is a virtual object surface ($j = 0$), as in Fig. 18.3(d).

If a surface does not satisfy these conditions, we have a boundary condition violation.

Edge Optical Thickness

The aperture of a lens is equal to or greater than any of the values of $(\bar{y} + y)$ at the two surfaces of a lens. With this aperture value and the lens curvatures, the edge optical thickness nt_{edge} has to be greater than a predetermined minimum value.

Average Lens Thickness

The average thickness of a lens must be greater than a certain minimum that depends on the lens diameter.

These are not the only possible boundary conditions, but many others may be used if desired, e.g., the maximum length of the optical system or the maximum thickness of a lens. The final choice of boundary conditions to be satisfied depends on the type of optical system being designed.

18.6 MODERN TRENDS IN OPTICAL DESIGN

The optimization methods are quite numerous, but the most widely used are the constrained damped least squares algorithms just described. However, there are some new methods that have recently received some attention. Among these, the most notable is global optimization by using a simulated annealing procedure, as next described.

18.6.1 Global Optimization and Simulated Annealing

As explained at the beginning of this chapter, the goal of any lens optimization program is to locate in variable space a point for which the so-called merit function has a minimum value. If the region being considered is small, perhaps only one minimum is there. However, if the space is large enough, many minima for the function may exist. Not all these minima will have the same value for the merit function, but some may be smaller than others. Thus, it is desirable to find the smallest minimum and not the closest to the starting point. This procedure, called *global optimization* (Kuper and Harris, 1992), has the advantage that the solution does not depend on the starting point. There are many procedures to perform global optimization, but only quite recently is there computer power available for this task.

A simple global optimization method is the *grid search*, in which the merit function is evaluated at equidistant points on a regular grid on the variables space. Once the point with the minimum merit function is located, the minimum on this region is found.

Another global optimization method is the *simulated annealing* algorithm, first used by Bohachevsky et al. (1984). In this algorithm the variables space is sampled with a controlled random search. The controlling parameter T is called "temperature" because it is analogous to the physical temperature in the thermal annealing process.

The main problem now with global optimization is that the computing time may become extremely large. This computing time grows very rapidly with the number of variables, making it feasible only for a relatively small number of variables. There is no doubt, however, that in the near future this method will become practical and more widely used.

18.7 FLOW CHART FOR A LENS OPTIMIZATION PROGRAM

The flow chart for a lens optimization program may have many different forms. It may also be very simple or extremely complex. As an example, Fig. 18.4 shows a more or less typical one.

18.8 LENS DESIGN AND EVALUATION PROGRAMS

Lens data input in lens evaluation programs shares many common procedures. Here, we will describe a few of these techniques. Regarding the signs, we have the following standard conventions:

1. The sign of the curvature or the radius of curvature of an optical surface is positive if the center of curvature is to the right of the surface and negative otherwise. This sign selection is independent of the direction in which the light is traveling.
2. The sign of a lens thickness or spacing between two of them is positive if the next surface is to the right of the preceding surface and negative otherwise. The next surface is not necessarily in the direction in which the light is traveling. In other words, it may be a virtual surface.
3. The sign of the refractive index medium is positive if the light travels in this medium from left to right and negative otherwise. Thus, at each reflecting surface the sign of the refractive index has to be changed.
4. The sign of any ray angle with respect to the optical axis is positive if the slope is positive and negative otherwise.
5. The sign convention for the angles of any ray with respect to the normal to the surface, as described in Section 1.3 and Fig. 1.14, are positive if the slope of the ray is greater that the slope of the normal to the surface and negative otherwise.

The curvature or the radius of curvature of an optical surface can be specified in several alternative ways:

1. Providing the value of the radius of curvature.
2. Specifying the value of the curvature, which is the inverse of the radius of curvature.

Lens Design Optimization

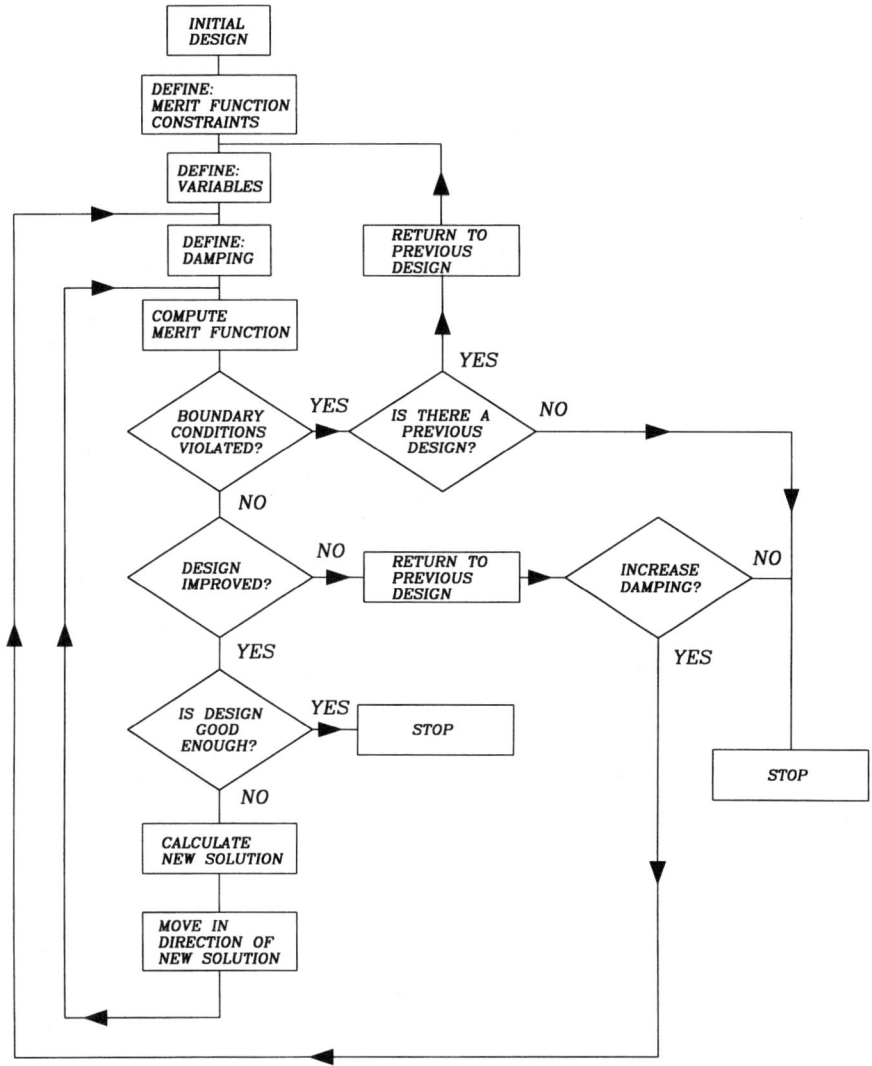

Figure 18.4 Flow chart for a typical lens design program.

3. By specification of the dioptric power of the surface.
4. Setting the value of the angle that the meridional ray must have after refraction or reflection at this surface. This method is useful in the last surface to set this curvature to a value that produces the desired effective focal length for the system. This value is

$u'_k = -D/(2F)$, where D is the entrance pupil diameter and F is the effective focal length.
5. With the height that the meridional ray must have when arriving at the next surface.
6. By making this curvature equal in magnitude and sign or equal in magnitude but with different sign to a previous curvature in the system.

The thickness or separation of a surface from the next one can be specified:

1. By the desired value of this parameter.
2. By setting the height of the meridional ray at the next surface. If we set this height equal to zero at the last surface, its position will be at the paraxial focus.
3. This separation can be made equal in magnitude, with the same or opposite sign, to the separation for a previously defined surface.

There are some useful tricks that lens designers apply when using a lens evaluation program. A few can be mentioned, for example:

1. The lens system is frequently oriented so that the light enters the system thorough the longest conjugate. Thus, if the object is closer to the system than the image, the system is reversed.
2. If the system is afocal, like a terrestrial telescope, a focusing element is placed at the end of the system, so that the light is focused at a finite distance. A possibility is to use a spherical mirror. To avoid introducing spherical aberration the radius of curvature should be much longer that the exit pupil of the system. Its selection depends on the resolution required from the system. To avoid off-axis aberrations the exit pupil of the system under evaluation is located at the center of curvature of the focusing spherical mirror.

With most commercial programs all the image analysis procedures and aberration plots described in Chap. 9 can be performed.

18.9 SOME COMMERCIAL LENS DESIGN PROGRAMS

Commercial lens design and evaluation programs are appearing quite frequently. Some of them are very complete, flexible, and sophisticated, but others are simple. A few of these programs have been reviewed in the proceedings of the International Lens Design Conference (Lawrence, 1990). A partial list of commercially available raytracing, lens design, and evaluation programs is as follows:

ACCOS
Optikos, 7796 Victor-Mendon Road, Victor, NY, 14564, U.S.A.
BEAM FOUR
Stellar Software, P.O. Box 10183, Berkeley, CA 94709, U.S.A.
CODE V (Version 9.20 for Microsoft Windows)
Optical Research Associates, 550 N. Rosemead Blvd., Pasadena, CA 91107, U.S.A.
EIKONAL
Juan Rayces Consulting, Inc., 22802 Montalbo Rd., Laguna Niguel, CA 92677, U.S.A.
GENII
Genesee Computer Center, 20 University Avenue, Rochester, NY 14605, U.S.A.
LASL
Los Alamos program, available to public users. Berlyn Brixner or Morris Klein, Los Alamos National Laboratories, Los Alamos, NM 87545, U.S.A.
OPTIX
P. O. Box 5243, 3637 U.S. 19N, Palm Harbor, FL 34684, U.S.A.
OSDP
Gibson Optics, 655 Oneida Drive, Sunnyvale, CA 94087, U.S.A.
OSLO
Sinclair Optics, Inc.,6780 Palmyra Rd., Fairport, NY 14450, U.S.A.
SIGMA PC
Kidger Optics Ltd., Sussex House, Farmingham Rd., Jarvis Brook, Crowborough, East Sussex TN6 2JP, U.K.
SCIOPT
Sciopt Enterprises, P.O. Box 20637, San Jose, CA 95160.
SOLORD
Lord Ingenierie Mediterranee, Ze de la Farlède, Rue Parmentier, B. P. 275, F-83078 Toulon, Cedex 9, France.
SYNOPSYS
Optical Systems Design Inc., P.O. East Bothbay, ME 04544-0247, U.S.A.
ZEMAX
Focusoft, Inc., P. O. Box 756, Pleasanton, CA 94566, U.S.A.

These programs can solve almost any lens design problem, but they require an experienced person with considerable design background to use them properly. The programs are so flexible that there is a great possibility of making mistakes if the operator does not have a good optics knowledge.

REFERENCES

Bohachevsky, I. O., Viswanathan, V. K., and Woodfin, G., "An Intelligent Optical Design Program," *Proc. SPIE*, **485**, 104–112 (1984).
Brixner, B., "Automatic Lens Design for Nonexperts," *Appl. Opt.*, **2**, 1281–1286 (1963a).
Brixner, B., "The Symposium Lens Improved," *Appl. Opt.*, **2**, 1331–1332 (1963b).
Brixner, B., "Automatic Lens Design: Further Notes for Optical Engineers," *J. Soc. Mot. Pic. Tel. Eng.*, **73**, 314–320 (1964a).
Brixner, B., "The Symposium Lenses—A Performance Evaluation," *Appl. Opt.*, **3**, 780–781 (1964b).
Brixner, B., "Automatic Lens Design Illustrated by a 600 mm $f/2.0$, 24 Field Lens," *J. Soc. Mot. Pic. Tel. Eng.*, **73**, 654–657 (1964c).
Buchele, D. R., "Damping Factor for the Least-Squares Method of Optical Design," *Appl. Opt.*, **7**, 2433–2435 (1968).
Cox, A., *A System of Optical Design*, Focal Press, New York, 1964.
Feder, D. P., "Optical Calculations with Automatic Computing Machinery," *J. Opt. Soc. Am.*, **41**, 630–635 (1951).
Feder, D. P., "Automatic Lens Design Methods," *J. Opt. Soc. Am.*, **47**, 902–912 (1957a).
Feder, D. P., "Calculation of an Optical Merit Function and Its Derivative with Respect to System Parameters," *J. Opt. Soc. Am.*, **47**, 913–925 (1957b).
Feder, D. P., Automatic Lens Design with a HighSpeed Computer," *J. Opt. Soc. Am.*, **52**, 177–183 (1962).
Feder, D. P., "Automatic Optical Design," *Appl. Opt.*, **2**, 1209–1226 (1963).
Forbes, G. W., "Optical System Assessment for Design: Numerical Ray Tracing in the Gaussian Pupil," *J. Opt. Soc. Am. A*, **5**, 1943 (1988).
Glatzel, E., "Ein Neues Verfahren zur Autmatschen Korrection Optischer Systeme mit Electronischen Rechenmaschinen," *Optik*, **18**, 577–580 (1961).
Glatzel, E. and Wilson, R., "Adaptive Automatic Correction in Optical Design," *Appl. Opt.*, **7**, 265–276, (1968).
Grey, D. S., "Aberration Theories for Semiautomatic Lens Design by Electronic Computer: I. Preliminary Remarks," *J. Opt. Soc. Am.*, **53**, 672–676 (1963a).
Grey, D. S., "Aberration Theories for Semiautomatic Lens Design by Electronic Computer: II. A Specific Computer Program," *J. Opt. Soc. Am.*, **53**, 677–680 (1963b).
Hayford, M. J., "Optimization Methodology," *Proc. SPIE*, **531**, 68–80 (1985).

Hilbert, R. S., Ford, E. H., and Hayford, M. J., "A Tutorial on Selection and Creation of Starting Points for Optical Design," *OSA Annual Meeting*, Boston, MA, 1990.

Holladay, J. C., "Computer Design of Optical Lens Systems," in *Computer Applications*, B. Mittman and A. Unger, eds., Macmillan, New York, 1960.

Hopkins, R. E., "Re-Evaluation of the Problem of Optical Design," *J. Opt. Soc. Am.*, **52**, 1218–1222 (1962a).

Hopkins, R. E., "Method of Lens Design," in *Military Standardization Handbook: Optical Design, MILHDBK 141*, U.S. Defense Supply Agency, Washington, DC, 1962b.

Hopkins, R. E., "An Application of the Method of Lens Design," in *Military Standardization Handbook: Optical Design, MILHDBK 141*, U.S. Defense Supply Agency, Washington, DC, 1962c.

Hopkins, R. E. and Spencer, G., "Creative Thinking and Computing Machines in Optical Design," *J. Opt. Soc. Am.*, **52**, 172–176 (1962).

Kuper, T. G. and Harris, T. I., "A New Look at Global Optimization for Optical Design," *Photonics Spectra*, January (1992).

Laikin, M., *Lens Design*, Marcel Dekker, New York, 1990.

Lawrence, G. N., ed., International Lens Design Conference. SPIE Proceedings, *Vol. 1354*, Bellingham, WA, 1990.

Meiron, J. "Automatic Lens Design by the Least Squares Method," *J. Opt. Soc. Am.*, **19**, 293–298 (1959).

Meiron, J., "Damped Least-Squares Method for Automatic Lens Design," *J. Opt. Soc. Am.*, **55**, 1105–1109 (1965).

Meiron, J. and Volinez, G., "Parabolic Approximation Method for Automatic Lens Design," *J. Opt. Soc. Am.*, **50**, 207–211 (1960).

Peck, W. G., "Automated Lens Design," in *Applied Optics and Optical Engineering, Vol. VIII*, R. R. Shannon and J. C. Wyant, eds., Academic Press, San Diego, CA, 1980.

Rayces, J. L. and Lebich, L., "Experiments on Constrained Optimization with Spencer's Method," *Opt. Eng.*, **27**, 1031–1034 (1988).

Rimmer, M. P., Bruegge T. J., and Kuper, T. G., "MTF Optimization in Lens Design," *Proc. SPIE*, **1354**, 83–91 (1990).

Robb, P. N., "Accelerating Convergence in Automating Lens Design," *Appl. Opt.*, **18**, 4191–4194 (1979).

Rosen, S. and Chung, A., "Application of the Least Squares Method," *J. Opt. Soc. Am.*, **46**, 223–226 (1956).

Rosen, S. and Eldert, C., "Least Squares Method of Optical Correction," *J. Opt. Soc. Am.*, **44**, 250–252 (1954).

Smith, W. J. and Genesee Optics Software, Inc., *Modern Lens Design. A Resource Manual*, McGraw Hill, New York, 1992.

Spencer, G., *A Computer Oriented Automatic Lens Correction Procedure*, PhD Thesis, The University of Rochester, Rochester, NY, 1963a.

Spencer, G., "A Flexible Automatic Lens Correction Proceedure," *Appl. Opt.*, **2**, 1257–1264 (1963b).

Stavroudis, O., "Automatic Optical Design", in *Advances in Computers, Vol. 5*, Alt and Rubinoff, eds., Academic Press, New York, 1964.

Wynne, C. G. and Wormell, P. M. J. H., "Lens Design by Computer," *Appl. Opt.*, **2**, 1233–1238 (1963).

Appendix 1
Notation and Primary Aberration Coefficients Summary

A1.1 NOTATION

The paraxial variables follow the notation in Table A1.1. Unprimed variables are used before refraction and primed variables are used after refraction on the optical surface. When the next surface is to be considered, a subscript $+1$ is used.

The are several kinds of focal lengths, as shown in Table A1.2. For example, one has a different value in the object space (lens illuminated with a collimated beam from right to left) than in the image space (lens illuminated with a collimated beam from left to right). In the first case an unprimed variable is used and in the second case a primed variable is used. When the object and image medium is the same, generally air, the two focal lengths have the same value. Then, the focal length is unprimed.

Table A1.1 Notation for Some Paraxial Variables

	At surface j		At surface $j+1$ (Before refraction)
	Before refraction	After refraction	
Meridional rays	i	i'	i_{+1}
	u	u'	u_{+1}
	l	l'	l_{+1}
	y	y'	y_{+1}
	Q	Q'	Q_{+1}
Principal rays	$\bar{i}$	$\bar{i}'$	$\bar{i}_{+1}$
	$\bar{u}$	$\bar{u}'$	$\bar{u}_{+1}$
	$\bar{l}$	$\bar{l}'$	$\bar{l}_{+1}$
	$\bar{y}$	$\bar{y}'$	$\bar{y}_{+1}$
	$\bar{Q}$	$\bar{Q}'$	$\bar{Q}_{+1}$

Table A1.2 Notation for Focal Lengths

		Object space	Image space	Same object and image medium
Thin lens or mirror	Axial Focal Length	f	f'	f
	Marginal Focal Length	f_M	f'_M	f_M
Thick lens or system	Effective Focal Length	F	F'	F
	Back Focal Length	F_B	f'_B	F_B
	Marginal Focal Length	F_M	f'_M	F_M

The focal length f for a thin lens or mirror is represented with lower case. The effective focal length F for a thick lens or a complete system is represented with upper case. The back focal length for a thick lens or system is represented with a subscript B.

The focal length as measured from the focus to the principal surface, along the optical axis, is used without any subscript. If this focal length is measured from the focus to the principal surface, along the meridional ray, a subscript M is used.

The focal ratio (or f-number), is represented by FN and defined as follows:

$$FN = \frac{F}{\text{Diameter of entrance pupil}} \tag{A1.1}$$

The numerical aperture for an object at a finite distance is

$$NA = n_0 \sin U_0 \tag{A1.2}$$

where n_0 is the refractive index in the object medium.

The primary aberration coefficients are represented by a short abreviation of its name. These names closely resemble those of Conrady. However, there are some important differences. To avoid confusion with the concept of longitudinal and transverse aberrations, the chromatic aberrations are named axial chromatic and magnification chromatic aberrations. A second important thing to notice is that some aberrations like the spherical aberration and astigmatism may be evaluated by their transversal or longitudinal extent. A letter T for transverse or a letter L for longitudinal is added to the name of these aberrations.

The aberrations due to only one surface or to a complete system are represented with the same symbol, asumming that one surface may be

Notation and Primary Aberration Coefficients Summary

Table A1.3 Notation for Primary Aberration Coefficients

Aberration	Total		Surface contribution	
	Longitudinal	Transverse	Longitudinal	Transverse
Spherical aberration	$SphL$	$SphT$	$SphLC$	$SphTC$
Coma (sagittal)	—	$Coma_S$	—	$ComaC_S$
Coma (tangential)	—	$Coma_T$	—	$ComaC_T$
Astigmatism (sagittal)	$AstL_S$	$AstT_S$	$AstL_SC$	$AstT_SC$
Astigmatism (tangential)	$AstL_T$	$AstT_T$	$AstL_TC$	$AstT_TC$
Distortion	—	$Dist$	—	$DistC$
Petzval curvature	Ptz	—	$PtzC$	—
Axial chromatic	$AchrL$	$AchrT$	$AchrLC$	$AchrTC$
Magnification chromatic	—	$Mchr$	—	$MchrC$

Table A1.4 Notation for Ray and Wave Aberrations

Exact aberration	Longitudinal	Transverse	Wave aberration
General (off-axis)	LA	TA	W
x Component (off-axis)	LA_x	TA_y	—
y Component (off-axis)	LA_x	TA_y	—
On-axis	LA_0	TA_0	W_0

considered as a system with only one refracting surface. The contribution of a surface to the total aberration in the system is represented by adding a letter C as usual. A subscript is sometimes used to indicate the surface to which it applies. The symbol (without primas, as in Conrady's notation) represents the aberration after refraction on the surface. The aberration before refraction would be represented by a subscript -1, which stands for the previous surface. Thus, the aberration in the object space (before surface 1 in the system) is represented with the subscript 0. The object is the surface number zero in the optical system. The aberration after the last surface (k) in the system is represented by the subscript k. Table A1.3 shows the symbols used to represent these aberrations.

When doing exact ray tracing, the aberration measured in a direction parallel to the optical axis is called the longitudinal aberration LA. The value of the aberration in a perpendicular direction to the optical axis is called the transverse aberration TA. The wavefront deformations are represented by W. These symbols are shown in Table A1.4.

A1.2 SUMMARY OF PRIMARY ABERRATION COEFFICIENTS

A1.2.1 Conrady's Form

This is the form of the coefficients as derived by Conrady, but with our sign notation.

Spherical aberration

$$SphTC = \frac{y(n/n')(n-n')(i+u')i^2}{2n'_k u'_k} \quad (A1.3)$$

and the contribution of the aspheric deformation is

$$SphTC_{asph} = -(8A_1 + Kc^3)\left(\frac{n-n'}{2}\right)\left(\frac{y^4}{n'_k u'_k}\right) \quad (A1.4)$$

Coma

$$Coma_S C = SphTC\left(\frac{\bar{i}}{i}\right) \quad (A1.5)$$

the aspheric contribution is represented by $Coma_{S\,asph}$ and given by

$$Coma_S C_{asph} = SphTC_{asph}\left(\frac{\bar{y}}{y}\right) \quad (A1.6)$$

Astigmatism

$$AstT_s C = SphTC\left(\frac{\bar{i}}{i}\right)^2 \quad (A1.7)$$

the aspheric contribution is

$$AstL_S C_{asph} = SphLC_{asph}\left(\frac{\bar{y}}{y}\right)^2 \quad (A1.8)$$

Petzval curvature

$$PtzC = \frac{h'^2_k n'_k}{2}\left(\frac{n'-n}{nn'r}\right) \quad (A1.9)$$

Distortion

$$DistC = Coma_S C\left(\frac{\bar{i}}{i}\right)^2 PtzC\left(\frac{\bar{i}}{i}\right) u'_k \qquad (A1.10)$$

the contribution introduced by the aspheric deformation is

$$DistC_{asph} = SphTC_{asph}\left(\frac{\bar{y}}{y}\right)^3 \qquad (A1.11)$$

Axial chromatic aberration

$$AchrTC = \frac{yni}{n'_k u'_k}\left(\frac{n_F - n_C}{n} - \frac{n'_F - n'_C}{n'}\right) \qquad (A1.12)$$

Magnification chromatic aberration

$$MchrC = AchrTC\left(\frac{\bar{i}}{i}\right) \qquad (A1.13)$$

A1.2.2 For Numerical Calculation

The following slightly different set of equations have been recommended by many authors for use in electronic computers.

Spherical aberration

$$SphTC = \sigma i^2 \qquad (A1.14)$$

where

$$\sigma = \frac{y(n/n')(n - n')(i + u')}{2n'_k u'_k} \qquad (A1.15)$$

the contribution of the aspheric deformation is

$$SphTC_{asph} = -(8A_1 + Kc^3)\left(\frac{n - n'}{2}\right)\left(\frac{y^4}{n'_k u'_k}\right) \qquad (A1.16)$$

Coma

$$Coma_S C = \sigma \bar{i} i \quad \text{(A1.17)}$$

The aspheric contribution is

$$Coma_S C_{\text{asph}} = SphTC_{\text{asph}} \left(\frac{\bar{y}}{y}\right) \quad \text{(A1.18)}$$

Astigmatism

$$Ast T_S C = \sigma \bar{i}^2 \quad \text{(A1.19)}$$

the aspheric contribution is

$$Ast L_S C_{\text{asph}} = SphLC_{\text{asph}} \left(\frac{\bar{y}}{y}\right)^2 \quad \text{(A1.20)}$$

Petzval curvature

$$PtzC = -\frac{{h'_k}^2 n'_k}{2}\left(\frac{n'-n}{n n' r}\right) \quad \text{(A1.21)}$$

Distortion

$$DistC = \bar{\sigma}\bar{i}i + \frac{h'_k}{2}(\bar{u}'^2 - \bar{u}^2) \quad \text{(A1.22)}$$

where

$$\bar{\sigma} = \frac{\bar{y}(n/n')(n-n')(\bar{i}+\bar{u}')}{2n'_k u'_k} \quad \text{(A1.23)}$$

the contribution to the aspheric deformation is

$$DistC_{\text{asph}} = SphTC_{\text{asph}} \left(\frac{\bar{y}}{y}\right)^3 \quad \text{(A1.24)}$$

Axial chromatic aberration

$$AchrTC = \frac{yni}{n'_k u'_k} \left(\frac{n'_F - n'_C}{n'} - \frac{n'_F - n'_C}{n'} \right) \qquad (A1.25)$$

Magnification chromatic aberration

$$MchrC = AchrTC \left(\frac{\bar{i}}{i} \right) \qquad (A1.26)$$

The magnitude of the aberrations depends both on the lens aperture and on the image height. Table A1.5 shows how each of the primary aberrations depend on these two parameters.

Table A1.5 Functional Dependence of Primary Aberrations on Aperture and Image Height

		Semiaperture y	Image height h'
Spherical aberration	Longitudinal	y^2	None
	Transverse	y^3	
	Wavefront	S^4	
Coma	Transverse	y^2	h'
	Wavefront	$S^2 y = (x^2 + y^2) y$	
Astigmatism	Longitudinal	None	h'^2
	Transverse	y	
	Wavefront	$S^2 + 2 y^2 = x^2 + 3 y^2$	
Petzval curvature	Longitudinal	None	h'^2
	Wavefront	S^2	
Distortion	Transverse	None	h'^3
	Wavefront	y	
Axial chromatic aberration	Longitudinal	None	None
	Transverse	y	
	Wavefront	S^2	
Magnification chromatic aberration	Transverse	None	h'
	Wavefront	y	

Appendix 2
Mathematical Representation of Optical Surfaces

A2.1 SPHERICAL AND ASPHERICAL SURFACES

An optical surface may have many shapes (Herzberger and Hoadley, 1946; Mertz, 1979a,b; Shannon, 1980; Schulz, 1988; Malacara, 1992), but the most common is spherical, whose sagitta for a radius of curvature r and a semidiameter $S = x^2 + y^2$ may be written as

$$Z = r - \sqrt{r^2 - S^2} \tag{A2.1}$$

However, this representation fails for flat surfaces. A better form is

$$Z = \frac{cS^2}{1 + \sqrt{1 - c^2 S^2}} \tag{A2.2}$$

where, as usual, $c = 1/r$, and $S^2 = x^2 + y^2$.

A conic surface is characterized by its eccentricity e. If we define a conic constant $K = -e^2$, then the expression for a conic of revolution may be written as

$$Z = \frac{1}{K+1}\left[r - \sqrt{r^2 - (K+1)S^2}\right] \tag{A2.3}$$

which works for all conics except the paraboloid. It also fails for flat surfaces, so a better representation is

$$Z = \frac{cS^2}{1 + \sqrt{1 - (K+1)c^2 S^2}} \tag{A2.4}$$

The conic constant defines the type of conic, according to the Table A2.1. It is easy to see that the conic constant is not defined for a flat surface. Figure A2.1 shows the shape of some conic surfaces.

To the equation for the conic of revolution we may add some aspheric deformation terms as follows:

$$Z = \frac{cS^2}{1 + \sqrt{1-(K+1)c^2S^2}} + A_1 S^4 + A_2 S^6 + A_3 S^8 + A_4 S^{10} \qquad (A2.5)$$

An axicon (McLeod, 1954, 1960), which has the conical shape illustrated in Fig. A2.2, may be represented by means of a hyperboloid with

Table A2.1 Values of Conic Constants for Conicoid Surfaces

Type of conic	Conic constant value
Hyperboloid	$K < -1$
Paraboloid	$K = -1$
Ellipse rotated about its major axis (prolate spheroid or ellipsoid)	$-1 < K < 0$
Sphere	$K = 0$
Ellipse rotated about its minor axis (oblate spheroid)	$K > 0$

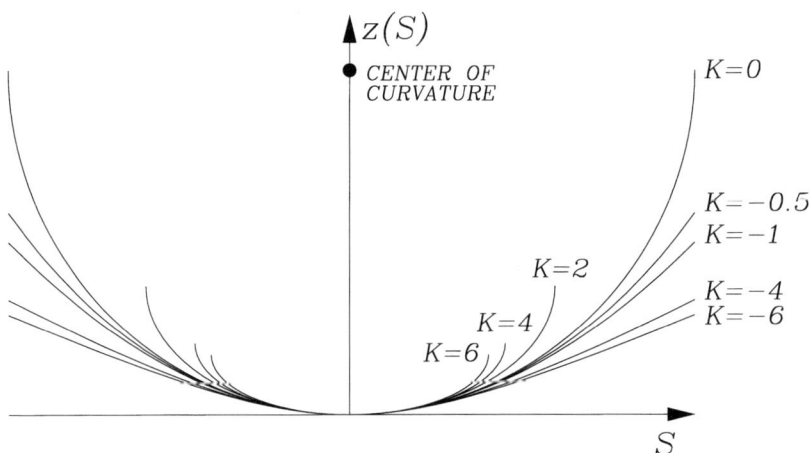

Figure A2.1 Shape of some conic surfaces.

Mathematics of Optical Surfaces

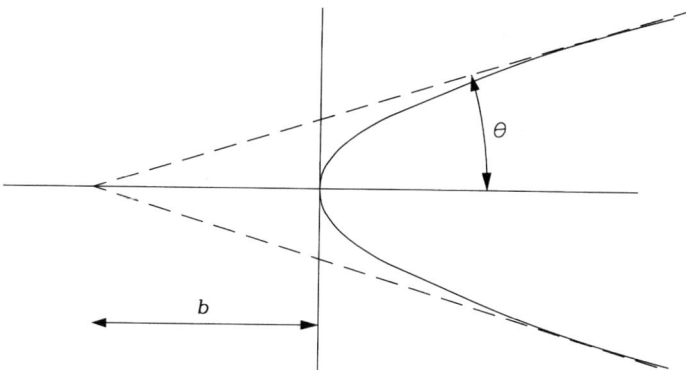

Figure A2.2 An axicon.

an extremely large curvature, obtaining

$$K = -(1 + \tan^2 \theta) < -1 \tag{A2.6}$$

and

$$c = \frac{1}{(K+1)b} \tag{A2.7}$$

Sometimes it may be interesting to express the optical surface as a spherical surface plus some aspheric deformation terms that include the effect of the conic shape. Then, we may find that

$$Z = \frac{cS^2}{1 + \sqrt{1 - c^2 S^2}} + B_1 S^4 + B_2 S^6 + B_3 S^8 + B_4 S^{10} \tag{A2.8}$$

where

$$B_1 = A_1 + \frac{[(K+1) - 1]c^3}{8} \tag{A2.9}$$

$$B_2 = A_2 + \frac{[(K+1)^2 - 1]c^5}{16} \tag{A2.10}$$

$$B_3 = A_3 + \frac{5[(K+1)^3 - 1]c^7}{128} \tag{A2.11}$$

and

$$B_4 = A_4 + \frac{7[(K+1)^4 - 1]c^9}{256} \tag{A2.12}$$

A2.1.1 Aberrations of Normals to Aspheric Surface

A normal to the aspheric optical surface intersects the optical axis at a distance Z_n from the center of curvature. Sometimes it is important to know the value of this distance, called aberration of the normals. To compute its value, we first find the derivative of Z with respect to S, as follows:

$$\frac{dZ}{dS} = \frac{cS}{\sqrt{1-(K+1)c^2S^2}} + 4A_1S^3 + 6A_2S^5 + 8A_3S^7 + 10A_4S^9 \tag{A2.13}$$

Then, the distance L_n as shown in Fig. A2.3 is

$$L_n = \frac{S}{dZ/dS} + Z \tag{A2.14}$$

which as shown by Buchroeder et al. (1972), for conic surfaces becomes

$$L_n = \frac{1}{c} - KZ \tag{A2.15}$$

The envelope of the caustic produced by the normals to the aspheric surface is called the *evolute* in analytic geometry. It is interesting to see that

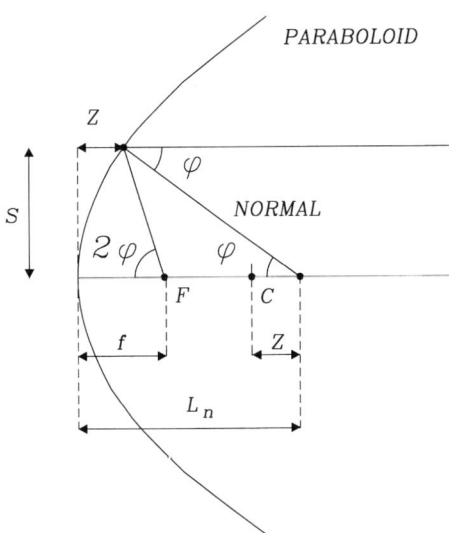

Figure A2.3 Some parameters for conic surfaces.

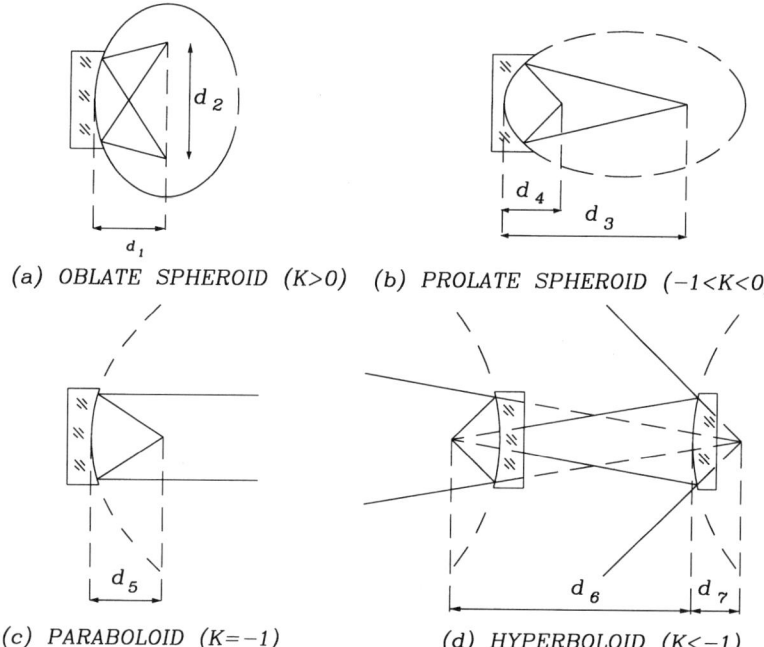

Figure A2.4 Aberration of the normals to the aspheric surface: (a) oblate spheroid ($K > 0$); (b) prolate spheroid ($-1 < K < 0$); (c) paraboloid ($K = -1$); (d) hyperboloid ($K > -1$).

for the case of a paraboloid ($K = -1$), as shown in Fig. A2.4 this aberration of the normals becomes

$$L_n = \frac{1}{c} + Z = \frac{1}{c} + f\tan^2\varphi \tag{A2.16}$$

where the angle φ is the angle between the normal to the surface and the optical axis, as illustrated in this figure, and f is the focal length of the paraboloid. We may see that, for this case of the paraboloid, the distance Z_n from the center of curvature to the intersection of the normal with the optical axis is equal to the sagitta Z, as shown in Fig. A2.3. In the general case of aspheric surfaces, the intersection of the normals may be approximated by

$$L_n = \frac{1}{c} - \frac{(Kc^3 + 8A_1)S^2}{2c^2} \tag{A2.17}$$

Sometimes, it is desirable to express a nonplane aspherical surface in terms of the angle φ between the normal to the surface and the optical axis instead of the ray height S. In this case the following relation can be used in Eq. (A2.3):

$$c^2 S^2 = \frac{\sin^2 \varphi}{1 + K \sin^2 \varphi} \tag{A2.18}$$

A2.1.2 Some Parameters for Conic Surfaces

The positions for the foci of the conic surfaces as functions of the radius of curvature r and the conic constant K, as illustrated in Fig. A2.4, are

$$d_1 = \frac{r}{(K+1)} \tag{A2.19}$$

$$d_2 = \frac{r}{(K+1)}(2\sqrt{K}) \tag{A2.20}$$

$$d_3, d_4 = \frac{r}{(K+1)}(1 \pm \sqrt{-K}) \tag{A2.21}$$

$$d_5 = \frac{r}{2} \tag{A2.22}$$

and

$$d_6, d_7 = \frac{r}{(K+1)}(\sqrt{-K} \pm 1) \tag{A2.23}$$

It is important to point out that the oblate spheroid is not an optical system with symmetry of revolution, since the object and image are off-axis. Thus, the image is astigmatic.

A2.1.3 Off-Axis Paraboloids

Figure A2.5 shows an off-axis paraboloid tilted an angle θ with respect to the axis of the paraboloid. The line perpendicular to the center of the off-axis paraboloid is defined as the optical axis. If the diameter of this surface is small compared with its radius of curvature, it may be approximated by a toroidal surface. Then, the tangential curvature, c_t, defined as the curvature along a circle centered on the axis of the paraboloid, as shown by Malacara (1991), is

$$c_t = \frac{\cos^3 \theta}{2f} \tag{A2.24}$$

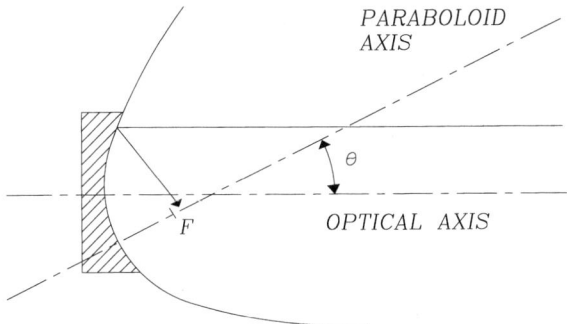

Figure A2.5 Off-axis paraboloid.

where f is the focal length of the paraboloid, and θ is the angle between the axis of the paraboloid and the optical axis, as in Fig. A2.5. The sagittal curvature c_s, defined as the curvature along a radial direction, is

$$c_s = \frac{\cos\theta}{2f} \qquad (A2.25)$$

The on-axis vertex curvature of the paraboloid is

$$c = \frac{1}{2f} \qquad (A2.26)$$

hence, we may find that

$$c_t c^2 = c_s^3 \qquad (A2.27)$$

which, as shown by Menchaca and Malacara (1984), is true for any conic, not just for paraboloids.

As shown by (Malacara, 1991) the shape of the off-axis paraboloid in the new system of coordinates rotated by an angle θ is given by

$$Z(x,y) = \frac{(X^2 + Y^2 \cos^2\theta + Z^2 \sin^2\theta)\cos\theta}{4f(1 + Y\sin\theta\cos^2\theta/2f)} \qquad (A2.28)$$

When the diameter of the paraboloid is relatively small, the surface may be approximated by

$$Z(x,y) = \frac{c_x X^2}{2} + \frac{c_y Y^2}{2} - \frac{c^2}{4}\cos^3\theta\sin\theta(1 + 3\cos^2\theta)(X^2 + Y^2)Y$$
$$- \frac{c^2}{4}\cos^3\theta\sin^3\theta(3X^2 - Y)Y \qquad (A2.29)$$

This surface has the shape of a toroid (represented by primary astigmatism) as indicated by the first two terms. An additional comatic deformation is represented by the third term. As should in Plloptd 5, a comatic shape is like that of a spoon, with a nonconstant increasing curvature along one diameter and another, constant curvature along the other perpendicular diameter. With somewhat larger diameters triangular astigmatism appears, as shown by the last term. This triangular astigmatism is the shape obtained by placing a semiflexible disk plate on top of three supports located at its edge, separated by 120°.

A2.1.4 Toroidal and Spherocylindrical Surfaces

An astigmatic surface is one that has two different curvatures along two orthogonal axes. For example, a toroidal surface, as described before, an ellipsoid of revolution, and an off-axis paraboloid are astigmatic. If we restrict our definition only to surfaces that have bilateral symmetry about these two orthogonal axes the off-axis paraboloid is out. Let us assume that the two orthogonal axes of symmetry are along the x and y axes. Then, the two orthogonal curvatures are given by

$$c_x = \frac{1}{r_x} + \left(\frac{\partial^2 Z(x,y)}{\partial x^2}\right) \tag{A2.30}$$

and

$$c_y = \frac{1}{r_y} + \left(\frac{\partial^2 Z(x,y)}{\partial y^2}\right) \tag{A2.31}$$

and the curvature c_θ in any arbitrary direction at an angle θ with respect to the x axis is given by

$$c_\theta = \frac{1}{r_\theta} = c_x \cos^2\theta + c_y \sin^2\theta \tag{A2.32}$$

If we further restrict our definition of astigmatic surfaces to surfaces where the cross-sections along the symmetry axes are circles we still have an infinite number of possibilities (Malacara-Doblado et al., 1996). The most common of these surfaces are the toroidal and the spherocylindrical surfaces. Sasian (1997) has shown that an astigmatic surface can sometimes replace an off-axis paraboloid, which is more difficult to manufacture.

Mathematics of Optical Surfaces 493

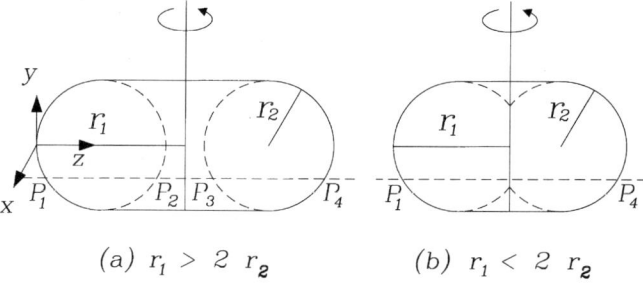

Figure A2.6 Toroidal surface parameters: (a) $r_1 > 2r_2$; (b) $r_1 < 2r_2$.

A toroidal surface, shown in Fig. A2.6, may be generated in many ways (Malacara and Malacara, 1971). It has the shape of a donut and is represented by

$$Z = \left(\left[\sqrt{(r_y^2 - Y^2)} + r_x - r_y \right]^2 - X^2 \right)^{1/2} + r_x \qquad (A2.33)$$

where r_x is the radius of curvature on the x–z plane (large radius) and r_y is the radius of curvature in the y–z plane (small radius). We may see that this expression is not symmetrical in X and Y because the axis of symmetry of the toroid is parallel to the y axis ($X=0$, $Z=r_x$), but does not have any symmetry about any axis parallel to the x axis.

As we may see in Fig. A2.6(a), observing the dotted line crossing the toroid, there are four solutions ($\mathbf{P}_1$, $\mathbf{P}_2$, $\mathbf{P}_3$, $\mathbf{P}_4$) for Z, given a pair of values of X and Y. This is obvious if we notice that we have two square roots, one inside the other. In Fig. A2.6(b) when $r_y < r_x$, two of the four solutions are imaginary.

Another similar surface, called a spherocylindrical surface (Menchaca and Malacara, 1986), is illustrated in Fig. A2.7 and expressed by

$$Z = \frac{c_x X^2 + c_y Y^2}{1 + \left[1 - (c_x X^2 + c_y Y^2)^2 / (X^2 + Y^2) \right]^{1/2}} \qquad (A2.34)$$

where c_x and c_y are the curvatures along the x and y axes, respectively. This surface is symmetric in X and Y. What these two surfaces have in common is that their cross-sections in the planes x–z and y–z are circles. If the clear apertures of these two types of surfaces, the toroidal and the spherocylindrical, are small compared with their radii of curvature, they become identical for all practical purposes.

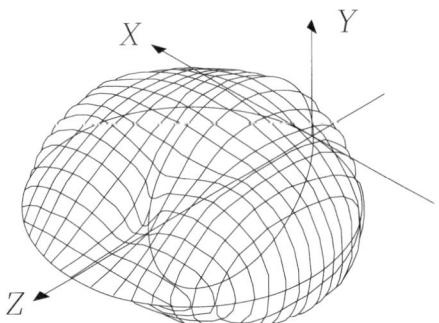

Figure A2.7 Spherocylindrical surface.

An important difference between the toroidal and the spherocylindrical surface is that the second has only two possible solutions for Z, since there is only one square root.

REFERENCES

Buchroeder, R. A., Elmore, L. H., Shack, R. V., and Slater, P. N., "The Design, Construction and Testing of the Optics for the 147-cm-Aperture Telescope," *Optical Sciences Center Technical Report No. 79*, University of Arizona, Tucson, AZ, 1972.

Herzberger, M. and Hoadley, H. O., "The Calculation of Aspherical Correcting Surfaces," *J. Opt. Soc. Am.*, **36**, 334–340 (1946).

Malacara, D., "Some Parameters and Characteristics of an Off Axis Paraboloid," *Opt. Eng.*, **30**, 1277–1280 (1991).

Malacara, D., "An Optical Surface and Its Characteristics," Appendix 1 in *Optical Shop Testing*, D. Malacara, ed., John Wiley, New York, 1992.

Malacara, D. and Malacara, Z., "Diamond Tool Generation of Toroidal Surfaces," *Appl. Opt.*, **10**, 975–977 (1971).

Malacara-Doblado D., Malacara-Hernández, D., and García-Márquez, J., "Axially Astigmatic Surfaces: Different Types and Their Properties," *Opt. Eng.*, **35**, 3422–3426 (1996).

McLeod, J. H., "The Axicon: A New Type of Optical Element," *J. Opt. Soc. Am.*, **44**, 592–597 (1954).

McLeod, J. H., "Axicons and Their Uses," *J. Opt. Soc. Am.*, **50**, 166–169 (1960).

Menchaca, C. and Malacara, D., "Directional Curvatures in a Conic Surface," *Appl. Opt.*, **23**, 3258–3260 (1984).

Menchaca, C. and Malacara, D., "Toroidal and Sphero-Cylindrical Surfaces," *Appl. Opt.*, **25**, 3008–3009 (1986).

Mertz, L., "Geometrical Design for Aspheric Reflecting Systems," *Appl. Opt.*, **18**, 4182–4186 (1979a).

Mertz, L., "Aspheric Potpourri," *Appl. Opt.*, **20**, 1127–1131 (1979b).
Sasian, J. M., "Double Curvature Surfaces in Mirror System Design," *Opt. Eng.*, **36**, 183–188 (1997).
Schulz, G., "Aspheric Surfaces," in *Progress in Optics*, Vol. XXV, E. Wolf, ed., Chap. IV, North Holland, Amsterdam, 1988.
Shannon, R. R., "Aspheric Surfaces," in *Applied Optics and Optical Engineering*, Vol. VIII, R. Shannon and J. C. Wyant, eds., Academic Press, San Diego, CA, 1980.

Appendix 3
Optical Materials

A3.1 OPTICAL GLASSES

Optical glass is mainly determined by their value of two constants, namely, the refractive index and the Abbe constant. A diagram of the Abbe number V_d versus the refractive index n_d for Schott glasses is shown in Fig. A3.1. The glasses with a letter "K" at the end of the glass type name are crown glasses and those with a letter "F" are flint glasses. Besides the refractive index for the d line, several other quantities define the main refractive characteristics of the glass. The difference $(n_F - n_C)$ is called the principal dispersion. The Abbe value expresses the way in which the refractive index changes with wavelength. The Abbe value V_d for the d line is defined as

$$V_{dSphT}C = \frac{n_d - 1}{n_F - n_C}\sigma i^2 \tag{A3.1}$$

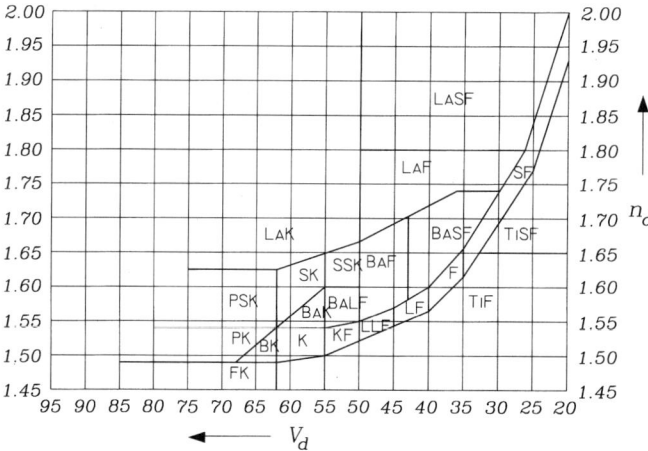

Figure A3.1 Abbe number versus refractive index chart for optical glasses.

497

The secondary spectrum produced by an optical glass is determined by the partial dispersion of the glass. The partial dispersion $P_{g,F}$ for the lines g and F is defined as

$$PV_{g,Fd} = \frac{n_{gd} - n_{Fl}}{n_F - n_C} \tag{A3.2}$$

There is such a large variety of optical glasses that to have a complete stock of all types in any optical shop is impossible. Many lens designers have attempted to reduce the list to the most important glasses, taking into consideration important factors, like optical characteristics, availability, and price. A list of some of the most commonly used optical glasses is given in Table A3.1.

Table A3.1 Some Schott Optical Glasses

Name	V_d	n_C	n_d	n_F	n_g
BaF4	43.93	1.60153	1.60562	1.61532	1.62318
BaFN10	47.11	1.66579	1.67003	1.68001	1.68804
BaK4	56.13	1.56576	1.56883	1.57590	1.58146
BaLF5	53.63	1.54432	1.54739	1.55452	1.56017
BK7	64.17	1.51432	1.51680	1.52238	1.52668
F2	36.37	1.61503	1.62004	1.63208	1.64202
K4	57.40	1.51620	1.51895	1.52524	1.53017
K5	59.48	1.51982	1.52249	1.52860	1.53338
KzFSN4	44.29	1.60924	1.61340	1.62309	1.63085
LaF2	44.72	1.73905	1.74400	1.75568	1.76510
LF5	40.85	1.57723	1.58144	1.59146	1.59964
LaK9	54.71	1.68716	1.69100	1.69979	1.70667
LLF1	45.75	1.54457	1.54814	1.55655	1.56333
PK51A	76.98	1.52646	1.52855	1.53333	1.53704
SF1	29.51	1.71032	1.71736	1.73463	1.74916
SF2	33.85	1.64210	1.64769	1.66123	1.67249
SF5	32.21	1.66661	1.67270	1.68750	1.69985
SF8	31.18	1.68250	1.68893	1.70460	1.71773
SF10	28.41	1.72085	1.72825	1.74648	1.76198
SF15	30.07	1.69221	1.69895	1.71546	1.72939
SF56A	26.08	1.77605	1.78470	1.80615	1.82449
SK4	58.63	1.60954	1.61272	1.62000	1.62569
SK6	56.40	1.61046	1.61375	1.62134	1.62731
SK16	60.32	1.61727	1.62041	1.62756	1.63312
SK18A	55.42	1.63505	1.63854	1.64657	1.65290
SSKN5	50.88	1.65455	1.65844	1.66749	1.67471

Optical Materials

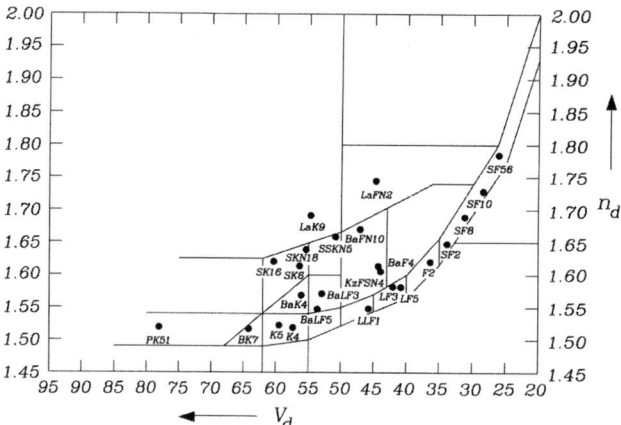

Figure A3.2 Some common optical glasses.

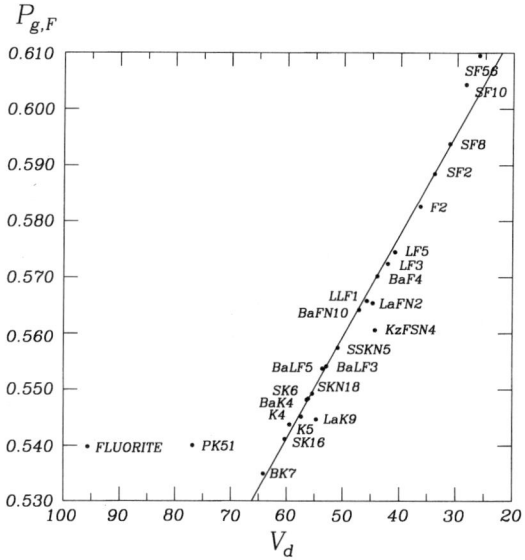

Figure A3.3 Abbe number versus relative partial dispersion of optical glasses.

The location of these glasses in a diagram of the Abbe number V_d versus the refractive index n_d is shown in Fig. A3.2. Figure A3.3 shows a plot of the partial dispersion $P_{g,F}$ versus the Abbe number V_d.

Ophthalmic glasses are also widely used. Table A3.2 lists some of these glasses.

Table A3.2 Some Schott Ophthalmic Glasses

Glass type		V_d	n_C	n_d	n_F	n_g	Density (g/ml)
Crown	(D 0391)	58.6	1.5203	1.5230	1.5292	1.5341	2.55
Flint	(D 0290)	44.1	1.5967	1.6008	1.6103	1.6181	2.67
Flint	(D 0389)	42.9	1.5965	1.6007	1.6105	1.6185	2.67
Flint	(D 0785)	35.0	1.7880	1.7946	1.8107	1.8239	3.60
Flint	(D 0082)	30.6	1.8776	1.8860	1.9066	1.9238	4.02
Low density flint	(D 0088)	30.8	1.6915	1.7010	1.7154	1.7224	2.99

Table A3.3 Other Optical Isotropic Materials

Material	V_d	n_C	n_d	n_F	n_g
Fused rock crystal	67.6	1.45646	1.45857	1.46324	1.46679
Synthetic fused silica	67.7	1.45637	1.45847	1.46314	1.46669
Fluorite	95.3	1.43249	1.43384	1.43704	1.43950

Table A3.4 Some Optical Plastics

Material	V_d	n_C	n_d	n_F	n_g
Acrylic	57.2	1.488	1.491	1.497	1.5000
Polystyrene	30.8	1.584	1.590	1.604	1.6109
Polycarbonate	30.1	1.577	1.583	1.604	1.6039
CR-39	60.0	1.495	1.498	1.504	1.5070

Finally, Table A3.3 lists some other optical isotropic materials used in optical elements.

A3.2 OPTICAL PLASTICS

There is a large variety of plastics, with many different properties, used to make optical components, but some of the most common ones are listed in Table A3.4.

A3.3 INFRARED AND ULTRAVIOLET MATERIALS

Most glasses are opaque to infrared and ultraviolet radiation. If a lens has to be transparent at these wavelengths special materials have to be selected.

The subject of these special materials is so wide that it cannot be treated in this book due to a lack of space. Instead, some references are given for the interested reader.

BIBLIOGRAPHY

Barnes, W. P., Jr., "Optical Materials—Reflective," in *Applied Optics and Optical Engineering*, Vol. VII, R. R. Shannon and J. C. Wyant, eds., Academic Press, San Diego, CA, 1979.

Kavanagh, A. J., "Optical Material," in *Military Standardization Handbook: Optical Design, MIL-HDBK 141*, U.S. Defense Supply Agency, Washington, DC, 1962.

Kreidl, N. J. and Rood, J. L., "Optical Materials," in *Applied Optics and Optical Engineering*, Vol. I, R. Kingslake, ed., Academic Press, San Diego, CA, 1965.

Malitson, I. H., "A Redetermination of Some Optical Properties of Calcium Fluoride," *Appl. Opt.*, **2**, 1103–1107 (1963).

McCarthy, D. E., "The Reflection and Transmission of Infrared Materials, Part 1. Spectra from 2 µm to 50 µm," *Appl. Opt.*, **2**, 591–595 (1963).

McCarthy, D. E., "The Reflection and Transmission of Infrared Materials, Part 2. Bibliography," *Appl. Opt.*, **2**, 596–603 (1963).

McCarthy, D. E., "The Reflection and Transmission of Infrared Materials, Part 3. Spectra from 2 µm to 50 µm," *Appl. Opt.*, **4**, 317–320 (1965).

McCarthy, D. E., "The Reflection and Transmission of Infrared Materials, Part 4. Bibliography," *Appl. Opt.*, **4**, 507–511 (1965).

McCarthy, D. E., "The Reflection and Transmission of Infrared Materials, Part 5. Spectra from 2 µm to 50 µm," *Appl. Opt.*, **7**, 1997–2000 (1965).

McCarthy, D. E., "The Reflection and Transmission of Infrared Materials, Part 6. Bibliography," *Appl. Opt.*, **7**, 2221–2225 (1965).

Parker, C. J., "Optical Materials—Refractive," in *Applied Optics and Optical Engineering*, Vol. VII, R. R. Shannon and J. C. Wyant, eds., Academic Press, San Diego, CA, 1983.

Pellicori, S. F., "Transmittances of Some Optical Materials for Use Between 1900 and 3400 Å," *Appl. Opt.*, **3**, 361–366 (1964).

Welham, B., "Plastic Optical Components," in *Applied Optics and Optical Engineering*, Vol. VII, R. R. Shannon and J. C. Wyant, eds., Academic Press, San Diego, CA, 1979.

Appendix 4
Exact Ray Tracing of Skew Rays

A4.1 EXACT RAY TRACING

Ray-tracing procedures have been described many times in the literature (Herzberger and Hoadley, 1946; Herzberger, 1951, 1957; Allen and Snyder, 1952; Lessing, 1962; Spencer and Murty, 1962; Malacara, 1965; Feder, 1968; Cornejo-Rodríguez and Cordero-Dávila, 1979). These methods are basically simple, in the sense that only elementary geometry is needed. However, tracing of skew rays through aspherical surfaces is quite involved from an algebraic point of view. This is the reason why these methods are not well described in many optical design books. Nevertheless, the practical importance of ray-tracing procedures is great, especially if a computer program is to be used or understood.

We will derive now the necessary equations to trace skew rays through aspherical surfaces, using a procedure described by Hopkins and Hanau (1962). This method is formed by the following four basic steps:

1. Transfer from first surface to plane tangent to next surface
2. Transfer from tangent plane to osculating sphere
3. Transfer from osculating sphere to aspheric surface
4. Refraction at aspheric surface

The rays are defined by the intersection coordinates X, Y, and Z on the first surface and their direction cosines multiplied by the refractive indices, K, L, and M. We will now study in some detail these steps.

A4.1.1 Transfer from First Surface to Plane Tangent to Next Surface

To begin the derivation of the formulas to trace skew rays, let us consider Fig. A4.1. The origin of coordinates is at the vertex of the optical surface. The starting point for the ray are the coordinates X_{-1}, Y_{-1}, and Z_{-1} on the preceding surface. The ray direction is given by the direction cosines

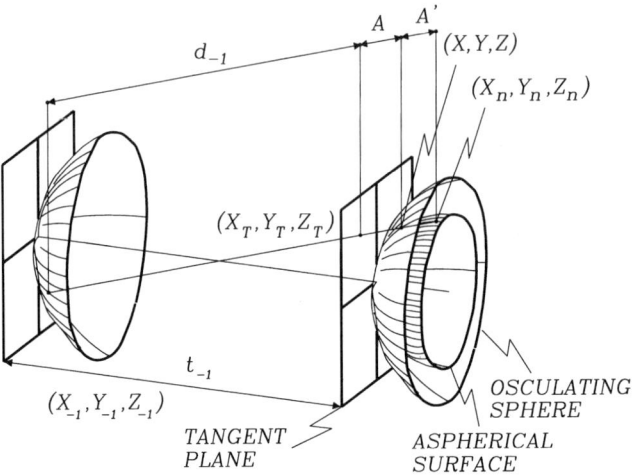

Figure A4.1 Ray tracing through an optical surface.

K_{-1}/n_{-1}, L_{-1}/n_{-1}, and M_{-1}/n_{-1}. The distance d, along the ray, from the starting point to the intersection of the ray with the plane tangent to the next surface is given by the definition of M_{-1} as

$$\frac{d_{-1}}{n_{-1}} = \frac{t_{-1} - Z_{-1}}{M_{-1}} \tag{A4.1}$$

then, using this value and the definitions of L and M, the coordinates X_T and Y_T on the tangent plane are given by

$$X_T = X_{-1} + \left(\frac{d_{-1}}{n_{-1}}\right) K_{-1} \tag{A4.2}$$

and

$$Y_T = Y_{-1} + \left(\frac{d_{-1}}{n_{-1}}\right) L_{-1} \tag{A4.3}$$

A4.1.2 Transfer from Tangent Plane to Osculating Sphere

A sphere is said to be osculating to an aspherical surface when they are tangents at their vertices and have the same radii of curvature at that point. Let us find now the intersection of the ray with the osculating (*osculum* is the latin word for kiss) sphere. If A is the distance along the ray, from the point on the tangent plane to the intersection with the sphere, the coordinates of this intersection are

$$X = X_T + \left(\frac{A}{n_{-1}}\right) K_{-1} \tag{A4.4}$$

$$Y = Y_T + \left(\frac{A}{n_{-1}}\right) L_{-1} \tag{A4.5}$$

and

$$Z = \left(\frac{A}{n_{-1}}\right) M_{-1} \tag{A4.6}$$

However, before computing these coordinates we need to know the value of the distance A. Then, the first step is to find this distance, illustrated in Fig. A4.1. From Fig. A4.2 we may see that

$$\begin{aligned} Z &= r - \sqrt{r^2 - (X^2 + Y^2)} \\ &= \frac{1}{c} - \frac{1}{c}\sqrt{1 - c^2(X^2 + Y^2)} \end{aligned} \tag{A4.7}$$

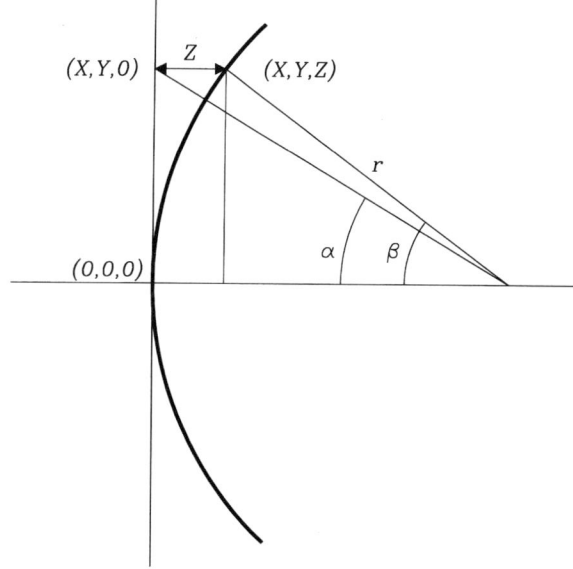

Figure A4.2 Some parameters in ray tracing.

and then, transposing and squaring we obtain

$$c^2(X^2 + Y^2 + Z^2) - 2cZ = 0 \tag{A4.8}$$

then, we substitute here the values of X, Y, and Z given by Eqs. (A4.4)–(A4.6):

$$\left(\frac{A}{n_{-1}}\right)^2 c(K_{-1}^2 + L_{-1}^2 + M_{-1}^2) - 2\left(\frac{A}{n_{-1}}\right)[M_{-1} - c(Y_T L_{-1} + X_T K_{-1})]$$
$$+ c(X_T^2 + Y_T^2) = 0 \tag{A4.9}$$

where we divided by c, assuming that c is not zero. Since the sum of the squares of the direction cosines is one, we may write

$$cn_{-1}^2 \left(\frac{A}{n_{-1}}\right)^2 - 2B\left(\frac{A}{n_{-1}}\right) + H = 0 \tag{A4.10}$$

where we have defined:

$$B = [M_{-1} - c(Y_T L_{-1} + X_T K_{-1})] \tag{A4.11}$$

and

$$H = c(X_T^2 + Y_T^2) = r\left(\frac{X_T^2 + Y_T^2}{r^2}\right) = r\tan^2\beta \tag{A4.12}$$

where the angle β is shown in Fig. A4.2. To obtain the desired value of A we must find the roots of the second-degree equation (A4.10), as follows:

$$\left(\frac{A}{n_{-1}}\right) = \frac{B \pm n_{-1}\sqrt{\left(\frac{B}{n_{-1}}\right)^2 - cH}}{cn_{-1}^2} \tag{A4.13}$$

Let us now consider the case of a plane surface ($c = 0$). When the value of c approaches zero, the value of A must also go to zero, as we may see in Fig. A4.1. This is possible only if we take the negative sign in expression (A4.13). Now let us find an alternative expression for the square root. Considering now Fig. A4.3 and using the cosine law, we may find that the segment D has a length given by

$$D^2 = X_T^2 + Y_T^2 + r^2 = A^2 + r^2 + 2Ar\cos I \tag{A4.14}$$

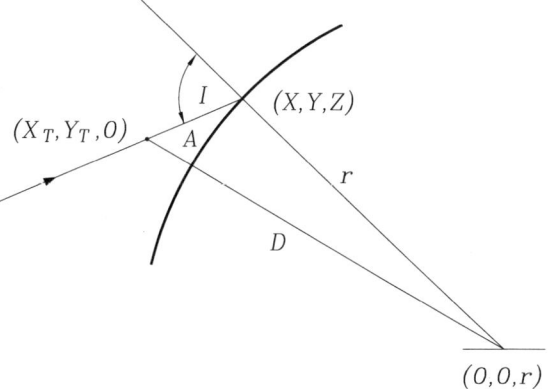

Figure A4.3 Ray refraction at the optical surface.

thus, solving for cos I and using the value of H in Eq. (A4.12), we obtain

$$n_{-1} \cos I = \frac{H - cn_{-1}^2 (A/n_{-1})^2}{2(A/n_{-1})} \tag{A4.15}$$

Now, we substitute here the value of (A/n_{-1}) given in Eq. (A4.13) (using the minus sign, as pointed out before) and, after some algebraic manipulation, we may find that

$$n_{-1} \cos I = n_{-1} \sqrt{\left(\frac{B}{n_{-1}}\right)^2 - cH} \tag{A4.16}$$

then, we substitute this result into Eq. (A4.13), with the minus sign in front of the square root, obtaining

$$\left(\frac{A}{n_{-1}}\right) = \frac{B - n_{-1} \cos I}{cn_{-1}^2} \tag{A4.17}$$

On the other hand, from Eq. (A4.16) we also may find that

$$cn_{-1}^2 = \frac{B^2 - n_{-1}^2 \cos^2 I}{H}$$
$$= \frac{(B + n_{-1} \cos I)(B - n_{-1} \cos I)}{H} \tag{A4.18}$$

Then, substituting this expression into Eq. (A4.17), the result for (A/n_{-1}) is

$$\left(\frac{A}{n_{-1}}\right) = \frac{H}{B + n_{-1} \cos I} \qquad (A4.19)$$

In conclusion, first the values of B and H are calculated with Eqs. (A4.11) and (A4.12), respectively. Then, the value of $n_{-1} \cos I$ is obtained with Eq. (A4.16) and substituted into Eq. (A4.19) to obtain the desired value of (A/n_{-1}).

A4.1.3 Transfer from Osculating Sphere to Aspheric Surface

We proceed here in a similar way as in the last section. The first part is to compute the distance A' (Fig. A4.1) from the intersection point of the ray with the osculating sphere to the intersection point with the aspherical surface. The direct method is extremely complicated and it is better to obtain this value in an iterative manner, as illustrated in Fig. A4.4. Let us assume that the coordinates of the ray intersection with the osculating spherical surface (point $\mathbf{a}_1$) are (X_1, Y_1, Z_1). The procedure is now as follows:

1. The point $\mathbf{b}_1$, with the same X_1 and Y_1 coordinates, at the aspherical surface is found and the distance $-F_1$ is computed.
2. We find the plane tangent to the aspherical surface at this point.
3. The intersection $\mathbf{a}_2$ of the ray with this plane is calculated.

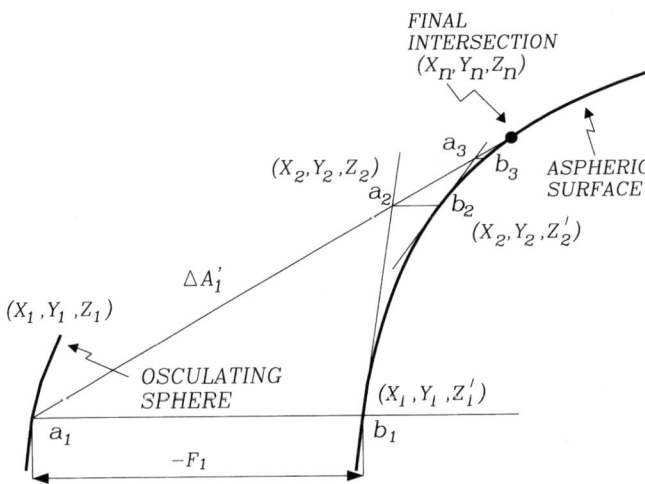

Figure A4.4 Transfer to the aspheric surface.

Exact Ray Tracing of Skew Rays

4. The same procedure is continued for points $\mathbf{b}_2$, $\mathbf{a}_3$, $\mathbf{b}_3$, and so on, until the error becomes small enough.

For any iteration step, let (X_n, Y_n, Z_n) represent the coordinates at the point $\mathbf{a}_n$, and (X_n, Y_n, Z'_n) the coordinates at the point $\mathbf{b}_n$, where this number n is the order of approximation. The radial distance S_n from the optical axis to the point $\mathbf{a}_n$ is

$$S_n^2 = X_n^2 + Y_n^2 \tag{A4.20}$$

then, we define W_n as the square root in expression (A2.8) for the aspherical surface:

$$W_n = \sqrt{1 - c^2 S_n^2} \tag{A4.21}$$

Now the distance $-F_n$ from $\mathbf{a}_n$ to $\mathbf{b}_n$ is given by

$$\begin{aligned} F_n &= Z_n - \left[\frac{cS_n^2}{1 + W_n} + B_1 S_n^4 + B_2 S_n^6 + B_3 S_n^8 + B_4 S_n^{10} \right] \\ &= Z_n - Z'_n \end{aligned} \tag{A4.22}$$

The coordinates of the point $\mathbf{a}_{n+1}$ may be found only after we calculate the length $\Delta A'_n$, as illustrated in Fig. A4.4, but first we need the equation of the plane tangent to the aspheric surface at point $\mathbf{b}_n$. To find the equation of this plane let us consider the equation of the aspheric surface (eliminating for simplicity the subscript n), which is

$$\Psi(X, Y, Z) = Z - \left[\frac{cS^2}{1 + W} + B_1 S^4 + B_2 S^6 + B_3 S^8 + B_4 S^{10} \right] = 0 \tag{A4.23}$$

then the equation of the tangent plane is

$$\Psi(X_n, Y_n, Z'_n) + (X - X_n)\left(\frac{\partial \Psi}{\partial X}\right)_{X_n, Y_n, Z'_n} + (Y - Y_n)\left(\frac{\partial \Psi}{\partial Y}\right)_{X_n, Y_n, Z'_n}$$
$$+ (Z - Z'_n)\left(\frac{\partial \Psi}{\partial Z}\right)_{X_n, Y_n, Z'_n} \tag{A4.24}$$

The next step is to compute these partial derivatives as follows

$$\frac{\partial \Psi}{\partial X} = \frac{\partial \Psi}{\partial S} \frac{\partial S}{\partial X} = \frac{\partial \Psi}{\partial S} \frac{X}{S} \tag{A4.25}$$

$$\frac{\partial \Psi}{\partial Y} = \frac{\partial \Psi}{\partial S} \frac{Y}{S} \tag{A4.26}$$

and

$$\frac{\partial \Psi}{\partial Z} = 1 \qquad (A4.27)$$

Differentiating Eq. (A4.23), and with the definition of W in Eq. (A4.21), we obtain

$$\frac{\partial \Psi}{\partial S} = -\frac{S}{W}c - S[4B_1S^2 + 6B_2S^4 + 8B_3S^6 + 10B_4S^8]$$

$$= -\frac{S}{W}E \qquad (A4.28)$$

where

$$E = c + W[4B_1S^2 + 6B_2S^4 + 8B_3S^6 + 10B_4S^8] \qquad (A4.29)$$

With these results, after substituting into Eq. (A4.24) we obtain the following equation of the plane:

$$-(X - X_n)\frac{X_n}{W_n}E - (Y - Y_n)\frac{Y_n}{W_n}E + (Z - Z'_n) + Z_n - Z'_n = 0 \qquad (A4.30)$$

but defining

$$U_n = -X_n E_n \qquad (A4.31)$$

$$V_n = -Y_n E_n \qquad (A4.32)$$

and using $F_n = Z_n - Z'_n$, the equation of the plane may be found to be

$$(X - X_n)U_n + (Y - Y_n)V_n + (Z - Z_n)W_n = -F_n W_n \qquad (A4.33)$$

If we now take the particular values X_{n+1}, X_{n+1}, X_{n+1}, for the coordinates X, Y, Z we may find that

$$(X_{n+1} - X_n)U_n + (Y_{n+1} - Y_n)V_n + (Z_{n+1} - Z_n)W_n = -F_n W_n \qquad (A3.34)$$

Similarly to Eqs. A4.4, A4.5, and A4.6, we may write for the coordinates X_{n+1}, Y_{n+1}, Z_{n+1}:

$$X_{n+1} = X_n + \left(\frac{\Delta A'}{n-1}\right)K_{-1} \qquad (A4.35)$$

$$Y_{n+1} = Y_n + \left(\frac{\Delta A'}{n-1}\right)L_{-1} \qquad (A4.36)$$

and

$$Z_{n+1} = Z_n + \left(\frac{\Delta A'}{n_{-1}}\right) M_{-1} \tag{A4.37}$$

Hence, substituting these values into Eq. (A4.33) and solving for $\Delta A'/n_{-1}$ we find that

$$D_{-1} = d_{-1} + A + A' \tag{A4.38}$$

This iterative loop ends when the desired tolerance in the value of $\Delta A'/n_{-1}$ has been obtained. Finally, from Fig. A4.1 we see that

$$\frac{\Delta A'}{n_{-1}} = -\frac{F_n W_n}{K_{-1} U_n + L_{-1} V_n + M_{-1} W_n} \tag{A4.39}$$

A4.1.4 Refraction at Aspheric Surface

From Eq. (A4.33) we see that the direction cosines of the normal to the plane (or to the aspherical surface) are U/G, V/G, W/G, where

$$G^2 = U^2 + V^2 + W^2 \tag{A4.40}$$

and the subscript n has been eliminated, since the iteration has been finished. Thus, the unit vector normal to the surface $\mathbf{S}_1$ is

$$\mathbf{S}_1 = \left(\frac{U}{G}, \frac{V}{G}, \frac{W}{G}\right) \tag{A4.41}$$

Then, the scalar product of the unit normal vector and the unit vector along the ray is the cosine of the angle between the two. Thus,

$$\cos I = \frac{K_{-1}}{n_{-1}} \frac{U}{G} + \frac{L_{-1}}{n_{-1}} \frac{V}{G} + \frac{M_{-1}}{n_{-1}} \frac{W}{G} \tag{A4.42}$$

which may be rewritten as

$$G n_{-1} \cos I = K_{-1} U + L_{-1} V + M_{-1} W \tag{A4.43}$$

From Eq. (1.17), the vectorial law of refraction is given by

$$\mathbf{S}_2 = \mathbf{S}_1 - \Gamma \mathbf{p} \tag{A4.44}$$

where Γ is given by Eq. (1.18) as

$$\Gamma = n\cos I' - n_{-1}\cos I \tag{A4.45}$$

which may be rewritten as

$$P = \frac{\Gamma}{G} = \frac{Gn\cos I' - Gn_{-1}\cos I}{G^2} \tag{A4.46}$$

In the same expression (1.18) we also have that

$$n\cos I' = n\left[\left(\frac{n_{-1}}{n}\cos I\right)^2 - \left(\frac{n_{-1}}{n}\right)^2 + 1\right]^{1/2} \tag{A4.47}$$

thus, multiplying by G we obtain

$$Gn\cos I' = n\left[\left(G\frac{n_{-1}}{n}\cos I\right)^2 - G^2\left(\frac{n_{-1}}{n}\right)^2 + G^2\right]^{1/2} \tag{A4.48}$$

Finally, the vectorial law of refraction may be written with three separate expressions, as

$$K = K_{-1} + UP \tag{A4.49}$$

$$L = L_{-1} + VP \tag{A4.50}$$

and

$$M = M_{-1} + WP \tag{A4.51}$$

A4.1.5 Refraction at Toroidal or Spherocylindrical Surfaces

To trace rays through toroidal (Murra, 1954; Spencer and Murty, 1962) or spherocylindrical surfaces (Menchaca and Malacara, 1986), we may follow basically the same procedure we used for the rotationally symmetric aspherical surfaces. For spherocylindrical surfaces the method described by Menchaca and Malacara (1986) may be used. In this case, the following equations must be used to find the values of U and V. First, we define the parameters:

$$Q = c_1 X^2 + c_2 Y^2 \qquad (A4.52)$$

and

$$R = \sqrt{1 - \frac{Q^2}{S^2}} \qquad (A4.53)$$

Then, the value of F defined in Eq. (A4.22) is now given by

$$F = Z_n - \frac{Q}{(1+R)} \qquad (A4.54)$$

The value of W is the same given in Eq. (A4.21) and the values of U and V become

$$U = W \frac{-\left[2c_1 X_n(1+R) + 2\left((c_1 X_n^2/S^4)(2c_1 - Q)Q\right)\right]}{(1+R)^2} \qquad (A4.55)$$

and

$$V = W \frac{-\left[2c_2 Y_n(1+R) + 2\left((c_2 Y_n^2/S^4)(2c_2 - Q)Q\right)\right]}{(1+R)^2} \qquad (A4.56)$$

A4.2 SUMMARY OF RAY TRACING RESULTS

The final set of expressions for tracing rays through an aspherical surface with rotational symmetry is now listed in the order in which they are to be used. The ray to be traced is defined by the intersection coordinates X_{-1}, Y_{-1}, and Z_{-1} on the first surface and its direction cosines multiplied by the refractive indices, K_{-1}, L_{-1}, and M_{-1}. First, to trace the ray from the previous surface to the plane tangent to the surface being considered:

$$\frac{d_{-1}}{n_{-1}} = \frac{t_{-1} - Z_{-1}}{M_{-1}} \qquad (A4.57)$$

$$X_T = X_{-1} + \left(\frac{d_{-1}}{n_{-1}}\right) K_{-1} \qquad (A4.58)$$

$$Y_T = Y_{-1} + \left(\frac{d_{-1}}{n_{-1}}\right) L_{-1} \qquad (A4.59)$$

Then, the ray is traced from the tangent plane to the spherical osculating surface as follows:

$$H = c(X_T^2 + Y_T^2) \tag{A4.60}$$

$$B = M_{-1} - c(Y_T L_{-1} + X_T K_{-1}) \tag{A4.61}$$

$$n_{-1} \cos I = n_{-1} \sqrt{\left(\frac{B}{n_{-1}}\right)^2 - cH} \tag{A4.62}$$

If the argument of this square root is negative, the ray does not intersect the spherical surface. Next, we calculate

$$\left(\frac{A}{n_{-1}}\right) = \frac{H}{B + n_{-1} \cos I} \tag{A4.63}$$

$$X = X_T + \left(\frac{A}{n_{-1}}\right) K_{-1} \tag{A4.64}$$

$$Y = Y_T + \left(\frac{A}{n_{-1}}\right) L_{-1} \tag{A4.65}$$

$$Z = \left(\frac{A}{n_{-1}}\right) M_{-1} \tag{A4.66}$$

We have calculated the coordinates of the ray on the osculating sphere. Now, we begin the iterative process to calculate the ray coordinates on the aspheric surface:

$$S_n^2 = X_n^2 + Y_n^2 \tag{A4.67}$$

$$W_n = \sqrt{1 - c^2 S_n^2} \tag{A4.68}$$

where, if the argument of this square root is negative, the ray is not crossing the aspherical surface. Then, we calculate

$$F_n = Z_n - \left[\frac{cS_n^2}{1 + W_n} + B_1 S_n^4 + B_2 S_n^6 + B_3 S_n^8 + B_4 S_n^{10}\right] \tag{A4.69}$$

When the optical surface is a conic, the coefficients B_i are computed with Eqs. (A2.9)–(A2.12). Then,

$$E_n = c + W_n[4B_1 S_n^2 + 6B_2 S_n^4 + 8B_3 S_n^6 + 10B_4 S_n^8] \tag{A4.70}$$

$$U_n = -X_n E_n \tag{A4.71}$$

Exact Ray Tracing of Skew Rays

$$V_n = -Y_n E_n \tag{A4.72}$$

$$\frac{\Delta A'}{n_{-1}} = -\frac{F_n W_n}{K_{-1} U_n + L_{-1} V_n + M_{-1} W_n} \tag{A4.73}$$

$$X_{n+1} = X_n + \left(\frac{\Delta A'}{n_{-1}}\right) K_{-1} \tag{A4.74}$$

$$Y_{n+1} = Y_n + \left(\frac{\Delta A'}{n_{-1}}\right) L_{-1} \tag{A4.75}$$

$$Z_{n+1} = Z_n + \left(\frac{\Delta A'}{n_{-1}}\right) M_{-1} \tag{A4.76}$$

If the magnitude of $\Delta A'$ is greater than the tolerance (typically about $\lambda/20$ or less), another iteration is performed by going again to the first equation (A4.67). Then, with the final values we continue by calculating

$$G^2 = U^2 + V^2 + W^2 \tag{A4.77}$$

$$G n_{-1} \cos I = K_{-1} U + L_{-1} V + M_{-1} W \tag{A4.78}$$

$$G n \cos I' = n \left[\left(G \frac{n_{-1}}{n} \cos I\right)^2 - G^2 \left(\frac{n_{-1}}{n}\right)^2 + G^2 \right]^{1/2} \tag{A4.79}$$

but if the argument of this square root is negative, the ray is not refracted, but totally reflected internally. Then, we continue with

$$P = \frac{G n \cos I' - G n_{-1} \cos I}{G^2} \tag{A4.80}$$

$$K = K_{-1} + UP \tag{A4.81}$$

$$L = L_{-1} + VP \tag{A4.82}$$

$$M = M_{-1} + WP \tag{A4.83}$$

This ends the ray-tracing procedure for the rotationally symmetric aspheric surface.

A4.3 TRACING THROUGH TILTED OR DECENTERED OPTICAL SURFACES

An optical surface may be tilted or decentered (Allen and Snyder, 1952) with respect to the optical axis of the system. In other words, there may not be a single common optical axis for all surfaces. Let us take a system of coordinates, as shown in Fig. A4.5, with its origin at the vertex of the

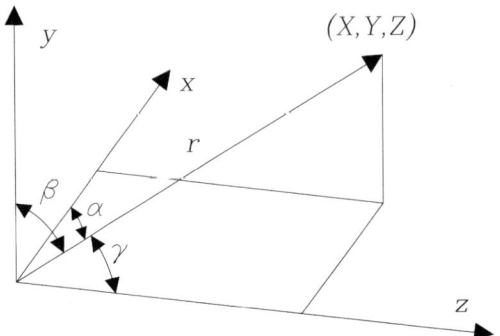

Figure A4.5 Tilting angles for a tilted optical surface.

surface under consideration, before tilting and/or decentering. Then, the z axis is aligned with the z axis of the previous surface and the y axis is parallel to the y axis of the previous surface. The optical axis of a surface may be inclined with respect to the optical axis of the previous surface by a rotation of this system of coordinates by an angle θ_x about the x axis or by a rotation by an angle θ_y about the y axis. A rotation by an angle θ_z about the z axis may be also important if the surface does not have rotational symmetry, as in a toroid. When tracing rays through a tilted or decentered surface a transformation with the desired rotations and decentration, with pivot at the origin (vertex of the surface), must be performed. This transformation is always made with respect to the previous surface, as shown in Fig. A4.6. So, if only one surface is tilted or decentered, a second transformation must be made to bring the optical axis to its previous position. The parameters to be transformed are the position from which the ray starts (X, Y, and, Z for the intersection of the ray with the previous surface) and the ray direction (cosine directors multiplied by refractive index values K, L, and M). These transformations for the three possible rotations are

$$\begin{aligned} X'_{-1} &= X_{-1} \\ Y'_{-1} &= -(Z_{-1} - t_{-1})\sin\theta_x + Y_{-1}\cos\theta_x + t_{-1} \\ Z'_{-1} &= (Z_{-1} - t_{-1})\cos\theta_x + Y_{-1}\sin\theta_x + t_{-1} \end{aligned} \qquad (A4.84)$$

and for the ray direction:

$$\begin{aligned} K'_{-1} &= K_{-1} \\ L'_{-1} &= -M_{-1}\sin\theta_x + L_{-1}\cos\theta_x \\ M'_{-1} &= M_{-1}\cos\theta_x + L_{-1}\sin\theta_x \end{aligned} \qquad (A4.85)$$

Exact Ray Tracing of Skew Rays

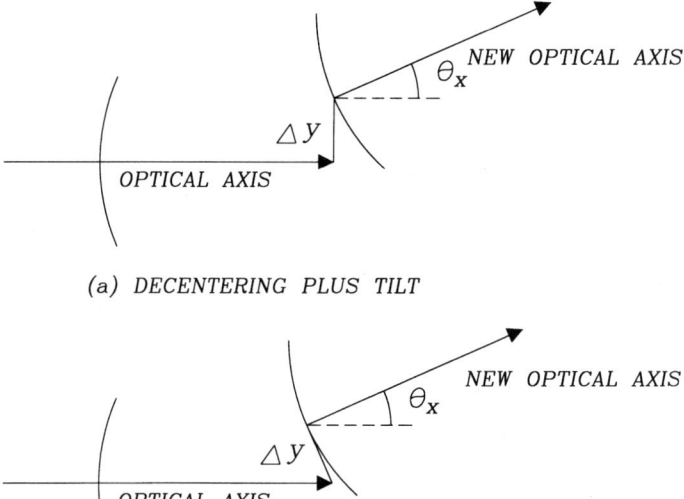

(a) DECENTERING PLUS TILT

(b) TILT PLUS DECENTERING

Figure A4.6 Tilting and decentering of optical surfaces: (a) decentering plus tilt; (b) tilt plus decentering.

The decentration is performed by means of the transformation

$$X'_{-1} = X_{-1} - \Delta X$$
$$Y'_{-1} = Y_{-1} - \Delta Y$$
(A4.86)

It is important to notice that the operations of tilting and decentering are not commutative, i.e., their order is important, as shown in Fig. A4.6.

After a surface has been decentered or tilted, the new optical axis for the following surfaces may have one of three different orientations, as shown in Fig. A4.7. These new possible orientations are:

1. The optical axis of the surface is tilted and/or decentered, as in Fig. A4.7(a). This is useful if not only one surface is tilted, but also several like a lens or system of lenses.
2. The refracted or reflected optical axis is as shown in Fig. A4.7(b). This is the case, e.g., when the tilted surface is a mirror and the new system elements have to be aligned with the reflected light beam.

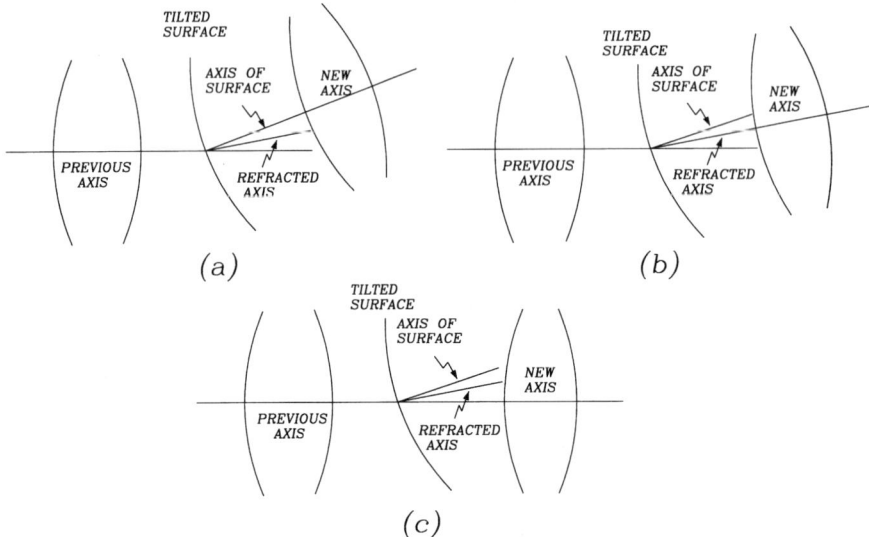

Figure A4.7 Selecting a new axis when tilting and/or decentering an optical surface.

3. The previous axis remains unchanged, as shown in Fig. A4.7(c). This is the case if only one surface, between two surfaces with a common axis, has been tilted and/or decentered.

REFERENCES

Allen, W. A. and Snyder, J. R., "Ray Tracing Through Uncentered and Aspheric Surfaces," *J. Opt. Soc. Am.*, **42**, 243–249 (1952).

Cornejo-Rodríguez, A. and Cordero-Dávila, A., "Graphical Ray Tracing for Conic Surfaces," *Appl. Opt.*, **18**, 3075 (1979).

Feder, D. P., "Differentiation of Ray Tracing Equations with Respect to Construction Parameters of Rotationally Symmetric Optics," *J. Opt. Soc. Am.*, **58**, 1494 (1968).

Herzberger, M. "Some Remarks on Ray Tracing," *J. Opt. Soc. Am.*, **41**, 805–807 (1951).

Herzberger, M. "Automatic Ray Tracing," *J. Opt. Soc. Am.*, **47**, 736–739 (1957).

Herzberger, M. and Hoadley, H. O., "The Calculation of Aspherical Correcting Surfaces," *J. Opt. Soc. Am.*, **36**, 334–340 (1946).

Hopkins, R. E. and Hanau, R., "Fundamentals Methods of Ray Tracing," in *Military Standardization Handbook: Optical Design, MIL-HDBK 141*, U.S. Defense Supply Agency, Washington, DC, 1962.

Lessing, N. V. D. W., "Cylindrical Ray-Tracing Equations for Electronic Computers," *J. Opt. Soc. Am.*, **52**, 472–473 (1962).

Malacara, D., "Geometrical Ronchi Test of Aspherical Mirrors," *Appl. Opt.*, **4**, 1371–1374 (1965).

Menchaca, C. and Malacara, D., "Toroidal and Sphero-Cylindrical Surfaces," *Appl. Opt.*, **25**, 3008–3009 (1986).

Murra, A. E., "A Toric Skew Ray Trace," *J. Opt. Soc. Am.*, **14**, 672–676 (1954).

Spencer, G. and Murty, M. V. R. K., "Generalized Ray-Tracing Procedure," *J. Opt. Soc. Am.*, **52**, 672–678 (1962).

Appendix 5
General Bibliography on Lens Design

Books

Born, M. and Wolf, E., *Principles of Optics*, Macmillan, New York, 1964.
Boutry, G. A., *Instrumental Optics*, Wiley-Interscience, New York, 1962.
Bouwers, A., *Achievements in Optics*, Elsevier, Amsterdam, 1946.
Brouwer, W., *Matrix Methods in Optical Instrument Design*, Benjamin, New York, 1964.
Chrètien, H., *Calcul des Combinaisons Optiques*, l'Ecole Superieure d'Optique, Paris, 1957.
Conrady, A. E., *Applied Optics and Optical Design, Part I*, Dover Publications, New York, 1957.
Conrady, A. E., *Applied Optics and Optical Design, Part II*, Dover Publications, New York, 1960.
Cox, A., *A System of Optical Design*, Focal Press, New York, 1964.
Habel, K. J. and Cox, A., *Engineering Optics*, Pitman, London, 1948.
Herzberger, M., *Modern Geometrical Optics*, Interscience, New York, 1958.
Hopkins, H. H., *Wave Theory of Aberrations*, Oxford University Press, London, 1950.
Hopkins, R. E., Hanau, R., Osterberg, H., Richards, O. W., Kavanagh, A. J., Wight, R., Rosin, S., Baumeister, P., and Nennett, A., *Military Standardization Handbook: Optical Design, MIL-HDBK 141*, U.S. Defense Supply Agency, Washington, DC, 1962.
Johnson, B. K., *Optical Design and Lens Computation*, Hatton, London, 1948.
Johnson, B. K., *Optics and Optical Instruments* (*Practical Optics*), Dover Publications, New York, 1960.
Kingslake, R., *Lens Design Fundamentals*, Academic Press, San Diego, CA, 1978.
Kingslake, R., *A History of the Photographic Lens*, Academic Press, San Diego, CA, 1989.
Korsch, D., *Reflective Optics*, Academic Press, Boston, MA, 1991.
Laikin, M., *Lens Design*, Marcel Decker, New York, 1990.
O'Shea, D. C., *Elements of Modern Optical Design*, John Wiley, New York, 1985.
Smith, W. J., *Modern Optical Engineering*, 2nd ed., McGraw Hill, New York, 1990.

Smith, W. J., *Lens Design, Critical Reviews of Science and Technology*, Vol. CR41, SPIE Optical Engineering Press, Bellingham, WA, 1992.

Smith, W. J. and Genesee Optics Software, Inc., *Modern Lens Design. A Resource Manual*, McGraw Hill, New York, 1992.

Welford, W. T., *Geometrical Optics and Optical Instrumentation*, North Holland, Amsterdam, 1962.

Welford, W. T., *Geometrical Optics*, North-Holland, Amsterdam, 1962.

Welford, W. T., *Aberrations of the Symmetric Optical System*, Academic Press, San Diego, CA, 1974.

Welford, W. T., *Aberrations of Optical Systems*, Adam Hilger, Bristol and Boston, 1986.

Review Articles or Chapters in Books

Hopkins, R. E., "Geometrical Optics," in *Methods of Experimental Physics, Geometrical and Instrumental Optics*, Vol. 25, D. Malacara, ed., Academic Press, San Diego, CA, 1988.

Hopkins, R. E. and Malacara, D., "Optics and Optical Methods," in *Methods of Experimental Physics, Geometrical and Instrumental Optics*, Vol. 25, D. Malacara, ed., Academic Press, San Diego, CA, 1988.

Kingslake, R., "Basic Geometrical Optics," in *Applied Optics and Optical Engineering*, Vol. I, R. Kingslake, ed., Academic Press, San Diego, CA, 1965.

Kingslake, R., "Lens Design," in *Applied Optics and Optical Engineering*, Vol. III, R. Kingslake, ed., Academic Press, San Diego, CA, 1965.

Welford, W. T., "Aplanatism and Isoplanatism," in *Progress in Optics*, Vol. XIII, E. Wolf, ed., North Holland, Amsterdam, 1976.

Index

Abbe, 29
  condenser, 436
  number, 149, 159, 497
  points, 89
  prism, 250
  surfaces, 128
Aberrated wavefront, 98
Aberration polynomial, 91, 171, 183
Aberration and wavefront
      deformations, 135
Aberrations
  balancing, 96
  chromatic, 145
  eye, 317
  fifth order, 173
  first order, 173
  high order, 123
  of normal to aspheric surface,
    488
  primary, 173
  Seidel, 103
  spherical, 73
  third order, 173
  transverse, 183
Achromatic
  condenser, 436
  double lens, 284
  doublet, 149, 151
  landscape lens, 283
  system with spaced elements,
    151
Achromatization method by
      Conrady, 156

Achromats, microscope objective,
    422
Active optics, 367
Adaptive optics, 367
Afocal
  five lens zoom system, 311
  systems, 373, 379
  three lens zoom system, 309
  zoom lens, 307
Airy function, 198
Amici
  objectives, 426
  prism, 244
Amplitude point spread function, 204
Anamorphic
  projection, 445
  prism system, 446
Anastigmat systems, symmetrical 301
Anastigmats, 291
Angular magnification, apparent, 261
Angles U and U', 14
Angles I and I', 14
Angular magnification, 34, 376
Anisotropic medium, 6
Annular aperture, 199
Aplanatic
  Abbe points, 89
  front, 427
  surfaces, 128
Apochromatic lenses, 159, 391
Apochromatism, 159
Apochromats, microscope objective,
    423

Apparent angular magnification, 261
Aqueous humor, 315
Ashtray astigmatism, 177
Aspherical
　off-axis aberrations, in aspheric surfaces, 132
　ophthalmic lenses, 326
　surface, 86, 132, 326, 511
Aspheric deformation terms, 86
Astigmatic curves, 122
Astigmatic image, 123, 126
Astigmatism, 108, 116, 121, 138, 177, 480, 482
　ashtray 177
　primary, 138
　primary longitudinal, 116
　sagittal longitudinal, 108
　tangential longitudinal, 108
　triangular, 177
Astronomical telescopes, 333
Asymmetrical systems, 292
Atmospheric turbulence, 333
Atmospheric seeing limited system, 235
Auxiliary axis, 107
Axial chromatic aberration 146, 148, 481, 483
Axial gradient
　index, 8
　lens, 100
Axial ray plots, 183
Axicon, 486
Axis orientation of spherocylindrical lenses, 330

Barrel distortion, 131
Basic telescopic system, 376
Beam-splitting prisms, 251
Bendings, 60
Best focus position, 95
Binocular beam-splitting system, 253
Biocular magnifiers, 263, 266
Boundary conditions in lens design, 467, 468

Bouwers camera, 339
Broken contact doublet, 387

Calcite, 6
Cardinal points, 53, 54
Cassegrain
　field correctors, 357
　light shields, 357
　telescope, 294, 346
Catadioptric systems, 156, 285, 337
　Bouwers, 339
　concentric, 340
　Maksutov, 341
　Schmidt, 337
　telescopes, 362
Cauchy interpolation formula, 145
Caustic 93, 96
Caustic waist, 95
CDM, 161
Cemented aplanatic doublet, 391
Center of curvature, 12
Characteristic focal line, 126
Chromatic difference of magnification, 161
Chromatic aberration, 145
　axial, 146, 148
　magnification, 160, 163, 165
　with one glass, 152, 165
Chromatic dispersion, 145
Circular aperture, diffraction of, 198
Coddington
　equations, 119
　magnifier, 264
Coherence effects in projectors, 440
Collimating a gaussian beam, 213
Color aberration, secondary, 156
Color averaging in optimization, 464
Coma, 112, 480, 482
　contribution, 114
　primary, 138
　sagittal, 108
　in single lens, 16
　tangential, 107
Comatic image, 123, 126
Compact disk objectives, 430

Index

Compound microscope, 415
Computer evaluation of systems, 217
Concave mirror, 271
Concave paraboloidal mirror, 276
Concave spherical mirror, 273
Concentric camera, 340
Condenser, 442
   Abbe, 436
   achromatic, 436
   for projectors, 443
Conic constant, 86, 485
Conicoid surfaces, 486
Conic surface, 89, 485, 490
Conjugate gradient method, 452
Conrady
   D-d method, 156
   interpolation formula, 145
   method to compute wavefront, 228
Constant deviation prism, 245, 255
Constrained damped least squares optimization method, 455
Convergent lens, 39
Convergent lenses, image formation, 45
Convex spherical mirror, 279
Cooke triplet, 294
Cornea, 315
Corrector plate, Schmidt, 339
Coudé focus configuration, 356
Cover glass, 427
Crystalline lens, 315
Crystals, 6
Curvature tolerances, 235
Curved or tilted object, 55
Cylindrical power of sphero-cylindrical lenses, 330

Dagor lens, 302
Dall-Kirham telescope, 354
Damped least squares, 453
Deflecting a light beam, 239
Deflecting prisms, 244
Deformation $\bar{y}$ terms, aspheric, 86
Delano y-$\bar{y}$ diagram, 67
Delano's relation, 21

Descartes' ovoid, 89
Design of a terrestrial telescope objective, 397
Diffraction
   effects in telescopes, 333
   in optical systems, 191
   of a circular aperture, 198
   system limited by, 234
   Fraunhoffer, 191
   Fresnel, 191
   images with aberrations, 200
Diopter, 39
Distances L and L', 14
Distortion, 109, 129, 481, 482
   barrel, 131
   negative, 130
   pincushion, 131
   positive, 130
   primary, 138
Divergent lens, 39
Divergent lenses, image formation, 47
Double Gauss lens, 303
Double lens, achromatic, 284
Doublet
   broken contact, 387
   cemented aplanatic, 391
   nonaplanatic, 383
   parallel air-space, 390
Dove prism, 245, 248, 249
Duplex front of an oil immersion, 427
Dyson system, 286

Eccentricity, 485
Effective focal length, 49
Eikonal, 6
Electromagnetic wave, 2
Electronic endoscope, 412
Ellipsoids, 89
Empty magnification, 420
Encircled energy, 201
Endoscope, 410
   electronic, 412
   fiber optics, 411
   gradient index, 411

[Endoscope]
　Hopkins, 410
　traditional, 410
Entrance pupil, 25
Equilateral triangle prism, 248, 253
Erecting eyepiece, 402
Erfle eyepieces, 405
Error function, 451
Exact ray tracing, 503
Exit pupil location in visual optical
　　systems, 373
Exit pupil, 25, 315
Eye aberrations, 317
Eye lens, 315, 376
Eye relief, 377
Eyepiece, 376, 378, 405
　erecting, 402
　Erfle, 406, 407
　hi-point, 431, 434
　Huygens, 431, 400
　Kellner, 401, 403
　Kingslake, 405
　microscopes, 430
　orthoscopic, 403, 406
　Plössl, 403, 405
　Ramsden, 400
　symmetric, 403, 405
　telescopes, 398
　wide field, 431, 433, 435

F-number, 391
Fermat's principle, 1, 7
Fiber optics endoscope, 411
Field correctors, 357
　aspheric, 361
　field flattener, 358
　Ross, 358
　Winne corrector, 360
Field averaging, 466
Field flattener, 358
Fifth order aberrations, 173
First-order optics, 18
Flow chart for a lens optimization
　　program, 470
Fluorite microscope objectives, 425

Focal length, 39
　effective, 49
Focus
　marginal, 94
　medium, 127
　paraxial 94
　Petzval, 127
　sagittal, 126, 127
Focussing a gaussian beam, 213
Fraunhoffer diffraction, 191, 195
　of a slit, 197
Fraunhofer lines, 1, 2
Frequency of light, 1
Fresnel diffraction, 191, 192
Fresnel lenses, 288
Front stop landscape lens, 269
Frontal telecentric system, 28

Gabor plates, 288
Galilean telescopes, 394, 397, 308
Gauss formula, 18
Gauss law, 41
Gaussian beams, 211
Gaussian optics, 18, 105
Gaussian image, 105
Gaussian integration, 223
Gaussians, 180
Geometrical optical transfer
　　function, 232
Geometrical optics, 1, 5
Geometrical spot size, 219
Glass plate, 239
Glatzel adaptive method, 453
Global optimization and simulated
　　annealing, 469
Gradient index, 7, 99
　axial, 8, 100
　endoscopes, 411
　radial, 8
　spherical, 8
Gregory telescope, 355
Grey's optimization method, 453

Half-turn rotation prism, 249
Hamilton's theory, 7

Hastings magnifier, 264
Helmholtz, 29
He-Ne laser collimators, 392
Herschel invariant, 33
Hi-point eyepiece, 431, 434
High order aberrations, 123
  spherical aberration, 98
Homogeneous materials, 5
Homogeneous medium, 6
Hopkins endoscope, 410
Human eye, 315
Huygens eyepieces, 400, 431
Huygens–Fresnel Theory, 191, 196
Hyperboloids, 89, 489
Hypermetropia, 319

Illuminator
  critical, 434, 436
  for microscope, 434
  Köhler, 434
Image
  convergent lenses, 45
  divergent lenses, 47
  formation, 22
  magnifications, 33
  ray, 67
  rear, 25
  thin lenses, 42
  virtual, 25
Imaging lens, simple, 265
Index of refraction, 145
  gradient, 7
Indirect ophthalmoscope, 408
Infrared and ultraviolet materials, 500
Inhomogeneous medium, 6
Interferometric quality, 234
Interpolation formula
  Cauchi, 145
  Conrady, 145
  Selmeier, 145
Invariant
  Herschel, 33
  Lagrange 33, 31, 63
Inverting prisms, 248

Isotropic medium, 3, 5, 6

Kellner eyepiece, 401, 403
Kingslake and symmetric telescope
  eyepieces, 405
Köhler illuminator, 434, 439

L-U method, meridional ray
  tracing, 16
Lagrange invariant, 31, 63, 112, 137
Lagrange multipliers in optimization, 457
Lagrange theorem, 31, 63, 112, 137
Landscape lens, 269
  achromatic, 283
Laser light collimators, 391, 392
Lateral magnification, 33, 44
Least squares method, 453
Leman prism, 250
Lens
  bendings, 60
  convergent or positive, 39
  design programs, 472
  design optimization, 451
  divergent negative, 39
  evaluation programs, 470
  power, 39
  thick, 49, 57
  thin, 39
Light gathering power of telescopes, 333
Light ray, 3, 5
Linear coma, 289
Line spread function, 230
Lister objectives, 425
Longitudinal astigmatism
  sagittal, 108
  tangential, 108
Longitudinal magnification, 33
Longitudinal spherical aberration, 75
Low-power objectives, 425

Magnification, 33
  angular, 34
  apparent angular, 261

[Magnification]
  chromatic aberration, 160, 163, 165, 481, 483
  lateral, 33, 44
  longitudinal, 33
  of pupils, 34
Magnifier, 260
  biocular, 263, 266
  Coddfington, 264
  Hastings, 264
  plano-convex, 264
  power, 261
  Ramsden, 264
  triplet, 264
Magnifying power, 376
  ophthalmic lenses, 321
  single lens, 415
Magnitude, visual limiting, 334
Maksutov camera, 341
Maksutov–Cassegrain telescope, 365
Malus law, 6
Manguin mirror, 285
Manufacturing tolerances, 235
Marginal focus, 94
Marginal sagittal surface, 105
Marginal spherical aberration, 94
Marginal tangential focus, 105
Marginal tangential ray, 107
Mean wavefront deformation, 175
Mechanical tube length, 418
Medium focus, 127
Meridional ray, 13
  tracing, 12, 217
    by L-U method, 16
    by Q-U method, 17
    plots, 183, 185
Merit function, 451, 463
Microscope, 415
  arrangement, 416
  compound, 415
  cover glass, 427
  critical illuminator, 434, 436
  eyepieces, 430
  empty magnification, 420
  illuminators, 434

[Microscope]
  mechanical tube length, 418
  numerical aperture of objective, 419
  optical tube length, 416
  resolving power, 418
  shoulder height, 416
  types of objectives, 428
  working distance, 418
Microscope objective, 421
  achromats, 422
  Amici, 426
  apochromats, 423
  duplex front of an oil immersion, 427
  fluorite, 422, 425
  Lister, 425
  low-power, 425
  oil immersion, 427
  parfocal, 418
  plan-achromats, 423
  plan-apochromats, 423
  reflecting, 428
  semiapochromats, 422
Mirror
  concave, 271
    paraboloidal, 276
    spherical, 273
  convex spherical, 279
  Manguin, 285
  paraboloidal, 270
  spherical, 39, 270
Modern trends in optical design, 469
Modulation transfer function, 205
Monochromatic off-axis aberrations, 103
Movie projectors, 439
Multiple mirror telescopes, 365
Myopia, 319

Nasmythe focus configuration, 356
Negative distorton, 130
Negative lens, 39
Newton's formula, 43
Newton telescope, 341

Index

Nodal points of thin lens, 44
Nonafocal three-lens zoom lens, 311
Nonaplanatic doublet, 383
Nondeflecting chromatic dispersing prism, 256, 257
Nondeflecting transforming prisms, 248
Normalized radius of curvature, 92
Numerical aperture of a microscope objective, 419

Objective, 376
Object ray, 67
Object, real, 23
Object shifting, 64, 71
Object, virtual, 23
Oblate spheroid, 489
Oblique rays, 13, 103, 104
Ocular (see also Eyepieces), 378
Off-axis aberrations
  in aspherical surfaces, 132
  monochromatic, 103
  for spherical mirrors, 272
Off-axis paraboloids, 490
Offense against the sine condition, 112
Offner system, 287
Oil inmmersion objectives, 427
Ophthalmic glasses, 499
Ophthalmic lenses, 315, 318, 322
  aspheric, 326
  magnifier power, 321
Opthalmoscope, 408
Optical axis, 13, 39
Optical glasses, 497
Optical materials, 4, 497
Optical models of the human eye, 375
Optical path, 7, 225
Optical plastics, 500
Optical sine theorem, 29, 31
Optical surface
  aspherical, 86, 132, 326, 511
  mathematical representation, 485
  spherocylindrical, 512
  toroidal, 512

Optical transfer function, 204, 208, 232
Optical tube length, 416
Optimization, lens design, 451
  conjugate gradient method, 452
  constrained damped least squares method, 455
  damped least squares method, 453
  Glatzel adaptive, 453
  global and simulated annealing, 469
  Grey's method, 453
  least squares method, 453
  optimum gradient method, 452
  steepest descent method, 452
  zig-zag method, 452
Orthogonality condition, 176
Orthoscopic eyepiece, 403
OSC, 113
Osculating sphere to aspheric surface, 508
OTF, 204
Other types of objectives, 428
Overhead projectors, 446

Plano-convex magnifier, 264
Paraboloid, 489
Paraboloidal mirror, concave, 270, 276
Parallel air-space doublet, 390
Parallel faces glass plate, 239
Paraxial approximation, 31
Paraxial axial image, 104
Paraxial focus, 94
Paraxial lateral magnification, 44
Paraxial ray, 13
  tracing, y-nu method, 21
Parfocal microscope objective, 418
Partial dispersion, 159, 160
Pechan prism, 249
Pentaprism prism, 244, 246
Periscope lens, 281
Petzval curvature, 109, 121, 138, 480, 482
Petzval focus, 127

Petzval lens, 292
Petzval projection lens, 444
Petzval surface, 105, 111
Petzval sum, 123
Petzval theorem, 109
Phase transfer function, 205
Photographic lens, 259, 291
   anastigmats, 291
   Cooke triplet, 292
   Dagor, 302
   double Gauss, 303
   landscape, 269
   Petzval, 292
   telephoto, 292
   Tessar, 300
Pincushion distortion, 131
Pinhole camera, 194
Plan-achromats, microscope objective, 423
Plan-apochromats, microscope objective, 423
Plane-parallel plate, 84
Plössl eyepiece, 403
Point characteristic, 7
Point spread function, 204, 230
Porro prism, 252
Portable television projector, 449
Positive distortion, 130
Positive lens, 39
Power of a lens, 39
Primary aberrations, 173
   astigmatism, 138
   coefficients, 477
   coma, 138
   distortion, 138
   longitudinal sagittal astigmatism, 116
   spherical aberration, 77, 108, 137
Principal surface, 114
Principal planes, 49
   for two separated thin lenses, 62
Principal ray, 25, 107
Prismatic lenses, 328
Prisms, 239
   Abbe, 250

[Prisms]
   Amici, 244
   anamorphic system, 446
   beam-splitting, 251
   constant deviation, 245, 255
   dove, 245, 248, 249
   deflecting, 244
   equilateral triangle, 248, 253
   half-turn, 249
   inverting, 248, 250
   Leman, 250
   nondeflectic chromatic dispersing, 256, 257
   nondeflecting transforming, 248
   Pechan, 249
   pentaprism, 244, 246
   Porro, 252
   rectangular retroreflecting, 247
   retroreflecting, 245
   reverting, 248, 250
   right angle, 244
   rotating, 249
   Schmidt-Pechan, 250, 252
   transforming, 244
   triangular dispersing, 253
   Wollaston, 244, 246
Profile projectors, 447
Projecting lens, 443
   Petzval, 444
Projection systems, 439
Projectors
   coherence effects in, 440
   components, 441
   condensers for, 443
   movie, 439
   overhead, 446
   profile, 447
   slide, 439
   television, 448
   telecentric system, 447
Prolate spheroid, 489
Pupils, 25, 315
   entrance, 25
   exit, 25
   magnification, 34

Index

Q-U method for meridional ray tracing, 17
Quartz, 6

Radau integration, 223
Radial energy distribution, 224
Radial gradient index, 8
Radius of curvature, 12, 14
 normalized, 92
Ramsden magnifier, 264
Ray aberrations, 34, 35
Rayces formulas, 35
Rayleigh resolution criterium, 210
Ray plots
 axial, 183
 meridional, 12, 183, 185
 sagittal, 183, 188
Ray tracing, exact, 503
Real object, 23
Real image, 25
Rear stop landscape lens, 269
Rear telecentric system, 28
Rectangular retroreflecting prism, 247
Reference sphere, 94, 98
Reflecting objectives, 428
Reflection laws, 9, 10
Reflective conic surfaces, 89
Refracting objectives, 382
Refraction laws, vectorial form, 11
Refractive index, 3, 4, 15, 497
Relays and periscopes, 407
Representation of optical surfaces, 485
Resolution criteria, 209
 Sparrow, 210
 Rayleigh, 210
Resolution of telescopes, 333
Resolving power of a microscope, 418
Retina, 316
Retroreflecting systems, 242, 245, 247
Reverting prisms, 248, 250
Right angle prism, 244
Ross corrector, 358

Rotating prisms, 249

Sagittal coma, 108
Sagittal focus, 126, 127
Sagittal image, 120
Sagittal longitudinal astigmatism, 108
Sagittal plane, 13, 27
Sagittal ray plots, 183, 188
Sagittal rays, 107, 121
Sagittal surface, marginal, 105
Schmidt camera, 337
Schmidt-Cassegrain telescope, 362
Schmidt corrector plate, 339
Schmidt-Pechan prism, 250, 252
Scintillation, 333
Secondary color aberration, 158
Seeing, 333
Seidel aberrations, 103
Selmeier interpolation formula, 145
Semiapochromats, microscope objective, 422
Shifting of the object, 64, 71
Shifting of the stop, 64, 71
Shoulder height, 416
Simple imaging lens, 265
Sine condition, offense against the, 112
Sine theorem 29, 31
Single lens, coma in a, 116
Single lenses, 260
Single magnifier, 261
Single microscope, 261
Skew ray, 13
Slide projectors, 439, 442
Snell's law, 11
Spaced elements, 151
Sparrow resolution criterium, 210
Speed of propagation of light, 1
Spherical aberration, 73, 480, 481
 aberration polynomial, 91
 of aspherical surfaces, 86
 longitudinal, 75
 marginal 94
 primary, 77, 108, 137
 surfaces without, 87

[Spherical aberration]
  of thin lens, 80
  transverse, 78, 94
Spherical and aspherical surfaces, 485
Spherical gradient index, 8
Spherical mirrors, 39, 270, 272
  concave, 273
  convex, 279
Spherical power of spherocylindrical
    lenses, 330
Spherical refracting surface, 12
Spherochromatism, 155
Spherocylindrical lenses, 329, 330
  axis orientation, 330
  cylindrical power, 330
  spherical power, 330
Spherocylindrical surfaces, 512
Spot diagram, 219
Spot size, 219
Spread function
  amplitude point, 204
  line, 230
  point, 204, 230
Stagnation in optimization, 462
Steepest descent method, 452
Stellar magnitude, 335
Stop, 25
Stop position analysis, 217
Stop-shift equations, 139, 163
Stop shifting, 64, 71
Strehl ratio, 202, 208
Surfaces without spherical
    aberration, 87
Symmetrical anastigmat systems, 301
Symmetrical principle, 138
Symmetric eyepiece, 403
System stop, 25
Systems of lenses, 49, 51
System of thin lenses, 60, 84
System of two flat mirrors, 240
System of two separated thin
    lenses, 61

Tangential astigmatism, 108
Tangential coma, 107
Tangential focus, marginal, 105
Tangential image, 119
Tangential plane, 13, 27
Tangential rays, 107, 119
Telecentric, rear, 28
Telecentric systems, 28
  projection system, 447
Telephoto lens, 294
Telescopes
  astronomical, 333
  Cassegrain, 346
  catadioptric, 362
  Dall–Kirham, 354
  diffraction effects in, 333
  Erfle eyepiece for, 406, 470
  eyepieces, 398
  Gregory, 355
  light gathering power, 333
  limiting magnitude for visual,
      334
  Maksutov–Cassegrain, 365
  multiple mirror, 365
  Newton, 341
  orthoscopic eyepiece for, 406
  resolution, 333
  Schmidt–Cassegrain, 362
  symmetric eyepieces for, 405
  two-mirror, 342
  visual, 373
Television projection,
    Schmidt system, 448
Television projector, 448
Terrestrial telescope objective, 398
Tessar lens, 300
Thick lenses, 49, 51, 57
Thickness, 15
  tolerances, 235
Thick optical system, 53
Thin lenses, 39
  axial chromatic aberration of, 148
  image formation, 42
  nodal points, 44
  spherical aberration of, 80
  system of, 60, 84
Third order aberrations, 116, 173

Index 533

Tilted or curved object, 55
Tilted or decentered optical surfaces, 515
Tilt of the wavefront, 138
Tolerance aberrations, 234
Tolerances, curvature, 235
Tolerances, thickness, 235
Toroidal surfaces, 492, 512
Traditional endoscope, 410
Transfer function, 204
   geometrical, 232
   modulation, 205
   optical, 232
   phase, 205
Transformation of an image, 242
   inversion, 242
   reflection, 242
   reversion, 242
   rotation, 242
Transforming prisms, 244
Transverse aberrations, 35, 183
Transverse spherical aberration, 78, 94
Triangular astigmatism, 177
Triangular dispersing prism, 253
Triplet magnifier, 264
Tscherning ellipses, 325
Tunnel diagram, 239
Twinkling, 333
Two-mirror afocal systems, 381
Two-mirror telescopes, 342

Varifocal lenses, 306
Vector addition of prisms, 329, 331
Vectorial form of refraction laws, 11
Vertex 12

Vignetting, 27
Virtual image, 25
Virtual object, 23
Visual and terrestrial telescopes, 393
Visual limiting magnitude, 334
Visual quality system, 234
Visual systems, 373
Visual telescopes, 373
Vitreous humor, 316

Wave aberration, 34, 35
   polynomial, 171
Wavefront, 3
   aberrated, 98
   deformations, 35, 135, 224, 228
   deviation, 94
   representation by gaussians, 180
   tilt, 138
   variance, 176
Wavelength of light, 1
Weight factors in optimization, 462
Wide field eyepiece, 431, 433, 435
Winne corrector, 360
Wollaston prism, 244, 246
Working distance, 418

y-$\bar{y}$ diagram, 67
y-nu method, paraxial ray tracing, 21

Zernike polynomials, 175
Zig-zag method, 452
Zoom lenses, 306
   afocal, 307
   five lens, 311
   nonafocal three-lens, 311
   three lens, 309